Nanostructured Materials for Solar Energy Conversion

Nanostructured Materials for Solar Energy Conversion

Edited By

Tetsuo Soga

Department of Environmental Technology and Urban Planning
Nagoya Institute of Technology
Nagoya, Japan

ELSEVIER   Amsterdam • Boston • Heidelberg • London • New York • Oxford
Paris • San Diego • San Francisco • Singapore • Sydney • Tokyo

Elsevier
Radarweg 29, PO Box 211, 1000 AE Amsterdam, The Netherlands
The Boulevard, Langford Lane, Kidlington, Oxford OX5 1GB, UK

First edition 2006

Library of Congress Cataloging-in-Publication Data
A catalog record for this book is available from the Library of Congress

British Library Cataloguing in Publication Data
A catalogue record for this book is available from the British Library

ISBN-13: 978-0-444-52844-5
ISBN-10: 0-444-52844-X

For information on all Elsevier publications
visit our website at books.elsevier.com

Printed and bound in The Netherlands

06 07 08 09 10   10 9 8 7 6 5 4 3 2 1

Preface

Our society is based on coal, oil and natural gas, but these fossil fuels will be depleted someday in the future because they are limited. Carbon dioxide is produced in the combustion of fossil fuels and the rapid increase of carbon dioxide concentration has affected the consequence of climate, resulting in the global warming effect. Under these circumstances, interest in photovoltaic (PV) solar cell is increasing rapidly as an alternative and clean energy source.

Photovoltaic solar cells provide clean electrical energy because the solar energy is directly converted into electrical energy without emitting carbon dioxide. The solar energy is not limited, free of charge and distributed uniformly to all human beings. Crystalline silicon solar cell has been extensively studied and used for practical terrestrial applications. However, the expensive material cost and lots of energy necessary for manufacturing have caused high cost and long energy payback time, which have prevented the large spread of PV power generation.

Recently, thin film solar cells using silicon or compound semiconductors have been actively studied instead of the bulk silicon solar cell. But the solar cells are still too expensive to compete with public electricity charge. In 2004 New Energy and Industrial Technology Development Organization (NEDO), Ministry of Economy, Trade and Industry, Japan, announced the "PV Roadmap Toward 2030 (PV2030)" in which the target of production cost for PV module is 50 yen/W in 2030. It is expected that the PV power generation can supply approximately 50% of residential electricity consumption (approximately 10% of total electricity consumption) in 2030. But it would be difficult to reach this goal only by the conventional technologies. One important concept to reduce the solar cell cost and to increase the conversion efficiency is to use NANOTECHNOLOGY, i.e., to use the nanostructured material in solar cell. Nanostructured materials are largely divided into inorganic materials and organic materials. In spite of the common interest and common purpose, two kinds of materials have been discussed in different conferences and different communities until now. There is no book that offers a comprehensive overview of the nanostructured inorganic and organic materials for solar energy conversion.

The aim of this book is to overview the nanostructured materials for solar energy conversion covering a wide variety of materials and device types from inorganic materials to organic materials. This book is divided into five parts: fundamentals of nanostructured solar cells, nanostructures in conventional thin film solar cells, dye-sensitized solar cells, organic and carbon based solar cells and other nanostructures. Authors are all specialists in their fields. But I must apologize that the important nanostructured materials are missing in this book because of the limit of my ability. This book was intended for researchers, scientists, engineers, graduate students and undergraduate students, majoring in electrical engineering, chemical engineering, material science, physics, etc., who are interested in the nanostructured solar cells. The content of my chapter is the subject of a graduate course in our department, Department of Environmental Technology and Urban Planning, devoted for the beginner of PV. I strongly hope that you will get some hints for the development of the solar cell from this book and contribute to the progress of PV.

Tetsuo Soga
Nagoya, Spring 2006

Introduction

Tetsuo Soga

Department of Environmental Technology and Urban Planning
Nagoya Institute of Technology
Gokiso-cho, Showa-ku, Nagoya 466-8555, Japan

Energy conversion in solar cell consists of generation of electron–hole pairs in semiconductors by the absorption of light and separation of electrons and holes by an internal electric field. Charge carriers collected by two electrodes give rise to a photocurrent when the two terminals are connected externally. When a resistance load is connected to the two terminals, the separation of the charge carriers sets up a potential difference.

Most of the solar cells used in the terrestrial applications are bulk-type single- or multi-crystalline silicon solar cells. The typical cell structure is a thin (less than 1 μm) n-type emitter layer on a thick (about 300 μm) p-type substrate. Photo-generated electrons and holes diffuse to the space charge region at the interface where they are separated by the internal electric field. The effective charge separation results from long diffusion length of electrons and holes in crystalline silicon. Although it is aimed to reduce the solar cell module manufacturing cost, the drastic reduction of cell cost and increase of the conversion efficiency cannot be expected by using the conventional materials and solar cell structures. Moreover, the shortage of the feedstock of high-purity silicon is predicted in the near future although it depends on off-spec silicon of electronics industry. Therefore, research and development of solar cells with low production cost, high conversion efficiency and low feedstock consumption are required.

An important concept to reach this goal is to use nanostructured materials instead of bulk materials. The motivations to employ nanostructures in solar cells are largely divided into three categories as follows:

1. To improve the performance of conventional solar cells.
2. To obtain relatively high conversion efficiency from low grade (inexpensive) materials with low production cost and low-energy consumption.

3.   To obtain a conversion efficiency higher than the theoretical limit
     of conventional p–n junction solar cell.

This book brings out an overview of the organic and inorganic nano-structured materials for solar energy conversion. The book comprises of five parts as follows:

PART I. FUNDAMENTALS OF NANOSTRUCTURED SOLAR CELLS

The fundamental issues to deal with nanostructured solar cells are described on device modeling, optical and electrical modeling and modeling of refractive index and reflectivity of quantum solar cells. The chapter on basic properties of semiconductor materials and the conventional p–n junction solar cells deals with nanostructured solar cells.

PART II. NANOSTRUCTURES IN CONVENTIONAL THIN FILM SOLAR CELLS

Nanostructures of conventional thin film solar cells such as silicon solar cells, chalcopyrite-based solar cells, CdS-based solar cells and CdTe-based solar cells are described. Amorphous silicon has attracted attention to reduce the manufacturing cost compared with bulk-type crystalline silicon. But there still remains a problem of stability. Recently, microcrystalline thin film silicon solar cells made up of nano-sized crystallites with the material properties between amorphous and bulk have been studied actively. It is expected to obtain very high conversion efficiency (more than 15%) by employing amorphous silicon/microcrystalline silicon tandem solar cells. It also describes that it is possible to improve the performance and reduce the cost of thin film solar cells based on chalcopyrite-based materials, CdS, CdTe and Cu_2S.

PART III. DYE-SENSITIZED SOLAR CELLS

The principle and the current status of dye-sensitized solar cells are described. In the conventional p–n junction solar cells, only the electrons and holes that can diffuse to the space charge region can be collected as a current. In order to get a long diffusion length, the purity of semiconductors should be increased and the defect concentration should be decreased, resulting in the expensive solar cell materials. In a dye-sensitized solar cell, a photon absorbed by a dye molecule gives rise to electron injection into the conduction band of nanocrystalline oxide semiconductors such as TiO_2 or ZnO. Because of the high surface area, relatively high photocurrent can be obtained in spite of the simple process. The dye is regenerated by electron transfer from a redox species in solution. A chapter on solid-state dye-sensitized solar cells in which the liquid electrolyte is replaced by p-type semiconductor is also dealt with.

PART IV. ORGANIC- AND CARBON-BASED SOLAR CELLS

The principle and the current status of organic solar cell and fullerene-based solar cell are described. Organic solar cells are attractive as solar cell materials because of high throughput manufacture process, ultra-thin film, flexible, lightweight and inexpensive. Organic materials differ from inorganic materials since the excited carriers exist as excitons, excitons are separated into electrons and holes at the interface, charge carrier transport is followed by hopping, etc. In order to increase the efficiency bulk, heterojunction solar cells using conjugated polymers and small molecule organic materials such as phthalocyanine have been investigated. It is important to understand the properties of fullerenes because it is often used as an organic solar cell. The photosynthetic materials are also studied as solar cell materials and a solid state cell is demonstrated.

PART V. OTHER NANOSTRUCTURES

Solar cells using other semiconductor nanostructures are overviewed. The concept of ETA (extremely thin absorber) is similar to that of dye-sensitized solar cells except that the ETA solar cell is completely made up of inorganic semiconductors. The concept of quantum structures is very important because there is a possibility to achieve the conversion efficiency higher than the theoretical limit of conventional p–n junction solar cells by employing quantum well or quantum dot structures. The idea is to extend the optical absorption to longer wavelengths by quantum wells, to use the carrier multiplication which produces the quantum efficiency exceeding unity, to use intermediate bands made of quantum dots, etc. It is also expected that single wall carbon nanotubes can improve the transport properties of polymer-based solar cells.

Nanostructured Materials for Solar Energy Conversion
T. Soga (editor)
©2006 Elsevier B.V. All rights reserved
ISBN-10: 0-444-52844-X/ISBN-13: 978-0-444-52844-5

Corrigendum

The following equations should replace those on the respective pages
of *Chapter 1 "Fundamentals of Solar Cell"* by *Tetsuo Soga*.

Page 5:

$$np = N_v N_c e^{\frac{E_v - E_c}{kT}} = N_v N_c e^{\frac{E_g}{kT}} = n_i^2 \;\rightarrow$$

$$p = 2(\frac{2\pi m_p kT}{h^2})^{3/2} \exp(-\frac{E_F - E_v}{kT}) = N_v \exp(-\frac{E_F - E_v}{kT})$$

$$np = N_v N_c e^{E_v}\, e^{E_\chi / \kappa T} = N_v N_c e^{\frac{E_g}{kT}} = n_i^2 \;\rightarrow$$

$$np = N_v N_c e^{\frac{E_v - E_c}{kT}} = N_v N_c e^{\frac{E_g}{kT}} = n_i^2$$

Page 24:

$$J_0 = \frac{qD_p P_{n0}}{l_p} + \frac{qD_n n_{p0}}{L_n} = \frac{qD_p n_i^2}{L_p N_D} + \frac{qD_n n_i^2}{L_n N_A} \;\rightarrow$$

$$J_0 = \frac{qD_p P_{n0}}{L_p} + \frac{qD_n n_{p0}}{L_n} = \frac{qD_p n_i^2}{L_p N_D} + \frac{qD_n n_i^2}{L_n N_A}$$

Page 31:

$$J_n + J_p + J_d = qF(1-R)\frac{\alpha L}{(\alpha L)^2 - 1}\frac{1}{\cosh\frac{x_j}{L}}\left\{\alpha L - e^{\frac{\alpha L - 1}{L} x_j}\right\} \;\rightarrow$$

$$J_n + J_p + J_d = qF(1-R)\frac{\alpha L}{(\alpha L)^2 - 1}\frac{1}{\cosh\frac{x_j}{L}}\left\{\alpha L - e^{-\frac{\alpha L - 1}{L} x_j}\right\}$$

We apologise for any inconvenience caused.

Table of Contents

PART III. DYE-SENSITIZED SOLAR CELLS

PART IV. ORGANIC- AND CARBON-BASED SOLAR CELLS

PART V. OTHER NANOSTRUCTURES

PART I

FUNDAMENTALS OF NANOSTRUCTURED SOLAR CELLS

Nanostructured Materials for Solar Energy Conversion
T. Soga (editor)
© 2006 Elsevier B.V. All rights reserved.

Chapter 1

Fundamentals of Solar Cell

Tetsuo Soga

Department of Environmental Technology and Urban Planning, Nagoya Institute of Technology, Gokiso-cho, Showa-ku, Nagoya 466-8555, Japan

1. INTRODUCTION

Solar cell is a key device that converts the light energy into the electrical energy in photovoltaic energy conversion. In most cases, semiconductor is used for solar cell material. The energy conversion consists of absorption of light (photon) energy producing electron–hole pairs in a semiconductor and charge carrier separation. A p–n junction is used for charge carrier separation in most cases. It is important to learn the basic properties of semiconductor and the principle of conventional p–n junction solar cell to understand not only the conventional solar cell but also the new type of solar cell. The comprehension of the p–n junction solar cell will give you hints to improve solar cells regarding efficiency, manufacturing cost, consuming energy for the fabrication, etc. This chapter begins with the basic semiconductor physics, which is necessary to understand the operation of p–n junction solar cell, and then describes the basic principles of p–n junction solar cell. It ends with the concepts of solar cell using nanocrystalline materials. Because the solar cells based on nanocrystalline materials are complicated compared with the conventional p–n junction solar cell, the fundamental phenomena are reviewed.

2. FUNDAMENTAL PROPERTIES OF SEMICONDUCTORS [1–5]

2.1. Energy Band and Carrier Concentration

The electrons of an isolated atom have discrete energy levels. When atoms approach to form crystals, the energy levels split into separate but closely spaced levels because of atomic interaction, which results in a continuous energy band. Between the two bands – the lower called the *valence*

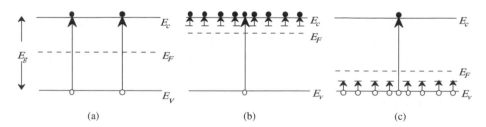

(a)   (b)   (c)

Fig. 1. (a) Energy band representations of intrinsic semiconductor, (b) extrinsic semi-conductor with donors and (c) extrinsic semiconductor with acceptors.

band and the upper called the *conduction band* – there is an energy gap called *band gap*, E_g, which is an important parameter in solar cell. All the energy levels in the valence band are occupied by electrons and those in the conduction band are empty at a temperature of 0 K. Some bonds are broken by the thermal vibrations at room temperature because the band gap is in the range of 0.5–3 eV. This results in the creation of electrons in the conduction band and holes in the valence band. The representation of energy band for semiconductor is shown in Fig. 1(a). E_c and E_v are designated as the bottom of the conduction band and the top of the valence band, respectively. The kinetic energy of electron is measured upward from E_c, whereas that of hole is measured downward from E_v, because a hole has a charge opposite to electron. The electrons in conduction band and holes in valence band can contribute to the current flow. On the contrary, in an insulator, the band gap is so large ($E_g > 5$ eV) that the conduction band is empty even at room temperature. In a conductor, the conduction band is partially filled with electrons or overlaps the valence band. Consequently, there is no band gap and the resistivity is very small.

2.1.1. Intrinsic Semiconductor

When electrons and holes generated from impurities are much smaller than thermally generated electrons and holes, they are called intrinsic semiconductors. The number of electrons in the conduction band per unit volume and that of holes in the valence band per unit volume are represented as n and p, respectively, and can be derived from the density of state and the distribution function. The electron concentration in the conduction band is expressed by

$$n = \int_0^{E_{top}} (\text{density of state}) \times (\text{probability that an electron state is occupied}) dE$$

where $E = 0$ means the energy of the bottom of the conduction band and E_{top} is the energy of the top of the conduction band. Assuming that the density of state is equal to $4\pi(2m_n/h^2)^{3/2}E^{1/2}$ and the probability of an energy level being occupied is given by Fermi-dirac distribution function, $\dfrac{1}{1 + e^{E - E_F/kT}}$, n is calculated to be

$$n = 2\left(\frac{2\pi m_n kT}{h^2}\right)^{3/2} \exp\left(-\frac{E_c - E_F}{kT}\right) = N_c \exp\left(-\frac{E_c - E_F}{kT}\right)$$

where E_F is the Fermi level, k the Boltzmann constant, T the absolute temperature, m_n the effective mass of the electrons, h the Planck's constant, and N_c the effective density of states of electrons in the conduction band. Similarly, the number of holes in the valence band can be calculated to be

$$np = N_v N_c e^{\frac{E_v - E_c}{kT}} = N_v N_c e^{\frac{E_g}{kT}} = n_i^2$$

where m_p is the effective mass of the holes and N_v the effective density of holes in the valence band.

For the ideal intrinsic semiconductor, the number of electrons in the conduction band is equal to that of holes in the valence band at a moderate temperature, that is, $n = p = n_i$, where n_i is the intrinsic carrier concentration. If we take a product of n and p, we get the following equation:

$$np = N_v N_c e^{E_v - E_x/\kappa T} = N_v N_c e^{E_g/kT} = n_i^2$$

It is obvious that the intrinsic carrier concentration decreases if the band gap becomes larger. The Fermi level of an intrinsic semiconductor is calculated to be

$$E_F = \frac{E_c + E_v}{2} + \frac{kT}{2}\ln\frac{N_v}{N_c}$$

Because the second term is much smaller than the first, the Fermi level of an intrinsic semiconductor lies close to the middle of band gap as shown in Fig. 1(a).

2.1.2. Extrinsic Semiconductor

When electrons and holes generated by impurity are not negligible, the semiconductor is called extrinsic semiconductor. Let us consider the carrier concentration in the case of Si. When group V atoms such as phosphorus (P) are doped as impurity, the phosphorous atom forms covalent bonds with its four neighboring Si atoms. The fifth electron is bound with P atom very loosely, and therefore ionized even at room temperature. Consequently, it becomes a conduction electron with negative charge. In this case, Si becomes n-type semiconductor and the phosphorous atom is called a donor. Under complete ionization condition, the electron (majority carrier) concentration is expressed as $n = N_D$, where N_D is the donor concentration. The donor is an immobile atom with positive charge. Because the equation $np = n_i^2$ is valid for extrinsic semiconductor under a thermal equilibrium, the hole (minority carrier) concentration is expressed as $p = n_i^2/N_D$. The Fermi level is expressed as

$$E_c - E_F = kT \ln \frac{N_c}{N_D}$$

The Fermi level can be controlled by the donor concentration and is close to the bottom of the conduction band. The schematic energy band representation of extrinsic semiconductor with donor is shown in Fig. 1(b).

Similarly, when group III atoms such as boron (B) are doped as impurity into silicon (Si), the boron atom forms covalent bonds with its four neighboring Si atoms. The positively charged conduction hole is created. This is a p-type semiconductor and the boron atom is called an acceptor. Under complete ionization condition, the hole (majority carrier) concentration is expressed as $p = N_A$, where N_A is the acceptor concentration. The acceptor is an immobile atom with negative charge. The electron (minority carrier) concentration is expressed as $n = n_i^2/N_A$. The Fermi level is expressed as

$$E_F - E_v = kT \ln \frac{N_v}{N_A}$$

In this case, the Fermi level moves closer to the top of the valence band. The schematic energy band representation of extrinsic semiconductor with acceptor is shown in Fig. 1(c).

2.2. Carrier Transport in Semiconductor

2.2.1. Mobility

The electrons in semiconductor move randomly in all directions by the thermal energy. After a short distance, the electrons collide with a lattice atom or an impurity atom, or other scattering center. This scattering process causes the electron to lose the energy taken from the electric field. The kinetic energy is transferred to the lattice in the form of heat. The average time between collisions is called the mean free time, τ_c. The random motion of electrons leads to the average net displacement to be zero. When a small electric field E is applied to the semiconductor, the electron experiences a force $-qE$ and gets accelerated toward the opposite direction of the field, where q is an electric charge (1.6×10^{-19}C). The velocity component produced by the electric field is called the drift velocity, v_n. The change in momentum of an electron in a mean free time is given by

$$m_n v_n = -qE\tau_c$$

so that

$$v_n = -\frac{q\tau_c}{m_n}E = -\mu_n E$$

where μ_n is called electron mobility. Similarly, the hole drift velocity is expressed by

$$v_p = \frac{q\tau_c}{m_p}E = \mu_p E$$

where μ_p is called hole mobility. The difference of sign is because of the direction accelerated by the electric field.

2.2.2. Drift Current

Let us consider an n-type semiconductor with a cross-sectional area of A and a carrier concentration of n as shown in Fig. 2. When an electric field E is applied to the sample, the electron current density is given by

$$J_n = \frac{I_n}{A} = -nqv_n = nq\mu_n E$$

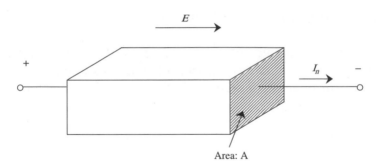

Fig. 2. Semiconductor sample to consider electron current density.

where I_n is the electron current. Similarly, the hole current density is given by

$$J_p = pqv_p = pq\mu_p E$$

The total current density due to the electric field is known as drift current density and given by

$$J = J_n + J_p = (nq\mu_n + pq\mu_p)E = \sigma E$$

where σ is called conductivity which is reciprocal of resistivity.

We consider the drift current using the energy band diagram. When the electric field is applied to the semiconductor, as shown in Fig. 3, the gradient of the conduction band and the valence band takes place and the electrons and holes flow to reduce the potential energy. It should be noted that the electrons and the holes move toward opposite directions, but the direction of current is the same.

2.2.3. Diffusion Current

When there is a spatial variation of electron concentration in the semiconductor sample, the electrons move from the region of higher concentration to that of lower concentration. This current is called the diffusion current. Let us consider the electron concentration with one-dimensional gradient in x-direction as shown in Fig. 4(a). The electron flows from right to left, and the electron flow rate per unit area is given by

$$F = -D_n \frac{dn}{dx}$$

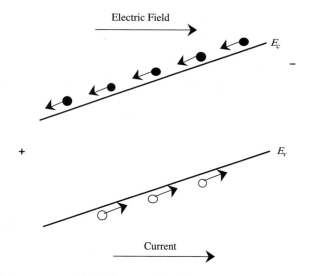

Fig. 3. Drift of electrons and holes in a semiconductor.

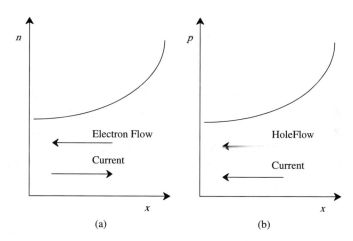

Fig. 4. (a) Diffusion of electrons and (b) holes.

where D_n is called diffusion coefficient of the electron. Therefore, the diffusion current density of electron is expressed by

$$J_n = qD_n \frac{dn}{dx}$$

When the electron concentration increases with x, the electrons diffuse toward the negative x-direction, resulting in the current flow toward the x-direction. Similarly, when the hole concentration increases with x, the holes diffuse toward the negative x-direction. It should be noted that the direction of current is opposite compared to electron. The diffusion current density by hole is given by

$$J_p = -qD_p \frac{dp}{dx}$$

where D_p is the diffusion coefficient of hole. When both an electric field and a concentration gradient are present, the total current density is given by

$$J_n = nq\mu_n E + qD_n \frac{dn}{dx}$$

for electrons and

$$J_p = pq\mu_p E - qD_p \frac{dp}{dx}$$

for holes.

We treated the diffusion and the drift phenomena separately, but there is a relationship between two phenomena. The diffusion coefficient is expressed by

$$D_n = \frac{kT}{q} \mu_n$$

which is known as Einstein relation.

2.3. Optical Absorption and Recombination in Semiconductor

2.3.1. Optical Absorption

The energy of a photon is hv, where h is Planck's constant and v is the frequency of the light. The relationship between photon energy and the wavelength λ is given by

$$\lambda \, (\mu m) = \frac{c}{v} = \frac{hc}{hv} = \frac{1.2398}{hv \, (eV)}$$

where c is the speed of light in vacuum. The spectrum of the solar light energy spreads from the ultraviolet region (0.3 µm) to the infrared region (3 µm). Let us assume that a semiconductor is illuminated with solar light. When the photon energy is less than band gap of the semiconductor, the light is transmitted through the material, that is, the semiconductor is transparent to the light. When the photon energy is larger than band gap, the electrons in the valence band are excited to the conduction band. It means that a photon is absorbed to create an electron–hole pair. This process is called intrinsic transition or band-to-band transition. The cutoff wavelength λ_0 is very important to choose the solar cell material because the light with wavelength longer than cutoff wavelength cannot be used for solar energy conversion.

The transition of electron by the optical absorption is shown in Fig. 5. When hv is bigger than E_g, an electron–hole pair is created, whereas the excess energy $hv - E_g$ gives the electron, or hole, additional kinetic energy that is dissipated as heat in semiconductor.

Assuming that a semiconductor is illuminated by a light with a photon flux F_0 normal to the surface in units of photons per unit area per unit time, the number of photons absorbed within a depth of x and $x + \Delta x$ is given by

$$F(x+\Delta x) - F(x) = -\alpha F(x)\Delta x$$

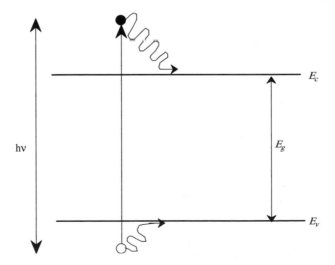

Fig. 5. Optical absorption process in semiconductor.

where α is the absorption coefficient. If the initial condition is given by $F(0) = (1-R)F_0$, the flux of photons at the depth of x is

$$F(x) = (1-R)F_0 e^{-\alpha x}$$

assuming that R is the reflectivity of the surface to normally incident light. At the distance of $1/\alpha$, the photon flux that exits is $1/e$ of the initial value. If the absorption coefficient is large, the photons are absorbed in a short distance, but the long distance is necessary for the photons to be absorbed when the absorption coefficient is small. It is important to note that the absorption coefficient is a strong function of photon energy. The absorption coefficients are approximately expressed as

$$A*(h\nu - E_g)^{1/2}$$

for direct gap semiconductor and

$$\frac{A**(h\nu - E_g + E_p)^2}{e^{E_p/kT} - 1} + \frac{A**(h\nu - E_g - E_p)^2}{1 - e^{-E_p/kT}}$$

for indirect gap semiconductor, where $A*$ and $A**$ are material dependent constants and E_p is the phonon energy associated at the absorption [6]. When $h\nu - E_g$ is much larger than E_p, and E_p is much smaller than kT, E_p can be neglected and the absorption coefficient of indirect gap semiconductor is proportional to $(h\nu - E_g)^2$. Generally, the absorption coefficient of an indirect gap semiconductor is much smaller than that of direct gap semiconductor because the absorption or the emission of phonon is involved in accompanying a change in momentum of electrons at the absorption of photons.

2.3.2. Recombination in Semiconductor

The excess charge carriers created in a semiconductor by absorption of light are annihilated after the source light is turned off. This process is called recombination. The recombination phenomena in bulk is classified into direct recombination (Fig. 6(a)), indirect recombination via localized energy states in the forbidden energy gap (Fig. 6(b)), Auger recombination (Fig. 6(c)), etc. First, let us consider direct recombination, which usually dominates in direct band gap semiconductors; this process is an inverse of absorption. When an electron makes a transition from the conduction band

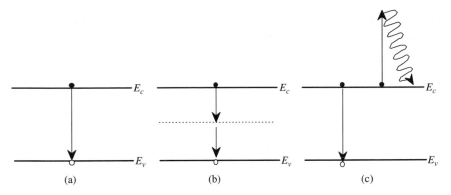

Fig. 6. Recombination processes in semiconductors: (a) direct recombination, (b) indirect recombination, and (c) Auger recombination.

to the valence band, an electron-hole pair is annihilated, resulting in the emission of photon. Under thermal equilibrium conditions, the recombination rate is $\beta n_{n0} p_{n0}$ for an n-type semiconductor and is equal to the generation rate (the number of electron–hole pairs generated per unit volume per unit time) by thermal vibration, where β is the proportionality constant and n_{n0} and p_{n0} are electron and hole concentrations, respectively, in n-type semiconductor at thermal equilibrium. When excess carriers are introduced by light illumination, the recombination rate is increased to βnp because the recombination is proportional to the number of electrons in conduction band and that of holes is valence band. In the case of low-injection level, the net recombination rate is given by

$$U = \beta np - \beta n_{n0} p_{n0} = \beta(n_{n0} + \Delta n)(p_{n0} + \Delta p) - \beta n_{n0} p_{n0}$$
$$\cong \beta n_{n0} \Delta p = \frac{\Delta p}{1/\beta n_{n0}} = \frac{\Delta p}{\tau_p}$$

where Δn and Δp are the excess electron and hole concentrations by light illumination. It means that the net recombination rate is proportional to the excess minority carrier concentration. $\tau_p = 1/\beta n_{n0}$ is called the minority carrier (hole) lifetime. Similarly, in the case of p-type semiconductor, the net recombination rate is expressed as $U = \Delta n/\tau_n$, where τ_n is the minority carrier (electron) lifetime expressed by $\tau_n = 1/\beta p_{p0}$.

Next, let us consider a semiconductor containing trap states near the midgap (Fig. 6(b)) with a concentration of N_t. This indirect recombination is very likely to be for indirect band gap semiconductor. For an n-type semiconductor, the minority carrier lifetime τ_p and the net recombination rate U are given by

$$\tau_p = \frac{1}{v_{th}\sigma_p N_t}$$

and

$$U = \frac{\Delta p}{\tau_p}$$

where v_{th} is the mean thermal velocity of hole and σ_p the capture cross section of the hole trap. Similarly, for a p-type semiconductor, the minority carrier lifetime and the recombination rate are given by

$$\tau_n = \frac{1}{v_{th}\sigma_n N_t}$$

and

$$U = \frac{\Delta n}{\tau_n}$$

In the case of indirect recombination, the minority carrier lifetime is independent of the majority carrier concentration, and is proportional to the inverse of the trap concentration.

In Auger recombination (Fig. 6(c)), one electron gives up its extra energy to another electron in the conduction band or the valence band during the recombination, resulting in the excitation of an electron to a higher energy level. The excited electron will give up this excess energy as heat when the excited electron relaxes to the band edge. Because the Auger process involves three particles, its recombination rate is expressed as $U = An^2p$ or $U = Ap^2n$ for electron–electron–hole process and hole–hole–electron process, respectively. A is the Auger constant which strongly depends on temperature. Auger process is important when the carrier concentration is high, especially in low band gap semiconductor.

Because a semiconductor is abruptly terminated, the disruption of the periodic potential function results in the energy states within the energy band gap at the surface. These states – surface states – enhance the recombination near the surface. They become very important with reducing the crystal size because the number of carriers recombining at the surface per unit volume is increased.

For a low-injection condition, the total number of carriers recombining at the surface per unit area and unit time is expressed as

$$U_s = S(p_s - p_{n0})$$

for an n-type semiconductor, where S is the surface recombination velocity and p_s is the hole concentration at the surface. When an n-type semiconductor is irradiated uniformly by the light to create excess carriers, the gradient of hole concentration yields a diffusion current, which is equal to the surface recombination current as shown in the following equation:

$$qD_p \frac{dp_n}{dx}\bigg|_{x=0} = qU_s = qS(p_s - p_{n0})$$

2.3.3. Continuity Equation

We treated the drift current, the diffusion current, the generation, and the recombination individually. But in the real semiconductor, all the processes occur simultaneously. In order to derive the relationship of these phenomena in one-dimensional form, we consider an infinitesimal slice with a thickness of dx and an area of A as shown in Fig. 7. Assuming that the electron current density at x is $J_n(x)$, the net increase of the electrons per unit time in this volume is the sum of the net flow into the slice and the net carrier generation in the slice, that is,

$$\frac{\partial n_p}{\partial t} Adx = \left[\frac{J_n(x)A}{-q} - \frac{J_n(x+dx)A}{-q} \right] + (G_n - R_n)Adx$$

where G_n and R_n are the generation rate and the recombination rate of electrons, respectively. Using the Taylor series, we obtain the following continuity equation for electrons in p-type semiconductor as follows:

$$\frac{\partial n_p}{\partial t} = \frac{1}{q} \frac{\partial J_n}{\partial x} + (G_n - R_n)$$

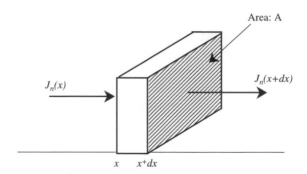

Fig. 7. Electron flow in small volume to explain the continuity equation.

Similarly, the one-dimensional continuity equation for holes in n-type semi-conductor is given by

$$\frac{\partial p_n}{\partial t} = \frac{1}{q}\frac{\partial J_p}{\partial x} + (G_p - R_p)$$

where J_p is the hole current density and G_p and R_p are the generation and the recombination rate of holes, respectively. Substituting the current expressions using the drift current and the diffusion current, the continuity equations are expressed as

$$\frac{\partial n_p}{\partial t} = n_p\mu_n\frac{\partial E}{\partial x} + \mu_n E\frac{\partial n_p}{\partial x} + D_n\frac{\partial^2 n_p}{\partial x^2} + G_n - \frac{n_p - n_{p0}}{\tau_n}$$

and

$$\frac{\partial p_n}{\partial t} = -p_n\mu_p\frac{\partial E}{\partial x} - \mu_p E\frac{\partial p_n}{\partial x} + D_p\frac{\partial^2 p_n}{\partial x^2} + G_p - \frac{p_n - p_{n0}}{\tau_p}$$

Let us calculate the steady state excess carrier concentration in a very simple case. Assuming that the excess generation occurs at $x = 0$ in an n-type semiconductor that is homogeneous and infinite in extent, the continuity equation is given by

$$D_p\frac{\partial^2 p_n}{\partial x^2} - \frac{p_n - p_{n0}}{\tau_p} = 0$$

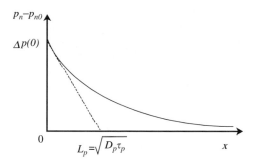

Fig. 8. Steady state excess carrier concentration profile when the excess carrier is generated at $x = 0$.

for zero-applied electric field. Because the excess carrier concentration should decay toward zero, the general solution is

$$p_n - p_{n0} = \Delta p(0)e^{-x/\sqrt{D_p \tau_p}} = \Delta p(0)e^{-x/L_p}$$

where $L_p = \sqrt{D_p \tau_p}$ is called the diffusion length for holes and $\Delta p(0)$ is the excess carrier concentration at $x = 0$. Similarly, $L_n = \sqrt{D_n \tau_n}$ is called the diffusion length for electrons. The distribution of the steady state excess carrier concentration is schematically shown in Fig. 8. The diffusion length is a measure of the average distance which a minority carrier can diffuse without recombination.

2.4. Photoconductive Effect

As described in Section 2.2., the conductivity of a semiconductor is given by

$$\sigma = n\, q\mu_n + p\, q\mu_p$$

where n and p are the electron concentration and the hole concentration, respectively, at thermal equilibrium. If the semiconductor is illuminated by light source to create electron–hole pairs, the conductivity of the semiconductor increases. This phenomenon is called photoconductive effect. For the practical p–n junction solar cell, the light-generated electrons and holes are separated by the internal electric field, which will be discussed in Section 3.2. Therefore, the photoconductivity gives very useful information on the performance of solar cell.

Let us calculate the photoconductivity of a semiconductor, which consists of a slab with Ohmic contacts at both ends, as shown in Fig. 9. Assuming that the slab of semiconductor is illuminated homogeneously, the conductivity is given by

$$\sigma_{ph} = (n + \Delta n)q\mu_n + (p + \Delta p)q\mu_p = \sigma + \Delta\sigma$$

at the steady state, where Δn and Δp are excess electron concentration and excess hole concentration, respectively. Since the generation rate is equal to the recombination rate, the continuity equations for electrons and holes are expressed as

$$\frac{\partial n}{\partial t} = G - \frac{\Delta n}{\tau_n} = 0$$

and

$$\frac{\partial p}{\partial t} = G - \frac{\Delta p}{\tau_p} = 0$$

Therefore, $\Delta\sigma$ is expressed by

$$\Delta\sigma = qG(\mu_n\tau_n + \mu_p\tau_p)$$

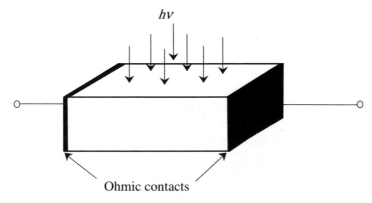

Fig. 9. Schematic diagram of photoconductor.

The photoconductive gain is given by

$$g = \frac{\tau_n}{t_n} + \frac{\tau_p}{t_p}$$

where t_n and t_p are the transit time of an electron and a hole between two Ohmic contacts, respectively. Usually, $\Delta\sigma/\sigma$ is a measure of how effectively the photogenerated electron–hole pairs arc collected at the external circuit.

3. BASIC PRINCIPLES OF p–n JUNCTION SOLAR CELL

3.1. Electric Properties [1–5]

3.1.1. Built-In Potential

The p–n junction is commonly used for solar cell. The important role of p–n junction is the charge separation of light-induced electrons and holes. Let us calculate the built-in potential of the p–n junction. Fig. 10 shows the energy band picture and majority carriers for n-type semiconductor and p-type semiconductor. When a p–n junction is formed, the large carrier concentration gradients cause the diffusion of carriers, that is, holes diffuse from p-type semiconductor to n-type semiconductor and electrons diffuse from n-type semiconductor to p-type semiconductor. Because of the ionized impurity atoms, a layer without mobile charge carriers is formed when the electrons

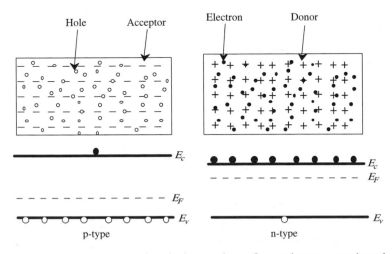

Fig. 10. Energy band pictures and majority carriers of n- and p-type semiconductors.

and holes diffuse across the junction. This space charge sets up an electric field, which opposes the diffusion across the junction as shown in Fig. 11(a). When the drift current due to the electric field is balanced by the diffusion current because of the carrier concentration gradient for each carrier, the thermal equilibrium is established. At this point, the Fermi levels of the p-type semiconductor and n-type semiconductor are equal as shown in Fig. 11(b).

The electrostatic potential difference between the p-type semiconductor and the n-type semiconductor at thermal equilibrium is called the built-in potential V_b. V_b is equal to the difference in the work function of p-side and n-side and is given by

$$V_b = \frac{kT}{q} \ln \frac{N_A N_D}{n_i^2}$$

where N_A and N_D are the concentrations of the acceptor and donor in p-type semiconductor and n-type semiconductor, respectively.

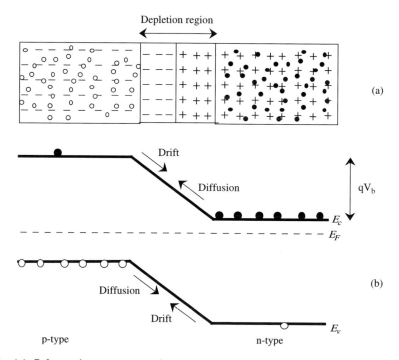

Fig. 11. (a) Schematic structures of p–n junction and (b) its energy band diagram in thermal equilibrium.

3.1.2. *Depletion Region*

At the p–n junction, a transition region free of mobile carriers called the depletion region is formed as shown in Fig. 11(a). The depletion region is charged by the ionized donor and acceptor ions, although the region beyond the depletion region is electrically neutral. Let us calculate the depletion layer width of an abrupt junction with a layer of p-type semiconductor of doping N_A for $x < 0$ and n-type semiconductor of doping N_D for $x > 0$ as shown in Fig. 12. x_p and x_n denote the depletion layer width of p- and n-side, respectively, ignoring the transition region. According to Poisson's equation, the electrostatic potential ϕ must obey

$$\frac{d^2\phi}{dx^2} = \frac{q}{\varepsilon}N_A \quad (-x_p \le x < 0)$$

and

$$\frac{d^2\phi}{dx^2} = -\frac{q}{\varepsilon}N_D \quad (0 < x \le x_n)$$

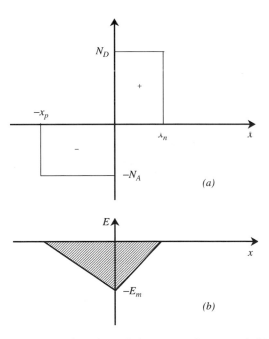

Fig. 12. (a) Rectangular approximation of the space charge and (b) the distribution of electric field at the p–n junction.

where ε is the dielectric constant of the semiconductor. By integrating these equations the electric field E is expressed as

$$E = -\frac{d\phi}{dx} = -\frac{qN_A(x+x_p)}{\varepsilon} \quad (-x_p \leq x < 0)$$

and

$$E = -\frac{d\phi}{dx} = -\frac{qN_D(x-x_n)}{\varepsilon} \quad (0 < x \leq x_n)$$

$E_m = qN_D/\varepsilon \, x_n = qN_A/\varepsilon \, x_p$ is the maximum electric field that exists at $x = 0$. The total potential difference, namely built-in potential, is given by

$$V_b = -\int_{-x_p}^{x_n} E\,dx = \frac{qN_A x_p^2}{2\varepsilon} + \frac{qN_D x_n^2}{2\varepsilon} = \frac{1}{2}E_m w$$

which is equal to the area of the field triangle shown in Fig. 12(b). The total depletion width w is given by

$$w = \sqrt{\frac{2\varepsilon}{q}\left(\frac{1}{N_A} + \frac{1}{N_D}\right)V_b}$$

The depletion layer width increases when either the donor concentration or the acceptor concentration is reduced. When the impurity concentration on one side is much higher than that of the other side, for example, in the case of p^+–n junction, where $N_A \gg N_D$, the total depletion layer width is given by

$$w = \sqrt{\frac{2\varepsilon}{qN_D}V_b}$$

3.1.3. Ideal Current–Voltage Characteristics under Dark

When a bias voltage V_F with the positive terminal to the p-side and the negative terminal to the n-side is applied, the applied voltage reduces the electrostatic potential across the depletion region as shown in Fig. 13(a).

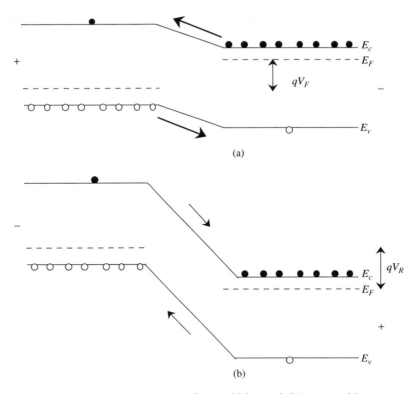

Fig. 13. Energy band diagram under (a) forward bias and (b) reverse bias.

This polarity is called the *forward bias*. In this case the drift current is reduced and the diffusion of electrons and holes increases from the n-side to the p-side and from the p-side to the n-side, respectively. Therefore, the minority carrier injection occurs, that is, electrons are injected into the p-side and holes are injected into the n-side.

At thermal equilibrium, the electron concentration in the n-side is expressed as

$$n_{n0} = n_{p_0}\, e^{qV_b/kT}$$

Therefore, when a forward bias V_F is applied to the junction, the electron concentration at the boundary of the depletion region in the n-side is expressed as

$$n_n = n_p\, e^{q(V_b - V_F)/kT}$$

using the electron concentration at the boundary of the depletion region in the p-side.

In the case of the low-injection condition ($n_n \approx n_{n0}$), n_p is expressed as

$$n_p = n_{p0} e^{qV_F/kT}$$

In the n-layer, the steady state continuity equation is

$$D_p \frac{d^2 p_n}{dx^2} - \frac{p_n - p_{n0}}{\tau_p} = 0$$

The solution of this differential equation is given by

$$p_n - p_{n0} = p_{n0}(e^{\frac{qV_F}{kT}} - 1) e^{-\frac{x - x_n}{L_p}}$$

where L_p is the diffusion length of holes in the n-layer. Therefore, the diffusion current density in n-side at $x = x_n$ is

$$J_p = -qD_p \frac{dp_n}{dx}\bigg|_{x=x_n} = \frac{qD_p p_{n0}}{L_p}(e^{qV_F/kT} - 1)$$

Similarly, the diffusion current density in p-side at $x = -x_p$ is

$$J_n = qD_n \frac{dn_p}{dx}\bigg|_{x=-x_p} = \frac{qD_n p_{p0}}{L_n}(e^{qV_F/kT} - 1)$$

where L_n is the diffusion length of electrons in the p-layer. Thus the total current density is the sum of the two, as follows:

$$J = J_n + J_p = \left(\frac{qD_p p_{n0}}{L_p} + \frac{qD_n n_{p0}}{L_n} \right)(e^{qV_F/kT} - 1) = J_0(e^{qV_F/kT} - 1)$$

J_0 is called the saturation current density and is expressed as

$$J_0 = \frac{qD_p p_{n0}}{L_p} + \frac{qD_n n_{p0}}{L_n} = \frac{qD_p n_i^2}{L_p N_D} + \frac{qD_n n_i^2}{L_n N_A}$$

When a reverse bias voltage V_R is applied to the p–n junction, the applied voltage increases the electrostatic voltage across the depletion region as shown in Fig. 13(b). Therefore, the diffusion current is suppressed. Similarly, the current–voltage characteristics under a reverse bias is given by

$$J = J_0(e^{-qV_R/kT} - 1)$$

3.1.4. Effects of Generation and Recombination

It is difficult to fabricate a p–n junction with an ideal current–voltage characteristic. In the practical p–n junction solar cell, it is necessary to consider the generation and the recombination of carriers in the depletion region. The electron and hole generations occur through the energy state in the forbidden energy gap in the case of reverse bias condition. At the forward bias, the carriers will recombine through the energy state at the forbidden energy gap. The recombination current is approximately given by

$$J_{rec} \propto e^{qV_F/2kT}$$

The forward current density of the practical p–n junction is represented empirically by

$$J = J_0(e^{qV_F/nkT} - 1)$$

where n is called the ideality factor. In the case of the ideal p–n junction, when the diffusion current dominates, $n = 1$ and when the recombination current dominates, $n = 2$, that is, n has a value between 1 and 2.

3.2. Photovoltaic Properties [7–10]

When the p–n junction is illuminated by the sunshine, electron–hole pair is generated by the photons that have energy greater than the band gap. The number of electron–hole pair is proportional to the light intensity. Because of the electric field in the depletion region due to the ionized impurity atoms, the drift of electrons toward n-side and that of holes toward p-side occur in the depletion region. This charge separation results in the current flow from n- to p-side when an external wire is short-circuited as shown in Fig. 14. The electron–hole pairs generated within a distance of diffusion length from the edge of the depletion region contribute to the photo current because of the diffusion of excess carriers up to the space charge region.

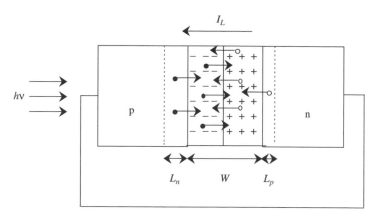

Fig. 14. Schematic illustration of carrier flow in illuminated p–n junction in the case of short-circuited.

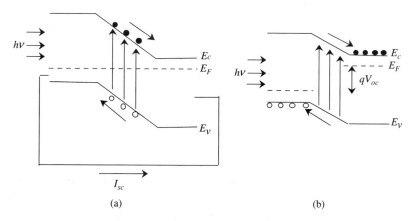

(a)   (b)

Fig. 15. Energy band diagrams of illuminated p–n junction in (a) the short-circuited and (b) open-circuited current.

When the p–n junction illuminated by sunshine is open-circuited, the voltage is generated due to the charge carrier separation. The energy band diagrams of p–n junction in the short-circuited and open-circuited current are shown in Figs 15(a) and (b), respectively. When the p- and n-side are short-circuited, the current is called the short-circuit current I_{sc} and equals to the photogenerated current I_L if the series resistance is zero. When the p- and the n-side are isolated, electrons move toward n-side and holes toward p-side, resulting in the generation of potential. The voltage developed is called the open-circuit voltage V_{oc}. The current-voltage characteristics of the p–n junction under illumination and dark are schematically

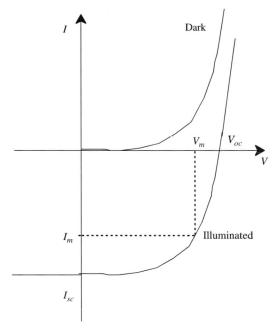

Fig. 16. Current–voltage characteristics of p–n junction under illumination and darkness.

shown in Fig. 16. Assuming that the area of the solar cell is unity, the current–voltage characteristic of the illuminated p–n junction is given by

$$I = I_0(e^{qV/nkT} - 1) - I_{s o}.$$

In the open-circuited, which is obtained for $I = 0$, the voltage is given by

$$V_{oc} = \frac{nkT}{q} \ln\left(\frac{I_{sc}}{I_0} + 1\right)$$

When the solar cell is operated under a condition that gives the maximum output power, the voltage V_m and the current I_m at the optimal operation point are shown in Fig. 16.

The fill factor *FF* of the solar cell is defined as

$$FF = \frac{V_m I_m}{V_{oc} I_{sc}}$$

The conversion efficiency of the solar cell η is defined as the ratio of the generated maximum electric output power to the total power of the incident light P_{in}.

$$\eta = \frac{V_m I_m}{P_{in}} = \frac{V_{oc} I_{sc} FF}{P_{in}}$$

The photovoltaic parameters are evaluated under standard test conditions: the air mass (AM) 1.5 spectrum with an incident power density of $1000\,W/m^2$ and a temperature of $25°C$. In order to improve the efficiency, it is necessary to maximize all the three photovoltaic parameters, such as V_{oc}, I_{sc} and FF.

3.3. Output Parameters of p–n Junction Solar Cell

Let us calculate the short-circuit current density, the open-circuit voltage, the fill factor and the energy conversion efficiency of real p–n junction solar cell. To derive analytical expressions of the output parameters, the simple p–n structure with the homogeneously doped abrupt junction, as shown in Fig. 17, is considered for the case of the low-injection level. It consists of a p-type emitter layer on the n-type base layer.

3.3.1. Short-Circuit Current

When light enters from the emitter side, the number of electrons and holes are generated at a distance x with the generation rate of

$$G = \alpha F (1 - R) e^{-\alpha x}$$

where α is the absorption coefficient of the light, R the reflectivity of the light at the surface, F the incident photon flux defined by the number of incident photons per unit area, unit time, and unit wavelength. All these three variables depend on the wavelength. Under steady state conditions, the continuity equation in the emitter layer is given by

$$D_n \frac{\partial^2 (n_p - n_{p0})}{\partial x^2} + \alpha F (1 - R) e^{-\alpha x} - \frac{n_p - n_{p0}}{\tau_n} = 0$$

The electrons generated in the p-layer reaching the edge of the depletion region by diffusion are immediately accelerated by the electric field to the

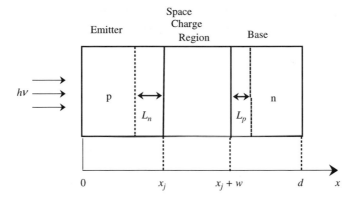

Fig. 17. Schematic cross-sectional view of p–n junction solar cell to calculate the solar cell parameters.

opposite side of the junction. So it is assumed that $n_p - n_{p0} = 0$ at $x = x_j$. Considering the surface recombination velocity S_n of the excess electron at the front surface

$$D_n \frac{\partial n_p}{\partial x} = S_n(n_p - n_{p0}) \quad \text{at} \quad x = 0$$

Under these boundary conditions, the electron current density in the emitter layer is given by

$$J_n = qD_n \left. \frac{\partial(n_p - n_{p0})}{\partial x} \right|_{x=x_j}$$

$$= qF(1-R)\frac{\alpha L_n}{(\alpha L_n)^2 - 1} \left[\frac{\dfrac{S_n L_n}{D_n} + \alpha L_n}{\dfrac{S_n L_n}{D_n} \sinh \dfrac{x_j}{L_n} + \cosh \dfrac{x_j}{L_n}} \right.$$

$$\left. - \left\{ \alpha L_n + \frac{\dfrac{S_n L_n}{D_n} \cosh \dfrac{x_j}{L_n} + \sinh \dfrac{x_j}{L_n}}{\dfrac{S_n L_n}{D_n} \sinh \dfrac{x_j}{L_n} + \cosh \dfrac{x_j}{L_n}} \right\} e^{-\alpha x_j} \right]$$

Similarly, the continuity equation in the base is given by

$$D_p \frac{\partial^2 (p_n - p_{n0})}{\partial x^2} + \alpha F(1-R)e^{-\alpha x} - \frac{p_n - p_{n0}}{\tau_p} = 0$$

Under the boundary conditions $p_n - p_{n0} = 0$ at $x = x_j + w$ and

$$D_p \frac{\partial p_n}{\partial x} = -S_p(p_n - p_{n0}) \quad \text{at} \quad x = d$$

where S_p is the surface recombination velocity of the excess holes at the back surface. The hole current density is given by

$$J_p = -qD_p \left. \frac{\partial(p_n - p_{n0})}{\partial x} \right|_{x=x_j+w}$$

$$= qF(1-R)\frac{\alpha L_p}{(\alpha L_p)^2 - 1} \left[\left[\alpha L_p - \frac{\frac{S_p L_p}{D_p}\cosh\frac{d}{L_p} + \sinh\frac{d}{L_p}}{\frac{S_p L_p}{D_p}\sinh\frac{d}{L_p} + \cosh\frac{d}{L_p}} \right] \right.$$

$$\left. - \left\{ \frac{\alpha L_p - \frac{S_p L_p}{D_p}}{\frac{S_p L_p}{D_p}\sinh\frac{d}{L_p} + \cosh\frac{d}{L_p}} \right\} e^{-\alpha d} \right] e^{-\alpha(x_j + w)}$$

Since the electric field in the space charge region is high, all the electrons and holes generated in this region are accelerated toward the opposite directions. The photocurrent density in the space charge region is given by

$$J_d = qF(1-R)(e^{-\alpha x_j} - e^{-\alpha(x_j + w)})$$

thereby ignoring the recombination.

The total short-circuit current density J_{sc} is calculated by integrating over the entire solar spectrum, so that

$$J_{sc} = \int_{\lambda_{min}}^{\lambda_{max}} (J_n + J_p + J_d) d\lambda$$

where λ_{min} is the smallest occurring wavelength and λ_{max} is the largest occurring wavelength. λ_{min} is around 0.3 μm for sunlight and λ_{max} is the wavelength corresponding to the absorption edge of the semiconductor. The photocurrent is proportional to the light intensity that emits to the semiconductor and strongly depends on the diffusion length and surface recombination velocity.

In the ideal p–n junction, it is possible to calculate the upper limit of the short-circuit current density. For simplicity, we assume that

$$d = \infty, \ w = 0, \ S_n = S_p = 0 \ \text{and} \ L_n = L_p = L$$

Then $J_n + J_p + J_d$ is expressed as

$$J_n + J_p + J_d = qF(1-R) \frac{\alpha L}{(\alpha L)^2} \frac{1}{\cosh \dfrac{x_j}{L}} \left\{ \alpha L - e^{\frac{\alpha L - 1}{L} x_j} \right\}$$

which is a function of the junction depth x_j. The maximum value of $J_n + J_p + J_d$ is

$$\frac{qF(1-R)}{\cosh \left(\dfrac{x_{j\,max}}{L} \right)} \quad \text{at} \quad x_{j\,max} \cong L \frac{\ln(\alpha L)}{\alpha L - 1}$$

If the diffusion length is long enough, that is, $\alpha L \gg 1$, the upper limit of $J_n + J_p + J_d$ is $qF(1-R)$. Therefore, the upper limit of the short-circuit current density is given by

$$J_{sc} = q \int_{\lambda_{min}}^{\lambda_{max}} F(1-R) d\lambda$$

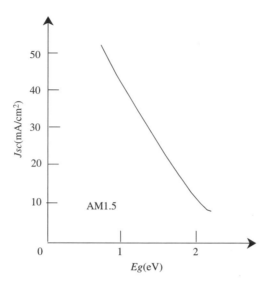

Fig. 18. Ideal short-circuit current density of p–n junction solar cell as a function of band gap.

Integrating from the lowest possible wavelength λ_{min} to the wavelength corresponding to the absorption edge $\lambda_{max}(\mu m) = 1.2398/E_g(eV)$, the upper limit of the short-circuit current can be calculated as a function of the band gap energy E_g. It is evident that the short-circuit current density increases with the reducing band gap energy. The relationship between the short-circuit current density and the band gap energy is shown in Fig. 18, assuming that R is zero.

3.3.2. Open-Circuit Voltage

The open-circuit voltage of the real p–n junction solar cell is given by

$$V_{oc} = \frac{nkT}{q} \ln\left(\frac{J_{sc}}{J_0} + 1\right)$$

where J_0 is the saturation current density expressed by

$$J_0 = qN_vN_c\left(\frac{1}{N_A}\sqrt{\frac{D_n}{\tau_n}} + \frac{1}{N_D}\sqrt{\frac{D_p}{\tau_p}}\right)e^{-E_g/kT}$$

It is evident that the saturation current should be small and the short-circuit current density should be large to get a large open-circuit voltage. To decrease the saturation current, the minority carrier lifetime should be long and the

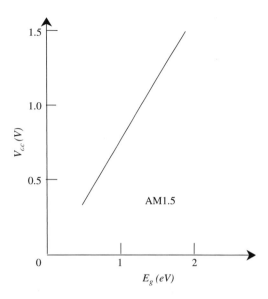

Fig. 19. Ideal open-circuit voltage of p–n junction solar cell as a function of band gap.

carrier concentration N_A and N_D should be high. The saturation current strongly depends on the choice of semiconductor. The upper limit of the open-circuit voltage can be approximately expressed as

$$V_{oc} \cong \frac{E_g}{q}\left(1 - \frac{T_0}{T_s}\right) + \frac{kT_0}{q}\ln\frac{T_s}{T_0} + \frac{kT_0}{q}\ln\frac{\Omega_{\text{inc}}}{4\pi}$$

using the thermodynamic engine with Carnot efficiency, where T_0 is the solar cell temperature, T_s the temperature of the sun, and Ω_{inc} the solid angle which the solar cell receive the incident irradiation from the sun [11]. The upper limit of the open-circuit voltage increases with the band gap energy as shown in Fig. 19.

3.3.3. Fill Factor and Conversion Efficiency

The fill factor is a function of the open-circuit voltage and is approximately expressed by

$$FF = \frac{\dfrac{qV_{oc}}{kT} - \ln\left(0.72 + \dfrac{qV_{oc}}{kT}\right)}{1 + \dfrac{qV_{oc}}{kT}}$$

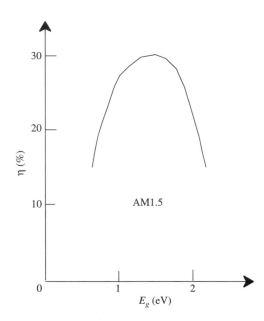

Fig. 20. Ideal conversion efficiency of p–n junction solar cell as a function of band gap.

The fill factor increases with the increase of the open-circuit voltage.

The upper limit of the conversion efficiency is a strong function of the band gap energy. The optimum efficiency of ~30% occurs when the band gap is between 1.4 eV and 1.6 eV, as shown in Fig. 20 at AM 1.5 and 1 sun. The band gap energy between 1 eV and 2 eV is suitable for solar cell to achieve relatively high efficiency.

3.4. Energy Losses of p–n Junction Solar Cell

3.4.1. Energy Losses of Ideal Solar Cell [12]

There are several reasons for the limited efficiency of an ideal p–n junction solar cell. First is the inability of the semiconductor to absorb photons below band gap energy. The photons with the energy less than the band gap energy ($hv < E_g$) are not absorbed in the semiconductor even if the thickness is sufficiently thick. The portion of this energy loss is indicated by A in Fig. 21, which is a solar energy spectrum assuming the black body radiation at 6000 K. Second is the excess energy loss, which occurs because the energy of the photon above the band gap energy is wasted in the form of heat. The excess energy of photons ($hv - E_g$) contributes to the lattice vibration (shown by B in Fig. 21). Even if the electrons and holes are created at

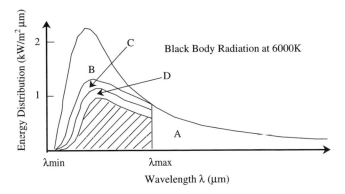

Fig. 21. Schematically illustrated spectral distribution of sunlight and energy losses. A corresponds to the energy which is not absorbed by semiconductor, B to the excess energy loss, C to the voltage factor loss, and D to the fill factor loss.

the bottom of the conduction band and at the top of the valence band, respectively, the open-circuit voltage is always smaller than the band gap energy because the Fermi level is located inside the energy band gap. This is the third reason for the loss that occurs because a p–n junction is incapable of fully utilizing the maximum voltage (shown by C in Fig. 21). Fourth is caused because the fill factor is smaller than unity. When the maximum power is extracted from the solar cell with the optimal load, the operation voltage is smaller than the open-circuit voltage. This loss is shown by D in Fig. 21. Therefore, only the energy shown by the shadow in Fig. 21 can be converted into electrical energy by p–n junction solar cell.

3.4.2. *Energy Losses of Real Solar Cell*

The energy conversion efficiency of real solar cell is usually lower than that of the ideal solar cell owing to various additional mechanisms. It is difficult to obtain the ideal material quality under limited procedure production costs. The three major factors that degrade the efficiency are: (1) The reflection loss. Because the reflectivity of bare Si wafer is ∼30%, only ∼70% of solar energy can be used for the energy conversion. To decrease the reflectivity, antireflection coating or texturing the surface are employed. The metal grid on the front surface also reduces the incident solar energy. (2) The recombination loss. The material parameters which affect the efficiency are the minority carrier lifetime and the mobility of carriers. Because the carriers generated in the depletion region and within the diffusion length from the edge of the depletion region can be collected as a photocurrent, some losses occur when the diffusion length is not sufficiently long. If deep levels or other lattice defects such as dislocations or grain boundary are present in the material, the diffusion length is reduced. High-impurity doping also reduces the diffusion length. The open-circuit voltage is degraded because of the increase of the saturation current by the lattice defects. The large surface recombination at the front and back surface degrades the open-circuit voltage and the short-circuit current. (3) The series and shunt-resistance losses. The origin of the series resistance is the resistance of semiconductor bulk, contact, interconnection, etc. The shunt resistance is caused by lattice defects and leakage current through the edge of the solar cell. When the thickness of the solar cell is not sufficiently thick, some portions of the photons are transmitted through the solar cell material, causing the loss. It is difficult to obtain an ideal material quality under limited procedure production costs. In order to increase the efficiency toward the ideal one, it is necessary to reduce the sum to losses.

4. BASIC PRINCIPLES OF SOLAR CELL USING NANOCRYSTALLINE MATERIALS

4.1. Fundamental Properties of Nanocrystalline Materials

A semiconductor indicates bulk properties when the crystal size is sufficiently large (Fig. 22(a)). But the size effects arise from the spatial confinement of charge carriers at very small semiconductor crystallites or structures where excitons are confined to two or three dimensions. Generally, there are two cluster regimes: (a) when the number of atoms is around 10^3–10^5, the bulk lattice structure exists but the electric properties are modified from the bulk and (b) when the number of atoms is ~30–10^3, the bulk lattice structure is not found and the properties are entirely molecules. The crossover point is a function of surface stabilization by external medium.

When the dimension of the semiconductor crystallite is smaller than the de Broglie wavelength of charge carriers of the semiconductor, the quantum size effect occurs. Therefore, the quantum size effects are dependent on the effective mass of the charge carriers. The electrons and holes are confined to the small crystallite, and discrete electronic states are created as shown in Fig. 22 (b). An analytical approximation for the lowest excited 1 s state is given by

$$E^* \approx E_g + \frac{h^2}{8r^2}\left(\frac{1}{m_e} + \frac{1}{m_h}\right) - \frac{1.8\,q^2}{\varepsilon r}$$

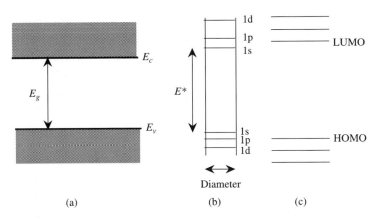

Fig. 22. (a) Electronic states for bulk semiconductor, (b) small crystallite, and (c) molecule.

where r is the crystal radius, m_e an effective mass of electrons and m_h an effective mass of holes [13, 14]. The second term shifts E^* to higher energy as r^{-2}, and the third term shifts E^* to lower energy as r^{-1}. Thus, E^* always increases for small r. Other than the increase of the effective band gap, the discrete electronic states are created in the valence and conduction bands. As the crystal size decreases, the charge–carrier confinement effects become noticeable and the discrete electronic states are produced, as shown in Fig. 22(b). Finally, with the decrease of crystal size the transition energy approaches the highest occupied molecular – lowest unoccupied molecular (HOMO–LUMO) levels which are observed in molecule as shown in Fig. 22(c). Fig. 22 shows the energy diagram of bulk semiconductor, small crystallite and molecule.

At the same time, the surface effects – which are usually neglected in the bulk semiconductor – become notable with the decrease of crystal size because the number of surface atoms becomes comparable to those in the crystallite. The surface dangling bond states exist as midgap traps that participate in surface reconstruction or chemical bonding to foreign atoms. The internal level structure is discrete and electrons occupy molecular orbital while satisfying the Pauli exclusion principle.

4.2. Energy Band Structure

When a bulk semiconductor is in contact with a different semiconductor, metal or electrolyte, a space charge region is formed at the interface to equalize the Fermi levels. The generation of space charge region is important to separate the photogenerated electrons and holes toward opposite

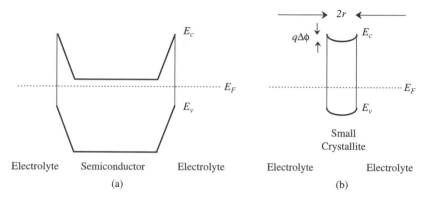

Fig. 23. Energy band diagrams of (a) large and (b) small semiconductor particles in electrolyte.

directions. An example of the energy band diagram for an n-type bulk semi-conductor contact with an electrolyte is shown in Fig. 23(a).

When the radius of the crystallite is smaller than the thickness of the space charge layer, the potential drop in the semiconductor becomes lim-ited. Under these conditions, all the donors are ionized and there are elec-trons left in the conduction band of the semiconductor. The potential drop between the center and the surface of the crystallite is given by

$$\Delta\phi = \frac{kT}{6q}\left(\frac{r}{L_{\mathrm{d}}}\right)^2$$

where r is the radius of the small crystallite and $L_d = \sqrt{\varepsilon kT/2q^2 N_D}$ is the Debye length, which depends on the donor concentration N_D [15–17]. The potential drop is very small compared to the band gap, and sometimes neg-ligible when the crystallite size is small. When the donor concentration is very high, the potential drop cannot be ignored even in the case of small crystallite.

4.3. Light-Induced Charge Separation

In a bulk semiconductor light-induced electrons and holes are separated effectively by the assistance of built-in electric field of space charge region. The minority carriers, which can diffuse to the edge of the depletion region before the recombination, also contribute to the photocurrent. In the case of small crystallite, the band bending is very small. Therefore, the light-induced

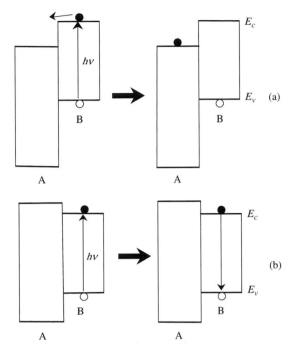

Fig. 24. Schematic illustrations of energy-level alignment for two cases. It is assumed that the band gap of A is larger than that of B and the photons are absorbed in B. (a) Charge carrier separation takes place, (b) the charge separation does not occur.

electrons diffuse toward the surface or recombine with holes or are captured by the trap levels. The average transit time from the interior of the semiconductor particle to the surface is approximately expressed as

$$\tau_d = \frac{r^2}{\pi^2 D}$$

from a random walk model, where D is the diffusion coefficient of electron [16]. Usually when the crystal size is small, the charge carriers can reach the surface before the recombination because τ_d is shorter than the relaxation time.

Let us assume that light-induced charge carriers are generated in small semiconductor crystallites B. When another semiconductor crystallite A (the band gap of A is larger than that of B), whose conduction band position is lower, is contacted to semiconductor crystallite B, the light-induced electrons are transferred from crystallite B to crystallite A as shown in Fig. 24(a).

In this figure the band bending was ignored. It means that the light-induced charge carriers were separated as is observed in the space charge region. However, if the conduction band position of the crystallite A is higher than that of crystallite B as shown in Fig. 24(b), the photogenerated electrons recombine with holes at the valence band. Therefore, the ability of light-induced charge separation is governed by the band position of the materials.

The light-induced charge separation takes place not only at semiconductor/semiconductor interface, but also at semiconductor/adsorbed molecule interface, organic material/organic material interface, semiconductor/quantum dot interface, etc. The difference with the conventional p–n junction is that the charge carrier separation is not governed by the electric field at the space charge region and minority carriers do not participate in the charge separation very much, whereas the photocurrent is strongly influenced by the minority carrier diffusion length in case of p–n junction solar cell. The effective interface area is strongly enhanced by the nanometer-sized small crystallites, resulting in the efficient absorption of the solar light.

The quantum yield of charge carrier injection is the fraction of the light-induced electrons into those injected in the conduction band of another material. The charge carrier injection is the competition with the radiative or nonradiative recombination. The quantum yield is given by

$$\phi_{inj} = \frac{k_{inj}}{k_{inj} + \tau^{-1}}$$

where k_{inj} and τ are the injection rate and the carrier lifetime, respectively [18]. To achieve a high-quantum yield close to unity, k_{inj} should be 100 times higher than τ^{-1} with reducing trap levels, interface levels, crystal size, etc.

4.4. Collection of Light-Induced Charge Carriers

The light-induced electrons and holes separated at the interface should reach each counter electrodes to be collected as a photocurrent. There are several losses during the transport of carriers. When the conductivity of the nanocrystalline material is not high, there is an energy loss during the transport of electrons in the high-resistive materials to the electrode. The back reaction of separated charge carriers toward the fundamental state also reduces the external quantum efficiency. The recombination with the localized energy levels created by defects or impurities causes current leakage. Therefore, it is necessary to reduce these losses for high energy conversion.

Examples of solar cell structures to collect the light-induced charge carriers using nanocrystalline semiconductors are shown in Fig. 25. In Figs 25(a)–(c) the light-induced electron and holes are separated at the interface of materials A and B. Electrons and holes reach each conducting layer via material A and hole conducting material, respectively (Fig. 25(a) and (b)), and materials A and B (Fig. 25(c)). In the case of organic solar cell where donor and acceptor materials are blended together, the charge carriers are separated at the interface by exciton dissociation and percolated toward each electrode through a continuous path of polymer network (Fig. 25(d)).

The open-circuit voltage is governed by the difference of the Fermi levels of two materials. An example of the energy band diagram for the whole device is shown in Fig. 26. The electron-conducting material and the

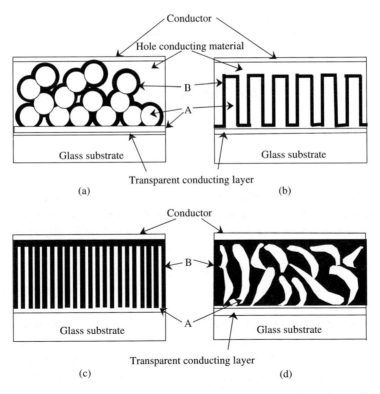

Fig. 25. (a–d) Schematic cross-sectional views of examples for solar cells based on nanocrystalline materials. A and B denote the materials A and B, respectively, shown in Fig. 24(a).

T. Soga

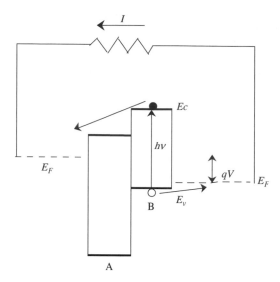

Fig. 26. Energy band profile of solar cell based on nanocrystalline materials.

hole-conducting material should be chosen, so that the energy level lineup encourages the electrons and holes to flow smoothly without energy loss.

REFERENCES

[1]   S. M. Sze, Physics of Semiconductor Devices, 2nd Edition, Wiley, New York, 1981.
[2]   S. M. Sze, Semiconductor Devices Physics and Technology, Wiley, New York, 1985.
[3]   M. A. Green, Solar Cells Operation Principle, Technology and System Application, The University of New South Wales, Kensington, 1992.
[4]   D. A. Neamen, Semiconductor Physics and Device, McGraw-Hill, New York, 2003.
[5]   D. K. Ferry and J. P. Bird, Electronic Materials and Devices, Academic Press, San Diego, 2001.
[6]   J. I. Pankove, Optical Processes in Semiconductors, Dover, New York, 1975.
[7]   H. J. Hovel, Semiconductors and Semimetals, Vol. 11 Solar Cells, Academic Press, New York, 1975.
[8]   A. Goetzberger, J. Knobloch and B. Vob, Crystalline Silicon Solar Cells, Wiley, New York, 1998.
[9]   H. J. Moller, Semiconductors for Solar Cells, Artech House, London, 1993.
[10] J. Nelson, The Physics of Solar Cells, Imperial College Press, London, 2003.
[11] W. Ruppel and P. Wurfel, IEEE Trans. ED-27 (1980) 877–882.
[12] M. Wolf, Proc. IRE, 40 (1960) 1246–1263.
[13] L. Brus, IEEE J. Quantum Electron., 22 (1986) 1909–1914.
[14] L. Brus, J. Phys. Chem., 90 (1986) 2555–2560.

[15] W. J. Albery and P. N. Bartlett, J. Electrochem. Soc., 131 (1984) 315–325.

[16] A. Hagfeldt and M. Graetzel, Chem. Rev., 95 (1995) 49–68.

[17] B. O'Regan, J. Mosser, M. Anderson and M. Graetzel, J. Phys. Chem., 94 (1990) 8720–8726.

[18] M. Graetzel, in Semiconductor Nanoclusters – Physical, Chemical, and Catalytic Aspects (P. V. Kamat and D. Meisel, eds.), pp. 353–375, Elsevier, Amsterdam, The Netherlands, 1997.

Nanostructured Materials for Solar Energy Conversion
T. Soga (editor)
© 2006 Elsevier B.V. All rights reserved.

Chapter 2

Device Modeling of Nano-Structured Solar Cells

M. Burgelman, B. Minnaert and C. Grasso

University of Gent, Department of Electronics and Information Systems (ELIS), Pietersnieuwstraat 41, B-9000 Gent, Belgium

1. INTRODUCTION

The aim of *device modeling* is to develop a link between materials properties and the electrical device characteristics of a nano-structured solar cell. This is in contrast to *materials modeling*, where materials parameters (like the optical absorption $\alpha(\lambda)$, the electrical mobilities μ_e, μ_h of electrons and holes, various relevant energy parameters, etc.) are studied and theoretically modeled based on physical and chemical phenomena and interactions. In device modeling, we suppose that all materials parameters needed are already known, either from quantitative materials modeling or from experiments. We also will assume that an optical analysis of the cell already has resulted in the knowledge of the optical generation of electron–hole pairs and excitons, both as a function of position and wavelength. Thus, the scope of this chapter is electrical device modeling of nano-structured solar cells.

The first goal of our device modeling is to simulate the *J–V* curve of a nano-structured solar cell, both in dark and under illumination. From this, the main photovoltaic parameters of the solar cell are deduced: the short-circuit current J_{sc}, the open-circuit voltage V_{oc}, the fill-factor, *FF* and finally the conversion efficiency η as the single most important photovoltaic parameter. Also more sophisticated measurements will be simulated: the quantum efficiency ($QE(\lambda)$) (often termed *IPCE* – incident photon to current conversion efficiency – in the context of nano-structured solar cells), admittance or impedance measurements, thus capacitance and conductance versus voltage or frequency. The benefits of device modeling are to gain insight in the internal cell processes, which are not available to measurement, to enable interpretation of measurements in terms of the internal cell physics and chemistry, to explain the conversion losses occurring in the cell, and if possible, to provide hints for improvement of the cell efficiency [1].

We will develop here a unified framework, which is applicable to a broad class of nano-structured solar cells, as explained in Section 2. We will then describe the characteristics of the 'unit cell' on a microscopic scale (Section 3). Next, two models will be presented to combine the multitude of unit cells to the macroscopic solar cells: the network model (NM) in Section 4, and the effective medium model (EMM) in Section 5. Both models are compared (Section 6), and their results are presented and discussed (Section 7). Finally, it is discussed how these models have to be adapted to include excitonic effects (Section 8).

2. NANO-STRUCTURED SOLAR CELLS

2.1. Common Characteristics of Nano-Structured Solar Cells

The internal of an idealized nano-structured solar cell consists of two phases or moieties. The first part is a material that conducts electrons, for example, TiO_2 or a polymer derived from fullerene, C_{60}. We will call this the *n*-part or *n*-phase. In organic solar cells, the term acceptor material is customary, which is a misleading term for a semiconductor physicist. The second part is a hole-conducting material, for example, $CuInS_2$ or a polymer derived from PPV, poly-phenyl-vinylene. We call it the *p*-part or *p*-phase, and in organic solar cells the term donor material is used. Each phase must form an electrically connected network by itself, which is contacted by one of the electrodes, and is extended over most of the cell thickness. The two phases must interpenetrate and be blended together intimately on a small distance scale, mostly far below 1 μm. As an example of a nano-structured solar cell, the structure of a polymer/fullerene solar cell is shown in Fig. 1.

The core phenomena of the solar cell are taking place on a microscopic (even 'nano-scopic') scale at the boundary between the two constituent phases (Fig. 1). A photon is absorbed, giving rise to a free electron–hole (*eh*) pair and/or a bound exciton pair. The *eh* pair is then separated into an electron in the electron conductor, and a hole in the hole conductor. Alternatively, the exciton is dissociated at the phase boundary into a free electron and a free hole, each in a separate material. We call the microscopic environment, where these phenomena take place, the *elementary* cell or *unit* cell. Often, a macroscopic solar cell is an assembly of a multitude of almost identical elementary cells; this assembly is in the form of two interpenetrating networks. This nano-structure is a common feature of all cell types studied here, though there may be, for example, differences in the distance scale. What actually discerns the various cell types is the nature of their unit cell.

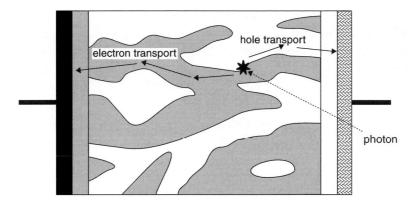

Fig. 1. An example of a nano-structured solar cell. The hatched phase is an electron transporting material, for example, the fullerene-derivate PCBM, and is only contacted by the left-hand-side electrode. The white phase is a hole-transporting material, for example, the PPV derivate MDMO-PPV, and is only contacted by the right-hand-side electrode. The phenomena of photon absorption, generation and separation of an electron and a hole, and transport of the carriers to the electrodes are shown schematically. The figure is not to scale: the width of the white and hatched channels is typically 1–50 nm, and the total cell width is 0.1–10 μm.

2.2. Classification of Nano-Structured Solar Cells

Nano-structured solar cells can be classified in several ways. In Fig. 2, the amount of inorganic versus organic materials in the cell is used as a criterion. As can be seen, a decrease in amount of inorganic materials goes together with an increase in the complexity of the cell. It often also goes with a decrease in production costs although this cannot be seen from the figure.

The 100% organic nano-structured cell is a mixture, or blend, on a nano-scopic scale; in such a 'bulk heterojunction' cell, the electron-conductor/hole-conductor mixture consists of a polymer/polymer blend (e.g., MDMO-PPV/PCNEPV [2]) or a polymer/organic molecule blend (e.g., MDMO-PPV/PCBM [3]). Less organic and less complex are the structures based on a nano-porous network of n-type oxide particles, for example, the widely used TiO_2 with particle size of 25–100 nm. In this, the oxide particles are covered with a thin, light-sensitive shell; this porous structure is immersed in a p-type hole conductor of various types. In case the absorber is a monomolecular layer of an organic dye, and the hole conductor a liquid electrolyte, the cells are called dye-sensitized solar cells (DSSCs) or Grätzel cells, developed by O'Regan et al. [4]. Practical advances are gained by replacement of

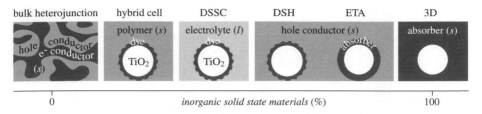

Fig. 2. Schematic representation of nano-structured solar-cell examples on unit cell scale: from an all organic cell to an all inorganic solid-state cell; the various cell types are described in the text.

the liquid electrolyte by an organic solid hole conductor like OMeTAD [5] or a polymer [6], obtaining a hybrid solar cell. When the electrolyte replacing material is an inorganic solid-state hole conductor like CuSCN [7] or CuI [8], the cells are called dye sensitized heterojunctions (DSHs), or interpenetrating heterojunctions (i-hets).

The last step on the way from a Grätzel cell to an all solid-state solar cell is the replacement of the organic dye by a solid-state absorber like CdTe [9], a-Si [10], or $CuInS_2$ [11]; the cells thus obtained are the so-called extremely thin absorber (ETA) cells [12]. In most of these cells however, the hole conductor is just the metal contact which penetrates the pores like a hole conductor would do. In other cells, the so-called 3-D cells, it is the absorber itself that fulfils the role of both absorber and hole conductor [13].

3. THE CONCEPT OF THE FLAT-BAND SOLAR CELL

3.1. Simplifying the Morphology

Both the *n*- and the *p*-phases of a nano-structured solar cell form an interconnected 3-D network and both networks are supposed to interpenetrate perfectly. In order to start the device modeling from a more clear arrangement, we simplify the geometric ordering of Fig. 1 (e.g., [14, 15]). In a Grätzel type cell, a well-conducting electrolyte, almost perfectly, contacts all individual TiO_2/dye grains. So, after idealizing the ordering of the TiO_2 particulates, the structure of Fig. 3(a) is obtained. For solid-state TiO_2-based cells (eta-cells, 3-D cells), the arrangement of Fig. 3(b) is more appropriate, as there is no separate hole conductor penetrating into the structure. The geometry of polymer bulk heterojunctions also can be simplified to the structure of Fig. 3(b). Along the dashed line of Fig. 3(b), there is, in this

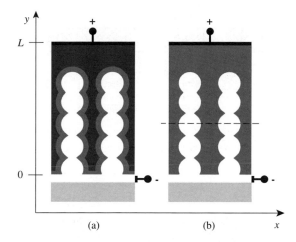

Fig. 3. Simplification of the geometry of the nano-structured solar cell of Fig. 1 into columns of grains; structure of a DSSC. (a) Each column is contacted by a well-conducting electrolyte; structure of a solid-state 3-Dcell (or ETA-cell in which the absorbing material is absorber and hole conductor in one), (b) Each column is contacted by a conducting solid-state material; at the dashed line there is a periodic ordering of TiO$_2$/absorber.

simplification, a periodic, nano-scale ordering: either of *n*-type TiO$_2$ grains and their *p*-type absorber shell (eta-cells, 3-D cells), or of the *n*- and *p*-type constituents of an organic bulk heterojunction solar cell.

3.2. A Solar Cell with Periodic Boundary Conditions

The morphology along the dashed line in Fig. 3 is further simplified in Fig. 4. Due to the periodicity of the structure, a unit cell can be defined between the dashed lines in Fig. 4. At these lines, Neumann-type boundary conditions apply for the electrostatic potential ϕ, that is, $\partial\phi/\partial x = 0$, and no current crosses these lines in the *x*-direction. The charge density $\rho(x)$ and the electrostatic potential $\phi(x)$ of such a periodic arrangement of a *p*-type material (here an inorganic absorber) and an *n*-type material (here TiO$_2$) are also sketched in Fig. 4, in equilibrium conditions, that is, at no applied voltage $V = 0$ and in dark. The energy band diagram is then obtained by vertically mirroring the $\phi(x)$ diagram and shifting it downward in each material over an energy amount χ and $\chi + E_g$, where χ is the electron affinity and E_g the bandgap of the materials:

$$E_C(x) = -q\phi(x) - \chi(x) \quad \text{and} \quad E_V(x) = -q\phi(x) - \chi(x) - E_g(x) \tag{1}$$

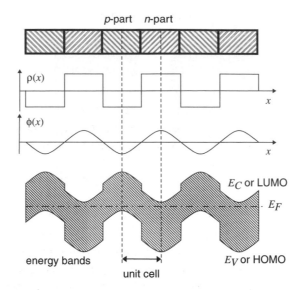

Fig. 4. A periodic ordering of n-type TiO$_2$ and a p-type absorber. The charge density $\rho(x)$, the electrostatic potential $\phi(x)$ and the equilibrium band diagram are sketched. The hatched region in the band diagram is the energy gap. A unit cell with periodic boundary conditions is indicated. Reprinted with permission from J. Appl. Phys., Vol. 95–94, M. Burgelman and C. Grasso, A network of flat-band solar cells as a model for solid-state nano-structured solar cells, pp. 2020–2024 ©2004, American Institute of Physics.

In organic structures, the energy E_C of the conduction band edge should be replaced by the lowest unoccupied molecular (LUMO) level, and the valence band edge E_V by the highest occupied molecular (HOMO) level.

The concept of unit cell becomes more complicated under nonequilibrium conditions ($V \neq 0$ and/or under illumination). A current J_d is now generated at the junction of the unit cell diode (Fig. 5). This current is carried away as a hole current J_h in the p-type subnetwork, and as an electron current J_e in the n-type subnetwork. As imposed by the periodical ordering of unit cells in the x-direction, no current can pass in x-direction through the borders of the unit cell: $J_{hx} = J_{ex} = 0$ at $x = -d$ and at $+d$. Instead, the current flows in the y-direction toward the contacts: $y = 0$ for the electrons and $y = L$ for the holes (the contacts lie outside Fig. 5, they are seen in Fig. 3).

Under nonequilibrium, our concept of the unit cell is clearly of a 2-D nature. There is however a large difference of scale between the x and y directions: typically, the dimension d of the unit cell is a few nanometer for a polymer cell to a few tens of a nanometer for a TiO$_2$-based cell, whereas the total cell thickness L is $\cong 100$ nm for a polymer cell and up to 10 μm for

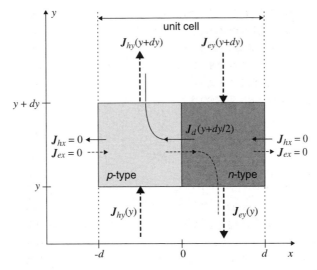

Fig. 5. A unit cell in a nano-structured solar cell under current-carrying conditions. J_e and J_h are the electron and hole particle currents. A lateral current (in x-direction) J_d is generated through the junction of the unit cell diode at $x = 0$, but $J_{ex} = J_{hx} = 0$ at the borders of the unit cell ($x = -d$ and $+d$). The hole current J_{hy} is collected at the top ($y = L$) of the p-type subnetwork, and the electron current J_{ey} is collected at the bottom ($y = 0$) of the n-type subnetwork.

a DSSC: the variations in the y-direction are thus at least an order of magnitude slower than in the x-direction. We will thus treat our unit cell as if there were no y-variations (thus as a 1-D cell), and then apply the result at each position y. This is analogous to Shockley's well-known gradual channel approximation, which is successfully used for many decades to describe the current in a MOS transistor: the electrostatic problem in the x-direction (into the depth of the transistor) is treated as being 1-D; the result is then used to describe the current in the y-direction (source to drain), see for example [16].

3.3. The Unit Cell is a Flat-Band Solar Cell

The first step in modeling the macroscopic cell is calculating the J–V characteristics of a single unit cell: this can be regarded as a 'local' internal J–V characteristic in the solar cell. This is difficult to handle analytically because, due to the small dimensions, the usual photovoltaic cell concepts do not apply (e.g., the concept of depletion layer, quasi-neutral region, diffusion of minority carriers, etc.). Also a numerical calculation is problematic for most common photovoltaic device simulators, including SCAPS [17, 18] and PC-1D [19], as no Neumann-type boundary conditions are implemented.

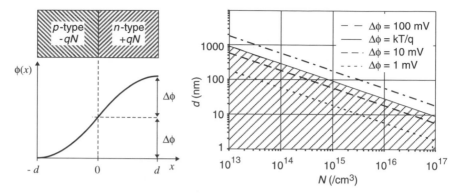

Fig. 6. Left-hand side: A unit cell with periodic boundary conditions ($\partial\phi/\partial x = 0$ at $x = \pm d$) and an electrostatic potential drop $\Delta\phi$ over each side. The p-side ($-d < x < 0$) carries a charge density $-qN$, and the n-side ($0 < x < d$) a charge density $+qN$. Right-hand side: The thickness/carrier density combinations (N,d) for a given potential drop $\Delta\phi$, calculated with Eq. (2) and $\varepsilon_s = 3$. All cells in the hatched region to the left-hand side and below the line $\Delta\phi = kT/q$ can be considered as flat-band solar cells.

Fortunately, an easy but excellent approximation exists, based on the fact that there is almost no electrostatic potential drop over the unit cell: the unit cell is in essence a 'flat-band cell' [15].

We will first make this statement plausible by considering a unit cell with a charge density $-qN$ in the p-part, and $+qN$ in the n-part of a symmetrical unit cell (Fig. 6, on left-hand side). When $\partial\phi/\partial x = 0$ at $x = -d$ and at $+d$, the potential drop $\Delta\phi$ over each part is given by

$$\Delta\phi = \frac{2Nd^2}{2\varepsilon_s\varepsilon_0} \tag{2}$$

where ε_s is the relative permittivity of both materials. In Fig. 6 (right-hand side), the (N, d) values giving rise to a fixed $\Delta\phi$ are drawn. When this potential drop over the unit cell is lower than the thermal voltage, $\Delta\phi < kT/q$, it can be neglected and the energy bands are essentially flat in the x-direction. All unit cells with (N, d) values in the hatched region of Fig. 6 (right-hand side) can thus be considered as flat-band cells. We anticipate that this is the case for most nano-structured solar cells.

This argument is now validated by numerical calculations of a CdTe/TiO$_2$ eta-cell with the device simulator SCAPS [17, 18]. As all other common solar-cell simulation tools available, SCAPS can only handle Dirichlet type boundary

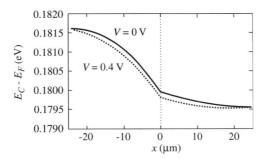

Fig. 7. SCAPS simulation of the electrostatic potential $\phi(x)$ of a thin CdTe/TiO$_2$ nano-structured unit cell. Solid curve: zero bias $V = 0$; dotted curve: forward bias ($V = 0.4$ V). To mimic the Neumann boundary conditions ($\partial\phi/\partial x = 0$ at the boundaries), specific values for the 'contact work function' are used: $\Phi^{\text{TiO}_2} = 4.6796$ eV and $\Phi^{\text{CdTe}} = 4.6816$ eV (at zero bias) or 5.0816 eV (at 0.4 V bias). Reprinted with permission from J. Appl. Phys., Vol. 95–94, M. Burgelman and C. Grasso, A network of flat-band solar cells as a model for solid-state nano-structured solar cells, pp. 2020–2024 ©2004, American Institute of Physics.

conditions, where the electrostatic potential ϕ, and hence the conduction band edge E_C are imposed at the boundary, for example, by specifying the contact work functions Φ_{m1} and Φ_{m2}. Simulation of Neumann type boundary conditions can however be done with SCAPS when appropriate fictitious values for both contact work functions are introduced, which are chosen such that $\partial\phi/\partial x = 0$ holds at the boundaries. These Φ_{m1} and Φ_{m2} values have to be adapted manually at each bias voltage V, which makes the numerical calculation quite laborious. The other simulation parameters are taken from Ref. 20. In Fig. 7, a periodic unit cell consisting of 25 nm p-type CdTe and 25 nm almost intrinsic TiO$_2$ is simulated. The kink at $x = 0$ is caused by the large difference in dielectric constant between TiO$_2$ ($\varepsilon_s = 40$ assumed) and CdTe ($\varepsilon_s = 10$). The total band bending is seen to be only 2 mV, which can be totally neglected. Hence, this nano-scale periodic unit cell in this case effectively behaves as a flat-band cell.

3.4. The Flat-Band Solar Cell: Energy Band Diagram

The recipe for constructing the energy band diagram of a flat-band solar cell is outlined in Fig. 8, where the terminology of an eta-cell (a very thin p-type absorber on a TiO$_2$ sphere) is used. In our flat-band cell approximation, the conduction and valence band are flat, apart from a discontinuity at the junction, as imposed by the difference in electron affinity $\Delta\chi$ and band gap ΔE_g between the two materials: this follows from Eq. (1). Here, we take $\Delta\chi = \chi^{\text{TiO}_2} - \chi^{\text{abs}} = 0.4$ eV, which is relevant for CdTe and TiO$_2$.

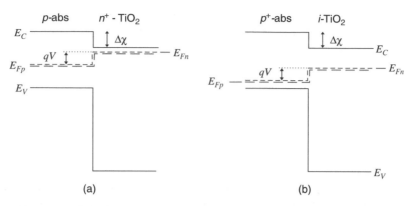

Fig. 8. Flat-band approximation of a unit cell in a periodic structure. (a) In case the n-region is doped more heavily. (b) In case the p-region is doped more heavily.

In the literature, even higher values of $\Delta\chi$, up to 0.7 eV were reported [20]. Also both Fermi levels E_{Fn} and E_{Fp} are flat, as we assumed no current in the x-direction in our 1-D flat-band cell, apart from the current J_d through the junction. This current is driven by a separation qV_d between the Fermi levels, where V_d is by definition the 'applied voltage' over the unit cell. We introduce a further simplification, that the p-absorber is a 'holes-only-material', where the electron concentration is negligible under all circumstances, and that TiO$_2$ is an 'electron-only-material', where the hole concentration can always be neglected. For TiO$_2$, this is justified by the large band gap $E_g > 3$ eV; for p-absorbers used in eta and 3-D cells, for example, CuInS$_2$ or CdTe, this is not so strongly justified since E_g is only $\cong 1.5$ eV. In organic photovoltaic materials including most polymers and fullerenes, the assumption is justified; indeed, ambipolar conduction is rarely reported in these materials. Thus, we assume that the position of E_{Fn}(abs) in the p-absorber and of E_{Fp}(TiO$_2$) in the TiO$_2$ is irrelevant: we only need to know E_{Fp}(abs) and E_{Fn}(TiO$_2$). We already know that

$$E_{Fn}(\text{TiO}_2) - E_{Fp}(\text{absorber}) = qV_d \tag{3}$$

To determine the position of one of the two relevant Fermi levels, we express that the total charge should be zero in a unit cell of a periodic structure.

We will illustrate this in the most simple case that both layers are equally thick ($d^{\text{TiO}_2} = d^{\text{abs}}$), and that the effective density of states is the same for both materials and for both carriers (thus, $N_V^{\text{TiO}_2} = N_C^{\text{TiO}_2} = N_V^{\text{abs}} = N_C^{\text{abs}}$). A more general formulation is not a problem, but it hardly brings anything new.

In a pn^+ cell (Fig. 8(a), with $N_D^{TiO_2} \gg N_A^{ab}$), holes in TiO_2 can be neglected, as well as electrons and holes in the absorber. The neutrality condition is then:

$$N_A^{abs} + n^{abs} - p^{abs} = N_D^{TiO_2} - n^{TiO_2} + p^{TiO_2}$$
$$N_A^{abs} \approx N_D^{TiO_2} - n^{TiO_2}$$
$$E_C(TiO_2) - E_{Fn}(TiO_2) = kT \ln \left(\frac{N_C^{TiO_2}}{N_D^{TiO_2} - N_A^{abs}} \right) \tag{4}$$

The position of $E_{Fn}(TiO_2)$ is easily determined from Eq. (4), and then $E_{Fp}(abs)$ is placed at an energy qV_d below $E_{Fn}(TiO_2)$. The p-absorber part is fully depleted, and the TiO_2 grain is neutralized by almost equal densities of ionized donors and electrons. The charge density in this structure approximately equals $-qN_A^{abs}$ at the left-hand side, and $+qN_A^{abs}$ at the right-hand side of the junction. It is thus determined by the lowest of the two doping densities $N_A^{abs}, N_D^{TiO_2}$. At high forward bias V, E_{Fp} approaches the valence band in the absorber, and holes become important (Fig. 9).

In a p^+n cell (with $N_A^{abs} \gg N_D^{TiO_2}$; Fig. 8(b) is presented for almost intrinsic TiO_2, thus for $N_D^{TiO_2} \to 0$), electrons in the absorber can be neglected, as well as electrons and holes in TiO_2. The neutrality condition now leads to:

$$N_A^{abs} + n^{abs} - p^{abs} = N_D^{TiO_2} - n^{TiO_2} + p^{TiO_2}$$
$$N_A^{abs} - p^{abs} \approx 0$$
$$E_{Fp}(abs) - E_V(abs) = kT \ln \left(\frac{N_V^{abs}}{N_A^{abs}} \right) \tag{5}$$

The position of $E_{Fp}(abs)$ is easily determined from Eq. (5), and then $E_{Fn}(TiO_2)$ is placed qV above E_{Fp}. Again the charge is related to the lowest of the two doping densities. At high forward bias V, E_{Fn} approaches the conduction band in TiO_2 (when $\Delta\chi > 0$), and electrons become important, as shown in Fig. 9.

Because the periodic boundary conditions, together with the small thickness, impose charge neutrality, the charge in both particles and hence the potential drop over them is very small. A net charge of 2×10^{15} cm^{-3} over two 25 nm thick grains indeed results in an electrical potential drop $\Delta\phi < 2$ mV over one unit cell. Remark that $\Delta\phi$ has to be compared with the built-in voltage V_{bi} (or diffusion voltage) in a thick planar diode. It should

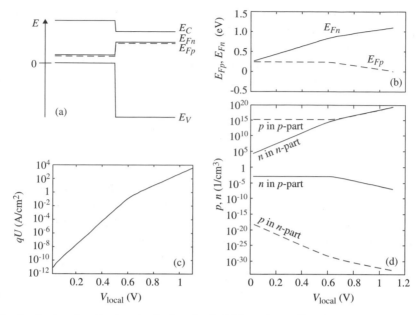

Fig. 9. Analysis of a flat-band p^+-absorber/n-TiO$_2$ solar cell. (a) Upper left-hand side: band diagram at $V_{local} = 0.5$ V with definition of energies, (b) Upper right-hand side: position of the Fermi levels $E_{Fp}(V_{local})$ and $E_{Fn}(V_{local})$, (c) Lower left-hand side: dark diode current $qU(V_{local})$, (d) Lower right-hand side: carrier concentrations $n(V_{local})$ and $p(V_{local})$. Parameter values: p-side: $N_A = 10^{15}$ cm^{-3} and $E_g = 1.5$ eV; n-side: $N_D = 10^{10}$ cm^{-3} and $E_g = 3$ eV; both sides $d = 25$ nm; junction: $\Delta\chi = 0.4$ eV; $\tau = 1$ ns; $S_i = 5$ 10^3 cm/s.

not be confused with a voltage drop V_d which could be measured at the external contacts, which is related to the splitting of the quasi-Fermi levels, Eq. (3).

Simulation shows that the rule of thumb expressed in Eqs. (4) and (5) also holds under illumination: the unit cells remain flat-band cells, that is, the electrostatic potential and the conduction-band level remain constant over the position x of the cell. The position of 'minority carrier' Fermi levels (E_{Fn} at the p-side and E_{Fp} at the n-side) however, are determined by the illumination intensity and the applied voltage V_d. In Fig. 8, both Fermi levels are drawn as coincident in each material, but the exact position of the minority Fermi levels is not relevant under our assumptions.

3.5. The Flat-Band Solar Cell: *J–V* Characteristics of the Unit Cell

In the dark, all carrier densities are easily obtained from the simple model. The dark diode current J_d in our flat-band solar cells is determined by recombination. Under low injection, all possible recombination mechanisms are proportional to the excess minority carrier density (excess means referred to the equilibrium concentration). Contact recombination is excluded, because in the periodic structure there are no contacts to the unit cells.

Bulk recombination is only important in the absorber material with its lower E_g. The relevant minority carriers are therefore the minority carriers in the absorber, which can be either electrons or holes depending on the doping ratio of the absorber to TiO_2: in a pn^+ cell (Fig. 8(a)), $n_{minor} = p^{abs}$, and in a p^+n cell (Fig. 8(b)), $n_{minor} = n^{abs}$, at a moderate bias voltage V_d.

The dominant path for interface recombination is between the conduction band of TiO_2 and the valence band of the absorber, when $\Delta\chi > 0$ (the case is shown in Fig. 8). The relevant minority concentration n_{minor} thus is either the hole concentration in the absorber or the electron concentration in TiO_2, whatever is the smallest. In a pn^+ cell (Fig. 8(a)) the minority concentration at the interface is $n_{minor} = p^{abs}$, and in a p^+n cell (Fig. 8(b)), $n_{minor} = n^{TiO_2}$.

In all cases, n_{minor} varies exponentially with applied voltage V_d (Fig. 9), and can easily be calculated from either Eq. (4) or (5). Hence recombination current J_d follows from:

$$J_d = \frac{qd_p}{\tau}[n_{minor}(V) - n_{eq}] \quad \text{bulk recombination, or}$$
$$J_d = qS_i[n_{minor}(V) - n_{eq}] \quad \text{interface recombination, with}$$
$$n_{minor}(V) = n_{eq} \exp\left(\frac{qV}{kT}\right)$$

$$(6)$$

Here d_p is the absorber thickness of the unit cell, and n_{eq} is the equilibrium minority concentration at the interface. As is customary, bulk recombination in the absorber is described by a lifetime τ, and interface recombination by an interface recombination velocity S_i. The local J–V curve calculated with Eq. (6) is shown in Fig. 9 for a p^+n cell. At $V = V_{kink} = 0.6$ V, there is a kink in the J–V curve. For a higher applied voltage, electrons and holes have equal concentrations (Fig. 9), and hence there are no longer minority carriers. The simple equations (Eq. (6)) cannot be used then. The application of the full Shockley–Read–Hall (SRH) expressions [16] results in an exponential J–V law with a decreased slope:

$$J_d \propto \exp\left(\frac{qV}{2kT}\right), \quad V \geq V_{kink}$$

$$(7)$$

It is not possible to truly verify this result (Eq. (6)) with a 1-D device simulator, as in the real 1-D periodic structure of Fig. 5 there is no current J_x in the x-direction. In the simulation of one unit cell however, a J_x is possible when a hole selective contact to the absorber, and an electron selective contact to TiO_2 are applied; the minority recombination at both 'contacts' is assumed to be zero. 1-D SCAPS simulations are carried out, adapting manually ϕ_{m1} and ϕ_{m2} for each bias voltage and for each illumination in order to obtain $\partial\phi/\partial x = 0$ at the boundaries (Section 3.3). They confirm (not shown) the gross features of the simple model described earlier: the dark characteristics have an exponential form, and the light J–V curve is obtained by shifting down the dark curve. We realize however that a unit cell in our periodic structure cannot be purely 1-D, because the current has to be drawn away as a lateral current J_y, as discussed in Section 3.2 and Fig. 5.

We stress that the absence of an electric field E in a flat-band unit cell does not render the collection of light-generated carriers impossible. It is well accepted and illustrated in the literature that there are other driving forces (besides E) for the separation of generated eh-pairs: the band edge discontinuities at the conduction band $\Delta E_C = \Delta\chi$ (favorable when it is positive, as in Fig. 4), and at the valence band $\Delta E_V = \Delta E_C + \Delta E_C$ (always favorable since TiO_2 is a wide band gap material), and the selectivity of the contacts [21–23]. Hence, we assume that all photons absorbed in the small unit cell contribute to the local light current J_L, which is added to the dark current J_d (note that, when $V_d > 0$, both currents have opposite direction).

Ideally (i.e., at $T = 0$) the flat-band unit cells should have a local open-circuit voltage V_{oc} corresponding to the absorber bandgap $E_g(abs)$ or to the 'interface bandgap' $E_g(abs) - \Delta\chi$, whatever the smallest. At $T > 0$, however, the dark recombination current lowers V_{oc} by an amount of $(kT/q) \ln(a/\tau J_L)$ or $(kT/q) \ln(S_i a/J_L)$, obtained by inserting Eq. (6) in the equation for V_{oc}; here, a is an appropriate constant. In what follows, we will treat V_{oc} as a parameter.

4. THE NETWORK MODEL (NM)

In a real nano-structured solar cell, the n-part forms an interconnected 3-D subnetwork which makes electrical contact with one electrode. The p-part forms a complementary subnetwork, which makes contact to the opposite electrode. In the following considerations, we simplify both subnetworks to one dimension to keep the model manageable.

In Fig. 10, each diode in the row stands for a periodic, flat-band unit cell as just described. Its local $J_d(V_d)$ law contains all the physical and chemical phenomena that occur in the nano-scopic unit cell. When more complex

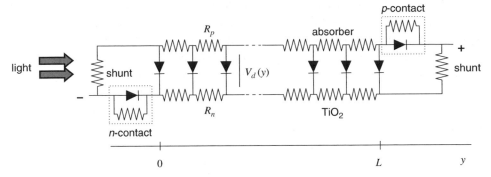

Fig. 10. A 1-D network connection of the unit cells used to simulate the macroscopic nano-structured solar cell. The terminology used in the figure applies to a solid-state eta or 3-D cell. Reprinted with permission from J. Appl. Phys., Vol. 95–94, M. Burgelman and C. Grasso, A network of flat-band solar cells as a model for solid-state nano-structured solar cells, pp. 2020–2024 ©2004, American Institute of Physics.

mechanisms are present than those described earlier, and give rise to a non-exponential local *J–V* law, a general $J_d(V_d)$ can be used for the network diodes. The resistors stand for the percolation in the *p*-network (top), and in the *n*-network (bottom). The contacts for majority carriers are described by a contact resistor in parallel with a reverse polarized diode representing a possible Schottky barrier at the electrode (enclosed in a dotted box in Fig. 10). Minority carriers making contact with the electrode are represented by a shunt resistance.

In physical terms, the voltage at the network nodes of the upper branch in Fig. 10 represents the hole Fermi level E_{Fp} in the *p*-part of the unit cells. The voltage at the nodes of the lower branch represent the electron Fermi level E_{Fn} in the *n*-part of the unit cells. We apply Eq. (3) to each position *y* along the network: the local voltage difference $V_d(y)$ over a network diode at position *y* is given by

$$V_d(y) = \frac{1}{q}(E_{Fn}^{\text{TiO}_2} - E_{Fp}^{\text{abs}})\big|_{\text{unit cell at position } y} \tag{8}$$

This constitutes the gradual channel approximation mentioned before. Constant resistors in the network representation stand for transport by drift and diffusion, since a voltage drop $V_d(y_1) - V_d(y_2)$ in the network stands for a Fermi-level gradient in the physical situation. Transport by space-charge limited currents would need appropriate *I–V* laws for the resistors (namely $I \propto V^2/\ell^3$ where ℓ is the length of a resistor). In the network description, all other physical quantities such as electrostatic potential ϕ, electron and hole concentration *n* and *p* within the unit cells, are lost and lumped in the current voltage law of the diode.

The cells are illuminated from one side, which implies that the unit cells generate less current and voltage as the light penetrates in the cell. Ideally, all generated currents are added to obtain the total light current. The open-circuit voltages however are not summed, they are rather averaged. This is an essential difference with classical solar cells. This argument will be developed quantitatively in Section 7.

5. THE EFFECTIVE MEDIUM MODEL (EMM)

Another way to describe the macroscopic cell structure of a nano-structured solar cell is with an EMM [24, 25]. In this model, the whole p–n nano-structure is represented by one single effective medium semiconductor layer (Fig. 11). The effective medium (EM) is characterized by an 'averaging' of the properties of the n- and the p-material. As the n-part of the nano-structured solar cell is considered as an 'electrons only' material, and the p-type as a 'holes only' material, all carrier-related properties of the EM should relate to the appropriate constituent. Thus the electron affinity χ (determining the position of the conduction band or the LUMO of the EM), the electron mobility μ_n, diffusion constant D_n, effective density of states in the conduction band N_C are all taken over from properties of the n-part. This is TiO$_2$ in most nano-structured cells derived from the DSSC, and for example, a fullerene derivate (e.g., PCBM) in a polymer/fullerene bulk heterojunction. Likewise, properties relating to the holes (including the position of the valence band or HOMO) are all taken over from properties of the p-part.

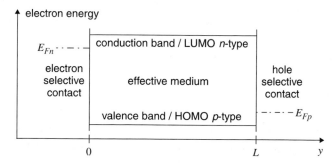

Fig. 11. In the EMM, the whole p–n nano-structure is represented by one single EM semiconductor layer. The EM has one conduction band namely the conduction band or LUMO of the n-type material and one valence band namely the valence band or HOMO of the p-type material. E_{Fn} and E_{Fp} represent the Fermi levels, respectively, of the electron and hole selective contact. For simplicity, the cell is shown in a flat-band condition.

This can be a dye/electrolyte combination in DSSC cells, an inorganic semiconductor in eta cells or 3-D cells, or a PPV derivate such as MDMO-PPV or MEH-PPV in a polymer/fullerene bulk heterojunction. When interface recombination is thought to be dominant, it is determined by both constituents. In particular, the interface band gap, which enters in the calculation of the minority carriers at the interface (Section 3.5) is determined by the lowest of the two conduction bands (or LUMOs), and the highest of the two valence bands (or HOMOs) at the interface. Noncarrier-related properties, such as the dielectric constant ε_s, the refractive index n, and the absorption constant α, are influenced by both materials. The precise way in which this happens depends strongly on the details and the size scale of the intermixing. For particles smaller than the wavelength of the illumination, a true effective medium theory can be used [26].

To calculate the external J–V characteristics of the effective medium cell of Fig. 11, it is easiest to feed all parameters into a standard solar-cell device simulator, for example, SCAPS [16]. In this way it is also possible to define several EM layers in the cell structure, or even to define a grading of the materials parameters of the EM in the y-direction of the cell structure. This way, one can define a cell structure with the pure materials at the contacts, as in Fig. 1, and a blend with gradual changing composition from one side toward the other [27].

6. THE EQUIVALENCE BETWEEN THE NETWORK MODEL AND THE EFFECTIVE MEDIUM MODEL

Before presenting any simulation results obtained from the two models studied earlier, first their meaning and mutual relationship is discussed in more detail. All differential and algebraic equations which govern the physical behavior of a semiconductor device can be quantitatively described by an infinite network. Indeed, the equations between the voltages and the currents of the network representation have exactly the same form as the discrete form of the differential semiconductor equations: this has been established by the work of C. T. Sah [28–30] almost forty years ago, and can be considered as belonging to the common background knowledge of semiconductor device engineers. Although the practical use of network calculations is on its way back due to the availability and ease of numerical device simulation software, which numerically solves the 'semiconductor equations', a network representation of simplified devices is still attractive and useful as it gives a pictorial and intuitive view of the internal working of a device (at least for those with some background in electrical networks). So, even if we have set up the network model in Section 4 as an almost visual image of the internal morphology, the network is more than a visual aid alone: it represents a full physical model, and is in principle equivalent with the effective medium model, as both are based on the same semiconductor equations. This being said, a few critical thoughts are in order.

Usually, a network representation is set up to describe the small signal ac behavior, thus to explain the capacitance and ac conductance of the device when a small ac voltage or illumination is applied superimposed to some steady state or dc-bias value: the linear network then represents the semiconductor equations linearized around the bias state. The value of all network elements then depends on the bias value of all physical properties. In particular, the resistor values R_n and R_p in Fig. 10 depend on the steady state electron and hole concentrations n and p. It is also legitimate to set up a network for the dc-problem, as we did in Section 4. But we have to admit that the solution of the problem (thus the values $n(y)$ and $p(y)$ at all positions) must be known before we can attribute a value to the network elements! In that sense, a dc-network model is often considered more as a vehicle to visualize and guide ideas than as a practical solver tool. In this case however, the situation is not so bad, as in important practical circumstances, all parameters are almost constant over the cell thickness, except in the very neighborhood of the contacts $y = 0$ and L. We will discuss in the next section that we did obtain relevant results even if we used a network with constant resistors over the whole cell thickness $0 \leq y \leq L$.

A second remark is that our network is a simplified version of Sah's full semiconductor device network: the NM thus only implements a limited number of physical phenomena. The EMM, on the other hand is solved with a standard numerical device simulator, which usually contains a lot of physical phenomena. We have to check thus carefully the correspondence between the two models.

We already commented on the meaning of the resistors in the network: they represent both diffusion and drift current. When ideal Shockley diodes are used in the NM, the diode current under illumination is given by

$$J_d(y) = J_s(y)\left[\exp\left(\frac{qV_d(y)}{kT}\right) - 1\right] - J_L(y) \tag{9}$$

where J_s is the dark saturation current and J_L the light current. The light current depends on the local optical generation $g_{eh}(x,\lambda)$ of eh pairs. Both in the

NM as in the EMM we simply derived $g_{eh}(x,\lambda)$ from a simple absorption law $\exp(-\alpha y)$ where y is the distance to the plane of light incidence and α the absorption constant. It is not a conceptual problem to use a more sophisticated calculation of $g_{eh}(x,\lambda)$, as was presented for example, by Ingenäs and coworkers [31]. In fact, SCAPS can accept any form of $g_{eh}(x,\lambda)$ as input from an external optical simulation program.

The first term in Eq. (9) refers to the net thermal recombination $U(y)$. Given the fact that

$$n(y).p(y) = n_i^2 . \exp\left(\frac{E_{Fn}(y) - E_{Fp}(y)}{kT}\right) \qquad (10)$$

the recombination term in Eq. (9) represents only physical recombination mechanisms that are proportional to $pn - n_i^2$ (with n_i the intrinsic carrier density). A recombination mechanism taking this functional form is radiative or direct recombination, sometimes also called monomolecular recombination. The net recombination is given by

$$U_{\text{rad}} = \alpha_r (pn - n_i^2) \quad \text{radiative recombination} \qquad (11)$$

where α_r is an appropriate constant.

Usually, device simulators contain other recombination mechanisms as well. One is Auger recombination, which is described by

$$U_{\text{Aug}} = C_n n^2 p + C_p np^2 \quad \text{Auger recombination} \qquad (12)$$

where C_n and C_p are constants. In silicon semiconductor devices, the dominant mechanism is often SRH recombination via trap centers. It is given by

$$U_{\text{SRH}} = \frac{pn - n_i^2}{c_p(n + n_t) + c_n(p + p_t)} \quad \text{SRH recombination} \qquad (13)$$

with c_n and c_p constants and with n_t and p_t the electron and hole concentration calculated as if the Fermi level were located at the trap level E_t [16]. Interface recombination is hardly present in standard Si devices, but might be the dominant mechanism in nano-structured solar cells. It is a complicated function of the electron and hole concentrations at either sides of the junction [32], but in practical cases it is well described by the SRH expressions

(Eq. (13)), where n is the electron concentration in the material with the lowest E_C or LUMO at the interface, and p the hole concentration in the material with the highest E_V or HOMO (thus, in the structure of Fig. 8, the values of n and p to be inserted in Eq. (13) each apply to a different material).

It is clear that the ideal Shockley diodes in the network model represent monomolecular recombination (Eq. (11)), but not Auger recombination (Eq. (12)). They also represent SRH recombination (bulk or interface) (Eq. (13)), but only if $n_t \gg n$ and $p_t \gg p$. This is the case if the trap centers are very close to the valence band or to the conduction band.

It is now clear that the network model and the EMM are in principle equivalent. In setting up a NM or an EMM for simulating a particular device, one has to take care that all physical and chemical mechanisms present and relevant in some particular cell, are well described in the model, and that the simulation does not take into account any phenomena which is not present, or deemed unimportant in the real cell. Both models can be used at choice (bearing in mind the remarks at the beginning of this section), but in setting up the parameter values, one has to take great care that both models describe the same mechanisms. This is illustrated in the next section.

7. SIMULATION RESULTS

7.1. Results with the Network Model (NM)

We assume that the J–V characteristics of a unit flat-band cell are purely exponential, with an ideality factor of 2. We take somewhat arbitrarily $J_L = 15$ mA/cm^2 and $V_{oc} = 0.8$ V, if all the light were absorbed in one unit cell, which is of course not realistic. The J–V characteristics of individual unit cells are scaled to account for the actual light input on each cell, which decreases as $\exp(-\alpha y)$, where y is the distance to the plane of light incidence, see Figs. 3 or 10. The value of α to be used is an effective value, which can account for possible optical pathway enhancement by, for example, scattering; this is however not considered here. We simplify the problem by assigning the same value to all resistors R_n and R_p in the network. The resistances of the two subnetworks add up to a total distributed series resistance. For the n-part (TiO$_2$) $R_n \cdot A = \rho_n d/3(1 - p)$, where ρ_n is the specific resistivity of the porous n-layer, p its porosity, and d the distance between the electrodes; the factor 1/3 is because the series resistance is distributed [22]. Likewise, the p-part is characterized by $R_p \cdot A$. We concentrate here on the influence of the absorption α and of the series resistances $R_p \cdot A$ and $R_n \cdot A$. The network can then be solved with any network solving program, for example, SPICE [33].

In Fig. 12, the *J–V* curves without and with series resistance are compared. Even when the total series resistance $R_s = 0$, V_{oc} does not reach its ideal value of 0.8 V. That is because all cells in the stack of Fig. 10 are forced to be at the same voltage. The unit cells at the illuminated side are then below their own local $V_{oc\,local}(y)$, the cells in the bulk and at the rear end of the stack are above their local $V_{oc\,local}(y)$ (Fig. 13). At open circuit of the whole cell, current flows from the rear cells to the front cells, and energy is internally dissipated. This is an inherent disadvantage of the geometrical ordering of the cells under discussion. We see in Fig. 12 that a nonzero resistance in the *n*-subnetwork improves V_{oc}. This is because cells deep in the stack, adversely contributing to V_{oc}, are effectively decoupled by a larger R_n. This is also illustrated in Fig. 13. The voltage $V_d(y)$ over the elementary cells is now nonuniform, the cells at the illuminated side carry a larger $V_d(y)$; this is favorable for the open-circuit voltage of the whole cell, which equals $V_{oc} = V_d(0)$ (see Fig. 10, with $R_p = 0$).

The resistances R_n and R_p have a different influence on the solar cell. TiO$_2$ based nano-structured solar cells are always illuminated from the TiO$_2$ side. Also, the largest fraction of the light is absorbed near the plane of incidence, especially at short wavelengths. Because light-generated electrons are drawn away to the front electrode (the shortest distance), and holes to the back contacts, a resistance R_p in the *p*-subnetwork is much more harmful: it

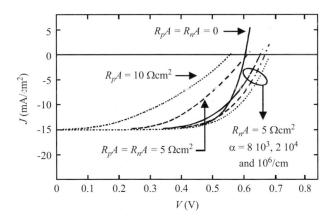

Fig. 12. Calculated *J–V* curves. The case without any series resistance is compared with the case with a total series resistance $R_s \cdot A = (R_n + R_p) \cdot A = 10\,\Omega\,\text{cm}^2$, where $R_s \cdot A$ is completely in the *n*-network, or completely in the *p*-network, or equally divided between both (see labels). In the first case, curves for three values of the absorption coefficient α are shown. Reprinted with permission from J. Appl. Phys., Vol. 95–94, M. Burgelman and C. Grasso, A network of flat-band solar cells as a model for solid-state nano-structured solar cells, pp. 2020–2024 © 2004, American Institute of Physics.

M. Burgelman et al.

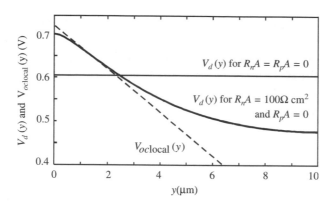

Fig. 13. Voltage $V_d(y)$ over the elementary unit cells as a function of position (see Fig. 10). The p-subnetwork is resistanceless ($R_p \cdot A = 0$), the n-subnetwork is resistanceless ($R_n \cdot A = 0$) or has a resistance of $R_n \cdot A = 100\,\Omega\,cm^2$. The local open-circuit voltage is also shown. Calculated with $\alpha = 10^4\,cm^{-1}$. Reprinted with permission from J. Appl. Phys., Vol. 95–94, M. Burgelman and C. Grasso, A network of flat-band solar cells as a model for solid-state nano-structured solar cells, pp. 2020–2024 © 2004, American Institute of Physics.

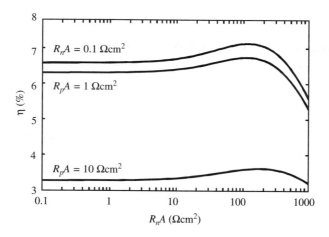

Fig. 14. Solar-cell efficiency η as a function of the resistance $R_n \cdot A$ in the n-subnetwork; the parameter is the resistance $R_p \cdot A$ in the p-subnetwork. Calculated with $\alpha = 10^5\,cm^{-1}$. Note the sensitivity of η to $R_p \cdot A$. Reprinted with permission from J. Appl. Phys., Vol. 95–94, M. Burgelman and C. Grasso, A network of flat-band solar cells as a model for solid-state nano-structured solar cells, pp. 2020–2024 © 2004, American Institute of Physics.

decouples the most illuminated cells at the front from the rear contact. With the parameters of Fig. 14, a value of $R_p \cdot A$ exceeding $1\,\Omega\,cm^2$ is detrimental, whereas value of $R_n \cdot A = 500\,\Omega\,cm^2$ can be tolerated. This could be an explanation for the substantially poorer behavior of 'dry' or all solid-state cells compared to 'wet' or Grätzel cells [4]: The ion conduction in the electrolyte is better than the hole conduction in the eta absorber or the solid-state p-conductor, and this is crucial for the cell performance, as we showed. The resistance in the n-TiO$_2$ network is not crucial, it is even beneficial for the V_{oc} if not too large. We note that the role of the p and n subnetwork will be interchanged when we would illuminate the cell from the p-contact side.

Further calculations are carried out to assess the efficiency enhancement $\Delta\eta$ resulting from the beneficial effect of the resistance in the n-network (Fig. 14). For the parameters used, the optical absorption should exceed $3 \times 10^4 \, \text{cm}^{-1}$ to see any positive $\Delta\eta$ at all. The effect is modest, for example, 0.7% (absolute) for a high value $\alpha = 10^5 \, \text{cm}^{-1}$, a moderate value $R_n \cdot A = 150 \, \Omega \, \text{cm}^2$ and a resistance less p-subnetwork ($R_p \cdot A = 0$) (Fig. 14). However, the mere fact that a moderately poor conduction in the TiO$_2$ network does not completely destroy the cell performance, is a remarkable result in itself.

7.2. Results with the Effective Medium Model (EMM)

The results obtained earlier with the NM are now verified with the EMM. We set up the parameters of the EM as follows. The cell thickness is $L = 10 \, \mu\text{m}$. For the EM, we take a band gap of $E_g = E_C^{\text{TiO}_2} - E_V^{\text{abs}} = 1.05 \, \text{eV}$, and an electron affinity of $\chi = 4.7 \, \text{eV}$. Again, this is relevant for a TiO$_2$/CdTe eta-cell. The majority carrier barrier ϕ_b at both contacts is taken to be $0.2 \, \text{eV}$. The built-in potential V_{bi} is then $(E_g - \phi_{b1} - \phi_{b2})/q = 0.65 \, \text{V}$. For the relative dielectric constant of the effective medium we take $\varepsilon = 30$, an average between the values of CdTe and TiO$_2$ (see Section 3.2). We assume that the dominant recombination will be monomolecular recombination described with Eq. (11). Only SRH recombination can be input in SCAPS, so we use a very shallow single neutral defect level at $0.05 \, \text{eV}$ above the valence band: then Eq. (13) reduces to Eq. (11). The cell is illuminated from the electron selective contact (n-contact side), and the absorption α is varied as a parameter. We assume that both constituents of the effective medium are nearly intrinsic (we take an arbitrary but sufficiently low value of $N_A = N_D = 10^4 \, \text{cm}^{-3}$). This means that all carriers (holes in the CdTe part and electrons in the TiO$_2$ part) are brought about by the external voltage V and/or by the illumination, but not by doping effects. In the NM simulations, the resistances R_n and R_p of the subnetworks were the parameters. In the EMM simulations of this section, the mobilities μ_n and μ_p of the carriers are treated as parameters. Simulations were done with the electron and hole mobilities varying in a wide range from 10^{-4} to $10^{+2} \, \text{cm}^2/\text{V s}$. These parameters are used as input in the device simulator SCAPS, and the results are discussed later.

M. Burgelman et al.

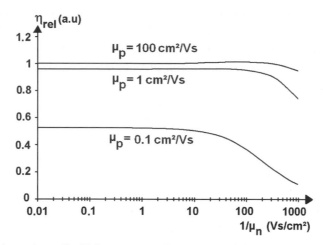

Fig. 15. Relative solar-cell efficiency η as a function of the inverse electron mobility $1/\mu_n$ calculated with the EMM. The parameter is the hole mobility μ_p. Calculated with $\alpha = 10^5\,\mathrm{cm}^{-1}$. Note the similarity with Fig. 14 obtained by the network model. Reprinted with permission from Comptes Rendus Chimie Vol. 9, B. Minnaert, C. Grasso and M. Burgelman, An effective medium model versus a network model for nanostructured solar cells pp. 735–741 © 2006, Elsevier.

In Fig. 15 the relative efficiency η is shown as a function of the inverse mobilities $1/\mu_n$ and $1/\mu_p$. This is done to emphasize the similarity with Fig. 14 obtained with the NM. The efficiency η deteriorates for carrier mobilities, but especially a low hole mobility μ_p is very detrimental (corresponding with high resistances R_p in the NM). It is remarkable that both results correspond, because the NM was calculated for values of R_n and R_p constant through the thickness of the cell, while in the EMM the carrier concentrations $n(y)$ and $p(y)$ and hence also the corresponding resistivities R_n and R_p can vary over up to seven orders of magnitude when going from one contact at $y = 0$ to the other at $y = L$.

A more detailed study shows that, if the absorption is high enough, the cell efficiency even rises for low electron mobilities, thanks to an improved open-circuit voltage V_{oc}; the cell efficiency does not suffer from a lower electron mobility μ_n unless it is below 10^{-3}–$10^{-4}\,\mathrm{cm}^2/\mathrm{V}$ s, depending on the absorption α. This result is similar to the NM where a rise in the resistance R_n of the n-network also improved cell efficiency, thanks to a better V_{oc}.

Simulations with the hole mobility μ_p as parameter show that the short current density J_{sc} quickly drops when the hole mobility is lower than ~ 1–$0.1\,\mathrm{cm}^2/\mathrm{V}$ s. This is the reason for the deterioration of the cell when the hole mobility is too low. Again, this result is similar to the NM where a small resistance R_p of the *p*-network drops the cell efficiency rapidly. Because both models are symmetrically set up, the role of the *p* and *n* subnetwork will be interchanged when we illuminate the cell from the *p*-contact side – simulations confirm this.

The efficiency enhancement $\Delta\eta$ resulting from the beneficial effect of resistance in the *n*-network is further quantified. For the EMM parameters used, the optical absorption α should exceed $10^5 \, \text{cm}^{-1}$ to see any positive $\Delta\eta$ at all, and even $\alpha > 2 \times 10^6 \, \text{cm}^{-1}$ is needed to increase the efficiency appreciably (a rise of 10% relative). With the NM, this efficiency increase was obtained already with $\alpha = 10^5 \, \text{cm}^{-1}$ (Section 4). The numerical results of the two models thus differ. This is because our NM was simply set up with constant resistances (assuming constant *n* and *p* concentrations over the cell), whereas the EMM calculates the $n(y)$ and $p(y)$ profiles, see Section 6. However, it is remarkable that the NM and the EMM both predict the same qualitative result: a moderately poor conduction in the *n*-network, together with a high optical absorption, can increase the cell efficiency. It is also clear that the cell efficiency can not be high if not enough photons are absorbed to create electron–hole pairs. As well as in the NM as in the EMM, the cell efficiency drops quickly if the absorption lowers from $5.10^3 \, \text{cm}^{-1}$.

7.3. Parameter Exploration with the Effective Medium Model (EMM)

We now apply the EMM to explore some important parameters in polymer solar cells. First, a parameter set was built up that describes measured *J–V* characteristics of a MEH-PPV/PCBM organic bulk heterojunction. The main parameters are listed in Ref. 34, and are explained here in brief.

The cell thickness *d* and the dielectric constant ε_s of the EM can be directly measured. The absorption constant $\alpha(\lambda)$ was calculated from the thickness *d* and the absorption *A*, where *A* was simply taken as $A(\lambda) = 1 - T(\lambda)$, with $T(\lambda)$ the measured transmission. This $\alpha(\lambda)$ file could also be substituted by $\alpha(\lambda)$ data from literature, based on more sophisticated measurements [35]. The work functions Φ_{m1} and Φ_{m2} of the electrode materials (ITO/PEDOT PSS and LiF/Al) come from literature [36, 37]. The electron affinity χ of the EM is determined by the LUMO level of the *n*-type constituent of the EM, thus the PCBM rich phase, which is taken from literature [38]. The value of $\chi + E_g$, where E_g is the band gap of the EM, is determined by the HOMO of the *p*-type constituent of the EM, thus the PPV rich phase, which is taken from literature [39]. Because the parameters Φ_{m1}, Φ_{m2}, χ and E_g can depend slightly on

M. Burgelman et al.

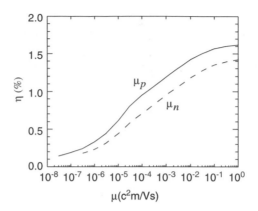

$\mu(c^2m/Vs)$

Fig. 16. Calculated influence of the mobilities μ_p and μ_n of the p and n material on the efficiency η. When one mobility was varied, the value of the other mobility, and of all other parameters, is taken from the standard parameter set for the MEH-PPV cell, see Table 1.

technological modifications, we allowed for some slight variation in the fitting procedure. The mobility μ_n of the EM is the electron mobility of the n-type constituent of the EM, and the mobility μ_p of the EM is the electron mobility of the p-type constituent of the EM. The mobilities measured on FET-structures [40] are taken as first guesses, and then adapted in the fitting procedure.

The EMM, with this parameter set, is fed into our simulation package SCAPS, and the parameters are adapted to obtain a reasonable fit with the measured characteristics of the MEH-PPV/PCBM blend cells.

The dark and illuminated J–V, and the quantum efficiency $QE(\lambda)$ measurements are fitted fairly well [34]. However, further refining of this parameter set is necessary to also (or better) simulate other measurements, especially impedance measurements (C-V and C-f), and all measurements at extended measurement conditions of illumination (intensity and wavelength) and temperature. This is not a trivial task at all.

This parameter was then used for a numerical parameter study.

Critical issues for cell performance are identified, and their influence quantified (based on the actual parameter set for MEH-PPV/PCBM cells). First, the cell performance is sensitive to the mobilities μ_n and μ_p in the materials: the efficiency η increases monotonously with increasing μ, over many decades (Fig. 16). The sensitivity is about 25% relative efficiency gain for an order of magnitude of increase of the mobilities. Second, the poor absorption of the solar illumination is a cause of weak performance. We

Table 1
Main parameters of the standard parameter set for MEH-PPV/PCBM cell # 3

Parameter	Symbol	Value	Unit
p-contact: ITO/PEDOT-PSS			
Work function	Φ_{m1}	4.7	eV
n-contact: Al/LiF			
Work function	Φ_{m2}	4.3	eV
Effective medium: MEH-PPV/PCBM			
Thickness	d	100	nm
Band gap	E_g	1.25	eV
Electron affinity	χ	3.75	eV
Electron mobility	μ_n	2×10^{-3}	cm^{2}V^{-1}s^{-1}
Hole mobility	μ_p	2×10^{-4}	cm^{2}V^{-1}s^{-1}
Dielectric constant (relative)	ε_s	3	
Maximum absorption constant	α_{max}	2.6×10^4	cm^{-1}
Wavelength of maximum absorption	λ_{max}	495	nm

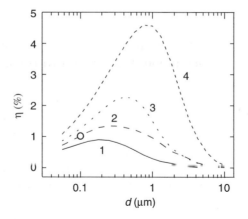

Fig. 17. Calculated influence of the cell thickness d on the efficiency η: curve 1: $\mu_n = \mu_p = 2 \times 10^{-4}$, curve 2: standard parameter set (Table 1); curve 3: $\mu_n = \mu_p = 2 \times 10^{-3}$ and curve 4: $\mu_n = \mu_p = 2 \times 10^{-2}$, all in units of cm^2/V s. The actual cell measurement, corresponding to the standard parameter set, is indicated with a circle.

investigated the influence on cell thickness d (Fig. 17, Table 1), and of the absorption characteristic $\alpha(\lambda)$. Making thicker cells is a technological problem, and not attractive from the fabrication viewpoint. Developing materials with smaller band gap, and a broader absorption $\alpha(\lambda)$ is a subject of active actual research [37]. Organic materials show a more narrow absorption $\alpha(\lambda)$ than inorganic semiconductors. A broader effective use of the solar spectrum could be obtained by combining organic materials with different band gap in a multijunction structure.

8. EXCITONIC SOLAR CELLS

The preceding models for nano-structured solar cells assumed the creation of electron-hole pairs that are separated immediately after their generation. In bulk heterojunction solar cells however, the process of charge separation does not take place spontaneously. The electron and the hole are bound by Coulomb forces, thus forming an exciton, until they are actually separated, or recombined. Most often, the charge separation only takes place at the boundary between two organic phases (molecules or polymers) driven by favorable energetic conditions. The process is then also called charge transfer or charge injection [35]. This section explores the influence of this specific feature on the performance of both crystalline cells and organic cells.

In what follows, we will denote electrons, holes and excitons with the subscript e, h and x, respectively. We will limit ourselves to a 1-D analysis.

Light with wavelength λ is absorbed in a solar-cell material with a total absorption constant $\alpha(\lambda)$, giving rise to a total optical absorption $G(\lambda,y)$. In a simple case,

$$G(\lambda,y) = \Phi_0 \alpha(\lambda) \exp[-y\alpha(\lambda)] \tag{14}$$

where ϕ_0 is the incident photon flux (e.g., in photons/cm^2 s) and y the distance to the plane of incidence. In real structures where multiple layers, interference and scattering phenomena can play a role, $G(\lambda,y)$ should be obtained by more advanced optical modeling, see Ref. 31 for example. A fraction f_{eh} of the absorbed light serves to generate electron–hole pairs, and a fraction f_x to generate a bound exciton pair. Other absorption mechanisms will be neglected here. The generation terms for eh pairs and excitons are thus:

$$g_{eh}(\lambda,y) = f_{eh}(\lambda)G(\lambda,y) \quad \text{and}$$
$$g_x(\lambda,y) = f_x(\lambda)G(\lambda,y) \quad \text{with} \quad f_{eh}(\lambda) + f_x(\lambda) = 1 \tag{15}$$

In inorganic semiconductors, $f_x \cong 0$, except at low temperatures in a narrow wavelength region around the band gap energy, $\lambda \lesssim \lambda_g = h\nu/E_g$. In organic materials, the dominant absorption is by excitons, and hence $f_x \cong 1$ for all absorbed wavelengths [35].

We will explain now how the electrical solar-cell device modeling should be enhanced to also include the effects of exciton. We start from the work of Green's group [41] and of Zhang [42], who presented a simple analytical model for excitonic effects in crystalline silicon cells. We then will extend this work for more realistic situations, and introduce the concept of exciton interface recombination and dissociation [43].

8.1. Simple Exciton Model and Extensions

We take a simple monomolecular form for the direct recombination (or annihilation) of excitons:

$$U_x = \frac{1}{\tau_x}(n_x - n_{x0}) \tag{16}$$

where τ_x is the exciton lifetime. Excitons also can dissociate and convert to a free electron–hole pair, with a net conversion rate $C_{x/eh}$ (this corresponds to $-U_{eh/x}$ in the notation of Refs. 41, 42).

$$C_{x/eh} = b(n^* n_x - n_e n_h) \tag{17}$$

where b (in $cm^3 s^{-1}$) describes the strength of the exciton binding and n^* is an appropriate constant, with the dimension of concentration (thus in cm^{-3}). In equilibrium, detailed balance requires this net rate $C_{x/eh}$ to be zero; this defines the equilibrium exciton concentration $n_{x0} = n_i^2/n^*$, occurring also in Eq. (16). Because excitons do not carry charge, their transport is by diffusion:

$$J_x = -D_x \frac{dn_x}{dx} \tag{18}$$

The simple model of Refs. 41, 42, is confined to the quasi neutral p region (p-QNR) of a one-sided n^+p silicon solar cell: both the heavy doped top n^+ layer and the space charge layer (SCL) in the p layer are neglected, and only electron diffusion is considered in the p-QNR; also, only the low-injection case is considered, that is, the recombination takes the simple monomolecular form (Eq. (11)) instead of the full SRH form (Eq. (13)). Furthermore, it is assumed that all excitons reaching the edge of the SCL are dissociated into eh pairs by the electric field in the SCL, and contribute to the electric current. We have introduced the following extensions [43]:

- We also consider the SCL in the p region. In the SCL, we assume that the electric field profile $E(y)$ is known from the abrupt depletion theory.

- In the SCL, we also consider electron flow by drift, and we use the full SHR recombination Eq. (13).
- We assume that the exciton dissociation rate b in Eq. (17) is field dependent: $b(E)$. Field-enhanced exciton dissociation is a well-documented phenomenon, see Ref. 44.
- We consider surface recombination of excitons at the boundaries (the contacts and the junction) by taking a surface formulation of the bulk Eq. (16), and surface exciton to eh dissociation at the boundaries, that is, a surface formulation of Eq. (17). Hence we introduce a boundary condition for excitons, for example, at a boundary $y = y_0$

$$J_x(y_0) = S_x(n_x(y_0) - n_{x0}) + b_s(n_x(y_0) - n_{x1})$$

(19)

where S_x is the exciton surface recombination velocity, b_s the exciton surface dissociation velocity and n_{x0} and n_{x1} appropriate constants. A similar boundary for electrons is used, where the term in b_s is a generation term instead of a recombination term [43].

With these assumptions and extensions, and with some care for numerical pitfalls and subtleties, the equations for the coupled electron–exciton problem are solved [43].

8.2. Results of Exciton Modeling in Crystalline Solar Cells

The cell structure modeled here [43] is a silicon cell with a thin, heavy doped n^+ region, the junction at $y = -W$; the p region is moderately doped ($N_A = 10^{15}\,cm^{-3}$) with the SCL of W in the range $-W \leq y \leq 0$ and the quasi-neutral p region in the range $0 \leq y \leq y_0$; the contact is at $y = y_0$. The calculation presented in Fig. 18 is for uniform generation (thus $\alpha \rightarrow 0$), only exciton generation, thus $f_x = 1$ and $f_e = 0$, and the incident photon flux adapted to give an ideal light current of $20\,mA/cm^2$.

Even if no electron–hole pairs are directly generated by the illumination, the light current can be decently high, provided that the exciton dissociation rate is high enough, that is at the right-hand side of Fig. 18. It is not so important where exactly the exciton dissociation takes place: it can be in the neutral bulk, in the SCL, or at an interface (the junction or the contacts). All cases 'work', provided that the appropriate dissociation parameter (b or b_s) is high enough. Threshold values for these parameters are deduced from the figure; they slightly vary with the other simulation parameters.

It is observed that the ideal $J_L = 20\,mA/cm^2$ is not obtained in the simulations of Fig. 18. This is because both diffusion lengths, L_e for electrons and

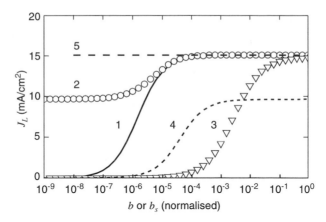

Fig. 18. Short-circuit current under uniform illumination ($\alpha \rightarrow 0$), generating excitons only ($f_x = 1$), for varying exciton dissociation parameters: surface dissociation at contact b_s: low = 10^{-2}/cms^{-1}, high = 10^7/cms^{-1}; bulk dissociation b: low = 10^{-16}/cm^{3}s^{-1}, high = 10^{-7}/cm^{3}s^{-1}; field-enhanced dissociation $b(E)$ in SCL: $b = b_{\text{low}}$ for $y > 0$ and exponentially increasing toward junction ($y = -W$). Curve (1): b varying and b_s = low. Curve (2): b varying and b_s = high. Curve (3): $b(E)$ varying and b_s = low. Curve (4): b_s varying and b = low. Curve (5): b_s varying and b = high. The horizontal axis is normalized to the high value of the varying parameter. Other parameters are: $L_e = L_x = x_0 = 50\,\mu$m and $S_e = S_x = 0$. Reprinted with permission from Thin Solid Films, Vol. 511–512, M. Burgelman and B. Minnaert, Including excitons in semiconductor solar cell modelling, pp. 214–218 © 2006, Elsevier.

L_x for excitons are too low, in this case equal to the thickness y_0 of the neutral part of the cell. Note that the diffusion lengths are given by $L_x = \sqrt{D_x \tau_x}$, and alike for electrons, where D_x is the diffusion constant in Eq. (18) and τ_x the lifetime in Eq. (16). In Fig. 19, the influence of the diffusion lengths is calculated.

It can be seen on Fig. 19 that both L_e and L_x must be rather high, that is, a few times the diode QNR thickness y_0, to obtain the ideal short-circuit current. This is plausible, because the excitons generated in the SCL and QNR must first diffuse to the back contact at y_0, where they dissociate, and then, as electrons, diffuse back to the SCL to be collected and contributed to the current. For the parameters used in Fig. 19, the electron diffusion is slightly more critical than the exciton diffusion. This applies to the situation of Fig. 19, where the exciton dissociation occurs at $y = y_0$; the arguments have to be adapted slightly when it occurs in the bulk, or near the junction ($y = -W$) (no illustration).

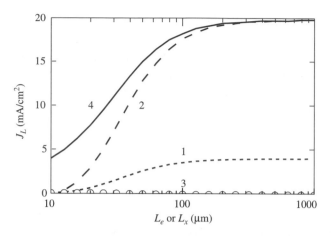

Fig. 19. Influence on J_L of the diffusion lengths L_e and L_x. The exciton surface dissociation at $y = y_0$ is high ($b_s = 10^7$ cm/s), and the bulk dissociation is low (uniform $b = 10^{-16}$ cm³/s in SCL and QNR). The other parameters are as in Fig. 18. Curve (1): L_e varying and $L_x = 10\,\mu$m. Curve (2): L_e varying and $L_x = 1000\,\mu$m. Curve (3): L_x varying and $L_e = 10\,\mu$m. Curve (4): L_x varying and $L_e = 1000\,\mu$m. Reprinted with permission from Thin Solid Films, Vol. 511–512, M. Burgelman and B. Minnaert, Including excitons in semiconductor solar cell modelling, pp. 214–218 © 2006, Elsevier.

8.3. Including Excitons in Electrical Device Modeling of Organic Solar Cells

An obvious difference between inorganic and organic semiconductors is the exciton binding energy Δ: it is small ($\Delta \cong 25$ meV $\cong kT$ at room temperature) in, for example, Si, but substantially larger for organic materials (e.g., $\Delta \cong 300$ meV $\cong 12\,kT$). As a result, in organics, the exciton is more stable [35]: excitons are the particles generated by illumination, and they only dissociate to free carriers in a high-field region, or at the contact with a suitable neighbor molecule, where one of the carriers is injected ('transferred'). These phenomena are described in our model by a field-dependent bulk dissociation constant $b(E)$ and by a surface dissociation rate b_s.

The increased value of Δ has a large influence on the other exciton-related parameters, as suggested in Refs. 41, 42: the parameter n^* in Eq. (17) is thermally activated with Δ, and thus would decrease by a factor of about 6×10^4, and the equilibrium exciton concentration n_{x0} would increase with the same factor. The enhancement of the bulk dissociation in an electric field is well documented (e.g., it increases with one or two orders of magnitude when E increases from 10^5 to 10^6 V/cm [44]), and also is the enhanced transfer of charge from an exciton dissociating at the tangent point of two neighboring molecules [44]. However, these phenomena are treated in different terms, and it is not straightforward to extract suitable values of the parameters $b(E)$ and b_s in organic molecules. Therefore, we have kept these as running parameters to show their influence in qualitative terms. When an organic bulk heterojunction solar cell is simulated with the NM, the concepts b and b_s retain their meaning in the 'unit cell'. When the EMM is used for the simulation, the surface dissociation b_s, in the nano-units that constitute the effective medium, will translate into an effective bulk dissociation rate b.

It can be expected that the major conclusion of our work on planar semiconductor cells will remain valid: the cells will work when there is enough exciton dissociation (wherever this be in the cell) and when both the electron and the exciton diffusion lengths exceed the unit cell thickness. There are indications that L_x can be very small in organic materials (down to a few nm), and thus the size of the unit cell, defined by the morphology of the donor/acceptor blend, will be extremely important.

9. CONCLUSION

Modeling of nano-structured solar cells is complicated for at least three features: (i) The geometry of the nano-structure, be it a structure based on TiO_2 nano-particles or a bulk heterojunction of organic materials, is very complicated and deviates far from the planar structure of bulk crystalline and even of polycrystalline thin film cells. (ii) The distance scale at which the essential phenomena take place is very small: the typical distance is well below a μm for an eta-cell, down to a few nm for an organic solar cell. This makes the application of nearly all concepts of the classical diode theory doubtful or even invalid. (iii) In organic solar cells, molecular processes of excitation, injection, charge transfer and recombination exceed the framework of solid-state semiconductor physics. Owing to these three features, the straightforward application of simulation programs available to the photovoltaic research community is rendered impossible.

Nonetheless, it is our strong conviction that the principles on which existing simulation software is based, do remain valid, as these principles are of a universal physical nature: electric charge is the cause of an electric field (Poisson's equation); particles of one kind can enter into an infinitesimal volume, they can recombine, be generated or converted in this volume, and the balance of these phenomena makes their concentration to change in time (conservation laws or continuity laws): these are called traditionally the 'semiconductor equations', but their validity is universal. It is however nonuniversal, even very specific to every cell system, to find suitable relations for the charge, the recombination, the current flow, etc. (these are called 'constitutive equations'), and to find reliable parameter values used in their description.

This chapter shows how excitonic phenomena (generation, dissociation into *eh* pairs, charge transfer) and various kinds of recombination at material boundaries can be handled in the framework of classical semiconductor device physics (feature (iii) discusses earlier). In very small sized 'elementary cells', the classical diode concepts can be replaced by the elegant concept of a flat-band cell (Section 3). Alternatively, full numerical calculations on nano-scale unit cells are possible to cope with feature (ii). Feature (i) maybe the most annoying one, as it has long time obscured the necessary link between physical and chemical phenomena at small scale, thus materials science, and the device characteristics of the complete macroscopic cell. Two way-outs are presented here. The network model is intuitive, but it is difficult to implement a variety of physical mechanisms in a correct way, and to attribute correct values to the network components. The EMM is easy to use in conjunction with a numerical device simulator, but also here great care has to be taken that the problem which one is setting up is indeed a good description of the solar cell – largely unknown – on the measurement table.

Numerical simulation of nano-structured solar cells of various kinds is a discipline in full development. Interesting and useful results are already obtained. But we also humbly admit that the status already obtained in simulating bulk crystalline and polycrystalline thin film solar cells still has to be reached in simulating nano-structured solar cells.

ACKNOWLEDGMENTS

The Research Fund of the University of Gent (BOF-GOA) (M.B., C.G.). The SBO-project 030220 'NANOSOLAR' funded by the Institute for the Promotion of Innovation by Science and Technology in Flanders (IWT) (B.M.). The European Research Training Network 'ETA' (HRPN-CT2000-0141) (C.G.).

REFERENCES

[1]   M. Burgelman, J. Verschraegen, S. Degrave and P. Nollet, Prog. Photovolt. Res. Appl., 12 (2004) 143–153.

[2]   S. Veenstra, W. Verhees, J. Kroon, M. Koetse, J. Sweelssen, J. Bastiaansen, H. Schoo, X. Yang, A. Alexeev, J. Loos, U. Schubert and M. Wienk, Chem. Mater., 16 (2004) 2503–2508.

[3]   S. Shaheen, C.J. Brabec, S. Sariciftci, F. Padinger, T. Fromherz and J. Hummelen, Appl. Phys. Lett., 78 (2001) 841.

[4]   B. O'Regan and M. Grätzel, Nature, 353 (1991) 737–740.

[5]   U. Bach, D. Lupo, P. Comte, J. Moser, F. Weissörtel, J. Salbeck, H. Spreitzer and M. Grätzel, Nature, 395 (1998) 583–585.

[6]   D. Gebeyehu, C. Brabec and N. Sariciftci, Thin Solid Films, 403–404 (2002) 271–274.

[7]   B. O'Regan and F. Lenzmann, J. Phys. Chem., B 108 (2004) 4342–4350.

[8]   C. Rost, K. Ernst, S. Siebentritt, R. Könenkamp and M. Lux-Steiner, Transparent p-type semiconductors for the ETA-solar cell with extremely thin absorber, Proc. of the 2nd World Photovoltaic Solar Energy Conference, pp. 212–215, Vienna, 1998.

[9]   K. Ernst, A. Balaidi and R. Könenkamp, Semiconductor Sci. Technol., 18 (2003) 475–479.

[10] A. Wahi, R. Engelhardt, P. Hoyer and R. Könenkamp, Interface characterisation of amorphous silicon on titaniumdioxide: towards a solid-state sensitizer cell, Proc. of the 11th Photovoltaic Solar Energy Conference, pp. 714–717, Montreux, 1992.

[11] I. Kaiser, K. Ernst, Ch. Fischer, R. Könenkamp, C. Rost, I. Sieber and M. Lux-Steiner, Sol. Energy Mater. Sol. Cells, 67 (2001) 89–96.

[12] S. Siebentritt, K. Ernst, C. Fischer, R. Könenkamp and M. Lux-Steiner, CdTe and CdS as extremely thin absorber materials in an η-solar cell, Proc. of the 14th European Photovoltaic Solar Energy Conference, pp. 1823–1826, Barcelona, 1997.

[13] M. Nanu, J. Schoonman and A. Goossens, Adv. Mater., 16 (2004) 453–456.

[14] Jörg Ferber, Elektrische und optische Modellierung von Farbstoffsolarzellen, Ph.D. Dissertation, Universität Freiburg (D), 1999 (in German).

[15] M. Burgelman and C. Grasso, J. Appl. Phys., 95 (2004) 2020–2024.

[16] S M Sze, Physics of Semiconductor Devices, 2nd ed., Wiley, New York, 1981.

[17] M. Burgelman, P. Nollet and S. Degrave, Thin Solid Films, 361–362 (2000) 527–532.

[18] A. Niemegeers, S. Gillis and M. Burgelman, A user program for realistic imulation of polycrystalline heterojunction solar cells: SCAPS-1D, Proc. 2nd World Conference on Photovoltaic Energy Conversion, pp. 672–675, Wien, Österreich, July 1998, JRC, European Commission, 1998.

[19] P. Basore, IEEE Trans. Electron Dev., 37 (1990) 337–343.

[20] C. Grasso, K. Ernst, R. Könenkamp, M. Burgelman and M.C. Lux-Steiner, Photo-electrical characterization and modeling of the eta-solar cell, Proc. of the 17th European Photovoltaic Conference, pp. 211–214, München, D, October 2001, WIP, München, 2002.

[21] M. Green, Silicon Solar Cells—Advanced Principles & Practice, The University of South Wales, 1995.

[22] S. Fonash, Solar Cell Device Physics, Academic Press, New York, 1981.

[23] S. Rühle, J. Bisquert, D. Cahen, G. Hodes and A. Zaban, Electrical and chemical potential distribution in dye-sensitized and similar solar cells in the dark and under illumination, QUANTSOL Conf., Bad Gastein, Austria, 2003.

[24] B. Minnaert, C. Grasso and M. Burgelman, Comptes Rendus Chimie, 9 (2006) 735–741.
[25] C. Brabec, in "Organic Photovoltaics: Concepts and Realization" (C. Brabec, V. Dyakonov, J. Parisi and N. Sariciftci, eds.), pp. 183–185, Springer, Berlin, 2003.
[26] L.H.K. Van Beek, in "Progress in Dielectrics 7" (J.B. Birks, ed.), p. 69, Heywood, London, 1967.
[27] J. Haerter, S. Chasteen and S. Carter, Appl. Phys. Lett., 86 (2005) 164101.
[28] C.T. Sah, Proc. IEEE 55–5 (1967), 654–671.
[29] C.T. Sah, Proc. IEEE 55–5 (1967), 672–684.
[30] C.T. Sah, Solid-State Electron., 13 (1970), 1547–1575.
[31] N.K. Persson and O. Ingenäs, in "Organic Photovoltaics" (S.S. Sun and N.S. Sariciftci, eds.), pp. 107–138, Taylor & Francis, Boca Raton, FL, 2005.
[32] H. Pauwels and G. Vanhoutte, J. Phys. D. Appl. Phys., 11 (1978) 649–667.
[33] G. Roberts and A. Sedra, SPICE, 2nd ed., Oxford University Press, New York, 1997.
[34] B. Minnaert and M. Burgelman, E-MRS Spring Conference (Symposium F: Thin Film and Nanostructured Materials for Photovoltaic THINC-PV2), Strasbourg, France, May 31–June 3, 2005.
[35] F.L. Zhang, M. Johansson, M.R. Andersson, J.C. Hummelen and O. Inganäs, Synth. Metals, 137 (2003) 1401–1402.
[36] C. Brabec, A. Cravino, D. Meissner, N. Sariciftci, T. Fromherz, M. Rispens, L. Sanchez and J. Hummelen, Adv. Funct. Mater., 11 (2001) 374–380.
[37] C. Winder and N.S. Sariciftci, J. Mater. Chem., 14 (2004) 1077–1086.
[38] S. Sensfuss, M. Al-Ibrahim, A. Konkin, G. Nazmutdinova, U. Zhokhavets, G. Gobsch, D·A.M. Egbe, E. Klemm and H.K. Roth, Proc. of SPIE, Vol. 5215, pp. 129–140, Bellingham, WA, 2004.
[39] S·A. McDonald, P.W. Cyr, L. Levina and E.H. Sargent, Appl. Phys. Lett., 85 (2004) 2089–2091.
[40] W. Geens, S.E. Shaheen, C.J. Brabec, J. Poortmans and N.S. Sariciftci, AIP Conference Proceedings, Vol. 544, pp. 516–520, Kirchberg, Austria, 2000.
[41] R. Corkish, D. Chan and M. Green, J. Appl. Phys., 79 (1996) 195–203.
[42] Y. Zhang, A. Mascarenhas and S. Deb, J. Appl. Phys., 84 (1998) 3966–3971.
[43] M. Burgelman and B. Minnaert, Thin Solid Films, 511–512 (2006) 214–218.
[44] V. Arkhipov and H. Bässler, Phys. Stat. Sol. A, 201–206 (2004) 1152–1187.

Nanostructured Materials for Solar Energy Conversion
T. Soga (editor)

Chapter 3

Optical and Electrical Modeling of Nanocrystalline Solar Cells

Akira Usami

Central Research Institute of Electric Power Industry, 2-11-1 Iwado Kita, Komae, Tokyo 201-8511, Japan

1. INTRODUCTION

A solar cell is a system that converts optical energy to electricity. Absorption of a photon generates an electron-hole pair, using its energy. However, in practical solar cells, nonnegligible electrons recombine with holes via several internal paths. The photocurrent as well as the photovoltage deteriorates because of this carrier loss. Thus, to attain a high-energy conversion efficiency, preventing the recombination is crucial. An approach, which consists of experimental cell build-ups and extensive testing, is costly and time consuming. Development of a theoretical model which would allow one to gain a better understanding of energy-conversion mechanisms is advantageous. This model is expected to suggest directions for improvements without the trial and error. The recent rapid progress of performance of computers has allowed an inexpensive personal computer to calculate a task performed conventionally within a workstation. Thus, a computer will be an indispensable tool for the device design soon as the device modeling is getting more and more importance.

This chapter focuses on theoretical modeling of nanocrystalline semiconductor films. Nanocrystalline semiconductor films have recently attracted considerable interest as a photoactive anode of an electrochemical solar cell. The recent most interesting example is dye-sensitized nanocrystalline solar cells [1–3]. A typical dye-sensitized solar cell is composed of a nanocrystalline titanium dioxide (TiO_2) film, dye-sensitizers, an electrolyte, a transparent conductive substrate, and a counter electrode; a dye-sensitized nanocrystalline TiO_2 film permeated by an electrolyte, and a bulk layer of the electrolyte are sandwiched between a transparent conductive front-substrate and a counter back-electrode. In a dye-sensitized solar cell, incident light is absorbed by a monolayer of the dye-sensitizers on the TiO_2 nanoparticles. Sufficient light absorption is achieved by nanocrystallization of TiO_2 because a large internal surface area of the nanocrystalline TiO_2 film significantly increases the dye concentration in the film. Photoinjected electrons travel through the TiO_2 network to the transparent conductive oxide (TCO) electrode. The transport of electrons occurs mainly by *diffusion* because of a negligible macroscopic electric field across the bulk of the film owing to screening by the high-ionic-strength electrolyte. The pores are completely filled with the electrolyte containing redox species. A photooxidized dye is regenerated by reduction by an electron donor in the electrolyte. The oxidized ion is diffused to the counter electrode, and then recycled to the original state by an electron captured from the electrode. The origin of the photovoltage is the energy of photons absorbed by the dye-sensitizers. A photovoltage loss stems from the following two driving forces: the electron injection from a dye-excited state into the TiO_2 conduction band, and the dye regeneration by redox species. The photocurrent decreases by the recombination of photoelectrons with photooxidized dyes and oxidized electrolytic ions. The recombination is significantly reduced by the employment of the "interfacial network heterojunction"; that is, photoelectrons are separated spatially from oxidized electron acceptors at an internal heterojunction. Similar improvement of photocurrent yields has been attained in organic photovoltaic cells [4].

A difficulty of optical and electrical modeling of the nanocrystalline electrode is complexity of the interfacial network heterojunction consisting of aggregates of a great number of nanosize particles. In principle, a straightforward method of conventional device modeling is applicable, dividing the electrode into spatial finite elements and formulating the Poisson and current equations. However, this is practically infeasible for the nanocrystalline electrode because of a required huge memory and a long computing time. To avoid this difficulty, a mean-field expression, such as a pseudo-homogeneous approximation, has been employed in a macroscopic electrical simulation of the dye-sensitized electrode. The first simulation of electric properties of a dye-sensitized solar cell has been presented by Södergren et al. [5] in 1994. They presented a model for photoelectron transport in a bare TiO_2 porous film, and the resultant model was applied to current–voltage characteristics simulations of a dye-sensitized solar cell. Neglecting redox species in the electrolyte, they modeled only electron transport in the semiconductor matrix. In addition, they treated the nanoporous electrode as an apparent continuum. As a result, an analytical formula was provided. Advanced models that consider redox species in the electrolyte were presented by Usami [6, 7] and Ferber et al. [8, 9] by adopting the pseudo-homogeneous approximation to not only the semiconductor matrix but also to the pore-filling electrolyte. However, they cannot be solved analytically, and thus, a numerical simulation with a computer is necessary.

Unfortunately, there are some simulations to which this approximation is inapplicable, that is, photoelectron random walk involving trapping/detrapping processes in nanoparticles. Especially, influences of the particle necking and coordinating number on the electron transport cannot be simulated without going into the geometric details of the porous photoelectrode [10–12]. This mezzoscopic electron behavior is essential for determination of the diffusion coefficient of electrons. For these simulations, Monte Carlo methods based on the continuous-time random walk (CTRW) [13, 14] or the multiple trapping (MT) [15, 16] have been employed. These simulations follow traces of electrons walking in an imaginary trap lattice created in a computer program with random numbers, like the backgammon. Monte Carlo simulations also have been applied to analyses of electron recombination dynamics, and satisfactory agreement with experimental data has been provided [13–16].

Electrons (and redox species) as well as photons play an essential role in the energy conversion processes. Thus, transmission of photons in the nanocrystalline electrode is also an important subject of modeling. The first simulation of light transmission in the nanocrystalline electrode was conducted for demonstration of improvement of the light absorption in the photoelectrode by light scattering of the TiO_2 particles [17]. For the same purpose, several simulations were presented [18–22]. These simulations have revealed that embedding relatively large TiO_2 particles into the semitransparent TiO_2 matrix increases the light absorption owing to optical length enlargement and optical confinement. Although these simulations considered multiple scattering effects, independent scattering was assumed. Thus, for relatively dense media of the scattering centers, a quantitative error becomes nonnegligible. As a result, light scattering of the nanocrystalline TiO_2 electrode itself, which is an extremely dense medium of the scattering centers, cannot be simulated with the independent scattering models. Usami and Ozaki [23] have presented a light scattering model for the nanocrystalline films.

In this chapter, theoretical simulation models and some examples of the simulations for electrical and optical properties of dye-sensitized nanocrystalline solar cells are outlined. Here, we restrict our discussion to descriptions of a basic concept; details are described in the cited literature. In Section 2, optical modeling of nanocrystalline TiO_2 films and application of light scattering to dye-sensitized solar cells are presented. In Section 3, electrical performance simulations of dye-sensitized solar cells are described as follows: first, formulation of an empirical model, that is, an equivalent circuit; second, Monte Carlo simulations of photoelectron transport in nanocrystalline particles; and third, formulation of differential equations expressing transport of electrons and redox species in the solar cells, and numerical solutions of these equations. In Section 4, on the basis of these models, prospects for formulation of advanced models are presented.

2. OPTICAL MODELING OF NANOCRYSTALLINE FILMS

2.1. Light Scattering Properties of Nanocrystalline TiO$_2$ Films

The study of light scattering and its applications is an old subject. Light scattering of discrete small dielectric particles has been widely studied, and these results have been published in books [24–26]. Light scattering by a dielectric particle much smaller than the light wavelength is referred to as the Rayleigh scattering. Physically light scattering by a dielectric particle is explained as follows. In a dielectric particle embedded in an electric field, electric charges due to dielectric polarization are induced. Since light is an electromagnetic wave, light has a spatially and temporally fluctuating electromagnetic field; thus, the induced charges fluctuate temporally. As this fluctuation of the charges induces a new electromagnetic wave radiation, the induced radiation is observed as light scattering. If the particle is much smaller than the incident light wavelength, the induced charges are approximately a dipole because the electromagnetic field of the incident light is assumed to be spatially uniform over the particle. This scattering field has a uniform angle distribution. This scattering based on a single discrete particle is applicable to scattering of plural discrete particles by simply summing the scattered intensity of each particle if distances between the particles are satisfactorily long: this is "independent scattering." However, for dense media of particles, interference between scattered fields is nonnegligible. For example, far fields of induced dipoles separated by the half wavelength cancel out, that is, "dependent scattering" must be considered in these dense media. In addition, "multiple scattering" – the scattered light that is re-scattered by other particles – has an essential influence on scattering properties in the dense media. Unfortunately, the dependent and multiple scattering cannot be easily simulated.

For a film consisting of randomly distributing small particles, the simplest scattering model is the independent scattering model. That is, the scattering field is described as the sum of the radiation fields from the randomly distributing dipoles induced in each particle. The scattering rate α for the simplest model has been already presented [26]:

$$\alpha = \beta \lambda^{-4} a^3 \tag{2.1}$$

where λ is the wavelength of incident light and a the radius of a particle. In Eq. (2.1), β depends on the refractive index, fractional volume, and distribution of the particles [26]; when the total scattering field is the sum of independent dipole fields (i.e., the optical interactions between the dipole fields are negligible), β is formulated analytically with the refractive indices of the particles n_s and the nanocrystalline film n, and the fractional particle volume f [24–26],

$$\beta = 2f(2\pi n)^4 \left| \frac{n_s^2 - n^2}{n_s^2 + 2n^2} \right|^2 \tag{2.2}$$

The light hemispherical ($=$ collimated $+$ diffused) transmittance T is related to α as $T = (1 - R)\exp(-\alpha d/2)$, where R is the specular reflectance on the film surface and d the film thickness. Here, since the forward scattering, which has the same intensity as the backward scattering owing to the symmetry in the Rayleigh scattering, is detected as diffuse transmission, an effective scattering rate is assumed to be $\alpha/2$. Consequently, the following dependence of T on λ, a and d is provided,

$$\ln(T) \propto \frac{\beta(f, n_s, n)}{2} \lambda^{-4} a^3 d \tag{2.3}$$

It should be noted that this scattering model is not precisely an independent model. n is an average refractive index of the medium into which the scattering centers are embedded. Here, the scattering centers are constituent particles of the nanocrystalline film. If we model the film as packing particles in the air, it is plausible that n is the refractive index of air; this is the literal independent scattering model. However, for dense media, it is well known that the scattering rate evaluated with this literal independent scattering model is much greater than experimental results [27, 28]. Further, a nanocrystalline film is not a simple dense medium of packing particles. In the applications, such as the solar cells, the charge transport in the film is essential. Annealing in the film preparation makes the particles interconnected, though the particle sizes do not change owing to an annealing temperature much lower than the melting point.

Ferrand and Romestain [29] have studied a light scattering loss from a medium of a porous silicon (Si) waveguide. They assumed that the light scattering is attributed to fluctuations of the refractive index of the light-propagating medium. Although the calculated scattering loss does not quantitatively agree with experimental results, these have the same order of magnitude. After their modeling, we assume that the medium surrounding the scattering centers is the nanocrystalline film having an effective refractive index.

Usami and Ozaki [23] have demonstrated experimentally this pseudo-independent model with light scattering in thin films consisting of TiO_2 nanoparticles smaller than approximately 50 nm. However, for intensified scattering in thick films, the internal multiple scattering between the scattering centers, which is neglected in this model, is nonnegligible. For the multiple scattering effects, $\ln(T) \propto \lambda^\gamma$ where γ increases from -4 in proportion to the film thickness is an alternative model for moderate thickness of films [23].

2.2. Application of Light Scattering to Dye-Sensitized Solar Cells

As discussed in Section 2.1, a nanocrystalline film has a light scattering ability. Practically, a nanocrystalline film prepared from a commercial powder P25, which is a nanopowder of approximately 50 nm diameter, is visually translucent because of the light scattering by the nanopowder. In a dye-sensitized solar cell, a nanocrystalline film is the photoactive electrode. Light scattering in the photoactive electrode is expected to increase the light absorption because of the increase of the optical length and optical confinement. The optical length depends on the paths of photons in the film. Without scattering, photons are transmitted straight into the film; thus, the absorbance is $\alpha_{abs}d$, where α_{abs} and d are the absorption coefficient and the film thickness, respectively. If photons are scattered in the film, the photons feel the film thicker than d; because of this increase of the optical length, the apparent absorbance is greater than $\alpha_{abs}d$. Optical confinement is also expected; because of a large average refractive index of the TiO_2 nanocrystalline film, nonnegligible scattered light is trapped in the film by the interfacial total reflection.

Although the scattering intensity increases in proportion to the particle diameter for the Rayleigh scattering, a nanocrystalline film of large particles has a small internal surface, leading to a decrease in the dye-sensitizer concentration. Thus, rather than increasing sizes of the film-constituting particles, a few large particles should be embedded into the nanocrystalline matrix of small particles. Even in this case, there is an optimal particle diameter to attain the greatest light absorbance, because forward scattering is intensified in large particles, decreasing the scattering effects. In order to estimate the optimal diameter and concentration of the scattering particles, and evaluate the improvement of the absorbance in the photoactive films, simulations have been carried out [18, 19, 21, 22].

In simulations of the light scattering by relatively large scattering particles embedded in the apparent transparent nanoparticles, multiple scattering effects were evaluated under the assumption of the independent scattering. A numerical solution of the radiative transport equation [18], a Monte Carlo simulation based on a ray-tracing method [21], and a solution of the four flax model employing empirical parameters [19] were presented. Here we describe the ray-tracing method, because through the explanation one can gain a better understanding of the increase of the light absorbance by the light scattering. Details of the radiative transport equation are also described by Ishimaru [24].

The ray-tracing Monte Carlo simulation has been widely applied to analyses of light transmission in a living body in infrared (IR) spectroscopy, such as the optical computer tomography [30]. Since a living body is also a strong light scattering medium, the same method is applicable to analyses of the light scattering in the nanocrystalline electrode. A scattering medium is characterized by the following parameters: the scattering coefficient, α_{sct}[m^{-1}], the absorption coefficient, α_{abs}[m^{-1}], and the angle distribution of the scattering intensity by a scattering center, $p(\theta)$. For simplicity, we assume a homogeneous medium and monochromatic light. Since the probability that a photon getting into the medium is transmitted without scattering or absorption through a length, l, is $\exp\{-(\alpha_{sct}+\alpha_{abs})l\}$, a transmission length, L, is determined by

$$L = -\frac{1}{\alpha_{sct} + \alpha_{abs}} \ln(Rnd[0,1]) \tag{2.4}$$

where $Rnd[0,1]$ is a random number between 0 and 1. Then, the photon transmitted by L is scattered or absorbed, scattering or absorption is determined as follows:

if $Rnd[0,1] < \alpha_{sct}/(\alpha_{sct}+\alpha_{abs}) \rightarrow$ scattering,

if $Rnd[0,1] < \alpha_{abs}/(\alpha_{sct}+\alpha_{abs}) \rightarrow$ absorption.

If absorbed, the location is recorded; if scattered, relative scattered direction (θ,φ) is

$$\theta = f^{-1}(Rnd[0,1]) \tag{2.5}$$

$$f(\theta) = \int_0^\theta p(\theta)d\theta, \quad f(\pi) = 1 \tag{2.6}$$

$$\varphi = 2\pi Rnd[0,1] \tag{2.7}$$

For the Rayleigh scattering, $p(\theta)$ is independent of θ. Here, $Rnd[0,1]$ is randomized at each step. The same procedure is repeated until the photon is absorbed. When sufficiently many paths of photons are simulated, the calculation is over.

Usami [21] has reported the light scattering simulations with the ray-tracing method considering the optical confinement effects. Ferber et al. [18] have presented similar simulations with the radiative transport equation. Qualitatively, these simulations agree with experimental results. However, a nonnegligible quantitative error is expected in a medium having dense scattering centers with the independent scattering models. Generally, in the dense media, real scattering intensities are smaller than simulation results based on the independent models [28].

3. SIMULATIONS OF PERFORMANCE OF DYE-SENSITIZED NANOCRYSTALLINE SOLAR CELLS

The conventional cell-designing method based on trial-and-error as a result of lacking a clear understanding of the cell operating mechanisms has prevented continuous improvement of the cell performance. As computer simulations provide a better understanding of cell operating processes, computer simulations are a very effective method for designing a high-performance cell. Until now, equivalent circuits, Monte Carlo simulations, and numerical analyses based on differential equations have been applied to dye-sensitized solar cells. Each method has its own advantages and disadvantages. Using an equivalent circuit, experimental current–voltage characteristics are successfully reproduced. However, the relation between the energy conversion processes and the parameters in the equivalent circuit is not clear. Monte Carlo simulations based on electron tracing are intuitive because the electron tracing is an imaginary experiment. For the electron recombination dynamics, satisfactory agreement with experimental results has also been reported. However, a static simulation of the current–voltage characteristics by tracing photocarriers is a heavy task for present computers. Numerical solution of differential equations is the primary candidate for a device simulator because the current–voltage characteristics are calculated from physical parameters, such as the diffusion coefficients. However, in the absence of a satisfactory understanding of the energy conversion mechanisms, there are discrepancies between simulation results and experimental current–voltage characteristics. In this section, we outline these three simulation methods.

3.1. Equivalent Circuit

A solar cell is an electrical device; this is expressed as an equivalent circuit. As conventional solar cells have a p–n junction, these behave as a diode in the dark (a schematic image is shown in Fig. 1). Under illumination, the output current is the sum of a reverse photocurrent and the forward diode rectification current. Thus, a general basic expression is a parallel connection of a constant current source and a diode. This electric circuit is represented as follows:

$$J = J_{sc} - J_0 \left[\exp\left(\frac{qV}{n\kappa_B T} \right) - 1 \right] \tag{3.1}$$

where J is the output current density, J_{sc} the short-circuit current density, J_0 the saturation current density, q the elementary charge, V the external applied voltage, κ_B the Boltzmann constant, and T the absolute temperature. Here, n is called "ideality factor." In the complete expression, series and shunt resistances are also considered; for simplicity, we use Eq. (3.1) here. As the exponential term in the right-hand side of Eq. (3.1) is much greater than 1 except extremely small V_s, where the bracket's term is negligiblc, 1 in the brackets is often neglected. In addition, from $J = 0$ at $V = V_{oc}$, Eq. (3.1) becomes

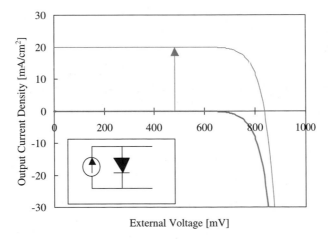

Fig. 1. Current–voltage characteristics and an equivalent circuit. Generally, a solar cell behaves as a diode in the dark. Under illumination, the photocurrent moves the dark current–voltage curve toward a high current. This is expressed as a parallel connection of a diode and a constant current source as shown in the inset.

$$J = J_{\text{sc}}\left[1 - \exp\left(\frac{q(V - V_{\text{oc}})}{n\kappa_{\text{B}}T}\right)\right] \tag{3.2}$$

Equation (3.2) indicates that the current–voltage characteristics are determined by J_{sc}, V_{oc} and n in a room temperature. Thus, not only J_{sc} and V_{oc} but also n is an important parameter. The diode ideality factor, n, has been studied for semiconductor p–n junctions [31–33]. According to the Sah–Noyce–Shockley theory, the ideality factor depends on the locus of the recombination of electron–hole pairs. If the recombination takes place in the neutral region of a p–n junction, this type gives an ideality factor of 1.0; the recombination in the space charge region, mediated by recombination centers in the band gap, results in an ideality factor of 2.0. Thus, ideality factors for p–n junctions are, in principle, between 1.0 and 2.0. However, some ideality factors greater than 2.0 have been reported [32, 33], and explained by tunneling effects [32], a series connection of diode elements [33], and so on.

Equation (3.2) is also applicable to dye-sensitized solar cells. It has been revealed that current–voltage characteristics of a wide variety of dye-sensitized solar cells are successfully reproduced by Eq. (3.2) [34]. However, as the dye-sensitized solar cells do not have a p–n junction, physical processes expressed in Eq. (3.2) are quite different from the conventional solar cells; for the dye-sensitized solar cells, Eq. (3.2) is based on the Butler–Volmer equation because the recombination takes place at the semiconductor/electrolyte interface. Ideality factors reported in relatively high efficiency dye-sensitized solar cells were around 2.0 [35]; practically, these are consistent with the Butler–Volmer equation. From electrical impedance spectroscopy spectra, van de Lagemaat et al. [36] explained an ideality factor of 1.0 by a barrier at the TCO/TiO$_2$ interface and an ideality factor of 2.0 by the electron transfer across the TiO$_2$/electrolyte interface, which depends on the Butler–Volmer equation; in fact, ideality factors in this range were also presented [37–40]. However, an ideality factor as great as 4.0 has been observed by increasing the light absorption at the TCO/TiO$_2$ interface [34]; this seems to be beyond the Butler–Volmer theory. Huang et al. [41] reported that the ideality factor was determined by the order in the electron density of the nanocrystalline electrode in the recombination reaction; however, the above variations of n cannot be explained. Thus, some other mechanisms than these explanations have nonnegligible influences on n of the dye-sensitized solar cells. Rau et al. [42] expressed the equivalent circuit as a series connection of two diodes. From significant variations of the fill factor with changing the work function of the front electrode, they modeled the TCO/TiO$_2$ interface as a Schottky diode. On the other hand, Rühle et al. [43] indicated the TCO/TiO$_2$ should be modeled as a tunnel barrier, rather than as a Schottky diode, and have recently demonstrated the presence of a tunnel barrier [44]. In both the Schottky and tunneling models, the TCO/TiO$_2$ has a significant influence on fill factor (FF), while V_{oc} is independent of the TCO/TiO$_2$. Usami et al. [34] also have revealed a significant influence of the TCO/TiO$_2$ on the idcality factor from experiments using light trapping in the TCO/glass front substrate; the ideality factor was a function of local photocarrier densities at the TCO/TiO$_2$ interface.

It should be noted that the equivalent circuit is not treated as a usual electrical circuit. As the constant current source is open in the dark, the current–voltage characteristics depend only on the diode. As shown in Fig. 1, when the constant current source is connected parallel to the diode, the external current is the sum of the source current and the diode current in a usual circuit. However, for dye-sensitized solar cells, under illumination, the current–voltage characteristics cannot be calculated from the photocurrent source and the dark diode property. Current–voltage characteristics under illumination have apparently deviated from calculation results of the sum of the photocurrent and the dark current. Some papers suggest that this is ascribed to potential drop at the series resistance [41, 45]. However, a definite answer has not been provided yet.

A disadvantage of the equivalent circuit is that the relation between the model and the energy conversion processes is unclear. Recent progress of fundamental studies on, in particular, the ideality factor described above makes this disadvantage less important. Since the equivalent circuit successfully reproduces experimental current–voltage characteristics, the equivalent circuit is expected to become a powerful tool, clarifying the relation between the model and the internal physical processes.

3.2. Monte Carlo Simulation

Monte Carlo simulation is a general term for simulations that use random numbers. For dye-sensitized solar cells, a trace of a photoelectron in a nanoparticle or a nanocrystalline film has been simulated with random numbers. With the Monte Carlo simulation, dynamics of the electron diffusion and recombination has been studied; static properties, such as the current–voltage characteristics, are not usually fit for the Monte Carlo simulation. Monte Carlo simulations of the macroscopic electron diffusion in a nanocrystalline film were carried out by van de Lagemaat and Frank [46]. A nanocrystalline film consists of nanoparticles and electrolyte-filled nanopores between the particles. In the simplest approximation, the nanocrystalline film is assumed to be a pseudo-homogeneous medium; that is, without going into exact geometric detail, the film is treated as the superposition of two continua, one representing the nanoparticle matrix and the other representing the solution. This approximation has been applied to theories for batteries [47]. They employed the pseudo-homogeneous approximation [46]. Here, we outline the simplest model. But in advanced models, the necking and coordinate number of the particles were considered; influences of these on the electron diffusion were simulated with Monte Carlo methods [10–12].

Since the Monte Carlo simulation of the electron diffusion is an imaginary experiment, details of the diffusion processes should be known beforehand. The electron transport in TiO_2 nanocrystals is similar to that in amorphous structures, rather than single crystals. Amorphous substances have no definite band gap, and tail states are exponentially distributed at lower energies of the conduction band. The tail states consist of localized trap levels, and "hopping" between the localized trap levels is the predominant electron-transport process (Fig. 2). The electron transport in the nanocrystalline TiO_2 is essentially the random walk hopping process between the tail states. In trap-dotting materials, electron transport has been explained by the following two models: CTRW and MT. While MT explicitly considers a difference between electrons transported in the conduction band and electrons localized in the trap levels, the difference is unclear in CTRW. In other words, electrons

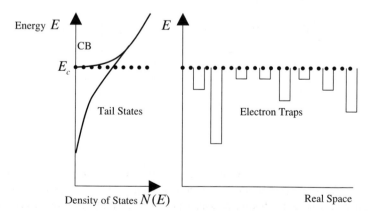

Fig. 2. Electronic states of TiO_2 nanoparticles. CB represents the conduction band. The tail states consist of the localized trap levels in the real space, and "hopping" between the localized trap levels is the predominant electron transport process.

wander about the traps in CTRW; electrons diffused in the conduction band are sometimes trapped in MT. These two models are essentially the same except for this difference [48]. In particular, there is essentially no difference between these two models for disordered materials like the TiO_2 nanocrystals where almost all electrons are trapped.

The tail states have the following energy distribution:

$$N(E) = \frac{N_{tot}}{m_c} \exp\left(\frac{E - E_c}{m_c}\right) \tag{3.3}$$

where $N(E)$ is the density of states at the energy, E, N_{tot} the total trap density, and E_c the energy of the conduction band edge. The characteristic energy $m_c = 60-100\,\text{meV}$ is reported experimentally for TiO_2 nanocrystals. As shown in Fig. 3, adopting the pseudo-homogeneous model, the nanocrystalline electrode is treated as a cubic lattice. Each node represents an electron trap. To reduce a computational load, the periodic boundary conditions are usually assumed (Fig. 3(B)); thus, the total nodes are much fewer than real traps in the nanocrystalline electrode. The trap energy depth, $E_T - E_c$, of a node should

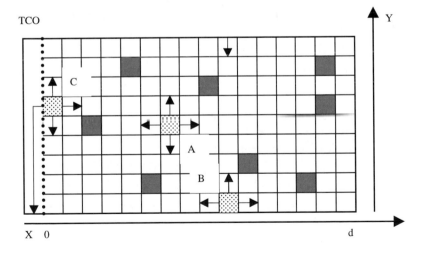

Fig. 3. An outline of the Monte Carlo simulation. Adopting the pseudo-homogeneous model, the nanocrystalline electrode is treated as a cubic lattice. Each node represents an electron trap. Shadows express electron-occupied traps. An electron in a trap can move to the neighboring traps if the destination is unoccupied (A). At the edges, the periodic boundary conditions are assumed (B). When an electron reaches the TCO, the electron becomes the external current (C).

be consistent with Eq. (3.3); thus, using a random number between 0 and 1, $Rnd[0,1]$, this is determined as follows:

$$E_T - E_c = m_c \ln(Rnd[0,1]) \tag{3.4}$$

If there are any static electrons in the electrode, such as under bias light in the transient photocurrent measurement, the preoccupation is determined by using a random number again,

$$f(E_T) = \frac{1}{1 + \exp(E_T - E_F/\kappa_B T)} \tag{3.5}$$

$Rnd[0,1] < f(E_T)$: the trap is occupied,

$Rnd[0,1] > f(E_T)$: the trap is empty.

Here, E_F is the quasi-Fermi level of the static electrons.

In the Monte Carlo simulation, an electron random walk between these nodes is followed. In each step, an electron attempts to move to a neighboring node. If the destination is empty, the electron moves to the node; if occupied, the electron passes the node. The trapped electron in a node waits the thermal detrapping there; the waiting time, τ_r, is assumed to depend on the trap energy depth,

$$\tau_r = \frac{\exp(E_T - E_c/\kappa_B T)}{\nu_{th}} \tag{3.6}$$

where ν_{th} is a frequency around $10^{12} - 10^{13}\, s^{-1}$, which relates to the trapping cross section and the free electron velocity.

In the Monte Carlo simulation of the electron diffusion in the nanocrystalline films, the recombination loss has been often neglected. However, some photoelectrons cannot be reached to the front electrode by the recombination. Monte Carlo simulations of these electron recombination dynamics were reported by Nelson et al. [13, 14, 49–52], and Barzykin and Tachiya [15, 16]. In order to simulate successfully the electron recombination dynamics with the Monte Carlo simulation, the rate-limiting process must be the electron diffusion rather than an electron transfer reaction with an acceptor at the particle surface. Durrant, Nelson, and coworkers [14, 49–52] have attempted to reproduce experimental recombination dynamics with the Monte Carlo simulation. Assuming the rate-limiting process is the electron diffusion in the nanocrystal,

Nelson has presented a Monte Carlo simulation model that an electron diffusing with the CTRW in a particle recombines promptly with a dye cation adsorbed on the surface of the particle when the electron occupies the nearest trap to the dye cation; satisfactory agreement of the simulation results with the experimental ones was provided for the recombination dynamics with both the oxidized ions of I_3^- and the dye cations of the usually employed Ru-complexes. This is also consistent with their contribution which has indicated that the recombination kinetics with $Ru(dcbpy)_2(NCS)_2$ (dcbpy = 2,2'-bipyridine-4,4'-dicarboxylate) where positive charge density of the cation state is localized on the Ru and NCS moieties is substantially the same as porphyrin dyes where the cation states are expected to be delocalized over the conjugated macrocycles [53]. However, a more detailed study has revealed that, by retarding the electron transfer time at the surface by spacing out the dye cations from the surface, the rate-limiting process varies from the mezzoscopic electron diffusion in the nanocrystal to the electron transfer reaction with the dye cations at the surface [51]. In any case, for the usually employed sensitizers, the microscopic electron diffusion in the nanocrystal is the dominant factor for the recombination kinetics with the dye cations, rather than the electron transfer reaction with the dye cations at the surface. Therefore, the electron recombination dynamics has been simulated by the electron random walk in the nanocrystal, though there are some other explanations for the rate-limiting step [54]. For the simulations of the electron recombination dynamics, the electron random walk in one nanoparticle, rather than the nanocrystalline film, is usually considered. Thus, the nodes are not the electron traps of the whole electrode but those in a particle. Since the traps stem from disorder of the crystal, the traps should be localized on the particle surface rather than on the bulk. However, the Monte Carlo simulation results were independent of the trap location [14].

3.3 Numerical Analysis Based on Differential Equations

In conventional solar cells based on a solid p–n junction, solar cell performance has been analyzed with computer simulation programs, such as "PC-1D" [55]. However, photogenerated charge carriers in a dye-sensitized solar cell have a different spatial distribution from those in the conventional solar cells. This leads to a fundamental difference between them as recently pointed out by Gregg and Hanna [56]. Thus, the computer programs such as PC-1D cannot be employed for dye-sensitized solar cells in which the carrier transport occurs by the diffusion, rather than the drift by a built-in potential, because of a negligible macroscopic electric field across the bulk of the photoanode owing to screening by the high-ionic-strength electrolyte. Thus, only the diffusion term is considered in the current equation at the bulk of the nanocrystalline photoelectrode. On the basis of a theory in batteries [47], the porous photoelectrode is treated as a superposition of two continua; as mentioned above, this is referred to as the "pseudo-homogeneous" approximation, that is, one represents a pore-filling solution and the other a solid semiconductor matrix. Current density i_1 and i_2 are introduced for electrons in the matrix and ions in the solution, respectively. The macroscopic current density in the electrode is obtained as the sum of i_1 and i_2.

The equations for the dye-sensitized solar cell are represented in one-dimensional form:

$$\frac{di_1}{dx} + \frac{di_2}{dx} = 0 \tag{3.7}$$

$$i_1 = \varepsilon \mu_n n \frac{d\varepsilon_F}{dx} \tag{3.8}$$

$$i_2 = \frac{2\varepsilon(\mu_3 - 3\mu_1)}{2 + \ln\gamma} C_3 \frac{d\mu}{dx} \tag{3.9}$$

$$\frac{di_1}{dx} = -e\Phi\alpha\exp(-\alpha x) + ef_{rec} \tag{3.10}$$

Here, the redox couple in the solution is I^-/I_3^-. Each notation represents the following: n, ε_F, C_3, and μ without a subscript are the electron density **in the** semiconductor matrix, the quasi-Fermi level in the semiconductor matrix, the triiodide ion concentration in the solution, and the chemical potential of the ions, respectively. μ with a subscript is the mobility; the subscripts of n, 1, and 3 in the mobility represent electron, iodide ion, and triiodide ion, respectively. γ is the activity coefficient of the ions, e the elementary charge, ε a correction factor, α the light absorption coefficient of dye-sensitizers, and Φ the light intensity. For simplicity, the incident light is assumed to be monochromatic.

The first equation represents the charge conservation. The second and third equations are the current equations for electrons and ions, respectively. In these equations, both the diffusion and drift terms are implicit. However, only the diffusion is taken into account because the macroscopic electric field across the electrode is negligible owing to screening by the high-ionic-strength electrolyte. The correction factor ε represents decrease of the mobility due to the nanocrystallization. In Eq. (3.9), from the net chemical reaction in the electrolyte: $I_3^- + 2e^- \leftrightarrow 3I^-$, the following is assumed: $dC_1:(-dC_3) = 3:1$, denoting the iodide ion concentration by C_1. The last equation represents the charge separation; here, the light absorption and the recombination are described. In the light absorption term, which is the first term on the right-hand side, the solar cell is illuminated from the interface between the nanocrystalline electrode and the TCO electrode ($x = 0$). The recombination term is represented as f_{rec}.

The simplest model of f_{rec} is first order in n: $f_{rec} = n/\tau$, where τ is the lifetime of electrons in the TiO$_2$ matrix. However, it is well known that the real recombination process in the dye-sensitized photoactive electrode is not so simple; several factors, such as electron transfer chain reactions in the electrolyte, electron-hopping transport in the TiO$_2$, and saturation of the electron density in a TiO$_2$ particle, have influences on f_{rec}. At present, a definite scheme for the electron recombination has not been provided. Here, in order to present some simulation results, the following f_{rec} is employed:

$$f_{rec} = f_{rec,low} + f_{rec,high} = \frac{n^{\beta_{low}} C_3}{\tau} + \alpha \Phi \left(\frac{n}{n_{r,d}} \right)^2 \exp(-\alpha x) \qquad (3.11)$$

where the recombination terms of $f_{rec,low}$ and $f_{rec,high}$ are for a low light intensity and a high light intensity, respectively. On the right-hand side in Eq. (3.11) the first term corresponds to $f_{rec,low}$, and the second term to $f_{rec,high}$. For the dye-sensitized solar cell, the electron diffusion coefficient decreases with lowering the light intensity. Since the diffusion length is the square of the product of the diffusion coefficient and the lifetime, this is partly compensated by an increase of the electron lifetime with lowering the light intensity, making the incident photon-to-current conversion efficiency (IPCE) less dependent on the light intensity (see Fig. 4 and Table 1). However, this compensation is not complete, and the IPCE decreases slightly with lowering the light intensity [57]. On the other hand, under high light intensities as well, the increase of the electron recombination rate was reported. When the electron density becomes greater than that at one electron per TiO$_2$ particle, significant acceleration of the recombination rate was observed [58]. Thus, in Eq. (3.11), both high and low light intensities are considered.

For normal testing conditions, the electron recombination path with the dye cations is reported to be negligible [50, 59, 60]. Here, the derivation

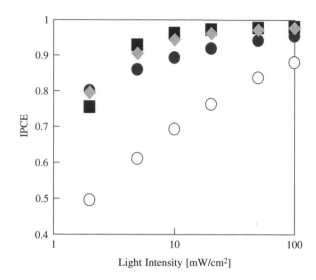

Fig. 4. Dependence of IPCE on low light intensity. Solid circles: $\beta_{low} = 0.5$, $\tau = 4.0 \times 10^{12}\,\mathrm{m}^{-1.5}\,\mathrm{s}$; solid diamonds: $\beta_{low} = 0$, $\tau = 25\,\mathrm{s}$; solid squares: $\beta_{low} = -0.5$, $\tau = 1.5 \times 10^{-10}\,\mathrm{m}^{1.5}\,\mathrm{s}$; open circles: $\beta_{low} = 0.5$, $\tau = 1.0 \times 10^{12}\,\mathrm{m}^{-1.5}\,\mathrm{s}$. The values of the parameters in Table 1 are employed. We calculate dependence of IPCEs on light intensities, 2–$100\,\mathrm{mWcm}^{-2}$, with the presented model. In low intensities, the recombination term $f_{rec,high}$ becomes very small because $n_{r,d}$ is expected to be greater than $10^{17}\,\mathrm{cm}^{-3}$. Although this term may be nonnegligible at relatively high light intensities, we disregard $f_{rec,high}$ to emphasize influences of the term $f_{rec,low}$ here. In Eq. (3.11), a definite value of τ is not known for real dye-sensitized solar cells. Thus, in the solid notations, diffuse lengths of electrons in the electrodes are assumed to be roughly 7–$10\,\mu\mathrm{m}$ at $1\,\mathrm{mWcm}^{-2}$. For comparison, the results of a fourth in τ at $\beta_{low} = 0.5$ are also shown as the open circles. The results indicate that the IPCE decreases with decreasing the incident light intensity; this reproduces experimental decrease of IPCE with decreasing illumination intensity [57].

of Eq. (3.11) considers only the recombination path with triiodide in the electrolyte. The net recombination reaction is the following:

$$I_3^- + 2e \rightarrow 3I^-$$

However, recent studies have revealed that the recombination is not a simple one-step reaction [54]. The electron diffusion involving the trap/detrap processes in TiO_2 also has a significant influence on the electron lifetime; this leads to a second-order recombination rate in n. Under relatively high light intensities, the electron acceptor in the recombination is a radical of I_2^- provided by the regeneration of the dye-sensitizer by iodide [58]. In this case, the concentration of I_2^- becomes in proportion to the product of the concentrations of the dye cation, $[D^+]$, and iodide, $[I^-]$: $[I_2^-] \propto [D^+][I^-]$. From

Table 1
Parameters of the theoretical model

Symbols	Employed values	Notes
λ	650 nm	Wavelength of incident light
$\alpha(\lambda)$	$1.95 \times 10^5 \, \mathrm{m^{-1}}$	Absorption coefficient of electrode
D_s	$5.0 \times 10^{-5} \, \mathrm{cm^2 \, s^{-1}}$	Diffusion coefficient of electrons in semi-conductor matrix
$D_{I^-}, D_{I_3^-}$	$2.6 \times 10^{-6} \, \mathrm{cm^2 \, s^{-1}}$	Diffusion coefficients of iodide and triiodide in electrolyte pores
d	10 μm	Nanocrystalline electrode thickness
n_i	$3.1 \times 10^{21} \, \mathrm{cm^{-3}}$	Effective density of states of conduction band

Notes: 100% light reflectance at the Pt-counter-electrode mirror is also assumed; concentrations of triiodide in the dark are 50 ± 1 mM.

the second order reaction in n and the electron acceptor of I_2^-, the recombination term becomes $f_{rec} \propto n^2[D^+]$, assuming that the concentration of iodide is a constant because of the high concentration of iodide. We also consider the following fact: when the electron density becomes greater than that at one electron per TiO_2 particle, significant acceleration of the recombination rate was observed [58]. Thus, normalizing this term by $n_{r,d}$ which is the electron density at one electron per TiO_2 particle, the second-order recombination term for $f_{rec,high}$ becomes the following: $f_{rec,high} = (n/n_{r,d})^2 \, [D^+] = \alpha\Phi(n/n_{r,d})^2\exp(-\alpha x)$. In $f_{rec,low}$ of Eq. (3.11), we assume that the rate-limiting process is the electron transfer with I_{ads}, $I_{ads} + e \rightarrow I^-$:$f_{rec,low} \propto n[I_{ads}]$ [52]; this term is modified by considering an influence of the trap/detrap diffusion process in TiO_2 that the electron lifetime depends on n:$f_{rec,low} \propto n^\beta[I_{ads}]$. Here, chain reactions of $I_3^- \leftrightarrow I_2 + I^-$ and $I_2 + e \rightarrow I_{ads} + I^-$ is the source of I_{ads} [52]. Thus, $[I_{ads}]$ become a function of C_3, C_1 and n. Since C_1 is generally an order of magnitude greater than C_3, for simplicity C_1 is assumed to be a constant. Thus, $f_{rec,low} \propto n^{\beta low}C_3$ is derived.

In addition to Eqs. (3.7)–(3.10), four boundary conditions must be considered. Three of these conditions are the following: $i_1(d) = 0$, $i_2(0) = 0$, and $i_1(0) = i_2(d)$, where d is the thickness of the nanocrystalline TiO_2 electrode. Here, the bulk of the electrolyte between the nanocrystalline electrode and the counter electrode is neglected. In the last condition, it is assumed that there is no macroscopic electric field within the simulation volume [61], and that the externally applied bias affects the current–voltage characteristics only by varying $n(0)$ with the Fermi–Dirac distribution. That is, any influence of the electric field is assumed to be included in this boundary.

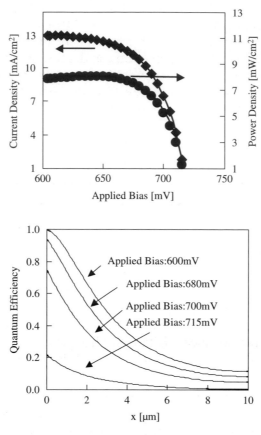

Fig. 5. Current–voltage characteristics and dependence of quantum efficiency on applied bias V for $n_{r,d} = 4 \times 10^{17}\,cm^{-3}$. The values of the parameters in Table 1 are employed. The other parameters were $\Phi = 50\,mWcm^{-2}$, $\beta_{low} = 0.5$, and $\tau = 4.0 \times 10^{12}\,m^{-1.5}\,s$. x is the distance from the TCO/TiO$_2$ interface.

Other factors, such as the recombination at the TCO, the resistance of the TCO, and the diffusion of the redox species in the electrolyte bulk layer, which are not included in the model, may influence the cell performance. For instance, with simple one-electron redox couple, such as ferrocenium/ferrocene, the oxidized redox species is reduced much faster on the SnO$_2$ surface than the TiO$_2$ surface [54]. In this case, a substantial loss of both the photocurrent and photovoltage has been observed because of the recombination at the TCO surface. Advanced models where these effects can be simulated are presented recently [62].

There has been a discussion about the last boundary condition [63]. In the widely accepted model, referred to as "kinetic model" or "interface model", the origin of the solar cell function in the dye-sensitized cell is practically interfacial exciton separation and subsequent charge diffusion [63]. In this chapter, we employ the widely accepted "kinetic model" for the last boundary condition. Several groups, on the other hand, have indicated an essential role of the built-in potential between the TCO substrate and the sensitized electrode, like the conventional p–n junction solar cells; this is referred to as "junction model" [64, 65]. Schottky or tunnel junctions at the FTO/nc-TiO$_2$ interface are also reported to have influences on the performance of the dye-sensitized solar cell [42–44]. As well as the formulation of f_{rec}, further fundamental studies are necessary to develop an accurate simulation model.

4. PROSPECTS

We present here some prospects for advances in the simulation in the dye-sensitized solar cell. As described above, each simulation method has its own advantages and disadvantages. Thus, we should select a suitable method for each job. For further advances, Kitao [66] has presented a strategy to make full use of these theoretical models; he called this "hierarchy and connectivity". The dye-sensitized solar cell is a composite of multiscale materials. Macroscopically, the solar cell performance relies on the charge diffusion in the several micron-scale pseudo-homogeneous media of the semiconductor matrix and the solution; the macroscopic charge diffusion has been modeled on the basis of the classical electromagnetic theory. On the other hand, the charge recombination depends on much smaller mezzoscopic scale processes; the rate-limiting process of the recombination is trapping/detrapping in a semiconductor nanoparticle. The mezzoscopic electron hopping has been simulated with the Monte Carlo method. The electron injection from the excited dyes to the semiconductor and the regeneration of these dyes are microscopic reactions at angstrom scales. The microscopic electron transfer reactions are a major subject in the field of quantum chemistry. Thus, it is very difficult to describe this multiscale energy-conversion mechanism with a computer code. The concept of hierarchy and connectivity is that a multi-scale problem is solved by providing parameter values in a large-scale model by small-scale simulations. For instance, the mezzoscopic trapping/detrapping process influences the simulation of the macroscopic diffusion by providing the diffusion coefficient in the current equation from the Monte Carlo simulations; modeling the electron transfer between the dyes and TiO_2 in the Monte Carlo simulation with the Marcus theory, the electron coupling constant is provided by a quantum chemistry code. The concept of hierarchy and connectivity allows us to simulate effectively the solar cell performance of the dye-sensitized cell consisting of the multiscale materials. Finally, it is still necessary to develop a sophisticated model for each simulation method. Unfortunately, with few exceptions, it is impossible to quantitatively reproduce experimental results with simulations.

REFERENCES

[1] A. Hagfeldt and M. Grätzel, Chem. Rev., 95 (1995) 49.
[2] A. Hagfeldt and M. Grätzel, Acc. Chem. Res., 33 (2000) 269.
[3] M. Grätzel, Nature, (London), 414 (2001) 338.
[4] G. Yu, J. Gao, J.C. Hummelen, F. Wudl and A.J. Heeger, Science, 270 (1995) 1789.
[5] S. Södergren, A. Hagfeldt, J. Olsson and S.-E. Lindquist, J. Phys. Chem., 98 (1994) 5552.
[6] A. Usami, Jpn. J. Appl. Phys., 36 (1997) L886.
[7] A. Usami, Chem. Phys. Lett., 292 (1998) 223.
[8] J. Ferber, R. Stangl and J. Luther, Sol. Energy Mater. Sol. Cells, 53 (1998) 29.
[9] R. Stangl, J. Ferber and J. Luther, Sol. Energy Mater. Sol. Cells, 54 (1998) 255.
[10] M.J. Cass, F.L. Qiu, A.B. Walker, A.C. Fisher and L.M. Peter, J. Phys. Chem. B, 107 (2003) 113.
[11] M.J. Cass, A.B. Walker, D. Martinez and L.M. Peter, J. Phys. Chem. B, 109 (2005) 5100.
[12] K.D. Benkstein, N. Kopidakis, J. van de Lagemaat and A.J. Frank, J. Phys. Chem. B, 107 (2003) 7759.
[13] J. Nelson, Phys. Rev. B, 59 (1999) 15374.
[14] J. Nelson, S.A. Haque, D.R. Klug and J.R. Durrant, Phys. Rev. B, 63 (2001) 205321.
[15] A.V. Barzykin and M. Tachiya, J. Phys. Chem. B, 106 (2002) 4356.
[16] A.V. Barzykin and M. Tachiya, J. Phys. Chem. B, 108 (2004) 8385.
[17] A. Usami, Chem. Phys. Lett., 277 (1997) 105.
[18] J. Ferber and J. Luther, Sol. Energy Mater. Sol. Cells, 54 (1998) 265.
[19] G. Rothenberger, P. Comte and M. Grätzel, Sol. Energy Mater. Sol. Cells, 58 (1999) 321.
[20] A. Usami, Sol. Energy Mater. Sol. Cells, 59 (1999) 163.
[21] A. Usami, Sol. Energy Mater. Sol. Cells, 64 (2000) 73.
[22] W.E. Vargas and G.A. Niklasson, Sol. Energy Mater. Sol. Cells, 69 (2001) 147.
[23] A. Usami and H. Ozaki, J. Phys. Chem. B, 109 (2005) 2591.
[24] A. Ishimaru, Wave Propagation and Scattering in Random Media, Academic Press, New York, 1978.
[25] C.F. Bohren and D.R. Huffman, Absorption and Scattering of Light by Small Particles, Wiley, New York, 1983.

[26] L. Tsang, J.A. Kong and R.T. Shin, Theory of Microwave Remote Sensing, Wiley, New York, 1985.

[27] C.L. Tien and B.L. Drolen, Ann. Rev., 1 (1987) 1.

[28] L.M. Zurk, L. Tsang, K.H. Ding and D.P. Winebrenner, J. Opt. Soc. Am. A, 12 (1995) 1772.

[29] P. Ferrand and R. Romestain, Appl. Phys. Lett., 77 (2000) 3535.

[30] M. Hiraoka, S.R. Arridge and D.T. Delpy, Kougaku, 24 (1995) 167.

[31] G.F. Cerofolini and M.L. Polignano, J. Appl. Phys., 64 (1988) 6349.

[32] A. Chitnis, A. Kumar, M. Shatalov, V. Adivarahan, A. Lunev, J.W. Yang, G. Simin and M. Asif Khan, Appl. Phys. Lett., 77 (2000) 3800.

[33] J.M. Shah, Y.-L. Li, Th. Gessmann and E.F. Schubert, J. Appl. Phys., 94 (2003) 2627.

[34] A. Usami, S. Seki, Y. Kobayashi and H. Miyashiro, CRIEPI Report Q05017 (2006).

[35] A. Usami, Electrochem. Solid-State Lett., 6 (2003) A236.

[36] J. van de Lagemaat, N.-G. Park and A.J. Frank, J. Phys. Chem. B, 104 (2000) 2044.

[37] Th. Dittrich, P. Beer, F. Koch, J. Weidmann and I. Lauermann, Appl. Phys. Lett., 73 (1998) 1901.

[38] M. Matsumoto, Y. Wada, T. Kitamura, K. Shigaki, T. Inoue, M. Ikeda and S. Yanagida, Bull. Chem. Soc. Jpn., 74 (2001) 387.

[39] M. Yanagida, T. Yamaguchi, M. Kurashige, K. Hara, R. Katoh, H. Sugihara and H. Arakawa, Inorg. Chem., 42 (2003) 7921.

[40] G. Kron, T. Egerter, J.H. Werner and U. Rau, J. Phys. Chem. B, 107 (2003) 3556.

[41] S.Y. Huang, G. Schlichthörl, A.J. Nozik, M. Grätzel and A.J. Frank, J. Phys. Chem. B, 101 (1997) 2576.

[42] G. Kron, U. Rau and J.H. Werner, J. Phys. Chem. B, 107 (2003) 13258.

[43] S. Rühle and D. Cahen, J. Phys. Chem. B, 108 (2004) 17946.

[44] S. Rühle and T. Dittrich, J. Phys. Chem. B, 109 (2005) 9522.

[45] L. Han, N. Koide, Y. Chiba and T. Mitate, Appl. Phys. Lett., 84 (2004) 2433.

[46] J. van de Lagemaat and A.J. Frank, J. Phys. Chem. B, 105 (2001) 11194.

[47] J. Newman and W. Tiedemann, AIChE. J., 21 (1975) 25.

[48] J. Bisquert, Phys. Rev. Lett., 91 (2003) 010602.

[49] R.L. Willis, C. Olson, B. O'Regan, T. Lutz, J. Nelson and J.R. Durrant, J. Phys. Chem. B, 106 (2002) 7605.

[50] I. Montanari, J. Nelson and J.R. Durrant, J. Phys. Chem. B, 106 (2002) 12203.

[51] J.N. Clifford, E. Palomares, Md. K. Nazeeruddin, M. Grätzel, J. Nelson, X. Li, N.J. Long and J.R. Durrant, J. Am. Chem. Soc., 126 (2004) 5225.

[52] A.N.M. Green, R.E. Chandler, S.A. Haque, J. Nelson and J.R. Durrant, J. Phys. Chem. B, 109 (2005) 142.

[53] Y. Tachibana, S.A. Haque, I.P. Mercer, J.R. Durrant and D.R. Klug, J. Phys. Chem. B, 104 (2000) 1198.

[54] A.J. Frank, N. Kopidakis and J. van de Lagemaat, Coord. Chem. Rev., 248 (2004) 1165.

[55] D.T. Rover, P.A. Basore and G.M. Torson, Proc. 18th IEEE Photovoltaic Specialists Conf., (1985) 703.

[56] B.A. Gregg and M.C. Hanna, J. Appl. Phys., 93 (2003) 3605.

[57] T. Trupke, P. Würfel and I. Uhlendorf, J. Phys. Chem. B, 104 (2000) 11484.

[58] C. Bauer, G. Boschloo, E. Mukhtar and A. Hagfeldt, J. Phys. Chem. B, 106 (2002) 12693.
[59] S.A. Haque, Y. Tachibana, D.R. Klug and J.R. Durrant, J. Phys. Chem. B, 102 (1998) 1745.
[60] S.A. Haque, Y. Tachibana, R.L. Willis, J.E. Moser, M. Grätzel, D.R. Klug and J.R. Durrant, J. Phys. Chem. B, 104 (2000) 538.
[61] A. Zaban, A. Meier and B.A. Gregg, J. Phys. Chem. B, 101 (1997) 7985.
[62] J.-J. Lee, G.M. Coia and N.S. Lewis, J. Phys. Chem. B, 108 (2004) 5269.
[63] B.A. Gregg, J. Phys. Chem. B, 107 (2003) 4688.
[64] K. Schwarzburg and F. Willig, J. Phys. Chem. B, 103 (1999) 5743.
[65] K. Schwarzburg and F. Willig, J. Phys. Chem. B, 107 (2003) 3552.
[66] O. Kitao, Finechemical, 34 (2005) 7.

Nanostructured Materials for Solar Energy Conversion
T. Soga (editor)

Chapter 4

Mathematical Modelling of the Refractive Index and Reflectivity of the Quantum Well Solar Cell

Francis K. Rault

Department of Electrical and Computer Systems Engineering, Monash University, PO Box 35, Victoria, Australia

1. INTRODUCTION

In the early 1990s, a new and revolutionary solar cell began development. Professor Keith W.J. Barnham of Imperial College of Science, Technology and Medicine, United Kingdom pioneered the development of the Quantum Well Solar Cell (QWSC) [1]. The simple concept was to introduce quantum wells (QW) into the intrinsic (i) region of a p–i–n solar cell. The quantum wells reduced the energy gap (E_g) and allowed the generation of electron–hole pairs by the longer wavelengths. These wavelengths would not normally be of sufficient energy to generate an electron–hole pair directly into the conduction and valence bands (surpass the energy band gap) (see Fig. 1).

The electrons or holes are confined within the respective conduction or valence quantum wells. These particles are freed from the wells by thermal excitations, tunnelling or photon-assisted excitation. Once freed from the wells the electric field across the p–i–n junction sweeps the particles away. The build up of charged particles and the connection of a load would cause a potential difference, therefore current conduction takes place.

The first quantum well solar cell showed remarkable promise. A standard AlGaAs/GaAs p–i–n solar cell was approximately 9% efficient at the time. The introduction of the quantum wells produced an efficiency of 14% [2]. An approximate doubling of efficiency was achieved. The new field of photovoltaics showed great promise with many areas to be investigated.

In this chapter, the main avenue of concentration will be mathematical and computational modelling. To give the reader an insight into the avenues currently under pursuit, initially a review of the science has been undertaken.

Francis K. Rault

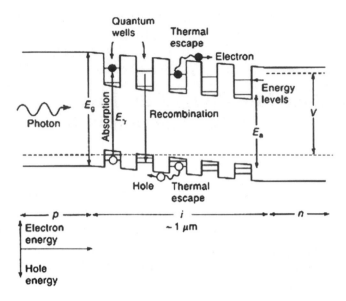

Fig. 1. Schematic of a quantum well solar cell.

Many researchers have performed the modelling of the QWSC [3–21]. The majority of authors concentrated on an electrical model of the cell. A combination of state density equations, generation rates with relations to incident photon flux has seen the formulation of the QWSC's current density versus voltage $(J-V)$ curves. From this the parameters such as open-circuit voltage (V_{OC}), short-circuit current density (J_{SC}), Maximum power (P_{MAX}) and efficiency are directly calculated. The work has proved to be vital in understanding the electrical phenomenon of the device.

Further work has seen the evaluation of the quasi-Fermi energy levels and the steady state escape of photo-carriers from quantum wells. This has achieved detailed modelling of dark and radiative currents [13, 22]. Other models have also investigated the photonic response of the QWSCs, which has led to the determination of the structure's quantum efficiency and spectral response [3–9].

The above two analyses have given great insights into the accuracy of the proposed models, as they yield insights into the structure's electrical performance. A model's accuracy in determining the structure's short-circuit current density, open-circuit voltage and quantum efficiency are all considered through the dark current analysis and spectral responses. To coincide with the theoretical models, experimental evaluations including electron beam induced currents and cathodoluminescence studies have been undertaken supporting the findings [14].

The effects of temperature on the QWSC properties have also been considered but with little depth. The author is aware of only three papers. These are experimentally and electrically based, respectively [17, 23, 24]. The experimental paper [23] consisted of the growth of various AlGaAs/GaAs samples and these were irradiated with the AM1.5G spectrum. The effects of temperature on QE, dark currents and efficiency were explored and documented. The electrically based paper [24] followed a similar line to others developing an electrical model with an emphasis on temperature effects on the $J-V$ curve and hence efficiency, dark currents and QE. The model in reference [17] was also electrically based with detailed investigations into temperature effects on the quasi-Fermi energy level of the QWSC.

The effects of cell dimensions such as QW and intrinsic region length on overall structure efficiency have been explored previously for InP-based QWSC [10, 11]. The paper consisted of "strained" structures grown with varying QW and intrinsic region lengths, experimentally evaluated under a solar spectrum. The findings showed lengths of 5 and 500 nm, respectively generating greatest efficiency (approximately 26%).

In more recent time, studies into growth techniques yielding reduction in dark currents through atomic hydrogen-assisted molecular beam epitaxy has also been conducted [15]. The results here have been promising in reducing the dark current and research continues.

The quantum photovoltaic group at Imperial College have extensively developed "strained" and "strain-balanced" devices. These devices have shown remarkable performance. The term strained relates to the mis-match of lattice properties of the semiconductor material causing changes in the semiconductor's properties.

Another recent pursuit has been investigations into the potential of quantum well solar cells for thermophotovoltaic applications [16]. It is still in its infancy but promising results are emerging.

These avenues of research have seen the efficiency of the QWSC rise to 25% and beyond.

From the above QWSC studies a key question has remained a controversial area. "Does the introduction of the wells into the cell produce an overall increase in cell performance exceeding that of single band gap structures?" It is clear that the introduction of quantum wells within the intrinsic region causes a short-circuit current increase, however there is an associated decrease in open-circuit voltage. This has been well discussed in Ref. 25. In short, there is a shift in the semiconductor's absorption edge which reduces the open-circuit voltage. Given this the underlying question remains, is there

still an overall performance increase and does it match or exceed performance levels of current single band gap devices?

The first detailed studies conducted were by Araujo et al. [19]. The
author presented a detailed balance theory and later followed with experimental results by Ragay et al. [21], which considered radiative recombination effects within the intrinsic region. From these studies they drew
conclusions that within the ideal limit all the recombination is radiative and
the quasi-Fermi energy levels are equal to the applied bias at all points. This
led to the conclusion that no enhancement in the limiting efficiency is
achievable with a single band gap limit of approximately 41%.

At the same time, studies conducted by the team at Imperial College [22,
25] showed through detailed electroluminescence and photoluminescence
experiments along with a generalised detailed balance theory, the recombination within the intrinsic region to have an ideality factor equal to unity indicating purely radiative recombination. Further work showed dark currents to be
dominated by Shockley, Read and Hall (SRH) recombination. Through the detailed balance approach taken, Nelson et al. [22] showed the quasi-Fermi level
to be less than the applied voltage. The requirement was the quasi-Fermi level
being significantly reduced indicating a reduced radiative recombination rate.

In more recent times research conducted by Bremner [21] investigated
the detailed balance efficiency limits through non-uniform quasi-Fermi levels. The approach taken here allows for the possibility of photons with energies below the QW energy gap to be absorbed by photocarriers in the
quantum wells, thus providing the energy required to escape the QW confinement. Further features of the approach includes effects of reduced carrier captured in the quantum wells due to hot carrier transport, this directly
relates to the reduction in radiative recombination and hence increases the
limiting efficiency to approximately 63%.

This work was followed by further studies conducted by Luque et al.
[21] regarding the thermodynamics of various multi-band solar cells. Their
studies acknowledged the parts of the Bremner work regarding the possibility of photon-assisted escape leading to an efficiency enhancement within
the QWSC. Further, they acknowledged the photon-assisted escape with
no hot carrier effects leading to an efficiency limit of approximately 63%.
Based on their analysis they disagreed with claims regarding incomplete
thermalisation of carriers yielding additional increase in limiting efficiency.
Furthermore, they disagree with the possibility of a simultaneous net flux of
carriers out of the QWs and suppression of the quasi-Fermi level in the well
with respect to the surrounding barrier material. This is the basis of the
experimental work conducted by the team at Imperial College.

On a final note, a recent review performed by Anderson [21] has placed forth the following comments. One can argue that the consensus has been effectively reached that QWSCs do provide a global enhancement given the analysis based on the Bremner's [21] work and the acknowledgement from the Luque et al.'s work [21] that photon-assisted carrier escape revises the limiting efficiencies to approximately 63%. This consensus comes with an argument into which mechanism enhances the performance levels. The photon-assisted escape approach, which provides the revision of the limiting efficiency relies on the process of intra-sub band optical absorption in the QW structure. This proves to be a weak process and the preferred electron–phonon interaction has been used to explain the extremely high photo-carrier escape from the wells [22]. This explains the process of the suppression of quasi-Fermi levels in the quantum wells with respect to barriers and applied voltage. The work of Luque et al. [21] argues on the thermodynamic grounds that without photon-assisted escape, the uniform quasi-Fermi level assumption represents the best scenario of the QWSC and hence no exceeding limiting efficiency is observed in comparison with a single band gap structure.

As can be seen this area still remains controversial today and in-depth investigation continues. The author believes this area of research is fundamental and of great importance, however, it does not play a major role in the work presented here. Primarily, the author is dealing with non-ideal structures to achieve performance enhancements.

The present author has tried to avoid this controversy and focussed his attention to developing models, which do not directly determine efficiency improvements in comparison with single band gap structures. The models presented are intended to be tools, which investigates the optical properties and develops a series of design parameters and hence a design configuration. The performance of this design configuration is determined using an independent model or experimental analysis. In this chapter, the author intends to present new models relating to the refractive index and reflectivity of quantum well solar cells.

2. MODELLING THE QWSC'S REFRACTIVE INDEX AND REFLECTIVITY WITH THE MODIFIED SINGLE EFFECTIVE OSCILLATOR MODEL

The determination of the refractive index of a semiconductor alloy can be achieved with the Kramer–Kronig relationship. This relationship relates to the complex dielectric constant. This is a complex integral formula, which considers the complete energy spectrum. This complex relationship can be approximated using the Modified Single Effective Oscillator (MSEO) model. Associated with the refractive index of the cell is the reflectivity or reflective loss. This is a loss common to all solar cells and the quantum well solar cell is not immune. For the case of silicon solar cells the losses can vary from 35% to 70% [26]. This is a considerable amount of photon energy being wasted. The combination of the refractive index model with the Fresnel formula allows the QWSC's reflective losses to be considered.

2.1. Refractive Index Modelling

Fundamentally, the refractive index is related to the well-known Kramers–Kronig relationship for the complex dielectric [20]

$$\varepsilon(\omega) = \varepsilon_1(\omega) + i\varepsilon_2(\omega) \tag{1}$$

where
$n = \sqrt{\varepsilon_1(\omega)}$ is the refractive index
$\varepsilon_2(\omega)$ is associated with the absorption coefficient.

This complex equation may be replicated through the MSEO model.

From Eq. (1) and through the Kramer–Kronig relationship it has been shown that [20]

$$\varepsilon_1 - 1 = \frac{2}{\pi} P \int_0^\infty \frac{E' \varepsilon_2(E') dE'}{E'^2 - E^2} \tag{2}$$

where P denotes the principal value of the integral.

Considering $\varepsilon_1(E)$ for the direct energy gap (E_Γ) and values below equalling zero, one can expand Eq. (2) to be

$$\varepsilon_1(E) - 1 = \chi(E) = \frac{2}{\pi} \int_{E_\Gamma}^\infty \varepsilon_2(E') \left[\frac{1}{E'} + \frac{E^2}{E'^3} + \frac{E^4}{E'^5} + \cdots \right] dE' \tag{3}$$

For $E < E_\Gamma \leq E'$
Integrating each term separately gives a power series

$$\chi(E) = M_{-1} + M_{-3} E^2 + M_{-5} E^4 + \cdots \tag{4}$$

where

$$M_i = \frac{2}{\pi} \int_{E_\Gamma}^{\infty} \varepsilon_2(E)E^i \, dE \qquad (5)$$

$$\varepsilon_2(E) = \eta E^4$$

For $E_\Gamma \leq E \leq E_f$

Otherwise, equal to zero for other energies.

Integrating Eq. (5) gives

$$M_{-1} = \frac{\eta}{2\pi}(E_f^4 - E_\Gamma^4)$$

$$M_{-3} = \frac{\eta}{\pi}(E_f^2 - E_\Gamma^2) \qquad (6)$$

It has been shown in previous work [20] that constraining Eqs. (3) and (5) to low energies the definition of the parameters E_f and η are obtained as functions of the dispersion energy (E_d) and the oscillator energy (E_o). This gives

$$E_f = (2E_o^2 - E_\Gamma^2)^{1/2}$$

$$\eta = \pi E_d / 2E_o^3(E_o^2 - E_\Gamma^2) \qquad (7)$$

Expanding Eq. (4) and substituting into (7) we obtain the complete refractive index given as [27]

$$n^2 = 1 + \frac{E_d}{E_o} + \frac{E_d}{E_o^3}E^2 + \frac{\eta}{\pi}E^4 \ln\left(\frac{2E_o^2 - E_g^2 - E^2}{E_\Gamma^2 - E^2}\right) \qquad (8)$$

where
$E = \hbar\omega$ is the photon energy (eV) and
$\hbar$ the Planck's constant divided by 2π

$$\omega = \frac{2\pi c}{\lambda}$$

c is the speed of light and
λ the wavelength of light.

The determination of the remaining parameters has been obtained through experiment [28], yielding the following equations.

For the oscillator energy

$$E_o = A + BE_\Gamma$$

where
E_Γ is the lowest direct band gap.

$$A \approx 2.6$$

(9)

$$B \approx 0.75$$

For the dispersion energy

$$E_d = \frac{F}{E_o}$$

(10)

where
F is related to the Kramer–Kronig relationship.

$$\varepsilon_1(\omega) - 1 = \omega_p^2 \sum_n \frac{f_n}{(\omega_n^2 - \omega^2)}$$

(11)

where

$$\omega_p^2 = \frac{4\pi q^2}{m_h} \frac{N}{\Omega} \quad \text{is the plasma frequency}$$

(12)

N/Ω is the number of valence electrons per unit volume.
q and m_h are the electron charge and effective mass of the valence band particle (hole), respectively.
f_n is the oscillator strength and ω_n the frequency of the corresponding transition.

It can be shown that Eq. (11) can be simplified to [28]

$$\varepsilon_1(\omega) - 1 \approx \frac{F}{E_o^2 - (\hbar\omega)^2}$$

(13)

where

$$F = \omega_p^2 f_n \tag{14}$$

For the purposes here, the oscillator strength was calculated using the equation proposed by Harrison [29]

$$f_n = \frac{2\hbar}{m^* \omega} \left| \left\langle \Psi_i \left| \frac{\partial \Psi_f}{\partial x} \right\rangle \right| \right|^2 \tag{15}$$

where
$m^* = m_e m_h / (m_e + m_h)$ is the effective mass of the particles
ω the transition frequency
Ψ_i and Ψ_f the wave functions of the respective energy levels associated with the transition.

From the above it becomes obvious that the oscillator strength (f_n) of a transition is dependent on the wave functions (Ψ) associated with the particles within the respective conduction and valence quantum wells. It is through this that the indirect effects of electric field within the structure may be considered. To understand this indirect link one must begin at its source "the quantum well".

In simple terms the quantum well is a structure formed through the process of epitaxial semiconductor growth. To form a quantum well a layer of pure semiconductor material is deposited between two layers of doped semiconductor material. An example of this would be "sandwiching" Gallium Arsenide (GaAs) between two layers of Aluminium Gallium Arsenide ($Al_x Ga_{1-x} As$).[1] Owing to the potential energy gap difference between the doped and pure material a "well like" region (quantum well) is formed. Within this well region particles may be confined.

These quantum wells are governed by the Schrödinger wave equation. Solving this equation yields discreet energy levels and their associated wave functions. The energy levels represent the energy of the confined particle and the wave functions are associated with the probability of finding the location of the particle within the well.

With the presence of an electric field perpendicular to the plane of the well regions the particle's wave functions and energy levels are affected.

---

[1]The molar concentration is defined as x.

The wave functions and energy become distorted and shifted within the quantum well. This phenomenon is called the quantum confined stark effect (QCSE) [29].

It is through this distortion and shifting of the particles properties that indirectly relates the electric field with the dispersion energy. The distortion in particle properties within the quantum well causes the oscillator strength to vary. This is observed through Eq. (15). Given Eq. (14) is dependent on oscillator strength (Eq. (15)), the effects of electric field on parameter F flows onto the parameter E_d through Eq. (10) and hence the indirect link to the refractive index is possible through Eq. (8).

For simplicity, the author has assumed the potential of the quantum wells to be significantly large so that they can be classified to be infinite wells. The barrier lengths are also assumed to be significantly large that "tunnelling effects" between the wells are not possible. Assuming the electric field (F below) to be perpendicular to the quantum well layers and the device to be quasi one-dimensional (x). This yields the boundary conditions of the wave functions within the well as

$$\Psi(x = 0) = 0$$

$$\Psi(x = L) = 0$$

It can be shown that the Schrödinger wave equation for such a configuration is given as [30]

$$\left(-\frac{\hbar^2}{2m} \frac{\partial^2}{\partial x^2} + |q| Fx - E_n \right) \Psi_{i/f} = 0 \qquad (16)$$

where
$\hbar$ is the Planck's constant divided by 2π; m the effective mass of the particle confined to the well; F the electric field; q the electronic charge; E_n the energy level of the particle within the well; and $\Psi_{i/f}$ the wave functions associated with the particles.

Eq. (16) can be solved using the variational technique proposed by Bastard et al. [31].

2.2. Approximating Quantum Wells

Determination of the refractive index of multiple quantum well structures has been pursued through various devices [32]. The matrix formalism is the preferred method of analysis [32]. The theory generates very accurate results, however it proves to be very time consuming through computation when there are a large number of layers. For this reason, approximation techniques have been developed [33]. The simple concept is the replacement of the multiple quantum well layers with a single homogenous layer yielding an average refractive index [33]. This technique allows the determination of the refractive index for transverse electric (TE) and transverse magnetic (TM). In the case of solar cells these two parameters are equal for the analysis. This is due to the fact that we assume the source of solar energy is directly above the solar cell.

Considering the layer configuration in Fig. 2

Following the Alman analysis for the case of a quantum well layer configuration the equations simplify to be [33]:

$$n^2 = \frac{\sum_j n_j^2 d_d}{\sum_j d_j} = \frac{n_1^2 l_1 + n_2^2 l_2}{l_1 + l_2} \tag{17}$$

for transverse electric (TE).

$$\frac{1}{n^2} = \frac{\sum_j \frac{1}{n_j^2} d_d}{\sum_j d_j} = \frac{\frac{l_1}{n_1^2} + \frac{l_2}{n_2^2}}{l_1 + l_2} \tag{18}$$

for transverse magnetic (TM).

The above equations form the "starting equations" for the investigations into the QWSC's refractive index.

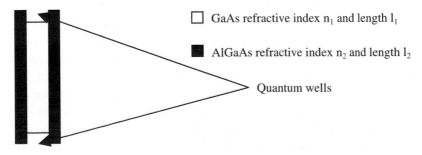

☐ GaAs refractive index n_1 and length l_1

■ AlGaAs refractive index n_2 and length l_2

Quantum wells

Fig. 2. Repeating layers.

2.3. Electric Field Effects on the Refractive Index of Multiple Quantum Wells

The QWSC is a p–i–n structure with the quantum wells embedded within the intrinsic region. The doped charged regions on either side shall produce an electric field perpendicular to the well layers. It is also possible to have induced electric fields due to piezoelectric effects [34]. These fields are induced by the strains in the semiconductor layers due to mis-matched lattice constants. This has been the backbone of the "strained" and "strain-balanced" structures briefly encountered earlier.

In this section the new approach documented above will be utilised to investigate the effects of electric field on the refractive index of the multiple quantum well solar cell. The analysis of induced electric field changes on the refractive index has been performed previously in other structures [27]. The approach has been to calculate the absorption coefficient of the material and from there the change in index is generated. These calculations are rather complex and are performed for a modulated electric field. This produces a positive and a negative change in the refractive index as a function of wavelength.

For the purpose of solar cells and in the case of QWSC, the assumption of the electric field generated by the charged regions on either side to be static has been made. Therefore, a non-fluctuating change in the index is predicted.

A detailed comparison between existing theoretical data and the new model is presented. This is performed for the wavelength spectrum relating to quantum well absorption for QWSC application (760–960 nm). The selection of this range for the modelling is clear from the next point.

From the original Barnham work [1], the purpose of the quantum wells within the QWSC is to utilise the region of spectrum, which would, typically not be adequate to produce a pair of photo carriers directly into the respective conduction and valence bands, but generate particles within the respective quantum wells.

It must be stressed that the device used for comparison is a quantum well modulator. This is due to the lack of data directly relating to the refractive index of the QWSC. The author has ensured the accuracy of the comparison by modelling the identical theoretical structure as documented in Ref. 35. Given the new model presented here is significantly different in approach to that used in Ref. 35 the author feels a comparison can be performed and conclusions drawn relating to the accuracy of the new model presented. As documented in Ref. 35 a structure of AlGaAs/GaAs ($x = 0.2$) with 25 quantum wells of 10 nm length and barriers of length 10 nm was considered.

Fig. 3 is a graph of the computationally modelled refractive index of the multiple quantum well modulator. From the figure all dashed lines (—) with the prefix "modelled" (e.g. Modelled AlGaAs with 0 kV/cm) are results generated by the author's new model. Those with asterisks (*) are results generated by the existing theoretical model [35].

Table 1 shows the comparison between the two models. Three main points corresponding to the two outer spectral wavelengths and one central wavelength of the energy spectrum have been considered.

From the table one can observe the accuracy of the new model for the various photonic energy and electric field levels. The new model does predict values, which are similar in magnitude to that of the existing theoretical model. Furthermore, there is an inaccuracy of approximately 0.3% for the relative energy levels. From the table one can observe the consistent exponential

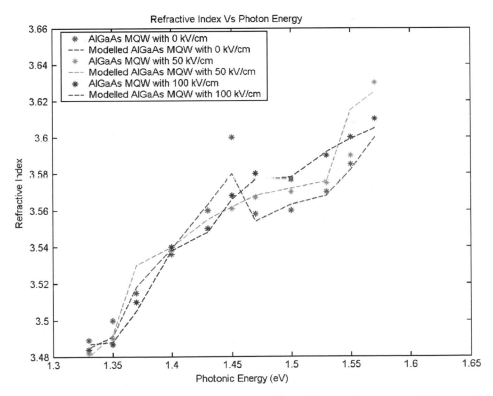

Fig. 3. Refractive index of quantum wells within a QW modulator. A comparison between model and theoretical data with an electric field of 0–100 kV/cm is considered.

Table 1
Refractive index versus photonic energy for a given electric field

Variable	Refractive index								
Photonic energy (eV)	1.35			1.45			1.55		
Electric field (kV/cm)	0	50	100	0	50	100	0	50	100
Existing model	3.500	3.490	3.478	3.580	3.560	3.580	3.585	3.60	3.600
New model	3.491	3.492	3.488	3.600	3.562	3.566	3.580	3.615	3.610
Inaccuracy (%)	0.20	0.05	0.3	0.50	0.05	0.4	0.1	0.04	0.2

increase in refractive index at the various photonic energy levels. This is consistent with the observations made in Ref. 18.

The application of the Bastard [31] solution for the quantum well wavelengths allows the new model to take into consideration the presence of excitonic effects. A large fluctuation at 1.45 eV energy mark is observed for the zero electric field curve. As the electric field increases, the excitonic energy significantly reduces. As the electric field reaches the 100 kV/cm value, the excitonic effect is near non-evident. The presence of excitonic effects is attributed to the absorption coefficient of the quantum wells becoming smaller with increase in electric field [35]. This has significant implications from a practical point of view for QWSC design.

Given that one wishes to utilise the quantum wells for their absorption of wavelengths below the band gap of the bulk semiconductor material, a large electric field across the intrinsic region would lead to a reduction in excitonic presence and hence quantum well absorption.

From a practical point, the electric field across the intrinsic region is governed by the p and n region charge along with the intrinsic region length. If we first consider doping concentrations levels, high-concentration levels ($N \geq 10^{18}/cm^3$) yielding high-electric fields, these are troublesome given that in the case of quantum wells there tends to be an increase in Auger recombination [12]. Increased Auger recombination causes an increase in phonon activity in the semiconductor lattice, leading to excess heat within the semiconductor and the cell efficiency is reduced [12].

Low-concentration levels ($N \leq 10^{17}$/cm^3) will increase the presence of excitonic effects. This will lead to high quantum well absorption; however, the cells operating bias is essential [21].

In order to minimise Auger recombination and to maintain the field across the intrinsic region at operating bias a doping level below ($N \leq 10^{15}$/cm^3) is recommended [21].

In relation to the intrinsic region length, a small intrinsic region length would be detrimental as electric fields would be large. Focus should be on utilising a medium to long region length as discussed in Refs. 10 and 11. The authors have shown the effects of well and intrinsic region length on performance. According to Refs. 10 and 11, QWSC designs with well lengths of the order of 60 Å and intrinsic regions of 300–500 nm produce conversion efficiencies approaching 26.50% for InP structures. It must be noted that this is a "strained" design. Furthermore, a physical limit on the intrinsic region length of 1000 nm has to be adhered to within strained structures [1]. This is to ensure correct operating bias.

Considering the effects of excitonic energy on the energy levels within the quantum wells. With the presence of the excitonic energy, the transition energy levels would require more energy for the particle to escape the confinement of the quantum well. This energy requirement would have to be sourced through thermal excitation, tunnelling or photon-assisted escape. In relation to the author's simulations, tunnelling is deemed negligible due to the simulation restrictions. Therefore, thermal excitation and photon-assisted escape are the only escape mechanisms.

Given the energy levels are inversely proportional to the well length. One draws the conclusion that photo carriers within a narrow well (≤ 10 nm) would require less energy to escape. This is based on the fact that the energy levels within a "wide well" will be lower to the base of the well requiring more energy to surpass the barrier. This is consistent with the work in [10, 11]. One of course is again restricted by the intrinsic region length and must keep this in mind during the design.

2.4. The Reflectivity of Quantum Well Solar Cells

Having established a refractive index model we are now in a position to explore the relationship refractive index and reflectivity have with one another. In order to achieve this one can combine the refractive index model with the Fresnel formula allowing the determination of the structure's reflective losses. The reflectivity (R) of the solar cell can be calculated using the Fresnel formula [36]

$$R = \frac{r_1^2 + r_2^2 + 2r_1r_2\cos2\Theta}{1 + r_1^2r_2^2 + 2r_1r_2\cos2\Theta}$$

with

$$r_1 = \frac{n_0 - n_1}{n_0 + n_1}$$

$$r_2 = \frac{n_1 - n_2}{n_1 + n_2}$$

$$\theta = \frac{2\pi n_1 d_1}{\lambda} \tag{19}$$

where
n_0 is the refractive index of the incident medium (air) = 1.00; n_1 the refractive index of the anti-reflection coating, and
n_2 the refractive index of the substrate; in this case the QWSC.

The refractive index of the QWSC was achieved through the work covered in Sections 2.1–2.2. These formulae allow the design parameters discussed above in Section 2.3 to be considered. d_1 the thickness of the anti-reflection layer (= 0.00 for no coating).

The thickness of the anti-reflection coating can be determined by minimising formula (19). It is common practice to have the minimum trough located at approximately 600–650 nm since this produces the lowest total reflection [26].

To show the model's capability and justification, comparative studies of designs are used in the modelling analysis. An established QWSC design for which experimental data are available is replicated and modelled [37] (Table 2).

Having established the model's ability to model existing structures, the author intends to utilise the new model and investigate the effects the refractive index has on the reflectivity. In order to achieve this the author intends to model an identical structure with the number of quantum wells within the intrinsic region varying for the values 0 (control cell), 5, 20, 50 and 100. This is documented as Designs 3–7 (Table 3).

Through this the effects of quantum well number on the refractive index and indirectly on the reflective losses will be observed and it is envisaged a preferred range of quantum well quantity within the structure will be achieved.

Table 2
Design parameters for QWSC

Region	ARC	P		Intrinsic		N	
Design 1							
Material	SiN ($n = 2.1$)	AlGaAs/GaAs		AlGaAs/GaAs		AlGaAs/GaAs	
		Doping	Molar	Doping	Molar	Doping	Molar
Concentration levels		1.34×10^{18}	0.40	I	0.00	6.0×10^{17}	0.20
Length (nm)	70	150		QW $= 50 \times 8.7$		600	

Note: Barrier length = 6.0 nm. A metallic back mirror with 95% reflectivity was not considered in this design.

Table 3
Design parameters for QWSC

Region	ARC	P		Intrinsic		N	
Designs 3–7							
Material	SiN ($n = 2.1$)	AlGaAs/GaAs		AlGaAs/GaAs		AlGaAs/GaAs	
		Doping	Molar	Doping	Molar	Doping	Molar
Concentration levels		2.0×10^{16}	0.33	I	0.33	2.0×10^{16}	0.33
Length (nm)	70	140		QW $= 5.5$		140	

Note: Barrier length = 17.0 nm. The number of quantum wells in each structure is 0 (Design 3), 5 (Design 4), 20 (Design 5), 50 (Design 6) and 100 (Design 7).

3.4.1. Solar Cell Designs

Fig. 4 shows the introduction of a SiN ARC of 70 nm in thickness. At short wavelengths (400 nm), the reflective loss of the QWSC is approximately 49% (model) compared to 48% (experimental). Moving through the spectrum, the reflection minimises at 650 nm with a value of 2% for both model and experimental values. Approaching the longer wavelength (900 nm) the QWSC has a value of approximately 13% (model) compared to 14% (experimental). It is clear from the comparative studies the results generated are in agreement with the experimental results obtained. In both cases the inaccuracy across the spectrum is approximately 2–7%. The discrepancies in the analysis are minimal and the author concludes the new model accurately determines values, which match with existing data.

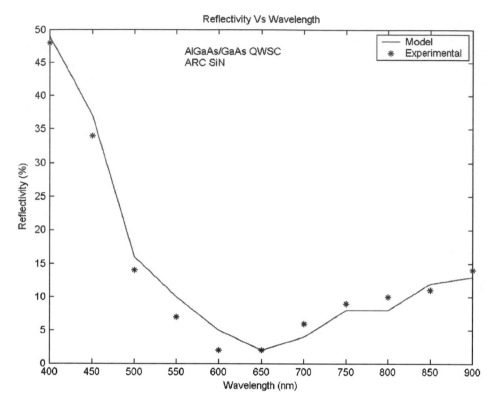

Fig. 4. Theoretical and experimental reflectivity of QWSC structure.

This new model can now be utilised to determine the effects of quantum well number variation on the refractive index and any "flow on" effects to the structure's reflective losses through Designs 3–7.

Fig. 5 shows the percentage change in the structure's reflectivity indirectly through the change in refractive index caused by the variation of quantum well number within the intrinsic region. The device's anti-reflection coating has been optimised for the 600 nm wavelength as discussed. Table 4 and Fig. 6 show a comparison between the changes of reflectivity (in percentage) for each structure when compared to the control cell.

From Table 4 and Fig. 6, it becomes evident that for a given wavelength the percentage change in reflectivity induced by refractive index change through variation in quantum well number is unique to wavelength and can be significant (fluctuating between 92% and 400%). It can be seen

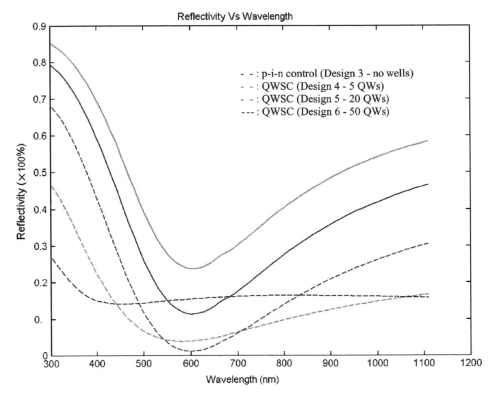

Fig. 5. Reflectivity of QWSC structures with varying quantity of quantum wells within the intrinsic region.

from the above curves the higher energy photons (300 and 600 nm) have a maximum change in reflectivity with a structure having 50 quantum wells when compared to the control structure.

In case of the lower energy photons (1100 nm) a minimum change is observed with a structure having 20 quantum wells when compared to the control cell. This suggests minimum reflective losses across the wavelength spectrum, which is achieved with structures consisting of quantum wells ranging between 20 and 50 in number.

Furthermore, the above finding suggests the trend of thought relating to only the optical properties of the anti-reflection coating affecting the reflective losses is not the case. Varying the cell's refractive index through well number significantly varies the cell's reflective losses.

Francis K. Rault

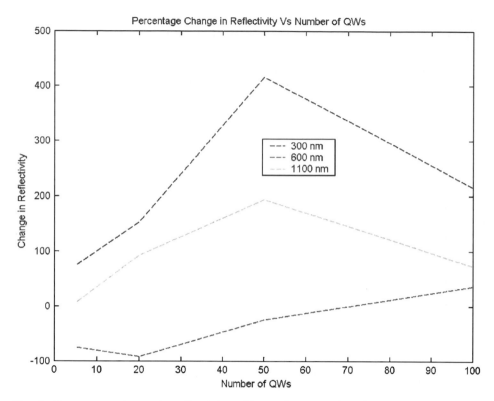

Fig. 6. Percentage change in reflectivity with varying quantity of quantum wells within the intrinsic region.

Table 4
Percentage change in reflectivity induced by the variation in quantum well number

Wavelength (nm)	Design 4 (5 QWs)	Design 5 (20 QWs)	Design 6 (50 QWs)	Design 7 (100 QWs)
300	73.39	152.44	415.45	215.61
600	−74.81	−92.07	−24.87	35.86
1100	5.54	92.37	193.38	72.81

3. QUANTUM WELL SOLAR CELLS DESIGNS

In this section the author presents the electrical performance results of some designs he has simulated based on the previous work. The electrical performance of these designs has been simulated using the Rimada–Hernendez model [5], along with the work of Lade [21]. The experimental results of

Table 5
Summary of design parameters

Molar concentration (x)	$x \leq 0.4$	Doping concentration (N)	$10^{15}\text{cm}^{-3} \leq N \leq 10^{18}\text{cm}^{-3}$
Cell dimension	See experimental data[a]	Intrinsic region $\leq 1000\,\text{nm}(\approx 600\,\text{nm})$	Well length $\geq 10\,\text{nm}(\approx 6\,\text{nm})$ Barrier length = 2 × well length
Temperature (K)	298	Well lengths (nm)	3–10
Spectrum	AM1.5G	Barrier (nm)	10
Cell area	1 cm × 1 cm		
Effective electron mass	$0.067\,m_\text{o}$	Interface superficial recombination	30–300 cm/s
Effective heavy hole mass	$0.8\,m_\text{o}$		
Material [6]	E_o	E_d	E_g
$\text{Al}_x\text{Ga}_{1-x}\text{As/GaAs}$	See section 3.3.3	$E_\text{o} = 2.6 + 0.75E_\text{g}$	$1.519 + 1.247x - \dfrac{dT^2}{T+204}$ $d = 5.4 \times 10^{-4}$ eV/K

Note: QW, quantum wells; QWL, quantum well length; BL, barrier length.
[a]The intrinsic region length and well length should be adhered to.

Aperathitis [23] have been utilised for comparison. Table 5 summarises the design parameters based on the findings of Section 2, along with the parameters of Rimada, Lade and Aperathitis.

Table 6 shows the results of the electrical performance compared to the experiement.

From the initial analysis, the values obtained are similar in magnitude to the experimental results. Furthermore, the discrepancies between model and experimental values are consistently similar in magnitude for each electrical parameter. These are good signs the Rimada model is generating reasonable results. Furthermore, we observe fractional improvements in performance of QWSC2 and QWSC3 when compared to the control (AlGaAs) structure, however none is observed when compared to the single band gap (GaAs) structure. The present author feels this latter result is ambigious as the QWSC structures of Aperathitis were grown without anti-reflection coatings while the GaAs control stucture had an MgF/ZnS anti-reflection coating. Clearly, an increase in photonic absorption would take place leading to

Table 6
QWSC cell performance under AM1.5G spectrum

Device	Cell perfomance			
	J_{SC} (mA/cm^2)	V_{OC} (V)	Fill factor	Efficiency (%)
p–i–n GaAs cell				
νExperimental	16	0.981	0.685	10.33
p–i–n AlGaAs (control) cell				
νModel	7.14	1.19	0.68	5.78
νExperimental	6.80	1.24	0.70	5.90
νModel–experimental descrepancy (%)	4.7	4.0	4.1	2.1
QWSC2				
νModel	10.68	0.96	0.73	7.56
νExperimental	10.2	1.009	0.713	7.34
νModel–experimental descrepancy (%)	4	5.0	3.9	2.9
νFractional improvement compared to QWSC 1 (%)	33	24	7.3	23
QWSC3				
νModel	9.86	0.97	0.768	7.21
νExperimental	9.40	1.004	0.735	6.93
νModel–experimental descrepancy (%)	4.66	3.51	4.2	4
νFractional improvement compared to QWSC 3 (%)	31	22.9	11.4	20

high-performance levels for the GaAs structure over the QWSCs. This forms part of the controversy of QWSC performance surpassing that of single band gap structures. With consistency during the physical structure growth possibly resolving the above ambiguity.

In relation to the fractional increase of short-circuit current between QWSC2 and QWSC3. The fractional increase in performance is approximately 8%, however open-circuit voltage and fill factor are near the same in magnitude. This increase in short-circuit current is likely due to the doubling in QW presence within the intrinsic region. Furthermore, the barrier length of QWSC3 being significantly smaller than QWSC2, leads to greater photo-carrier escape through the tunnelling effect, which according to the Aperathis work [23] is evident at any temperature and dominant at low temperatures (in our case room temperature).

From the above it is clear that reasonable results have been achieved. There is a clear improvement of cell performance over the p–i–n control (AlGaAs) cell. Although experimental growth of the structures would give full confirmation, the author feels confident that the results are accurate. This is due to the fact that the discrepancies relating to the accuracy of the Rimada model are significantly small. Furthermore, the electrical perform-ance analysis shows the discrepancies to be consistent for all the structures.

Over the past decade, the quantum well solar cell has developed into one of the many structures pioneering the development of third-generation photovoltaics. The next generation of structures are complex devices in both theoretical and practical perspectives. They show great promise in achiev-ing one common goal, harnessing a clean renewable source of electricity from the glowing beacon of our solar system, the Sun.

REFERENCES

[1] K.W.J. Barnham, "A novel approach to higher efficiency-the quantum well solar cell", 11th E.C. Photovoltaic Solar Energy Conference, Montreux, Switzerland, pp. 146–149, 1992.

[2] E. Aperathitis, Z. Hatzopoulos, M. Kayambaki and V. Foukaraki, "1 cm × 1 cm GaAs/AlGaAs MQW solar cells under one sun and concentrated sunlight", 28th IEEE Photovoltaics Specialists Conference, Anchorage, Alaska, USA, pp. 1142–1145, 2000.

[3] J.P. Connolly, J. Nelson, Keith W.J. Barnham and Ian Ballard, "Simulating multiple quantum well solar cells", 28th IEEE Photovoltaics Specialist Conference, Anchorage, Alaska, USA, pp. 1304–1307, 2000.

[4] R. Corkish and M.A. Green, "Recombination of carriers in quantum well solar cells", 23rd IEEE Photovoltaics Specialist Conference, Louisville, Kentucky, USA, pp. 675–680, 1993.

[5] J.C. Rimada and L. Hernandez, Microelectron. J., 32 (2001) 719–723.

[6] J.P. Connolly, J. Nelson, Ian Ballard, Keith W.J. Barnham, Carsten Rohr, Chris Button, John Roberts and Tom Foxon, "Modelling multiquantum well solar cell efficiency", 17th European Photovoltaic Solar Energy Conference, Munich, Germany, 2001.

[7] R. Corkish and C.B. Honsberg, "Numerical modelling of quantum well solar cells", 2nd World Conference and Exhibition P.V. Solar Energy Conversion, Vienna, Austria, 1998.

[8] N.G. Anderson, J. App. Phys., 78 (3) (1995) 1850–1861.

[9] H. Wenzel et al., IEEE JSTQE, 5 (3) (1999) 637–642.

[10] A. Freundlich, V. Rossignoi, M.F. Vilela and P. Renaud, "InP-based quantum well solar cells grown by chemical beam epitaxy", 1st World Conference on Photovoltaic Energy Conversion, HI, USA, pp. 1886–1889, December, 1994.

[11] P. Renaud, A. Freundlich, V. Rossignoi and M.F. Vilela, "Modelling p–i(multi-quan-tum well)–n solar cells: A contribution for a near optimum design" 1st World Conference on Photovoltaic Energy Conversion, HI, USA, pp. 1787–1790, December, 1994.

[12] M. Wolf, R. Brendel, J.H. Werner and H.J. Quessier, J. App. Phys., 83 (8) (1998) 4213–4221.

[13] N.J. Ekins-Daukes, J. Nelson, J. Barnes, K.W.J. Barnham, B.G. Kluftinger, E.S.M. Tsui, C.T. Foxon, T.S. Cheng and J.S. Roberts, Phys. E, 2 (1998) 171–176.

[14] R.J. Walters, G.P. Summers, S.R. Messenger, M.J. Romero, M.M. Al-Jassim, R. Garcia, D. Araujo, A. Freundlich, F. Newman and M.F. Vilela, J. App. Phys., 90 (6) (2001) 2840–2846.

[15] Y. Suzuki, T. Kikuchi, M. Kawabe and Y. Okada, J. App. Phys., 86 (10) (1999) 5858–5861.

[16] J.P. Connolly and C. Rohr, Semiconductor Sci., 18 (2003) S216–S220.

[17] B. Kluftinger, K.W.J. Barnham, J. Nelson, T. Foxon and T. Cheng, Sol. Energy Mater. Sol. Cells, 66 (2001) 501–509.

[18] F.K. Rault and A. Zahedi, Microelectron. J., 34 (4) (2003) 797–803.

[19] G. Araujo and A. Marti, Sol. Energy Mater. Sol. Cells, 33 (1994) 213–240.

[20] M.A. Afromowitz, Solid State Commun., 15 (1974) 59–63.

[21] S.J. Lade and A. Zahedi, Microelectron. J., 35 (5) (2004) 401–410.

[22] J. Nelson, M. Paxman, K.W.J. Barnham, J.S. Robert and C. Button, IEEE J. Quantum Electron., 29 (6) (1993) 1460–1467.

[23] E. Aperathitis, Z. Hatzopoulos, M. Kayambaki and V. Foukaraki, Mater. Sci. Eng., B, 51 (1998) 85–89.

[24] I. Ballard, K.W.J. Barnham, J. Nelson, J.P. Connolly, C. Roberts, J.S. Roberts and M.A. Pate, "The effects of temperature on the efficiency of multi-quantum well solar cells", 2nd World P.V. Conference, Vienna, Austria, pp. 3624–3626, 1998.

[25] J. Nelson, et al., "Quantum well solar cell dark currents", 12th European Photovoltaic Solar Energy Conference, Amsterdam, The Netherlands, pp. 1370–1373, April, 1994.

[26] D.J. Aiken, Sol. Energy Mater. Sol. Cells, 64 (2000) 393–400.

[27] I. Suemune, INSPEC Pub. EMIS, 15 (1996) 283–287.

[28] S.H. Wemple, Phy. Rev. B, 7 (8) (1973) 3767–3777.

[29] P. Harrison, Quantum Wells, Wires and Dots; Theoretical and Computational Physics, Wiley, West Sussex. England, p. 293, 2000.

[30] V.A. Sinyak, IEEE JQE, 2 (1996) 379–382.

[31] G. Bastard, E.E. Mendez, L.L. Chang and L. Esaki, Phy. Rev. B, 28 (6) (1983) 3241–3245.

[32] K.H. Schlereth and M. Tacke, IEEE JQE, 26 (4) (1990) 627–630.

[33] G.M. Alman, L.A. Molter, H. Shen and M. Dutta, IEEE JQE, 28 (3) (1992) 650–657.

[34] J.P. Loehr. INSPEC Pub. EMIS., 15 (1996) 71–73.

[35] Chih-Hsiang Lin, J.M. Meese, M.L. Wroge and Chun-Jen Weng, IEEE PTL, 6 (5) (1994) 623–625.

[36] A. Goetzberger, J. Knobloch and B. Voss, Crystalline Silicon Solar Cells, Wiley, London, pp. 14–19, 1998.

[37] J.P. Connolly, "Modelling and optimising GaAs/$Al_{1-x}Ga_xAs$ multiple quantum well solar cells", Ph.D. Thesis, University of London, Imperial College of Science, Technology and Medicine, p. 150, 1997.

PART II

NANOSTRUCTURES IN CONVENTIONAL THIN FILM SOLAR CELLS

Nanostructured Materials for Solar Energy Conversion
T. Soga (editor)

Chapter 5

Amorphous (Protocrystalline) and Microcrystalline Thin Film Silicon Solar Cells

R.E.I. Schropp

Faculty of Science, SID – Physics of Devices, Utrecht University, PO Box 80.000, 3508 TA Utrecht, The Netherlands

ABSTRACT

Thin film silicon, like no other thin film material, has been shown to be very useful in tandem and triple-junction solar cells. Such multijunction cells, due to their spectrum splitting capability, have true potential for high-conversion efficiency. Amorphous silicon has a band gap of 1.7–1.8 eV, and thus microcrystalline silicon, with a band gap of 1.1 eV, makes an ideal match to amorphous silicon in tandem cells. Since the first reports on practical microcrystalline cells in 1994, much research effort has been put worldwide into the development of both fundamental knowledge and technological skills that are needed to improve thin film silicon multijunction solar cells. The research challenges are to enhance the network ordering of amorphous semiconductors (leading to protocrystalline networks), mainly for improving the stability, to increase the deposition rate, in particular for microcrystalline silicon; to develop thin doped layers, compatible with the new, fast deposition techniques; to design light-trapping configurations, by utilizing textured surfaces and dielectric mirrors.

Examples of progress can be given for each one of these four challenges. To enhance the deposition rate, modifications of the widely used plasma enhanced chemical vapor deposition (PECVD) are studied as well as new, inexpensive and fast techniques. For instance, protocrystalline silicon with enhanced medium-range order (MRO), deposited at a rate of 10–30 Å/s has been tested both in p–i–n and in n–i–p solar cells, and these cells have exhibited remarkably stable performance. Microcrystalline silicon has undergone progressive optimization, for example, by the exploration of novel plasma-deposition regimes, which has led to higher open-circuit voltages and fill factors. For a single junction cell, a stabilized efficiency of $\eta = 10\%$ ($V_{oc} = 0.52$ V, FF $= 0.74$) on texture-etched ZnO:Al has been obtained. The deposition rate of this absorber material is currently being accelerated toward 8–10 nm/s at various institutes. The world record initial efficiency of 15% for thin film silicon photovoltaic (PV) technology has been obtained for triple-band gap triple-junction cells, using adequate optical enhancement techniques, such as textured back reflectors and intermediate partially reflecting layers.

In production, the trend is toward implementation of these multijunction structures in large area modules. Modules are presently produced batch wise or in-line on glass, or roll-to-roll on stainless steel and other foils, such as polymer plastics.

1. INTRODUCTION

The first semiconductor-based solar cells with energy-conversion efficiencies larger than 10% were made of silicon, in the years 1950–1960. At present, 85–90% of the solar PV modules produced worldwide are based on crystalline silicon (c-Si) wafers [1]. Thin film solar cells, on the other hand, are attractive because they can be produced at low cost, in an energy-efficient way, using only a small amount of abundantly available materials. Yet, on the rapidly growing PV market, the relative share of thin film solar cells is presently only 5–10%. Owing to lower production costs, there is an opportunity for thin film technologies to capture a larger market share. Among the available thin film technologies, silicon thin film technology is the most advanced and takes up 90–100% of the thin film PV market segment.

Whereas c-Si solar cells are commonly made from self-supporting slices sawn from a silicon ingot (with a typical thickness $d \sim 200$–300 µm), the active silicon layers in thin film silicon PV are deposited from the gas phase (using SiH_4 feedstock gas, often diluted in H_2) in a low-temperature process on a cheap substrate such as glass, steel or plastic.

Thin film silicon exists in different phases, ranging from amorphous via microcrystalline to single crystalline (the latter via epitaxy and transfer techniques, which is beyond the scope of this chapter). In contrast to the periodic lattice that characterizes the crystalline form, there is only very short-range order in amorphous silicon (a-Si:H). However, most atoms are fourfold coordinated and hydrogen plays an important role in the growth process as well as in the passivation of dangling bonds. Owing to variations from the ideal tetrahedral bond lengths and bond angles, electronic band tail states are present. They play an important role in the electronic density of states distribution and in the transport of carriers. In addition, due to remaining unpassivated dangling bonds, there is always a significant number of mid-gap states (defects) in a-Si:H.

In n-type or p-type doped a-Si:H, the dangling bond density is so high that photogenerated charge carriers in an abrupt p–n junction are lost by recombination. Therefore, a-Si:H-based solar cells are made in the p–i–n or in the n–i–p configuration. The vast majority of charge carriers is then generated in the intrinsic absorber layer (the i-layer) and drifts to the external contacts due to the field that is set up by the p- and n-type doped layers. Due to the long-range disorder in the amorphous silicon network, the momentum conservation law for electron excitation by photon absorption is relaxed. This makes a-Si:H behave like a material with a *direct band gap* (no additional phonon is required for the electronic transition). Therefore, the absorption coefficient of a-Si:H is much higher than that of c-Si and thus a-Si:H solar cells can be made very thin. The typical thickness for single junction cell is $d \sim 0.2$–$0.4\,\mu m$.

Based on this concept, the first experimental a-Si:H p–i–n solar cell, with an energy-conversion efficiency of 2.4%, was made by Carlson and Wronski [2] in 1976. Presently, the efficiency of such cells is well above 10%. The major disadvantage of a-Si:H, however, is that the dangling bond density in the i-layer reaches a higher concentration over time upon exposure of the solar cell to light (the so-called Staebler–Wronski effect [3]). It is generally assumed that the presence of H combined with that of weak Si–Si bonds plays a role in the microscopic process. Because of this light-induced defect creation the performance saturates at a level that is lower than the initial efficiency. Therefore, there has been a search for materials with a smaller H content, better MRO, and a lower concentration of dihydride bonding configurations (a parameter that frequently appears to be correlated with light-induced defect creation). One form of thin film silicon offering better resistance to light-induced defect creation is protocrystalline silicon [4]. The enhanced stability of this material is thought to be due to the enhanced MRO. This material has an optical band gap similar to amorphous silicon, or slightly higher (1.8 eV), and has all characteristics of "classic" amorphous silicon. Microcrystalline silicon, a mixed-phase material consisting of nanocrystals embedded in amorphous tissue, also appears to be stable when the volume fraction of nanocrystals is sufficiently large (>40%). The optical band gap of microcrystalline silicon is 1–1.1 eV.

Since the optical band gap can be altered by creating materials with different crystalline contents, thin film silicon, like no other thin film material, can relatively easily be used in tandem and triple-junction solar cells. Such multijunction cells, due to their spectrum splitting capability, have true potential for high-conversion efficiency. The two band gaps mentioned earlier together form an ideal match with the terrestrial solar spectrum. Starting from the first reports on practical microcrystalline cells in 1994, much research effort has been devoted worldwide to the development of both fundamental knowledge and technological skills that are needed to improve thin film silicon multijunction solar cells.

The challenges of the present research are (i) to enhance the network ordering of amorphous semiconductors, mainly for improving the stability, (ii) to increase the deposition rate, in particular for microcrystalline silicon; (iii) to develop thin doped layers, compatible with the new, fast deposition techniques; and (iv) to design light-trapping configurations, by utilizing textured surfaces and dielectric mirrors.

2. MATERIALS PROPERTIES

Irrespective of the specific plasma process used for deposition of silicon thin films (see Section 3), dilution of the silicon-containing (e.g., SiH_4) source gas with hydrogen results into silicon thin films with different ordering, morphologies, and textures than "classic" amorphous silicon. Using moderately hydrogen-diluted mixtures, a material is deposited which has a higher medium-range structural order than standard amorphous silicon. This material also has a higher stability against light-induced changes than amorphous silicon [4] and is often referred to as protocrystalline silicon [5]. At increasing hydrogen-dilution ratios, a transition from amorphous to crystalline growth is observed. In this transition regime, several material morphologies have been obtained, which are referred to as polymorphous silicon (pm-Si:H) [6], (hydrogenated) microcrystalline silicon (μc-Si:H) [7] or nanocrystalline silicon (nc-Si:H), quasi-amorphous [8] or nanoamorphous silicon (na-Si:H) [9] or heterogeneous silicon (het-Si:H) [10], and polycrystalline silicon (poly-Si) [11], depending on whether nanocrystals are present and on the crystalline fraction of the material.

2.1. Amorphous and Protocrystalline Silicon

Using conventional PECVD techniques, where hydrogen dilution of the silane (SiH_4) feedstock gas is necessary to produce protocrystalline silicon [4, 5, 12], it has thus far not been possible to reach a high deposition rate for such material. Rates higher than 3 Å/s using conventional PECVD have not been reported yet and typically, 1 Å/s is used. In addition, the deposition parameters for achieving a protocrystalline regime are dependent on the type of substrate as well as on film thickness [13, 14] due to structural evolution of the ordering along the growth direction.

An enhanced stability for thin silicon solar cells has also been reported for material with another structure, which is referred to as polymorphous (pm-Si:H) [6]. This material has high hydrogen content and is deposited in a regime that is close to that where powder formation occurs. Powder formation occurs under plasma conditions close to the transition regime that favor secondary reactions, such as high pressure, high power, large inter-electrode distance, and low substrate temperature. To keep the plasma in the transition regime, high dilutions of SiH_4 in H_2 ($R > 30$) are also required. Under this regime, small crystalline particles of 3–5 nm are formed in the plasma that can be embedded in the growing amorphous film.

A technique that is not plasma based, hot-wire chemical vapor deposition (HWCVD), has been shown to be capable of producing stable a-Si:H films in the protocrystalline regime (i.e., with enhanced MRO) [15]) at relatively high rate. In order to give an example of the characterization of protocrystalline silicon, the hot-wire deposited material is more extensively presented here.

In 1997, the deposition rate for these films was already high (5–8 Å/s), but a possible drawback was the high substrate temperature regime (360–425°C) in which the materials were obtained. X-ray diffraction measurements indicated that the most stable materials exhibited the narrowest widths of the first scattering peaks. This characteristic parameter can be considered a fingerprint for protocrystalline materials [16].

As the type of substrate has an influence on the degree of ordering in the deposited films, it has further been determined that PECVD conditions leading to *protocrystalline* silicon on n^+-type doped layers lead to *microcrystalline* silicon on plain stainless steel [17]. Such substrate-dependent effects were first observed for HWCVD layers in the substrate temperature domain of 360–425°C [17]. Thus, it was thought that for hot-wire deposited material either high substrate temperature or low-bonded hydrogen content was needed to obtain protocrystalline material. In 2004, however, it was reported [18] for the first time that materials fabricated by HWCVD at relatively low substrate temperature (250°C) were in fact protocrystalline Si:H materials. Moreover, materials with protocrystalline characteristics were achieved in a high deposition rate regime (>10 Å/s) without using any H_2 dilution [18].

Thus, high-quality intrinsic layers could be made from *undiluted* silane at 250°C and 0.02 μbar, using Ta filaments. To investigate whether there is any "template" effect on the structure of the deposited films, the stainless steel substrates were first coated with a-Si:H (sample 2-SS) or μc-Si:H (sample 3-SS and 3-SS') n-type layers (as in n–i–p solar cells). For reference, also bare stainless steel substrates were used (sample 2-SS and 2-SS'). The samples denoted without the prime had a 2 μm thick i-layer; those with a primed sample denotation had a 0.7 μm thick i-layer (in the order of the real thickness in solar cells).

Determination of the peak width of the FSP in X-ray diffraction experiments (XRD) is a sensitive tool for demonstrating protocrystallinity [16]. As the samples 1-SS' and 3-SS' are rather thin, long scan times of about 12 h were used and each film was run twice to ensure reproducibility. Also, the back of the stainless steel substrate was scanned in order to allow subtraction of the substrate signal.

Fig. 1 shows the XRD data after subtraction of the SS substrate. Table 1 lists the quantitative values of the full-width-at-half-maximum (FWHM), the position (P) of the first sharp peak (FSP) of the intrinsic a-Si:H layer, and total integrated intensity I_T of both a-Si:H peaks. In cases where the μc-Si:H template n-layer was present, the integrated intensities of the (111), (220), and (311) peaks amounted to a fraction of the total integrated intensity I_T that is consistent with the thickness ratio of 80 nm of the μc-Si layer to the total thickness of the silicon film. The XRD results show that the HWCVD a-Si:H have narrow FSP linewidth, consistent with that of protocrystalline silicon. This effect is enhanced if a μc-Si n-type template is present. Therefore, the i-layers as built-in in n–i–p solar cells have an improved MRO based on the narrower FWHM for i-layers grown on μc-Si:H film. The thinner films (0.7 μm) seem to have better MRO than the thicker films (2.0 μm). In solar cells, i-layers are usually even thinner ($\sim$0.4 μm).

Further experiments using small-angle X-ray scattering (SAXS) and cross-sectional transmission electron microscopy (TEM) revealed that the samples with a density of 2.195 g/cm^3 (which is not exceptionally dense) contained spherical (or ellipsoidal) voids with most probable sizes of only 2 or 3 nm. The voids are not interconnected and are arranged along the growth direction. It is conjectured that the voids allow the surrounding material to relax and thus to possess enhanced MRO.

Light soaking for over 1500 h under 100 mW/cm^2, AM1.5 conditions showed that cells incorporating such material have excellent stability (see Fig. 2). The changes in the FF are within 10%, while the absolute FF saturates at around 0.65 (see Fig. 2a).

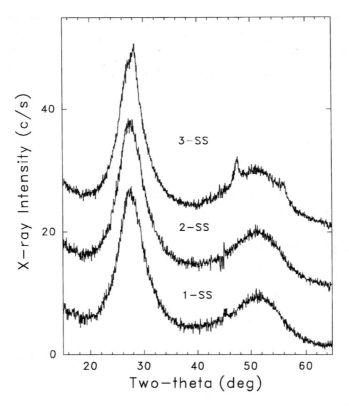

Fig. 1. XRD patterns from the three thick films made on SS (1-SS, 2-SS, 3-SS) after subtraction of the SS reference pattern. Sample 3-SS (on μc-Si:H n-layer) shows evidence of the (111), (220) and (311) peaks of c-Si.

Table 1
Quantitative XRD results. FWHM and P are the full-width-at-half-maximum and position of the first sharp peak of the intrinsic layer. IT is the total integrated intensity of both a-Si:H peaks. The indicated error is the statistical uncertainty

Sample no.	FWHM (2θ-deg)	P (2θ-deg)	I_T (deg c/s)
1-SS	5.59 ± 0.09	27.59 ± 0.02	229
2-SS	5.55 ± 0.09	27.58 ± 0.02	236
3-SS	5.29 ± 0.11	27.64 ± 0.04	234
1-SS′	5.29 ± 0.09	27.72 ± 0.06	73
3-SS′	5.10 ± 0.09	27.82 ± 0.08	77

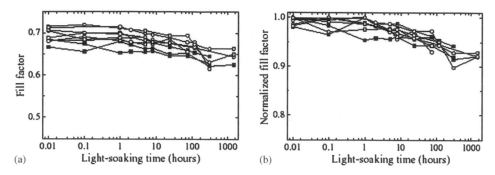

Fig. 2. (a) Fill factor and (b) normalized fill factor as a function of light-soaking time for single junction n–i–p solar cells with hot-wire deposited intrinsic silicon.

2.2. Microcrystalline Silicon

Microcrystalline silicon was first reported by Vepřek et al. in Europe in 1968 [19], and in Japan by Matsuda et al. [20] and Hamasaki et al. [21] in 1980.

Microcrystalline silicon is a mixed-phase material consisting of a-Si:H, embedded nanosized crystallites or conglomerates of nanocrystallites, and grain boundaries. During the growth of the layer, the formation of crystallites starts with a nucleation phase after an amorphous incubation phase. During continued layer deposition, clusters of crystallites grow (crystallization phase) until a saturated crystalline fraction is reached. These processes are very much dependent on the deposition conditions. In general, crystalline growth is enhanced by the presence of atomic hydrogen, which chemically interacts with the growing surface [7, 22, 23]. Different growth models for μc-Si:H exist. They are discussed in this chapter in Section 3.

A schematic representation of the incubation and crystallization phase as a function of the source gas dilution with hydrogen is shown in Fig. 3. The incubation phase and crystallization phase have a large influence on the optoelectronic material properties. The optical absorption in μc-Si:H is because of local absorption in both the amorphous and the crystalline part of the volume. Compared to a-Si:H, the absorption of high-energy photons ($h\nu \sim 1.7$ eV and higher energies) is low, due to the fact that the band gap in the crystalline volume fraction is indirect, whereas in the low-energy part ($h\nu < 1.7$ eV) the absorption of μc-Si:H is higher than in amorphous silicon, due to the smaller band gap of 1–1.1 eV. The optical absorption spectra of μc-Si:H, a-Si:H, and single crystal Si (c-Si) are compared in Fig. 4.

In mixed-phase material (μc-Si:H), the charge carrier transport mechanism is not yet completely understood. Whether the electrical properties are determined by the amorphous or the crystalline fraction of the volume

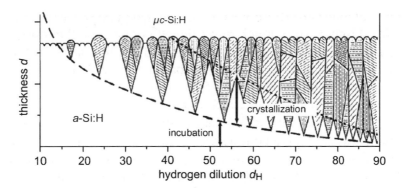

Fig. 3. Schematic representation of the cross-section of a µc-Si:H layer grown at different hydrogen dilution of the source gas (based on Ref. 22). The marked regions represent crystallites and the white region the amorphous volume.

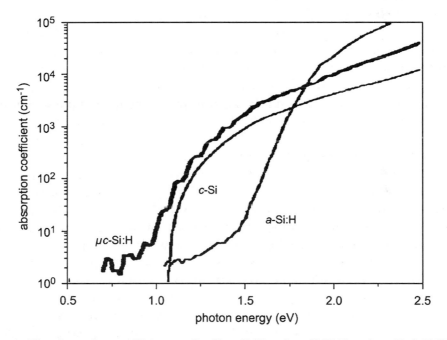

Fig. 4. The absorption coefficient α of c-Si, a-Si:H and µc-Si:H (data from Ref. [24]).

is dependent on the transport path. If the crystalline fraction is high enough, percolation takes place along interconnected paths through the crystallites. From geometrical considerations, the percolation threshold is ~33%. In n-type and p-type microcrystalline silicon, percolation is observed at even lower crystalline volume fractions. Of these two doping types, n-type material seems to exhibit percolation at the lowest crystalline content, in experiments at Utrecht University at only ~10%, much lower than theoretical predictions [25]. For intrinsic microcrystalline silicon, the threshold has instead been found at a crystalline content that is instead higher than predicted. For example, Kočka et al. [26] observe this threshold at 65%. Overall, it can be stated that the transport of charge carriers in μc-Si:H is influenced by (1) the crystalline fraction, (2) electronic barriers because of the difference in mobility gaps of crystalline grains and grain boundaries or amorphous tissues between grains, (3) doping due to impurities such as oxygen and (4) grain boundary defects.

Obviously, grain boundary properties play an important role in carrier transport in μc-Si:H. Matsui et al. [27] studied electron and hole transport in separate devices and concluded that μc-Si:H becomes more n-type at higher crystalline fractions, leading to low activation energies for the dark conductivity. Elsewhere, it has been found that the grain boundary defect density increases with the crystalline fraction [28].

The performance of solar cells based on transition-type microcrystalline silicon is based on the principle that a considerable amorphous silicon content passivates the grain boundary defects, whereas the crystalline fraction is high enough to bring about sufficient absorption in the near infrared part of the spectrum. This principle is fundamentally different from that in polycrystalline silicon [29, 30], which has no a-Si:H content but rather contains large crystals oriented in the [110] direction, with a low grain boundary defect density, extending from bottom to top in the growth direction.

The optoelectronic properties of μc-Si:H are thus sensitively dependent on the structure of the crystallites and the grain boundaries. Dangling bonds give rise to localized states that are effective recombination centers for charge carriers. Most of the defects are concentrated at the grain boundaries and in the amorphous phase. Dangling bond densities N_d can be as low as for compact a-Si:H, but in the low-deposition temperature regime (below $T_s = 180°C$) N_d is usually reported to be higher [31].

The presence of oxygen in concentrations higher than 1×10^{18} cm^{-3} may deteriorate the electronic properties of the layers considerably [32]. Interconnected voids adjacent to grain boundaries can act as a diffusion path for impurities (such as oxygen, acting as an n-type dopant), after deposition as well as during deposition [33, 34, 35]. The n-type doping effect leads to an unfavorable field distribution in p–i–n or n–i–p type solar cells. It is thus necessary to develop a material with high compactness, so that grain boundaries are fully passivated and impurities, such as oxygen, do not deteriorate the transport mechanism.

2.2.1. Doped Microcrystalline Layers

Microcrystalline silicon thin p- and n-type doped layers are used in silicon-based thin film solar cells because of their favorable optical and electrical properties: low optical absorption in the ultra-violet–visible–near-infrared (UV–VIS–NIR) range, and high conductivity and doping efficiency (fraction of the dopant atoms in the material that donates holes or electrons). As the optimal thickness of doped layers in solar-cell applications is very thin (around 20 nm), it is essential to reduce the thicknesses of the amorphous incubation and crystallization phases as much as possible to benefit from the favorable properties of μc-Si:H. Prasad et al. [36] observed in 1991 that the use of higher radio frequencies (rf) (VHF PECVD) lead to a thinner amorphous incubation layer for μc-Si:H doped layers. VHF PECVD is frequently used for the deposition of μc-Si:H to benefit from the high atomic hydrogen concentrations and the low ion energies [37, 38].

For a few cell types, particularly those in which an elevated substrate temperature is used to allow high-rate deposition of the absorber layer, the doped layers need to be temperature stable. Damage to the already deposited doped layers can be induced by enhanced diffusion of dopants at the high temperature at which i-layer deposition takes place and/or by excessive hydrogen diffusion into the doped layer [39]. It is observed that the hydrogen-diffusion coefficient in a silicon matrix is smaller for materials made at higher substrate temperatures [40]. This mechanism implies the need of doped microcrystalline layers deposited at a high substrate temperature.

In general, the optimum deposition temperature for microcrystalline silicon is still under discussion. Matsuda [41] claims that hydrogen coverage of the growth surface is necessary for crystallization. At high temperatures (above 350°C), the crystallinity would be limited by the reduced hydrogen coverage due to the hydrogen out-diffusion. Vepřek et al. [42], however, observe that if the material does not contain impurities, crystallization can take place up to very high temperatures. Both these statements illustrate that it is not straightforward to deposit microcrystalline silicon-doped layers (i.e., layers containing impurities) at high temperatures. Given the unavoidable presence of impurities, in practice the best doped microcrystalline silicon layers are obtained at temperatures down to 100°C. The optimum deposition temperature for doped layers in a solar-cell device with a high temperature, high deposition rate i-layer lies between 250°C and 400°C. An alternative deposition process to circumvent the crystallization problem at high temperatures is the layer-by-layer (LBL) process [43, 44].

3. DEPOSITION METHODS

The first amorphous silicon layers were deposited in a rf-driven glow discharge using silane [45], now usually called PECVD.

PECVD has become the workhorse of thin film semiconductor industry and is generally used for the deposition of thin film silicon solar cells. Some of the energy transferred to silane molecules in the collisions with electrons is radiated as visible light, hence the deposition method is also called glow discharge deposition. An important advantage of PECVD deposition is that the deposition temperature of device quality silicon thin films can be kept between 180°C and 250°C. This allows using a variety of low-cost materials as a substrate, such as glass, stainless steel and flexible plastic foils.

The rf glow discharge technique is based on the dissociation of silicon-containing gasses in a rf plasma, usually at a frequency of 13.56 MHz. The plasma is generated between two electrodes, and the substrate is attached to the grounded electrode. Inelastic collisions of energetic free electrons with the source gas molecules not only result in the production of silicon-containing radicals (Si_xH_y) and hydrogen radicals, but also in the production of ions. These species can react further with other radicals, ions and molecules in the gas phase. Radicals that diffuse to the substrate can contribute to the film growth. For the growth of a compact layer, it is important that the surface mobility of the growth precursors is high. This surface mobility is greatly improved by passivation of the dangling bonds at the growth surface by hydrogen [46, 47]. A growth radical reaching the surface of a growing film attaches to one of the hydrogen passivated surface silicon dangling bonds. In order to find an energetically favorable position, the radical has to diffuse over the surface by hopping over the hydrogenated surface atoms. Hydrogen can be removed by thermal excitation or by abstraction by a SiH_3 radical, in which the dangling bond and a SiH_4 molecule is formed [46]. Cross-linking with the neighboring silicon atoms finally results in film growth, under the release of molecular hydrogen. For the deposition of device quality a-Si:H, SiH_3 radicals are the preferred main growth precursors [46, 48]. Other radicals, such as SiH_2 and higher silane radicals, having higher sticking coefficients than SiH_3, can be directly incorporated into the hydrogen-terminated surface [49]. The contribution of these radicals to the growth, however, results in films with poor quality [33, 46], and therefore the presence of these radicals in the plasma should be avoided. This can be achieved by lowering the electron temperature in the plasma and/or by increasing the deposition temperature. A lower electron temperature in the plasma is achieved by using a higher plasma excitation frequency. A high deposition temperature of up to 350°C is required at high deposition rates in order to achieve sufficient diffusion of growth radicals on the surface. Such an elevated substrate temperature has the drawback that thermal damage to the previously deposited layers in the solar cell structure may occur; see Section 2.2.

Ions also contribute to the growth; their contribution is estimated to be 10% [46, 50]. The ions that are produced in the plasma are accelerated under the influence of the electric field between the plasma and the electrodes. The resulting ion bombardment on the growing surface strongly affects the material properties. The energy that is released when an ion is stopped and neutralized at the film surface improves the surface mobility of the growth precursors as well as the cross-linking process, resulting in a more dense silicon network. Therefore, ions are believed to play a beneficial role during growth of device quality thin film silicon [51]. On the other hand, energetic ions with a kinetic energy higher than 20 eV are thought to cause defects in the material [52, 53], particularly in μc-Si:H [54].

Thus, the deposition of thin film silicon from a silane plasma can be described as a four-step process [52]:

1. The primary reactions in the gas phase, in which SiH_4 molecules are decomposed by electron-impact excitation, generate various neutral radicals and molecules, positive and negative ions and electrons.
2. The secondary reactions in the plasma between molecules and ions and radicals that result in formation of reactive species and eventually in formation of large silicon–hydrogen clusters, which are described in literature as dust or powder particles. Neutral species diffuse to the substrate, positive ions bombard the growing film and negative ions are trapped within the plasma.
3. Interaction of radicals with the surface of the growing film, such as radical diffusion, chemical bonding, hydrogen sticking to the surface or abstraction from the surface.
4. The subsurface release of hydrogen and relaxation of the silicon network.

The PECVD technique still delivers the best cell efficiencies using a-Si:H, proto-Si:H or μc-Si:H intrinsic absorber layers. To obtain device-quality a-Si:H in a laboratory rf PECVD deposition system, the typical processing conditions that have generally been used are silane flow 20–50 sccm, process pressure 0.5–0.7 mbar, substrate temperature 200–250°C, rf power

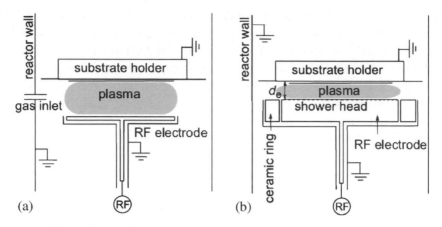

Fig. 5. Schematic diagram of an rf PECVD chamber. (a) A conventional electrode configuration is shown. (b) The configuration is shown for high-pressure conditions at low
interelectrode distance. At small interelectrode distances, the use of a shower head is
required. Figure from Ref. [100].

density 20–50 mW/cm^2, electrode distance 1–3 cm. A typical deposition rate
is 0.2 nm/s. To obtain protocrystalline or microcrystalline material, additional
hydrogen source gas is added to the silane feedstock gas during the plasma
process, which generally slows down the deposition rate even further.
Typically, H$_2$/SiH$_4$ gas flow ratios of 10 or higher [41] have been used.

Such deposition rates are widely considered to be too low for cost-
effective mass production of solar cells. For high-volume mass production
a deposition rate of 2–3 nm/s is required. A considerable research effort is
therefore devoted to increasing the deposition rate of protocrystalline silicon
and μc-Si:H while maintaining material quality.

A schematic diagram of an rf PECVD deposition chamber is given in
Fig. 5.

3.1. Deposition of Amorphous or Protocrystalline Silicon
3.1.1. Modifications of Direct PECVD Techniques
High pressure, high power regime. Using the high pressure, high power
(hphP) rf PECVD regime, device quality a-Si:H films have been fabricated
at a deposition rate of 1.2 nm/s [55]. The pressure and power density were
5–10 mbar and 270–530 mW/cm^2, respectively. By varying the silane to
hydrogen flow ratio or by changing the deposition pressure at a fixed flow
ratio, a transition from amorphous to microcrystalline silicon growth was

obtained. Single junction a-Si:H solar cells, having a-Si:H absorbers deposited at 190°C and $r_d = 1.2$ nm/s, reached a stabilized efficiency of 6.5%. The band gap for this high-rate a-Si:H material is about 50 meV higher than for "classic" a-Si:H i-layers prepared at the same deposition temperature of 190°C (low hydrogen dilution and low deposition rate (0.15 nm/s). The wide band gap and correspondingly high V_{oc} (typically 0.88–0.9 V) make these high rate a-Si:H cells interesting candidates for top cells in tandem solar cells.

Dusty regime. Similar widening of the band gap is observed for polymorphous (pm-Si:H) [6]. This material is deposited in a regime that is close to the powder formation. The powder formation is achieved by using plasma conditions that favor secondary reactions including the increase of pressure, power, interelectrode distance, and the decrease of the substrate temperature. Under this regime, small crystalline particles of 3–5 nm can be formed in the plasma, which are embedded in the growing amorphous film. To keep the plasma in the transition regime, high dilutions of SiH_4 in H_2 ($R > 30$) are used.

Very high-frequency PECVD. University of Neuchâtel investigated the influence of increasing the plasma excitation frequency in PECVD in the VHF range from 13.56 up to 150 MHz [56]. The group demonstrated that by increasing the excitation frequency from 13.56 MHz up to 70 MHz at constant plasma power, the deposition rate increases monotonously from 3 to 10 Å/s, while good quality of a-Si:H films is maintained [57]. The shift to higher excitation frequencies modifies the electron energy distribution function in the plasma, which leads to a faster dissociation rate of source gases, and thus in higher deposition rates. The higher operating frequencies allow applying higher power densities in the discharge, while staying in a powder-free operational regime [58].

3.1.2. Remote Techniques

Expanding thermal plasma CVD. In the ETP CVD [59] technique the plasma generation and the film deposition are spatially separated. The plasma is generated by a dc discharge in a cascaded arc in argon and/or hydrogen. The power to the plasma is typically 2–8 kW. The pressure in the plasma source is 200–700 mbar. The plasma expands supersonically into the deposition chamber, where the pressure is typically 0.1–0.3 mbar. Pure SiH_4 is injected into the plasma jet from an injection ring. The silane is dissociated by the reactive species from the plasma source. The deposition rate of a-Si:H layers is dependent on a number of factors, such as gas flows and arc current, and can reach 800 Å/s [59]. In order to deposit device quality a-Si:H at high deposition rates (>70 Å/s), elevated substrate temperatures of around

400°C have to be used [60]. Applying an external rf bias on the substrate during ETP CVD deposition of a-Si:H has proved to be a tool to provide extra energy to the surface due to a moderate ion bombardment of the growing surface and facilitates a reduction of the deposition temperature of ~100°C [60].

Hot-wire CVD. Since the first patent in 1979, the HWCVD technique has been further developed and presently is a viable method for the deposition of silicon-based thin films and solar cells [61–64]. This technique can be considered to be a remote technique, since it is based on the decomposition of silicon-containing gasses at a catalytic hot surface, while the substrate itself has no active role in generating the active precursors, unlike in the case of PECVD where it usually has at least a role as the grounded electrode. The absence of the requirement to achieve an equipotential plane at the substrate makes it easier to transport either rigid or foil type substrate materials during deposition and to scale up to large areas. Because source gases are catalytically decomposed, the method is often referred to as thermo-catalytic CVD (TCCVD) [65] or catalytic CVD (Cat-CVD) [66].

Usually, tungsten or tantalum filaments are used as catalyst, with filament temperatures roughly between 1400°C and 2100°C. Depending on the filament material and temperature, silane can be entirely dissociated into atomic Si and H at the catalyst surface. The created species react further with unreacted SiH_4 in the gas phase, so that various growth precursors are formed. As no ions are generated there during deposition, the gas-phase species and reactions that occur are different from those in conventional PECVD processes. Furthermore, the absence of any plasma ensures that no particles are trapped, thus eliminating one important source of dust.

Advantages of the HWCVD technique over the PECVD method are the high deposition rate, the scalability to large areas, and the relatively low equipment costs. Ultrahigh deposition rates of over 10 nm/s have been obtained in the case of a-Si:H [67]. In addition, it is possible to deposit amorphous silicon films with a low hydrogen content below 1 at.% that have a low defect density [63]. Solar cells incorporating such layers show less degradation upon light soaking than cells with a conventional PECVD deposited layer [68, 69, 70]. This work has initiated considerable interest in the HWCVD deposition method and today many groups study HWCVD thin film silicon and its alloys for various applications such as solar cells [71], passivation layers [72], and thin film transistors [73].

The reduced Staebler–Wronski effect has first been attributed to the very low hydrogen content of these hot-wire deposited layers (1–4 at.%) [63], but could also be explained by the presence of enhanced MRO [15, 18].

The absence of ions has the advantage of no defects being created as a result of energetic ion bombardment. However, as there are no low-impact species to densify the layer, it is still unclear how a low hydrogen content, device-quality amorphous silicon is deposited with the HWCVD method. One aspect is, however, that the H flux to the growing substrate surface is five to ten times higher than in PECVD, and that this H is active in subsurface reactions.

A drawback of the HWCVD method in comparison to PECVD is the control of the substrate temperature. Due to the heat radiation from the filaments, it is difficult to use substrate temperatures below 200°C. Artificial substrate cooling is a possibility, but the effects on material quality have not been fully investigated. Another aspect is that reactions of the source gasses with the filaments result in the formation of metal silicides on the filaments. These silicides change the catalytic properties and cause aging of the filaments. Eventually the filaments become brittle, particularly at the relatively "cold" ends of the filament. Any early breakage of the filaments can be prevented by physically shielding the filament ends or by flushing them with hydrogen. By choosing appropriate designs for both the deposition chamber geometry as well as the catalyst geometry, and by an appropriate (pre-) treatment of the filaments, filament lifetime issues can be overcome [74–76].

3.2. Deposition of Microcrystalline Silicon

In order to obtain good optoelectronic quality (low defect density) μc-Si:H, ion damage to the growing surface should be avoided, while keeping a high atomic H density in the gas phase to interact with the growing surface. In order to reach high deposition rates, new PECVD regimes featuring higher rf frequencies (very high frequency, "VHF" [56, 57, 58]), conditions of source gas depletion at high pressure (high-pressure depletion, "HPD" [33] or hphP [55]) have been explored. Further, alternative plasma excitation techniques such as microwave PECVD (MW-PECVD) [77] or the expanding thermal plasma CVD (ETP-CVD) technique [59, 78], and advanced electrode designs are explored. Another approach is to abandon the mechanism of plasma decomposition altogether and use HWCVD or Cat-CVD techniques [66, 79].

There are three interrelated models explaining μc-Si:H growth. Although the role of atomic H in these models is different, there is no controversy about it having an important role.

(i) The surface diffusion model [80], first proposed by Matsuda [41] in 1983, states that the chemically activated H diffusion at the surface as well as the coverage of the surface by hydrogen are important prerequisites for the growth precursors to find favorable sites for microcrystalline growth.

(ii)   The chemical transport model by Veprek et al. [81] in 1987 is based on the equilibrium between growth and etching at the surface. If the etching rate is high (comparable to the growth rate) microcrystalline silicon material is deposited. The selective etching model, proposed by several groups [82, 83, 84] is based on the chemical transport model and the difference in etching rate between the dense crystalline regions (lower etching rate), and the amorphous tissue (higher etching rate).

(iii) Various growth zone models [33] describe mechanisms that are responsible for crystal nucleation taking place in the subsurface region of the growing film. The responsible processes are based on interaction with in-diffused atomic hydrogen. Layer-by-layer growth of μc-Si:H seems to be an appropriate system to study the hydrogen-induced nucleation in the subsurface.

From optical emission spectroscopy (OES) measurements, it has become clear that there exists a pressure- and temperature-dependent threshold for the flux of atomic hydrogen to the growing surface [41]. Atomic hydrogen can originate from SiH_4 and from H_2. An important loss mechanism for atomic hydrogen is the recombination reaction with SiH_4, $H + SiH_4 \rightarrow H_2 + SiH_3$. Hydrogen dilution thus slows down the H annihilation reaction.

Another possible method to increase the atomic hydrogen density is to operate under silane depletion conditions, in which the silane is decomposed faster than that supplied by the flow of feedstock gas. High-density plasmas, where the silane depletion can be easily realized, have been applied for depositing microcrystalline silicon [85–89]. This has even succeeded using pure silane [85, 89].

As mentioned, ion bombardment above a certain ion energy ($\sim 20\,eV$) causes damage [53], particularly in μc-Si:H [54]. The magnitude of the ion damage can be expressed by the product of ion energy and ion flux density per deposited monolayer. If the rf power density is increased under depletion conditions, the ion damage per monolayer increases because there is no increase in deposition rate.

The HPD regime has been investigated [90, 91] as an approach to suppress the ion damage while maintaining a high atomic hydrogen density. The HPD method combined with the VHF technique improves the crystalline volume fraction in the very high deposition rate regime (higher than 2 nm/s).

In recent work, in addition, a small interelectrode distance was implemented that was made possible by using a showerhead electrode [92]. The cell results using this combined approach are highlighted in Section 4.

4. SINGLE JUNCTION CELL

This section first discusses the basic operation of a basic thin film silicon solar cell and then presents the thin film structure and technology. Thin film silicon solar cells were, in the first instance, merely based on a-Si:H as the single absorber material. Single junction solar cells of this type are now commercially available and the operation principles of thin film drift type cells are best explained with the help of this basic cell structure.

Intrinsic (undoped) amorphous silicon is a semiconductor with a mobility band gap of approximately $E_g = 1.7$–$1.8\,eV$ at room temperature. Within the band gap, band tail states and mid-gap defect states are present. In n-type or p-type-*doped* a-Si:H layers, the defect density is so high that excited carriers have effectively zero mean free path and thus photogenerated electrons and holes recombine locally within the doped layer. Therefore, high-quality (low defect density) intrinsic layers (i-layers) are used as absorber layers. The absorption of photons leads to the formation of mobile electron–hole pairs in the i-layer. The electrons in the conduction band and holes in the valence band move to the external contacts aided by the electrical field that is generated by the doped layers.

Because of the localized states within the band gap of the i-layer, charge carriers can intermittently be trapped in band tail states and may be lost due to recombination (through defect states). The diffusion length of charge carriers in a-Si:H is much shorter than in crystalline Si. Therefore, unlike the situation in c-Si, in a-Si:H, the main charge transport mechanism is not diffusion, but drift by the electrical field. For this reason, a-Si:H-based solar cells are often described as drift type solar cells.

Fig. 6 shows a schematic energy band diagram of a p–i–n solar cell under different external voltage conditions. The electric-field distribution over the i-layer is determined by the mobility band gap of the i-layer, the doping levels of the doped layers, the mobility band gaps of the doped layers, the defect density in the i-layer, its spatial distribution and its occupation, and the thickness of the i-layer. The electric field is not constant over the entire i-layer. Owing to charged defects at the p/i and i/n interface, as indicated in Fig. 6, strong band bending occurs near the p- and the n-layer, and thus the electric field in the middle of the i-layer is lower than that in the regions closer to the

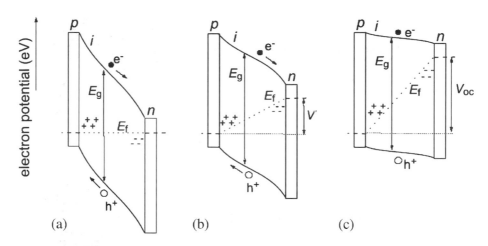

Fig. 6. Schematic band diagram (with electrons e⁻ in the valence band and holes h⁺ in the conduction band, and an imaginary Fermi–level connecting the Fermi levels at the p- and n-layer) of a p–i–n solar cell in (a) short-circuit condition, (b) operating condition and (c) open-circuit condition. Figure from Ref. [100].

doped layers. This is one of the reasons for the fact that thin film silicon solar cells cannot be made as thick as would optically be of interest.

Therefore, light trapping in the absorber layer is commonly used and is brought about by employing textured or rough substrates, for example textured transparent conductive oxides (TCOs). Because of light scattering at the rough substrate, a thin i-layer can absorb the same amount of light as a thicker layer on a flat substrate. In this configuration, also the collection of charge carriers is improved: the recombination rate is lower due to the larger overall electrical field over the thinner i-layer.

Also in μc-Si:H solar cells, the transport mechanism is predominantly drift driven. The charge carrier transport mechanism is still under discussion; many models and concepts that correlate the V_{oc} to material properties can be found in the literature. Kočka et al. [93] claim that the V_{oc} is decreased when large grain boundaries introduce barriers in the conduction path. Droz et al. [94] experimentally observe a strong correlation between the V_{oc} and the estimated crystalline fraction of the i-layer. Yan et al. [95] describe the V_{oc} with a two-diode model. Werner et al. [96] correlated the V_{oc} to the grain size and they conclude that the grain boundary recombination velocity for small grain sizes ($\sim$10 nm) is very low. Apart from the obvious need for further improvement of the charge transport properties of μc-Si:H, a recently published study, in which the limitations to the performance of μc-Si:H-based

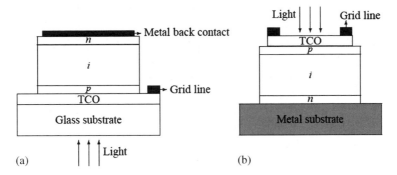

Fig. 7. Schematic cross section of (a) a p–i–n structured solar cell and (b) an n–i–p structured solar cell. The dimensions are not to scale. Figure from Ref. [100].

thin film solar cells are evaluated, concludes that the highest potential for improvement lies in light trapping [97].

At present, the focus of the research has shifted from a-Si:H to different forms of thin-film silicon such as protocrystalline Si, nanocrystalline Si, µc-Si:H (microcrystalline Si) and thin film poly-Si.

Thin film solar cells can be made in the substrate or in the superstrate mode (see Fig. 7). In superstrate type solar cells, the carrier on which the various thin film materials are deposited serves as a window to the cell. Usually in superstrate cells, glass is used as the carrier. A more technologically challenging version of a superstrate configuration is a cell on a transparent thin polymer foil. This type of cell is bound to have performance limitations as the limited temperature resistance of most transparent plastics dictates the maximum temperature allowable in further process steps. In substrate type solar cells, the carrier on which the various thin film materials are deposited forms the backside of the cell. They are usually made on a stainless-steel carrier that at the same time serves as the back contact. A highly flexible device can be made in the substrate configuration by employing a very thin metal carrier or a metal-coated polymer foil. As the polymer need not be transparent, a temperature-resistant type of polymer can be employed such as polymid.

The status of the technology of single junction cells in the laboratory is as follows. This summary is not meant to be exhaustive, but gives the highlights of the last few years.

4.1. Amorphous (Protocrystalline) Silicon

The University of Neuchâtel has presented a p–i–n a-Si:H cell with a stabilized efficiency of 9.5% [98] using the VHF PECVD technique.

The cell was deposited on a glass substrate coated with textured ZnO made by the Low Pressure Chemical Vapor Deposition (LPCVD) technique. The amount of light entering the i-layer was additionally enhanced by applying a multilayer antireflective coating (ARC). This p–i–n cell has an i-layer thickness of only ~0.25 μm and the short-circuit current density J_{sc} is very high (>17.5 mA/cm^2 stabilized and >18 mA/cm^2 initially). The deposition rate for the a-Si:H intrinsic layer was around 5 Å/s. This is the highest stabilized efficiency for a purely amorphous silicon p–i–n cell structure.

Solar cells in the n–i–p configuration incorporating HWCVD protocrystalline layers at a substrate temperature of 250°C have demonstrated excellent stability against light soaking, where the continuous exposure time was over 1500 h. The change in fill factor and overall performance was within 10% [79]. The stability of these layers against degradation has been attributed to a special void nature that allows enhanced MRO in the material itself.

The record single junction n-i-p *a*-Si:H solar cells with an HWCVD absorber layer deposited at 16.5 Å/s had an initial efficiency of 9.8% [99]. Also p–i–n single junction cells on glass/SnO$_2$:F substrates have been made using the HWCVD technique. Recently, such cells with HWCVD absorber layers deposited at high rates of 32 and 16 Å/s reached initial efficiencies of 7.5 and 8.5%, respectively [101].

Using the ETP CVD technique, a single junction a-Si:H solar cell, with the absorber deposited at 250°C at an $r_d = 11$ Å/s, has achieved an initial efficiency of 8% [102].

4.1.1. Microcrystalline (μc-Si:H) Silicon

In 1992, Faraji et al. [103] reported a thin film silicon solar cell with a μc-Si:H:O i-layer. The first solar cell with a μc-Si:H i-layer was reported in 1994 by Meier et al. [24] at IMT Neuchâtel, Switzerland with $\eta = 4.6\%$.

At Utrecht University, the concepts of discharge frequencies in the VHF PECVD regime [24], plasma conditions in the HPD regime [33], and gas distribution through a shower head gas inlet [33, 104] have been combined for the fabrication of high efficiency, stable, single junction μc-Si:H solar cells. These concepts have separately been studied by other groups in the following references: VHF PECVD [24, 105–108], HPD [33, 90, 91, 108] and shower head [33, 109].

In order to control the material properties in the growth direction, the hydrogen dilution of silane in the gas phase is graded using various gas ratio profiles with a parabolic shape as a function of deposition time.

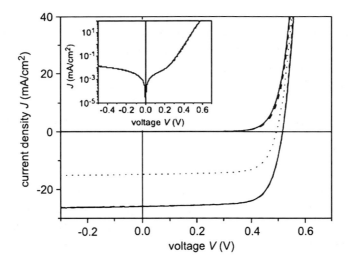

Fig. 8. Stable 10% μc-Si:H p–i–n cell.

By optimization of p–i–n solar cells based on the transition type μc-Si:H, an efficiency of 10% is obtained for a single junction solar cell on a texture-etched ZnO:Al front contact [110] (see Fig. 8). The μc-Si:H i-layer is 1.5 μm thick, deposited at a rate of 0.45 nm/s.

Using a hot-wire deposited p/i interface layer (buffer layer) in p–i–n μc-Si:H cells with a plasma-deposited bulk i-layer at a conventional rf frequency of 13.56 MHz in the high pressure, high-power regime, an initial efficiency of 10.3% was obtained at IPV, Jülich [111]. The buffer layers were made at 1 Å/s. The bulk layers were made at 11 Å/s. This result is primarily due to the enhancement of V_{oc} (to 568 mV) for the cell with the special HW buffer layer with respect to a cell without it (545 mV). The HW buffer layer also leads to an improvement from 8.7 to 9.4% of regular rf PECVD cells (V_{oc} from 505 to 549 mV). A possible mechanism for this improvement is the protection by the HW buffer of the p-type seed layer from energetic ions in the plasma. The 10.3% μc-Si:H single junction solar cell shows a small light-induced degradation of 5.8% relative to the initial efficiency, leading to a stabilized efficiency of 9.8%.

4.1.2. Higher Rate

Since microcrystalline silicon layers have to be made rather thick, especially when they are used in hybrid or micromorph a-Si/μc-Si tandem cells, the deposition rate is a major research challenge. Microcrystalline silicon materials with higher deposition rates are under development at many laboratories worldwide. At Utrecht University, combinations of a high rf discharge power and high total gas flow are chosen, such that the depletion condition is constant. At a deposition rate of 4.5 nm/s, a stabilized conversion efficiency of 6.7% is obtained for a single junction solar cell with a μc-Si:H i-layer of 1 μm. It is found that the defect density increases one order of magnitude upon the increase in deposition rate from 0.45 to 4.5 nm/s. This increase in defect density is partially attributed to the increased energy of the ion bombardment during plasma deposition. The solar cells with the high-rate μc-Si:H i-layers have been light soaked in AM1.5 light, filtered by an a-Si:H top cell in order to simulate the light exposure of the bottom cell in an a-Si:H/μc-Si:H tandem cell. It appeared that performance of the solar cells deposited at high rate actually *improve* upon light soaking (mainly due to the improving FF), contrary to the 5–10% degradation with light soaking of the transition materials reported in the literature. The reason of the improvement might be thermal equilibration of the fast-deposited material due to the elevated temperature of 50°C, or the improvement of the contacts due to this temperature.

Even higher deposition rates for μc-Si:H have been reported. T. Matsui et al. [113] and S. Goya et al. [114] have reported deposition rates of 23 and 22 Å/s, respectively. Matsui et al. use VHFCVD, at a frequency of 100 MHz, and an initial efficiency of 8.8% was obtained. This cell has 2.3 μm thickness. Goya et al. also use VHFCVD, but with a ladder-shaped electrode, and an initial efficiency of 8.3% was obtained.

Fig. 9 gives an overview of the efficiencies obtained for μc-Si:H-based single junction solar cells as a function of the deposition rate of the i-layer. Table 2 gives a brief overview of the achievements on single junction μc-Si:H solar cells.

5. MULTIBAND GAP CELLS (MICROMORPH OR HYBRID CELLS)

Microcrystalline intrinsic silicon (μc-Si:H) is very attractive for application in thin film silicon solar cells because, in combination with a-Si:H, it makes an ideal pair for use in multibandgap tandem solar cells. This combination not only is better matched to the solar spectrum than either a-Si:H or μc-Si:H by itself (see Fig. 10), but it is also the best possible combination. Theoretical calculations [115] show that the combination of band gap of a-Si:H (1.7–1.8 eV) and μc-Si:H (1–1.1 eV) is close to the optimal combination for

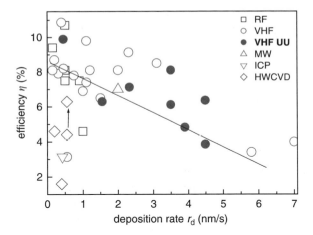

Fig. 9. Cell performance versus deposition rate for μc-Si:H based single junction solar cells. Solar cells from the present studies are compared to results published by other groups (based on Ref. [112] and recent reports).

Table 2
Overview of the achievements on single junction μc-Si:H solar cells (small area)

Laboratory	Deposition rate (nm/s)	Intitial efficiency (%)	Stabilized efficiency (%)	Reference
Utrecht university	0.45	9.9	10.0	[92]
IPV Jülich	1.1	10.3	9.8	[111]
AIST, Japan	2.2	8.8	–	[113]
MHI, Japan	2.3	8.3	–	[114]

maximum efficiency. Microcrystalline silicon doped layers are suitable as doped window layers because they have low optical absorption compared to a-Si:H at photon energies higher than $h\nu = 2\,\text{eV}$. In this energy range the use of amorphous doped layers in window layers can lead to a large loss of photons. The electrical properties of μc-Si:H doped layers are superior to those of a-Si:H, because microcrystalline material can be doped more efficiently. The activation energies of the dark conductivity are smaller in a-Si:H, because the Fermi level is closer to the valence or the conduction band for p- and n-type-doped layers, respectively. Therefore, the built-in voltage in solar cells is higher when μc-Si:H doped layers are used.

The concept of "micromorph" tandem cells, stacked cells with a *microcrystalline* bottom cell and an *amorphous* top cell, was first proposed in 1996

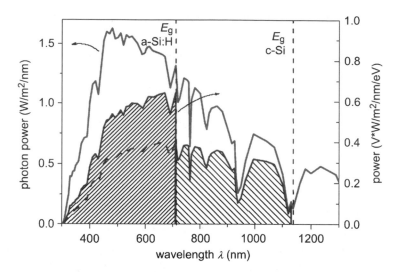

Fig. 10. AM1.5 irradiated solar power spectrum according to Ref. [116] (upper solid line) and the power that can be generated using a a-Si:H/μc-Si:H tandem solar cell (lower solid line), an a-Si:H single junction (lower solid line on the left-hand side) and a μc-Si:H single junction (dotted line on the left-hand side; solid line in the center). Assumptions: mobility gap E_g (μc-Si:H) = E_g (c-Si), V_{oc} = 0.85 V for the a-Si:H subcell and V_{oc} = 0.53 V for the μc-Si:H subcell, infinitely sharp band edges, and total absorption for photon energies above the band gap.

by the University of Neuchâtel [117]. In 2002, Meier et al. [118] published a micromorph tandem cell with a stabilized efficiency of 10.8% (12.3% initial) of which the bottom cell (with a thickness of d~2 μm) was deposited at a rate of r_d = 0.5 nm/s. The highest initial efficiency so far, for a thin film silicon tandem cell, is 14.7%, reported by Yamamoto et al. of Kaneka, Japan [119]. For triple junction cells, comprising of at least one microcrystalline silicon cell and one amorphous silicon cell, the highest efficiencies report to date are 15.1% [120] and 15%, Yamamoto et al. [121], respectively.

    To improve the stability behavior, micromorph or "hybrid" cells comprise a reflecting intermediate layer (IL) at the tunnel recombination junction between the amorphous and the microcrystalline cell. The IL allows one to keep the amorphous cell very thin (<0.3 μm, in tandem cells), while enough high current can be generated. Without an IL, the amorphous top cell would have to be made so thick that excessive Staebler–Wronski effect cannot be avoided. The optical effect of an IL is stronger in n–i–p type tandem cells than in p–i–n type tandem cells [122], and the required thickness (or rather the n.d product) is dependent on the roughness of the back reflector or the front TCO, respectively. Thicknesses have been varied between 30 and 110 nm. A low n value would be beneficial because of lower reflection loss [121].

Microcrystalline silicon with a sufficiently high crystalline volume fraction (>45% as determined by Raman spectroscopy) does not show light-induced degradation of optoelectronic properties [123, 124], whereas a-Si:H shows an increasing defect density upon light soaking (Staebler–Wronski effect [3]). Therefore, the microcrystalline bottom cell is often chosen to be current limiting, so that the tandem cell as a whole behaves stable. For outdoor application, however, in certain locations a top-cell limited design instead offers higher annual and daily yield than a bottom cell limited tandem cell [125]. This is because under practical outdoor circumstances the operation temperature and light intensity are correlated. This correlation is advantageous for the daily and annual energy yield of the top cell but adverse for the bottom cell.

In addition to the efficiency and stability, the manufacturing costs of a solar cell determine its feasibility as an industrial product. Therefore, the trend in research is toward higher deposition rates, while minimizing the loss in material quality and cell performance. Microcrystalline silicon single junction and μc-Si:H/a-Si:H tandem solar are now studied by numerous groups worldwide [31, 104, 105, 126, 127].

Further development and optimization of a-Si:H/μc-Si:H tandems will remain very important because it is expected that in the near future (first half of this century), its market share can be considerable. For example, in the European Roadmap for PV R&D [1], it is predicted that in 2020, the European market share for thin film silicon (most probably a-Si:H/μc-Si:H tandems) will be 30%. This shows the importance of thin film multibandgap cells as second-generation solar cells.

6. MODULES AND PRODUCTION

Within the silicon-based PV technologies, thin film silicon has important advantages over c-Si wafer technology. Among the thin film PV technologies it is the most mature, because commercial modules are available from multiple manufacturers. Silicon thin film solar cells can be made using only nontoxic components. Moreover, the constituent elements are abundant in the earth crust. The deposition processes can easily be scaled up to larger than a 1 m^2 area in batch processes, while in continuous roll-to-roll processes the area can even be in excess of 1000 m^2 (and is cut afterward in submodules of the desired size). The processing temperatures are generally low: for Si deposition, the temperature is less than 300°C and for the textured TCO layers (depending on the type), the temperatures are also in this range or at most 550°C. The energy payback times are short (depending on the location of the system, typically between one and three years). Thin film silicon PV is the only thin film technology so far that allows effective stacking of cells to form truly spectrum-splitting tandems or triple-junction cells. Monolithic series connection avoids soldering and tabbing (as in c-Si modules) and thus provides modules with the desired output voltage that are manufacturable completely in-line, while area losses due to these connections are as low as 5%. The efficiency of these thin film drift type devices shows a low T coefficient, while the fill factor and V_{oc} are only very weakly affected by the illumination intensity. Owing to this, the energy yield of thin film Si PV modules, in particular those consisting of multijunction cells [128], is ~10% higher than c-Si PV modules with the same power rating under standard test conditions (STC conditions: 100 mW/cm^2, AM1.5 spectrum and 25°C) [129].

There are several ongoing methods to obtain a high deposition rate on a large area. The principle challenges in applying VHF PECVD on an industrial scale are (i) the uniformity of the deposited layers on a large substrate, where the effect of standing electrical waves at high frequencies (>60 MHz) may limit its application; (ii) effective coupling of the power to the plasma. The remarkable features of the HWCVD process, such as the high deposition rate, the high gas utilization, the low-pressure process (avoiding dust formation and therefore avoiding the need for frequent chamber cleaning), the large-area deposition capability (achieved by the use of multiple filaments and/or filament grids [130]), and the decoupling of the gas dissociation from the deposition on substrate, are all very attractive for industrial production [79]. The issues regarding scaling-up of the HWCVD technique concern the uniform deposition over the large area, the filament lifetime and filament ageing. Recently, the Japanese company Anelva [130] introduced a HWCVD deposition system for large-area deposition, and Ulvac Inc. developed a large-area deposition apparatus with an effective deposition area of 150 × 85 cm^2 [131].

A few examples of successful demonstrations of solar cells on a large area and production activities are discussed here. The first three examples deal with deposition on glass, the latter three deal with roll-to-roll deposition on foil.

Oerlikon (previously Unaxis Solar) has adapted their KAI PECVD commercial deposition system, originally designed for deposition of the active matrix for liquid crystal displays, to allow deposition of a-Si:H and μc-Si:H silicon at a plasma frequency of 40 MHz [132]. In the KAI-M reactor (52×41 cm^2 substrate size), single junction 1 cm^2 area a-Si:H solar cells were fabricated on Asahi U-type substrates with an initial efficiency of 10.4% and a stabilized efficiency of 8.4%. A 10×10 cm^2 module had an initial efficiency of 9.6%.

Mitsubishi heavy industries (MHI) have developed a large-area VHF-PECVD production apparatus for depositing on 1.1×1.4 m^2 substrate implementing a ladder-shaped electrode and using a phase modulation method [133]. The base frequency in the process is 60 MHz and the modulation frequency is varied to minimize the thickness uniformity. The average deposition rate is 11 Å/s and a thickness uniformity within $\pm 18\%$ was obtained. Amorphous silicon p–i–n modules with stable aperture efficiency of 8% have been made. Since October 2002, a-Si:H modules have been commercially produced. The production level was 12 MW in 2005, and plans for tandem cell production are underway.

Kaneka Corporation is producing modules on glass of 4 mm thickness with an area of 91×45.5 cm^2 and 91×91 cm^2. In 1993, the goal for mass production was a stable efficiency of 8%, based on a single junction p–i–n structure made by large area CVD and precision laser patterning. Later, thin microcrystalline Si technology was introduced. In 1995, a 10% double junction was obtained on a large area. Light-soaking tests of a 91×45.5 cm^2 module showed a degradation of less than 8% (light-soaking conditions 0.75 suns, 38°C) [134].

The present goal is to obtain stable 12% on 91×91 cm^2. Production was started in 1999, and presently the average stable efficiency is 7.5% (>50.000 modules). The record module (91×45.5 cm^2) has an efficiency of 12.3% (47.1 W). On 91×91 cm^2 it is 11.6%.

United Solar Ovonic's first roll-to-roll facility was built in 1986. In 1991, it was modified to obtain a capacity of 2 MW/year. At present, United Solar has a roll-to-roll production capacity of 30 MW/year. In the production machine, six rolls of stainless steel are mounted vertically, and the foils run once through a 240 ft long machine. All layers of the triple-junction solar-cell structure are deposited automatically. The length of the stainless-steel foil is 1.5 mile (2.3 km) and its thickness is 5 mil (125 μm). This machine is being duplicated in Auburn Hills, Michigan, where manufacturing has started in 2006. Expansion into Europe and China is also part of the strategy

[135]. The stainless-steel roll is first washed in a roll-to-roll washing machine, then the back Al/ZnO reflector is deposited, then the triple-junction Si-based alloy solar cell, and finally the top contact, which also has antireflection properties is deposited. After a short passivation step, the electrodes are bonded, cells are cut and interconnected, followed by lamination, framing and finishing. The active area efficiency is 10.4% over the long run [135]. Typical modules have a power of ~40 W indoors, outdoors it is ~38 W. The light-induced degradation is 12% in 800 h, with a generally observed stable efficiency of 8%.

Fuji Electric Co. Ltd, after first developing thin film silicon tandem modules on $30 \times 40\,cm^2$ glass substrates, adopted a "stepping roll" deposition method on rolls of polyimide foil.

The products use the concept of the so-called series connection through apertures formed on film (SCAF) series connection. There are two kinds of holes: series connecting holes, and current collection holes.

As a part of the process sequence, basic steps are to punch series connecting holes, followed by a metallization step (sputtering), then current collection holes are punched, and then the solar cell structure including the transparent top contact is deposited and laser scribed. One roll is 1.5 km long and in one run ~30 kW can be produced. In R&D the stabilized efficiency of $40 \times 80\,cm^2$ a-Si/a-SiGe modules is 9% (initial aperture area efficiency 10.1%).

The completed modules are less than 1 mm thick and weigh approximately only $1\,kg/m^2$. Examples of first applications launched in the market, since October 2004, are power sources for mobile use and on the rooftop of water reservoir ponds at water purification plants. Another type of module is a 4 m long steel plate integrated module, which can be used for installation as metal rooftops of factories or public facilities. The low weight of such roof plates reduces the requirements with respect to the construction of the building. The annual capacity is presently 3 MW and will be gradually increased: a 10 MW plant has been announced. The typical tandem modules are based on a-Si/a-SiGe cells, $40 \times 80\,cm^2$, and have a power output of 24 W (DC 80 V) [136].

Helianthos b.v. (a Nuon Company) has introduced a manufacturing concept in which a-Si:H solar cells are produced in a roll-to-roll manner on a temporary metal substrate (temporary superstrate process) [137, 138]. The cell structure is first deposited on a SnO_2:F-coated metal foil. At one point in the process, the solar cell is transferred to a permanent plastic substrate on which cells are monolithically series connected to produce lightweight PV modules. In the pilot line modules with efficiency above 7% are obtained on 1 ft wide foil.

In order to increase efficiency, a-Si:H/μc-Si devices are incorporated in the temporary superstrate process. This allows for cost reduction in both the module production (high-throughput roll-to-roll). As a result from research, a-Si:H/μc-Si:H modules on $10 \times 10\,cm^2$ foils with an aperture area efficiency of 8.6% (initial 9%) have been presented. Single junction encapsulated modules from the roll-to-roll pilot line have an efficiency of 7% (stable 5.5–6%) [139].

ACKNOWLEDGMENTS

I would not only like to thank all colleagues and group members at Utrecht University, but also all colleagues worldwide, for their constant dedication to advance this field and providing the right atmosphere to do so. In particular, I am grateful to my nearest coworker Jatin Rath for his insights and contributions, and Miro Zeman at Delft University of Technology for stimulating discussions. Much of this overview has been made possible through studies performed by Ph.D. students, of which I would like to specifically mention Aad Gordijn and Marieke van Veen. The research at the Utrecht group is supported by the Foundation for Fundamental Research on Matter and by SenterNovem of The Netherlands.

REFERENCES

[1] Photovoltaic Technology Research Advisory Council (PV-TRAC), A Vision for Photovoltaic Technology for 2030 and Beyond, Photovoltaic Technology Research Advisory Council (PV-TRAC), European Communities, 2004.

[2] D.E. Carlson and C.R. Wronski, Appl. Phys. Lett., 28 (1976) 671.

[3] D.L. Staebler and C.R. Wronski, Appl. Phys. Lett., 31 (1977) 292.

[4] S. Guha, K.L. Narasimhan and S.M. Pietruszko, J. Appl. Phys., 52 (1981) 859.

[5] R. J. Koval, J. Koh, Z. Lu, L. Jiao, R. W. Collins and C. R. Wronski, Appl. Phys. Lett., 75 (1999) 1553.

[6] P. Roca i Cabarrocas, J. Non-Cryst. Sol., 31 (2000) 266–269.

[7] E. Vallat-Sauvain, U. Kroll, J. Meier, A. Shah and J. Pohl, J. Appl. Phys., 87 (6) (2000) 3137.

[8] A.R. Middya, S. Hamma, S. Hazra, S. Ray and C. Longeaud, Mater. Res. Soc. Proc., 664 (2001) A9.5.

[9] H. Yu, R. Cui, F. Meng, C. Zhao, H. Wang, H. Yang, H. Chen, P. Luo, D. Tang, S. Lin, Y. He, T. Sun and Q. Ye, 15th Int. PVSEC-15, p. 1048, Shanghai, China AD-9, 2005.

[10] R. E. I. Schropp, B. Stannowski, A. M. Brockhoff, P. A. T. T. van Veenendaal and J. K. Rath, Mater. Phys. Mech., 1 (2000) 73.

[11] R.E.I. Schropp and J.K. Rath, IEEE Trans. Electron Devices, 46 (1999) 2069.

[12] D.V. Tsu, B.S. Chao, S.R. Ovshinsky, S. Guha and J. Yang, Appl. Phys. Lett., 71 (1997) 1317.

[13] J.H. Koh, Y. Lee, H. Fujiwara, C.R. Wronski and R.W. Collins, Appl. Phys. Lett., 73 (1998) 1526.

[14] R.W. Collins and A.S. Ferlauto, Curr. Opin. Solid State and Mater. Sci., 6/5 (2002) 425.

[15] A.H. Mahan, D.L. Williamson and T.E. Furtak, Mater. Res. Soc. Symp. Proc., 467 (1997) 657.

[16] D.L. Williamson, Mat. Res. Soc. Symp. Proc., 557 (1999) 251.

[17] S. Guha, J. Yang, D.L. Williamson, Y. Lubianiker, J.D. Cohen, A.H. Mahan, Appl. Phys. Lett., 74 (1999) 1860.

[18] R.E.I. Schropp, M.K. van Veen, C.H.M. van der Werf, D.L. Williamson and A.H. Mahan, Mat. Res. Soc. Symp. Proc., 808 (2004) A8.4.1.

[19] S. Veprek and V. Marecek, Solid-State Electron., 11 (1968) 683.

[20] A. Matsuda, S. Yamasaki and H. Yamamoto, Jpn. J. Appl. Phys., 19 (1980) L305.

[21] T. Hamasaki, H. Kurata and Y. Osaka, Appl. Phys. Lett., 37 (1980) 1084.

[22] R.W. Collins, A.S. Ferlauto and C.R. Wronski, Sol. Energy Mater. Sol. Cells, 78 (2003) 143.

[23] G. M. Ferreira, A.S. Ferlauto, Chi Chen, R.J. Koval, J.M. Pearce, C. Ross, C.R. Wronski and R. W. Collins, J. Non-Cryst. Solids, 338–340 (2004) 13–18.

[24] J. Meier, R. Flückiger, H. Keppner and A. Shah, Appl. Phys. Lett., 65 (7) (1994) 860.

[25] A. Gordijn, J.K. Rath and R.E.I. Schropp, Mat. Res. Soc. Symp. Proc., 762 (2003) 637.

[26] J. Kočka, A. Fejfar and I. Pelant, Sol. Energ. Mat. Sol. C., 78 (2003) 493.

[27] T. Matsui, R. Muhida and Y. Hamakawa, Appl. Phys. Lett., 81 (25) (2002) 4751.

[28] A.L. Baia Neto, A. Lambertz and F. Finger, J. Non-Cryst. Solids, 274 (2000) 299–302.

[29] J.K. Rath, H. Meiling and R.E.I. Schropp, Jpn. J. Appl. Phys., 36 (1997) 5436.

[30] P.A.T.T. van Veenendaal, T.J. Savenije, J.K. Rath and R.E.I. Schropp, Thin Solid Films, 175 (2002) 403–404.

[31] A. Matsuda, J. Non-Cryst. Solids, 1 (2004) 338–340.

[32] T. Kamei and T. Wada, J. Appl. Phys., 96 (4) (2004) 2087.

[33] M. Kondo and A. Matsuda Curr. Opin. Solid State Mater., 6 (2002) 445.

[34] J. Kočka, H. Stuchlikova and A. Fejfar, J. Non-Cryst. Solids, 299–302, (2002) 355–359.

[35] M. Konagai, T.Tsushima and R. Asomoza, Thin Solid Films, 395 (2001) 152.

[36] K. Prasad, U. Kroll and M. Schubert, Mat. Res. Soc. Symp. Proc., 219, (1991) 469.

[37] R. Flückiger, J. Meier and A. Shah, Proc. 23rd IEEE PVSC, Louisville, KY 1993, p. 839, 1993.

[38] H. Keppner, U. Kroll and A. Shah, Proc. 25th IEEE Photovoltaics Specialist Conf. Washington, DC. p. 669, 1996.

[39] K.F. Feenstra, J.K. Rath and R.E.I. Schropp, 2nd World PVSEC p. 956, Vienna, 1998.

[40] W. Beyer, J. Non-Cryst. Solids, 40 (1996) 198–200.

[41] A. Matsuda, J. Non-Cryst. Solids, 767 (1983) 59–60.

[42] S. Veprek, F.-A. Sarott and M. Rückschloss, J. Non-Cryst. Solids, 733 (1991) 137–138.

[43] A. Asano, Appl. Phys. Lett., 56 (1990), 533.

[44] H. Shirai, J. Hanna and I. Shimizu, Jpn. J. Appl. Phys., 30 (1991) L881.

[45] H.F. Sterling, R.C.G. Swann, Solid-State Electron., 8 (1965) 653.

[46] A. Matsuda, K. Nomoto, Y. Takeuchi, A. Suzuki, A. Yuuki and J. Perrin, Surf. Sci., 227 (1990) 50.

[47] Y. Toyoshima, K. Arai, A. Matsuda and K. Tanaka, J. Non-Cryst. Solids, 137–138 (1991) 765–770.

[48] G. Bruno, P. Capezzuto and A. Madan, Plasma Deposition of Amorphous Silicon-Based Materials, Academic Press, San Diego, CA, USA 1995.

[49] R. A. Street, Hydrogenated Amorphous Silicon, Cambridge University Press, Cambridge, 1991.

[50] E.A.G. Hamers, J. Bezemer and W. F. van der Weg, Appl. Phys. Lett., 75 (5) (1999) 609–611.

[51] A. Matsuda, Thin Solid Films, 337 (1996) 1.

[52] R. E. I. Schropp and M. Zeman, Amorphous and Microcrystalline Silicon Solar Cells: Modeling, Materials and Device Technology, Kluwer, ISBN 0-7923-8317-6, Boston/Dordrecht/London, 1998.

[53] W. Shindoh and T. Ohmi, J. Appl. Phys., 79 (1996), 2347.

[54] M. Kondo, M. Fukawa, L. Guo and A. Matsuda, J. Non-Cryst. Solids, 84 (2000) 266–269.

[55] B. Rech, T. Roschek, J. Müller, S. Wieder and H. Wagner, Sol. Energy Mater. Solar Cells, 66 (2001) 267–273.

[56] H. Curtins, N. Wyrsch and A.V. Shah, Electron. Lett., 23 (5) (1987) 228–230.

[57] A. Shah, J. Dutta, N. Wyrsch, K. Prasad, H. Curtins, F. Finger, A. Howling and Ch. Hollenstein, Mater. Res. Soc. Symp. Proc., 258 (1992) 15–26.

[58] J.L. Dorier, C. Hollenstein, A.A. Howling and U. Kroll, J. Vac. Sci. Technol. A., 10 (4) (1992) 1048–1052.

[59] W.M.M. Kessels, R.J. Severens, A.H.M. Smets, B.A. Korevaar, G.J. Adriaenssens, D.C. Schram and M.C.M. van de Sanden, J. Appl. Phys., 89 (2001) 2404.

[60] A.H.M. Smets, W.M.M. Kessels and M.C.M. van de Sanden, Mater. Res. Soc. Symp. Proc., 808 (2004) 383.

[61] H. Wiesmann, A.K. Gosh, T. McMahon and M. Strongin, J. Appl. Phys., 50 (1979) 3752.

[62] H. Matsumura, J. Appl. Phys., 65 (1989) 4396 .

[63] A.H. Mahan, J. Carapella, B.P. Nelson, R.S. Crandall and I. Balberg, J. Appl. Phys., 69 (1991) 6728 .

[64] K.F. Feenstra, R.E.I. Schropp and W.F. van der Weg, J. Appl. Phys., 85 (1999) 6843.

[65] B. Schröder, U. Weber, H. Seitz, A. Ledermann and C. Mukherjee, Thin Solid Films, 395 (2001) 298.

[66] H. Matsumura, Jpn. J. Appl. Phys., 25 (12) (1986) L949.

[67] B.P. Nelson, E. Iwaniczko, A.H. Mahan, Q. Wang, Y. Xu, R.S. Crandall and H.M. Branz, Thin Solid Films, 395 (2001) 292.

[68] M. Vanecek, A.H. Mahan, B.P. Nelson and R.S. Crandall, in Proceedings of the 11th European Photovoltaic Solar Energy Conference (L. Guimaraes, W. Palz, C. De Reyff, H. Kiess and P. Helm, eds.), p. 96, Montreux, Switzerland, October 12–16, 1992.

[69] M. Vanecek, Z. Remes, J. Fric, R.S. Crandall and A.H. Mahan, in Proceedings of the 12th European Photovoltaic Solar Energy Conference (R. Hill, W. Palz and P. Helm, eds.), p. 354, Amsterdam, The Netherlands, April 11–15, 1994.

[70] A.H. Mahan, E. Iwaniczko, B.P. Nelson, R.C. Reedy Jr., R.S. Crandall, S. Guha and J. Yang, Proceedings of the 25th IEEE Photovoltaic Specialists Conference, p. 1065, Washington, D.C., USA, May 13–17, 1996.

[71] B. Schröder, Thin Solid Films, 430 (2003) 1–6.

[72] V. Verlaan, C.H.M. van der Werf, W.M. Arnoldbik, H.C. Rieffe, I.G. Romijn, W.J. Soppe, A.W. Weeber, H.D. Goldbach and R.E.I. Schropp, 20th European Photovoltaic Solar Energy Conference, p. 1434, Barcelona, Spain, June 6–10, 2005.

[73] H. Meiling and R.E.I. Schropp, Appl. Phys. Lett., 70 (20) (1997) 2681–2683.

[74] P. Alpuim, V. Chu and J.P. Conde, J. Non-Cryst. Sol., 110 (2000) 266–269.

[75] H. Matsumura, Thin Solid Films, 395 (2001) 1.

[76] M.K. van Veen and R.E.I. Schropp, Appl. Phys. Lett., 82 (2003) 287.

[77] K. Saito, M. Sano, K. Ogawa and I. Kajita, J. Non-Cryst. Solids, 164–166 (1993) 689–692.

[78] A.H.M. Smets, W.M.M. Kessels and M.C.M. van de Sanden, Appl. Phys. Lett., 82 (6) (2003) 865–867.

[79] R.E.I. Schropp, Thin Solid Films, 451–452 (2004) 455.

[80] T. Akasaka and I. Shimizu, Appl. Phys. Lett., 66 (1995) 3441.

[81] S. Veprek, F.A. Sarrott and Z. Iqbal, Phys. Rev., B 36 (1987) 3344.

[82] I. Solomon, B. Drevillon and N. Layadi, J. Non-Cryst. Solids, 164–166 (1991) 989.

[83] C.C. Tsai, G.B. Anderson and R. Thompson, Mat. Res. Soc. Symp. Proc., 192 (1990) 475.

[84] S. Kumar, B. Drevillon, C. Godet, J. Appl. Phys., 60 (1986) 1542.

[85] M. Scheib, B. Schröder and H. Oechsner, J. Non-Cryst. Solids, 895 (1996) 198–200.

[86] R. Nozawa, K. Murata, M. Ito, M. Hori and T. Goto, J. Vac. Sci. Technol., A 17 (1999) 2542.

[87] R. Knox, V. Dalal, B. Moradi and G. Chumanov, J. Vac. Sci. Technol., A 11 (1993) 1896.

[88] J.P. Conde, Y. Schotten, S. Arekat, P. Brogueira, R. Sousa, V. Chu, Jpn. J. Appl. Phys. 36, 38 (1997).

[89] K. Endo, M. Isomura, M. Taguchi, H. Tarui and S. Kiyama, Sol. Energ. Mat. Sol. C., 66 (2001) 283.

[90] L. Guo, M. Kondo, M. Fukawa, K. Saitoh and A. Matsuda, Jpn. J. Appl. Phys., 37 (1998) L1116.

[91] M. Fukawa, S. Suzuki, L. Guo, M. Kondo and A. Matsuda. Sol. Energy Mater. Sol. Cells, 66 (2001) 217.

[92] A. Gordijn, J.K. Rath and R.E.I. Schropp, Progress in Photovoltaics: Research and Appl. 13 (2005) 1–7.

[93] J. Kočka, J. Stuchlik and A. Fejfar, Appl. Phys. Lett., 79 (16) (2001) 2540.

[94]   C. Droz, E. Vallat-Sauvain and A. Shah, Sol. Energy Mater. Sol. Cells, 8 (2004) 61.

[95]   B. Yan, J. Yang, G. Yue, K. Lord and S. Guha, Proc. of 3rd World Conference on Photovoltaic Energy Conversion, p. 1627, Osaka, May 11–18, 2003.

[96]   J.H. Werner, K. Taretto and U. Rau, Solid State Phenomena, 80–81 (2001) 299.

[97]   A.V. Shah, J. Meier and J. Bailat, J. Non-Cryst. Sol., 338–340 (2004) 639.

[98]   J. Meier, J. Spitznagel, U. Kroll, C. Bucher, S. Faÿ, T. Moriarty and A. Shah, Proc. of the 3rd World Conference on Photovoltaic Energy Conversion, paper S20-B9-06, Osaka, 2003.

[99]   A.H. Mahan, R.C. Reedy, E. Iwaniczko, Q. Wang, B.P. Nelson, Y. Xu, A.C. Gallagher, H.M. Branz, R.S. Crandall, J. Yang and S. Guha, Mater. Res. Soc. Symp. Proc., 507 (1998) 119.

[100] A. Gordijn, Microcrystalline Silicon for Solar Cells, Ph.D. thesis, Utrecht University, Utrecht, The Netherlands 2005.

[101] R.H. Franken, C.H.M. van der Werf, J. Loeffler, J.K. Rath and R.E.I. Schropp, Thin Solid Films, 501 (2006) 47–50.

[102] A.M.H.N. Petit, V. Nadazdy, A.H.M. Smets, M. Zeman, M.C.M. van de Sanden and R.A.C.M.M. van Swaaij, 20th European Photovoltaic Solar Energy Conference, 1616, Barcelona, Spain, June 6–10, 2005.

[103] M. Faraji, S. Gokhale and S.V. Ghaisas, Appl. Phys. Lett., 60 (26), (1992) 3289.

[104] B. Rech, T. Roschek, T. Repmann, J. Müller, R. Schmitz and W. Appenzeller, Thin Solid Films, 427 (1–2) (2003)157–165.

[105] A.V. Shah, J. Meier and U. Graf, Sol. Energy Mater. Sol. Cells, 78 (2003) 469.

[106] A. Matsuda, J. Non-Cryst. Solids, 1 (2004) 338–340.

[107] M. Fukawa, S. Suzuki and A. Matsuda, Sol. Energy Mater. Sol. Cells, 66 (2001) 217.

[108] Y. Mai, S. Klein and F. Finger, Appl. Phys. Lett., 85 (14) (2004) 2839.

[109] B. Rech, T. Repmann, S. Wieder, M. Ruske, and U. Stephan, Thin Solid Films 502 (2006) 300–305.

[110] O. Kluth, B. Rech, L. Houben, S. Wieder, G. Schoepe, C. Beneking and H. Wagner, Thin Solid Films, 351 (1999) 247.

[111] S. Klein, Y. Mai, F. Finger, M.N. Donker and R. Carius, 15th International Photovoltaic Science and Engineering Conference (PVSEC-15), 49 (2), pp. 736–737, Shanghai, China, 2005 .

[112] J.K. Rath, Sol. Energ. Mat. Sol. C., 76 (2003) 431.

[113] T. Matsui, A. Matsuda and M. Kondo, Proc. of 19th European PVSEC, 1407, Paris, 2004.

[114] S. Goya, Y. Nakano, T. Watanabe, N. Yamashita and Y. Yonekura, Proc. of 19th European Photovoltaic Solar Energy Conference, 1407, Paris, 2004.

[115] K. Zweibel, Chem. Eng. News, 64 (1986) 34.

[116] R. Hulstorm, R. Bird and C. Riordan, Solar Cells, 15 (1985) 365.

[117] J. Meier et al., Mater. Res. Soc. Proc. Ser., 420 (1996) 3.

[118] J. Meier, S. Dubail and A. Shah, Sol. Energy Mater. Sol. Cells, 74 (2002) 457.

[119] K. Yamamoto, Proc. of 3rd World Conference on Photovoltaic Energy Conversion, S2O-B9-03, Osaka, 2003.

[120] B. Yan, G. Yue, J.M. Owens, J. Yang, S. Guha, 4th World Conference on Photovoltaic Energy Conversion, May 2006, Waikoloa, Hawaii.

[121] K. Yamamoto et al., 15th International Photovoltaic Science & Engineering Conference (PVSEC-15), pp. 529–530, Shanghai, China, 2005.

[122] D. Dominé, J. Steinhauser, L. Feitknecht, A. Shah and C. Ballif, 20th European Photovoltaic Solar Energy Conference, pp. 1600–1604, Barcelona, Spain. June 6–10, 2005.

[123] H. Keppner, J. Meier and A. Shah, Appl. Phys., A 69 (1999) 169.

[124] M. Fonrodona, D. Soler, J. Escarre, J.M. Asensi, J. Bertomeu and J. Andreu, J. Non-Cryst. Solids, 659 (2004) 338–340.

[125] A. Nakajima, M. Ichikawa, T. Sawada, M. Yoshimi and K. Yamamoto, Jpn. J. Appl. Phys., 43, 9A/B (2004) L1162–L1165.

[126] K. Yamamoto, A. Nakajima and Y. Tawada, Sol. Energy, 77 (6) (2004) 939.

[127] O. Vetterl, F. Finger and H. Wagner, Sol. Energy Mater. Sol. Cells, 62 (2000) 97.

[128] S. Guha, Mat. Res. Soc. Symp. Proc., 808 (2004) A6.4.1.

[129] R. Gottschalg, C.N. Jardine, R. Ruether, T.R. Betts, G.J. Conibeer, J. Close, D.G. Infield, M.J. Kearney, K.H. Lam, K. Lane, H.P. Pang and R. Tscharner, Proc. of the 29th IEEE PV Specialists Conference, New Orleans, LA, USA. pp. 1699–1702, 2002.

[130] K. Ishibashi, M. Karasawa, G. Xu, N. Yokokawa, M. Ikemoto, A. Masuda, H. Matsumura, Thin Solid Films, 430 (2003) 58–62.

[131] S. Osono, M. Kitazoe, H. Tsuboi, S. Asari and K. Saito, Thin Solid Films, 501 (2006) 61.

[132] J. Meier, U. Kroll, J. Spitznagel, S. Benagli, T. Roschek, G. Pfanner, Ch. Ellert, G. Androutsopoulos, A. Hügli, G. Büchel, D. Plesa, A. Büchel and A. Shah, Proc. of the 19th EU PVSEC, pp. 1328–1333, Paris, June 2004.

[133] H. Takatsuka, M. Noda, Y. Yonekura, Y. Takeuchi, Y. Yamauchi, Sol. Energy, 77 (2004) 951–960.

[134] Y. Tawada, 3rd World Conference on Photovoltaic Energy Conversion, 5PLD103, Osaka, May 11–18, 2003.

[135] S. Guha, Proc. of the 31st IEEE Photovoltaic Specialists Conference, pp. 12–16, Lake Buena Vista, FL, USA (2005).

[136] M. Shimosawa, K. Tabuchi, M. Uno, S. Kato, M. Tanda, Y. Takeda, S. Iwasaki, Y. Yokoyama, T. Wada, Y. Sakakibara, H. Nishihara, H.E. Nomoto, A. Takano and T. Kamoshita, 20th European Photovoltaic Solar Energy Conference, paper 3CO.7.1, p. 1533 Barcelona, Spain, June 6–10, 2005.

[137] R.E.I. Schropp, H. Meiling, C.H.M. van der Werf, E. Middelman, E. van Andel, P.M.G.M. Peters, L.V. de Jonge-Meschaninova, J. Winkeler, R.J. Severens, G.J. Jongerden, M. Zeman, M.C.M. van de Sanden, A. Kuipers and C.I.M.A. SPEE, 2nd World Conference and Exhibition on Photovoltaic Solar Energy Conversion, pp. 820–822, Vienna, Austria, July 6–10, 1998.

[138] E. van Andel, E. Middelman and R.E.I. Schropp, Method of Manufacturing a Photovoltaic Foil, Priority date 26/09/1996, Filing date 24/09/1997, Publication number, EP1066642, PCT publication number WO9949483.

[139] R. Schlatmann, B. Stannowski, E.A.G. Hamers, J.M. Lenssen, A.G. Talma, G.C. Dubbeldam and G.J. Jongerden, 15th International PV Science & Engineering Conference (PVSEC-15), 744, Shanghai, China, 2005.

Nanostructured Materials for Solar Energy Conversion
T. Soga (editor)

Chapter 6

Thin-Film Solar Cells Based on Nanostructured CdS, CIS, CdTe and Cu$_2$S

Vijay P. Singh, R.S. Singh and Karen E. Sampson

Department of Electrical and Computer Engineering & Center for Nanoscale Science and Engineering, University of Kentucky, Lexington, KY 40506, USA

Photovoltaic devices based on thin-film cadmium sulfide (CdS), copper indium diselenide (CIS) and cadmium telluride (CdTe) are leading contenders for large-scale production of solar cells at low cost. Recent developments in nanotechnology provide new exciting opportunities for further improvements in cell performance and cost reduction in manufacturing processes in these devices. These opportunities and some initial results are described below.

In Section 1, we present the historical background and basic operation of CdS/CdTe and CdS/CIS devices. This is followed, in Section 2, by the advantages offered by nanotechnology. Some promising device designs and experimental approaches are described in the sections that follow.

1. BACKGROUND

1.1. CdS/CIS Solar Cell

Thin-film CdS–CuInSe$_2$ Solar Cells have already been developed to a point where an efficiency of greater than 17% has been achieved [1–8]. However, the open-circuit voltage remains low and is ultimately limited by the relatively low band gap of 1.05 eV. Recent developments in nanotechnology can be used to achieve higher band gap and higher open-circuit voltage through quantum confinement.

CuInSe$_2$ (CIS) drew attention as a solar cell absorber material after the work of Wagner and coworkers [9] who demonstrated a 12% conversion efficiency using single crystals. Relatively high absorption coefficient of CIS offers a distinct advantage.

CdS/CuInSe$_2$ thin-film devices have been fabricated by several methods. Physical vapor deposition (PVD) and selenization of the deposited metals have yielded high efficiency devices [10–13]. Other device fabrication methods such as sputtering [14, 15], spray deposition [16], direct evaporation [17] and screen printing [18] have also been used. Recent studies have demonstrated that the band gap of semiconductors can be tuned by varying its particle size [19]. An illustrative energy band diagram [20] of the CIS/CdS heterojunction is sketched in Fig. 1 below.

Photons with energy less than 2.41 eV but greater than 1.05 eV pass through the CdS window layer, acting as the window material and are absorbed by the CIS layer. Absorption of light in the depletion region and within a diffusion length of the junction in CIS will create carriers which will be collected. Photons with energy greater than 2.41 eV will be absorbed by the CdS layer, and the carriers created in the depletion region and within a diffusion length of the junction in CdS will also be collected. The separation of the light-generated carriers across the CdS/CIS junction gives rise to the light-generated current I_L.

1.2. CdS/CdTe Solar Cell

CdTe has a near ideal band gap value of 1.5 eV [21], making it an ideal absorbing material in solar cells. Also, CdTe has a short absorption length relative to the grain size of a few micrometers typical of good devices. This

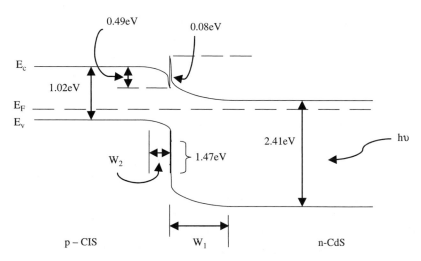

Fig. 1. Energy band diagram of CIS/CdS heterojunction.

results in a reduced recombination at grain boundaries, a problem encountered in other polycrystalline materials. Larger efficiencies are obtained when photo-generated carriers are collected in the depletion region. n-CdS has proved to be the most suitable heterojunction partner to p-CdTe because of, (i) the high quality of the CdTe layer that can be deposited on top of CdS and, (ii) beneficial inter-diffusion at the CdS-CdTe junction. CdS has a band gap of 2.41 eV and can be made thin enough for it to be an appropriate window material. However, there is still a need to make the CdS even thinner to avoid absorption at short wavelengths. Efficiencies greater than 16% have been achieved in thin-film p-CdTe/n-CdS solar cells on glass substrates coated with a transparent conducting oxide (TCO) [22]. However, efficiencies are still limited by issues at the CdS/CdTe junction interface [23].

Device designs on lightweight, flexible substrates have also been investigated [24, 25]. Due to reduced weight, these cells have space as well as terrestrial applications. Inexpensive methods of fabrication like spray pyrolysis, closed-space sublimation and dip coating have been demonstrated [26, 27]. Molybdenum is a suitable substrate because its coefficient of thermal expansion is close to that of CdTe, thus allowing for fewer problems in high-temperature processing. A low-resistance tunneling contact to CdTe was achieved. In addition, an increase in open-circuit voltage was achieved with CdS/CdTe devices on a molybdenum substrate by employing annealing techniques that resulted in junction improvement due to the formation of CdS_xTe_{1-x} interlayer between CdS and CdTe. This interlayer is thought to reduce localized interface states at the junction and thus reduce the reverse saturation, leading to an increase in open-circuit voltage. An open-circuit voltage of 820 mV was achieved with such an interlayer [28]. However, efficiency in cells on flexible substrates remains low and further improvements are needed. Issues that remain to be addressed for these devices are: a better understanding of the CdTe/CdS junction [28], a low resistance top contact to CdS and a conducting contact to p-type CdTe. Fig. 2 shows the ideal energy band diagram for a Mo/p-CdTe junction. Since the electron affinity of CdTe is 4.3 eV and its band gap is 1.5 eV, the work function of p-CdTe, (depending upon its acceptor concentration) lies between 5.05 and 5.8 eV. The work function of Mo, however, is less than 5 eV. Thus the Mo/p-CdTe junction is a Schottky diode as shown in Fig. 2.

The current transport in metal-semiconductor contacts is mainly due to majority carriers (holes in a p-type material). The current flow across the metal-semiconductor contacts can take five paths [29–31]. These are: (i) thermionic emission current from the semiconductor over the potential barrier

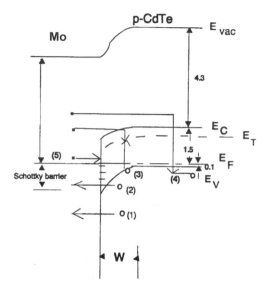

Fig. 2. Energy band diagram of Mo/p-CdTe junction.

into the metal, the dominant process for Schottky diodes with moderately doped semiconductor operated at a moderate temperature like 300 K, (ii) quantum-mechanical tunneling current through the barrier, the dominant process for contact with a heavily doped semiconductor, (iii) recombination current in the space-charge region, (iv) injection current from the metal to the semiconductor and (v) recombination current via interface states.

The depletion layer width W is given by

$$W = \sqrt{\frac{2\varepsilon_s(V_{bi} - V_1)}{qN_A}} \tag{1}$$

where N_A is the acceptor concentration at the CdTe surface, q the charge of an electron, ε_s the dielectric constant of CdTe, V_{bi} the junction potential of the contact in equilibrium and V_1 the forward bias. The doping concentration N_A can be measured from the capacitance–voltage characteristics.

$$N_A = \frac{2}{q\varepsilon_s}\left[-\frac{1}{d(1/C^2)/dV}\right] \tag{2}$$

where C is the capacitance.

Modeling the diode D, as a Schottky diode and assuming that path (1) thermionic emission current is the dominant current mechanism, its current–voltage characteristics can be expressed as [30, 31],

$$J_{te} = J_{ote}[e^{qV/nkT} - 1]$$ (3)

where

$$J_{ote} = BT^2 e^{-\phi_b/kT}$$ (4)

$$\phi_b = (\chi + E_g)_{CdTe} - W_{m'}$$ (5)

The barrier height ϕ_b depends on the band gap (E_g), electron affinity (χ) and the metal work function (W_m); n is the diode ideality factor; T the temperature; k the Boltzmann's constant; q the electron charge and B the effective Richardson constant. If the path (ii), quantum-mechanical tunneling current through the barrier becomes the dominant transport process for a heavily doped CdTe, then we can model MS junction as a good conductivity contact. In this case, the doping concentration N_A must be very large so that the depletion into the semiconductor (W) is small enough to allow carriers to tunnel across the barrier. The tunneling current can be increased by (i) reducing the Schottky barrier height and thus increasing the probability of majority carrier tunneling and (ii) increasing the number of majority carriers at the semiconductor surface.

1.3. Cu2S/CdS Solar Cells

Cu$_2$S/CdS solar cells were investigated by many research groups in the past but the interest dwindled because this cell structure suffered from degradation with time. However, it is still valuable as a vehicle for understanding nano-electronic effects in thin-film devices. Cu$_2$S/CdS thin-film solar cells are fabricated by a variety of techniques including a combination of spray pyrolysis and chemiplating [32, 33], a combination of evaporation and chemiplating [34, 35], all evaporation [36] and sputtering [37]. Chemiplated Cu$_2$S/evaporated Zn$_x$Cd$_{1-x}$S (with anti-reflective coating) thin-film solar cells on Zn/Cu substrates have exhibited the highest efficiencies [38]. These cells had a V_{oc} of 0.599 V, a J_{sc} of 18.5 mA cm^{-2}, a fill factor (FF) of 0.748 and an efficiency of 10.2%. Cu$_2$S/CdS (with anti-reflective coating) thin-film solar cells, fabricated by chemiplating an evaporated CdS layer, had a

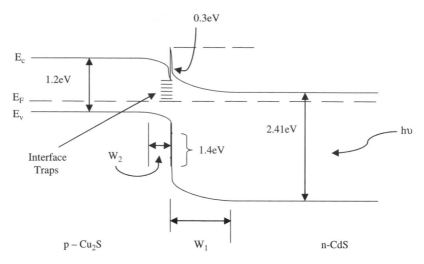

Fig. 3. Energy band diagram of Cu$_2$S/CdS heterojunction.

conversion efficiency of 9.15% with $V_{oc} = 0.516$, $J_{sc} = 21.8\,\text{mA cm}^{-2}$ and FF $= 0.714$ [35, 38]. Cu$_2$S/CdS solar cells have also been fabricated by an all evaporation process [20, 39]. CdS and CuCl are successively evaporated onto indium tin oxide (ITO)-coated glass substrate and the CuCl is converted into Cu$_2$S by annealing the device in vacuum. A thin layer of copper is evaporated on top of Cu$_2$S to correct its stoichiometry. A variety of contacts including copper, silver and a combination of chromium/lead have been evaporated to serve as top contacts for Cu$_2$S. Cu$_2$S/CdS cells are generally prone to degradation over time as the copper ions start migrating. Exposing the Cu$_x$S surface to a strong oxidizer such as oxygen can cause a substantial degradation in device performance [40]. The Cu atoms were found to move rapidly to both the Cu$_x$S surface and the CdS–Cu$_x$S interface on exposure to air at room temperature [41]. The degradation of the Cu$_2$S/CdS cells, attributed to the formation of surface oxides, can be significantly prevented by replacing the ambient air with argon [42]. The energy band diagram of the Cu$_2$S/CdS heterojunction is shown below in Fig. 3.

Under illumination, most of the light is absorbed in the p-type Cu$_2$S layer. As prepared Cu$_2$S/CdS cells consist of uncompensated CdS and nearly stoichiometric Cu$_2$S. The space charge would thus be narrow and ionization of deep levels near the junction on illumination causes further narrowing. Tunneling to interface states occurs, lowering the effective barrier height and hence the V_{oc}. Heat treatment of the cell allows oxygen and/or copper to reach the space-charge region, forming compensating acceptor states in the CdS. This process widens the space-charge region which restricts tunneling. With the increased barrier, an improvement in V_{oc} could be expected. A major factor affecting the photocurrent in Cu_2S solar cells is the high surface recombination velocity which results in most of the photo-generated carriers close to the surface to be captured. Surface recombination can be reduced by proper doping of the region closer to the surface, to produce a drift field to counteract the minority carrier diffusion to the surface [43].

2. ADVANTAGES OF NANOSTRUCTURES

Nanoscience and nanotechnology offer exciting, new approaches to addressing the challenges in the manufacture of thin-film solar cells. Large-scale production requires not only device designs for high efficiency and stability, but also an ability to produce large area, uniform and self-ordered films. For this, one needs a process for economically fabricating large periodic arrays of semiconductor nanostructures that will allow (i) the size and composition to be varied, (ii) encapsulation in a rugged host material, (iii) flexibility to use a variety of substrate materials and preferably, (iv) compatibility with standard silicon-based fabrication techniques. These issues are addressed by the nanostructured solar cell designs described below.

The key aspects to improving the efficiency of a solar cell lie in (1) improving light harvesting and (2) improving the charge transport of free carriers. In this context, *nanostructured layers in thin-film solar cells offer three important advantages*. First, due to scattering, the effective optical path for absorption is much larger than the actual film thickness. Second, light-generated electrons and holes need to travel over a much shorter path and thus recombination losses are greatly reduced. As a result the absorber layer thickness in nanostructured solar cells can be as thin as 150 nm instead of several micrometers in the traditional thin-film solar cells [44, 45]. Third, the energy band gap of various layers can be tailored to the desired design value by varying the size of the nanoparticles (quantum confinement). This allows the device engineer to better design the absorber and window layers in the solar cell.

In order to achieve the three enhancements listed above, a clear understanding of the physics of nanoscale heterojunctions is needed. At present, such an understanding is lacking. Nanoporous alumina templates offer a platform for the development of this basic understanding.

2.1. Advantages of Nano Template

A technique that can create nano-dots of uniform size and well-controlled inter-dot separation relies on some form of template, a pattern imposed on a surface that acts as a guide for the subsequently deposited material. In a particularly interesting example of such a technique, a nanoporous alumina film is used as a "negative" template for fabricating nano-dots and nanowires. This technique is ideally suited for creating metal or semiconductor nano-dots within an insulating alumina matrix. One can deposit semiconductors and metals inside nanoporous alumina and study their electro-optical characteristics and electron transport with the objective of developing an analytical model for nanoscale photovoltaic heterojunctions. The unique advantage of having an insulating template is that it allows us to (i) perform annealing treatments without losing nanocrystallinity and, (ii) study a single nanoscale heterojunction inside the pores. If the thickness of the deposited material is also in the order of a few nanometers, a "quantum dot", a structure with nanometer-range lengths in all three spatial dimensions is obtained. In short, this is an economical method for manufacturing large quantities of nanostructures. In summary, the principal features of this technique are: (i) uniform regular distribution of microscopic pores with sub-micrometer to nanometer diameter, (ii) arrangement of vertically directed pores with high aspect (depth/diameter) ratio at almost identical distance from each other, (iii) ability to control diameter of cells and pores by changing electrolyte composition and electrochemical processing regimes and (iv) high reproducibility of the film structure for samples of large sizes.

3. NANOSTRUCTURED CdS–CdTe AND CdS–CIS DEVICE DESIGNS

3.1. CdS–CuInSe$_2$ Devices

Thin-film CdS–CuInSe$_2$ solar cells have already been developed to a point where an efficiency of greater than 17% has been achieved [1]. However, CIS thickness is several micrometers and the band gap of CIS (1.1 eV) is not ideally matched with the solar spectrum. One can use quantum confinement to enhance the band gap and the open-circuit voltage of these cells. Furthermore, optical path can be enhanced through the use of nanoporous structures [44, 45]. Such enhancement will result in, increased short circuit current density and an ability to reduce the planar thickness (and hence the amount of the chemicals needed) in these cells.

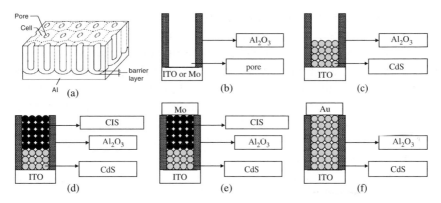

Fig. 4. Schematic of the nanoscale heterojunction fabrication. (a) Anodized alumina top view; (b) anodized alumina side view; (c) electro-deposit semiconductor A into the pores; (d) electro-deposit semiconductor B into the pores; (e) electro-deposit the top contact; (f) nanoscale Schottky diode.

To prepare glass/ITO/nanostructured porous Al_2O_3 with CdS/CIS cells in the nanopores, first, template structures with nanopores of a selected diameter (in the 8–100 nm range) are fabricated by anodization of aluminum coated on ITO-coated glass. Next, electro-deposition (ED) [46–48] is used to coat the nanopores first with CdS and later with CIS. In the deposition of CIS films for polycrystalline thin-film solar cells, difficulty in forming near-stoichiometric CIS films is often experienced for non-vacuum techniques (in- and out-diffusion is a common problem during thermal processing). This should be solvable either by initially depositing an In-rich compound, or by subsequent deposition of In onto the top surface of the CIS. Finally, the top contact can be electro-deposited to complete the device structure. Based on quantum confinement and LCAO calculations [49] one can design for an effective band gap of the order of 1.5 eV which is the optimal for the solar spectrum.

Steps in the fabrication process (see Fig. 4) are: (i) deposit a thin-film of Al on a conducting substrate, (ii) anodize the Al and form porous alumina (Al_2O_3 or AAO) on the above conductive substrate, (iii) electro-deposit semiconductor A (n-CdS) into the pores, (iv) electro-deposit semiconductor B (p-CIS) into the pores to make a junction, (v) deposit a top contact (molybdenum, for example) by electro-deposition, (vi) alternately, after Step (iii), a metal can be electro-deposited to make a Schottky contact. The resulting ITO/CdS/CIS, ITO/CdS/CdTe device structures with a vertical heterojunction are illustrated in Figs. 5(a) and (b).

We expect an increase in the effective energy band gap of CdS and CIS due to quantum confinement in nanoscale pores. Increase in the band gap of CdS will allow high-energy photons to pass through, making it a better window material, while an increase in the CIS band gap will make it optimal for the solar spectrum, and should lead to higher V_{oc} and higher efficiency. Since only half of the alumina template contains CdS/CIS, a reduction in current may be foreseen as a problem; however, this loss will be partially compensated by scattering of incident sunlight as long as the pores are deep enough so that a sufficient amount of CIS is available.

3.2. Nanostructured CdS–CdTe Solar Cells

The device structure is illustrated in Fig. 5(b). Here, we want the CdS window layer to be nanocrystalline. Quantum confinement of CdS leads to an effective band gap that is higher than the 2.4 eV, thus making nanocrystalline CdS a better window material than the bulk CdS [19]. The biggest challenge in traditional bulk CdS/CdTe devices has been the contact between the p–CdTe layer and the top electrode like graphite. An interlayer of nanocrystalline CdTe between bulk CdTe and top electrode will be worth investigating as a means of improving this contact.

Also, a homojunction device design consisting of nanocrystalline–n-CdTe/bulk-p-CdTe will be worth investigating. In this case, one could conceive of the nanocrystalline–n-CdTe layer as primarily a window layer and the bulk-p-CdTe layer as primarily an absorber layer. n-type CdTe can be produced by indium doping [50], temperature control of deviation from stoichiometry and annealing in a Cd atmosphere [51–53]. Issues involved in doping n-type CdTe nanoparticles are introducing a Cd atmosphere while annealing CdTe or controlling CdTe stoichiometry by forming Cd rich CdTe under appropriate temperatures or doping with indium.

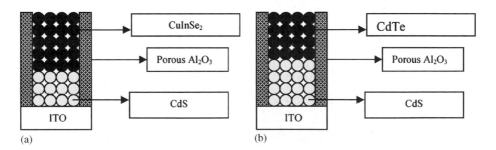

Fig. 5. (a) Schematic of an ITO/CdS/CIS heterojunction; (b) schematic of an ITO/CdS/CdTe heterojunction.

4. FILM AND DEVICE CHARACTERISTICS

4.1. Nanoporous Alumina

Films of porous alumina on ITO-coated glass substrates as well as free-standing porous alumina membranes were fabricated.

4.1.1. Films of Nanoporous Alumina

Porous alumina films were fabricated by anodizing aluminum film deposited on ITO glass. One micron of Aluminum was deposited on ITO-coated glass slides. Samples were annealed at 550°C for 90 min in flowing nitrogen. After cooling, samples were removed and anodized at 10°C in 0.3 M oxalic acid. The sample was anodized for two minutes, then the oxide layer was partially removed in a heated phosphoric/chromic acid mixture. The sample was re-anodized for approximately 20 min. The typical electron micrographs are shown below in Figs. 6(a) and (b). Annealing Al films at 550°C in flowing nitrogen before anodization resulted in a smoother surface and straighter pores.

4.1.2. Free-Standing Membranes of Nanoporous Alumina

Aluminum foil was cut in 1/2″×1″ pieces, de-greased in acetone for 2 h and anodized at 40 V in 0.3 M oxalic acid; anodization time ranged between 10 min (when thin AAO film was desired) and 25 min (when thicker AAO film was desired). Next, these samples were submerged in 45°C phosphoric acid for 3.5 min for pore widening. This resulted in a layer of AAO pores (2–25 μm thick) above an aluminum substrate. After protecting the front

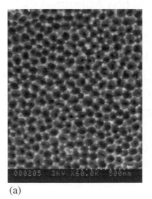

(a)   (b)

Fig. 6. (a) Top view of AAO on ITO; (b) cross-sectional view of AAO on ITO.

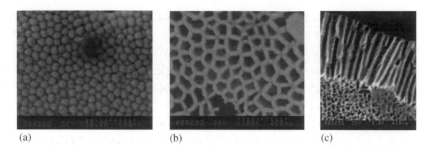

(a)   (b)   (c)

Fig. 7. (a) Bottom of pores before barrier layer removal; (b) bottom of pores after barrier layer removal in phosphoric acid for 27 sec; (c) cross-sectional view of the alumina foil.

with nail polish, the black oxide (Fig. 7(a)) was removed by submerging it in 45°C phosphoric acid for 30 min. The aluminum layer was removed by immersing in a mixture of hydrochloric acid and cupric chloride [29], and the barrier layer was removed by keeping the sample in concentrated phosphoric acid for 20–30 s as shown in Fig. 7(b).

Next, samples were mounted on a clean ITO-coated glass slide and the nail polish layer was dissolved in acetone, leaving a thin layer of AAO pores as shown Fig. 7(c). For the anodization times ranging from 10 to 25 min, a linear relation between thickness and time was observed.

4.2. Nanostructured CdS

Nanostructured CdS films were fabricated inside porous alumina as well as on ITO-coated glass substrates.

4.2.1. Nanostructured CdS Films in Porous Alumina

CdS nanocrystals were prepared by electro-deposition (ED) in dimethyl sulfoxide (DMSO) solution composed of 0.055 M $CdCl_2$ and 0.19 M elemental sulfur. The temperature was maintained at 120°C. A 10 V dc voltage was applied between the ITO/AAO foil working electrode and the platinum counter electrode. Initially the electro-deposition was carried out for 10 min. The scanning electron micrograph shows that CdS is deposited inside the pores of the template as shown in Fig. 8(a). The walls of the pores and top surface are insulating (aluminum oxide), hence the applied electric field directs the CdS particles toward ITO through the pores. The templates filled with CdS were annealed at 450°C for 30 min prior to their examination under a scanning electron microscope (SEM). The typical particle size for electro-deposited CdS was about 15 nm. The X-ray diffractograms of the

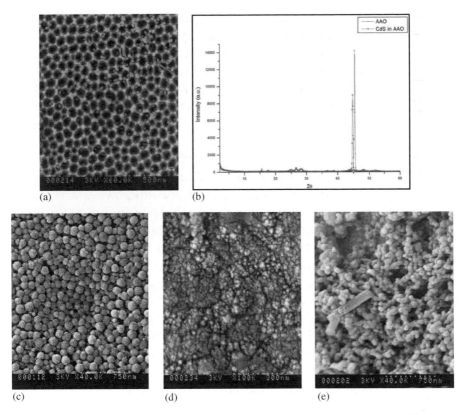

Fig. 8. (a) Electro-deposited CdS in AAO on ITO; (b) X-ray diffractogram of CdS in AAO; (c) CdS on ITO by CBD; (d) CdS on ITO by electro-deposition; (e) electro-deposited CdS in AAO on ITO (long duration).

porous alumina template and CdS-filled alumina template are shown in Fig. 8(b). Comparing the X-ray diffractogram with the standard JCPDS data indicated that the CdS deposited inside the template is hexagonal (JCPDS 41-1049).

In case of chemical bath-deposited CdS films, even after annealing at 450°C, the film did not show any coalescence of the particles. This confirms that the growth of CdS is through the pores and the pore walls prevent the adjacent particles from coalescing during annealing.

As test structures for comparison, we also deposited CdS films on ITO/glass substrate by chemical bath deposition (CBD) and electro-deposition (ED). Micrographs of these films are shown in Figs. 8(c) and (d). It is interesting to compare the morphology of CdS of Fig. 8(d) with that of Fig. 8(e) where the CdS was electrodeposited in alumina pores for long duration (30 min) and the pores got completely filled over.

4.2.2. Nanostructured CdS Films on ITO-Glass Substrates

CdS films were fabricated by three methods, namely (a) solution growth (b) sonochemical method and (c) microwave-assisted synthesis [54]. The films obtained were characterized by XRD, SEM, TEM, UV–VIS spectroscopy. The X-ray diffractograms indicate peaks at 2θ positions of 27°, 44° and 52° corresponding to the (111), (220) and (311) planes of the cubic CdS phase, respectively. Electron microscopic observations indicated a particle size of 12 nm for CdS films made by sonochemical and microwave-assisted synthesis. Optical absorption measurements revealed a blue shift in the absorption spectrum. An optical absorption excitation peak was seen at 334 nm, and a hyperbolic band model calculation gave an excitonic energy of 2.98 eV; this is 0.58 eV higher than that of the bulk CdS (2.4 eV). The increased effective band gap allows high-energy (blue) photons to go through making nanocrystalline CdS a more effective window material in photovoltaic applications like the CdS/CdTe solar cells.

Porous CdS Films   Fabrication of porous CdS involves irradiating the CdS solution with ultrasound waves during the process of dip coating [55]. ITO-coated glass and plastic (commercial transparency) were used as substrates. XRD studies performed on these powders show a phase corresponding to cubic CdS. The FE–SEM images of the films on plastic showed 80 nm uniform pores for all the three methods. The optical absorption results indicated a blue shift and also multiple peaks in the absorption curve. These films are useful for fabricating nanoporous heterojunction solar cells.

4.3. Nanostructured CIS

Several research groups have investigated nanostructured CIS particles and films and their applications to solar cells [56]. Nanu, Schoonman and Goossens [57] have obtained an energy conversion efficiency of 5% in a nanocomposite 3D device based on an interpenetrating network of titania and CIS. CIS was deposited by spraying inside a nanocrystalline film of titania. Graphite paste and tin oxide were used as conducting contacts to CIS and titania, respectively. In our studies, attempts to electro-deposit CIS films on top of CdS in the nanoporous alumina template have not yielded high-quality CIS films so far. It appears that the CIS films are not stoichiometric and a selenization step will be needed. These investigations are continuing.

4.4. Nanostructured CdTe

Nanocrystalline CdTe films are of interest because of their potential applications as n-type window layers in a p–n homojunction thin-film CdTe solar cell, contact to CdS/CdTe solar cells and in electroluminescent display devices. CdTe nanocrystals were prepared by microwave-assisted synthesis [58] and films were cast from colloidal solutions containing nano-CdTe. These films were characterized by optical absorption, photoluminescence spectroscopy, profilometer and SEM on glass and plastic substrates for various conditions. The typical particle size of CdTe nanocrystals was about 10 nm. The UV–VIS spectra recorded on this film shows a blue shift in absorption due to smaller particle size. Band gap estimate using Tauc's law indicates a value of about 2.8 eV as compared to a bulk CdTe band gap of 1.5 eV. This increased band gap makes CdTe a good candidate for electroluminescent display devices and for window layers in n-CdTe/p-CdTe homojunction solar cells.

5. HETEROJUNCTIONS AND SCHOTTKY DIODES IN POROUS ALUMINA

5.1. Schottky Diode in an AAO Membrane

AAO membranes with pores of 30 nm width and 4 μm height were prepared by the method described in Section 3. An ITO film was sputtered as a transparent back contact and a top contact (Cu or Au) was thermally evaporated over an area of $0.07 \, \text{cm}^2$. Current–voltage curves are as shown in Figs. 9(a) and (b) for the case of copper and gold, respectively.

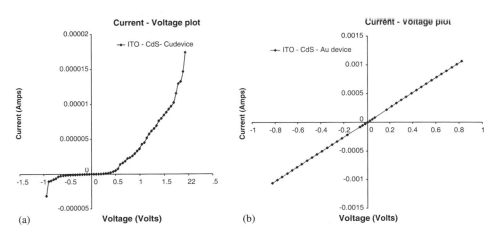

Fig. 9. (a) *I–V* of nanoscale Cu/CdS junction; (b) *I–V* of nanoscale Au/CdS junction.

The device with copper electrode shows a diode like behavior while the device with gold electrode behaves like a conducting contact. This difference can be understood in terms of a model where the CdS film in the AAO pore is highly conductive. This would lead to a very thin depletion layer at the Au/CdS interface resulting in excessive tunneling across the junction and hence a conductive contact. Copper on the other hand will be able to effectively dope the CdS surface to a reduced n-type conductivity leading to a wider depletion layer at the Cu/CdS junction and no tunneling; hence a diode-like behavior is observed.

5.2. Heterojunctions in AAO

5.2.1. CdS/Cu$_2$S Junctions in Nanoporous Alumina Film on ITO–Glass

Photovoltaic CdS/Cu$_2$S junctions were fabricated inside the nanopores of alumina and their characteristics were studied. As a starting absorber material for proof of concept, copper sulfide was used instead of CIS. Even though copper sulfide is not as stable as CIS, it is easier to fabricate [59] and serves as a reliable vehicle for testing the quantum confinement effects in thin-film heterojunction solar cells. One micron of aluminum was deposited on ITO-coated glass slides. Samples were annealed at 550°C for 90 min in flowing nitrogen. After cooling, samples were removed and anodized at 10°C in 0.3 M oxalic acid. The sample was anodized for 2 min, then the oxide layer was partially removed in a heated phosphoric/chromic acid mixture. The sample was re-anodized for approximately 20 min, using the process described by Chu et al. [60].

CdS was deposited by both electro-deposition and thermal evaporation on different samples. For electro-deposition, a solution composed of 0.055 M CdCl$_2$ and 0.19 M elemental sulfur was used. The temperature was maintained at 120°C. A 36 V dc voltage was applied between the working electrode and the platinum counter electrode. Electro-deposition was performed for different durations in order to calculate the deposition rate. SEM pictures showed CdS deposition at the bottom of the pores. The samples were then annealed at 350°C for 45 min in flowing nitrogen. Next, 1.1 μm of CuCl was evaporated, then annealed in vacuum at 195°C for 18 min. This anneal resulted in a reaction between CdS and CuCl to form copper sulfide and cadmium chloride. Samples were then rinsed in de-ionized water to remove the cadmium chloride and leave behind a nanoscale CdS/copper sulfide heterojunction in each alumina pore. Scanning electron micrographs showing the cross section of the device are shown below in Figs. 10(a) and (b); ITO layer is on top while the alumina pores filled with CdS and copper sulfide

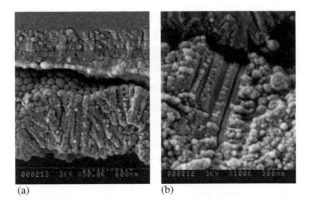

(a)   (b)

Fig. 10. (a) Cross-sectional view of electro-deposited CdS in AAO; (b) fracture section of electro-deposited CdS in AAO.

are at the bottom. The nature of the gap between these two is not fully understood yet.

Gold electrode was next deposited to complete the photovoltaic cell. An open-circuit voltage of 242 mV was measured under 1-sun illumination. The current, however, was too small to measure. We attribute this to an extremely high-series resistance probably due to a bad contact between ITO and CdS caused either by the gap mentioned above or an incomplete removal of the aluminum oxide barrier layer at the bottom of the alumina pore; this barrier layer is present in most anodized AAO films.

Several other nanoscale heterojunctions were prepared by the method described above except that CdS layer was prepared by vacuum evaporation instead of electro-deposition. These exhibited an open-circuit voltage of 300 mV. However, the currents were still very low.

5.2.2. CdS/Cu₂S Junctions in Nanoporous AAO Membranes

Nanostructured CdS/Cu_2S junctions were made inside a nanoporous alumina membrane. An all evaporation process was employed in the fabrication of these devices. Well-ordered porous alumina membranes with a pore width of 30 nm and a pore depth of 4 μm were used for this purpose. The membranes were mounted on a clean glass slide and CdS film of thickness 3.5 μm was evaporated onto the nano template. Next, samples were annealed at 450°C in a nitrogen atmosphere for 90 min. Partially filled porous alumina templates were turned upside down and mounted on a cleaned glass substrate. The glass substrate was placed inside the vacuum chamber and a 280 nm thick layer of ITO was sputter deposited to contact CdS. Next, the device was turned upside down again and bleached CuCl of thickness 3 μms was deposited onto CdS by thermal evaporation. The samples were then annealed in vacuum for 18 min at 195°C to form Cu_2S through a reaction between CuCl and CdS. Next, solar cells were rinsed in de-ionized water and a 10 nm thick layer of copper was thermally evaporated on top of Cu_2S followed by chromium and lead electrodes.

For comparison, bulk heterojunctions of Cu_2S/CdS were fabricated on ITO-coated glass substrates. CdS of thickness 4 μm and CuCl of thickness 0.5 μm were evaporated to form the junction. The vacuum anneal yielded a Cu_2S thickness of 300 nm and a CdS thickness of 3.7 μm. Chromium was evaporated as the top ohmic contact for Cu_2S and a 1-μm thick layer of lead was evaporated on top of chromium to improve its conductivity. This test cell yielded a V_{oc} of 445 mV, a J_{sc} of 20.3 mA cm^{-2}, FF of 0.635 and an efficiency of 5.73%. Cells formed inside the porous alumina fared worse, primarily due to extremely high-series resistance. The dark *I–V* characteristics for Cu_2S/CdS devices fabricated inside nanoporous alumina template is shown in Fig. 11. A series resistance of 1323 Ω was observed. Such a huge value for the series resistance indicates that the device suffers from a very poor ohmic contact. The light curve for the device is shown in Fig. 12. The nano heterojunction solar cell yielded a V_{oc} of 367 mV and a J_{sc} of 17.8 mA cm^{-2}. An in-depth current–voltage analysis revealed that the value of ideality factor was 4.7 and effective reverse saturation current was 0.225 μA cm^{-2}.

The bulk and the nano heterojunctions have comparable open-circuit voltages, but they differ greatly in the magnitudes of the light-generated currents. The bulk devices have light-generated currents of the order of tens of milliamperes while the nanoscale heterojunctions have currents of the order of tens of microamperes. The major reason for this difference is the poor ohmic contact to the CdS layer. The high-aspect ratio of the alumina template results in only partial filling of the pores by evaporated CdS. This is confirmed by cross-sectional SEM. When ITO is sputtered from the other side of the foil, it forms a poor contact. Such a poor contact between CdS and ITO is the cause of the large series resistance.

In theory, the band gap of Cu_2S is expected to increase from 1.2 to 1.5 eV, which is the ideal band gap for an absorber material to take full advantage of the solar spectrum. Since the band gap of CdTe is also 1.5 eV, the nano Cu_2S/CdS heterojunctions are expected to match the highest open-circuit voltage of 824 mV observed in the CdTe/CdS solar cells on Mo foil [25]. But such high open-circuit voltages were not observed in the nano

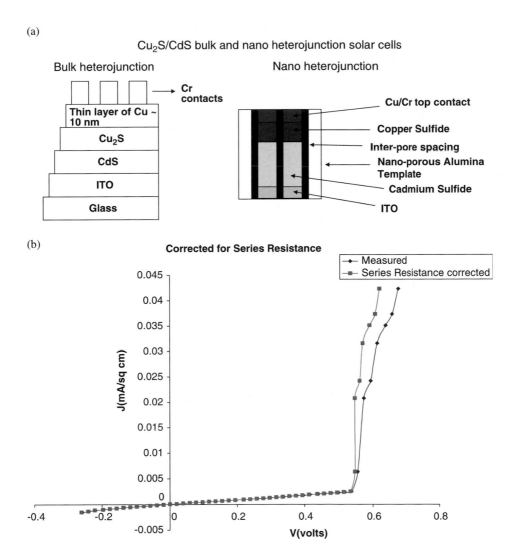

Fig. 11. (a) Schematic of the CdS/Cu$_2$S nano heterojunction in AAO; (b) *I–V* characteristics in the dark for a CdS/Cu$_2$S nano heterojunction in AAO with Cr/Pb top contact.

Cu$_2$S/CdS heterojunctions. The performance of these cells was greatly limited by series resistances arising out of poor contacts to the window material. These cells can be improved by reducing the thickness of the porous alumina foil. The reduction in the thickness of the foil reduces the aspect ratio. This would make it easier for the CdS to get through the pore depth completely. With CdS having gone all the way through the pore, better

Cause of Series Resistance in Cu$_2$S/CdS
hetero-junction inside nanoporous alumina template

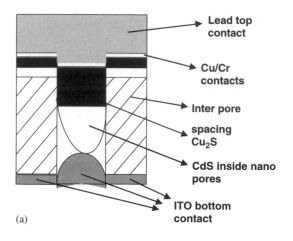

· CdS does not go all the way
  through and stops midway
· Cu$_2$S growth is supported inside
  CdS
· ITO from the bottom contacts CdS
  inside the pore
· 10 nm layer copper improves
  photovoltaic behavior of Cu$_2$S
· Cr/Pb act as top contact to Cu$_2$S

(a)

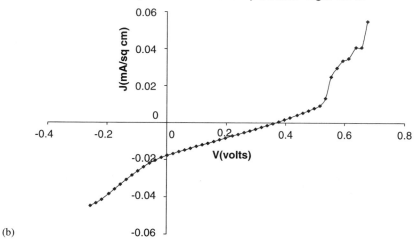

Nano heterojunction with Cr/Pb top contact - Light Curve

(b)

Fig. 12. (a) Schematic of the cause for high-series resistance; (b) *I–V* characteristics under illumination for a CdS/Cu$_2$S nano heterojunction in AAO with Cr/Pb top contact.

ohmic contacts can be made to it. These improved ohmic contacts would reduce the series resistance associated with the cell considerably and hence improve the J_{sc}, V_{oc} and FF for the cell. More experiments will be needed to optimize the thickness of the porous alumina films and to improve the performance of these cells.

6. CONCLUSION

Porous alumina templates were fabricated from a host of starting substrates including aluminum foil, tape, Al on ITO, Al on molybdenum and Al on glass. Methods for selectively electro-depositing and vacuum evaporating semiconductor arrays in an insulating matrix were developed and executed.

Nanoscale photovoltaic hetcrojunctions in porous alumina on ITO and in porous alumina membranes were produced and characterized. However, the currents in dark and in light were very small due to poor contact between CdS and ITO inside the pore. Further work is needed to solve this problem and optimize device parameters for high-efficiency solar cells.

Nanoscale Schottky diode junctions in porous alumina membranes were produced and characterized. However, the currents were very small due to poor contact between CdS and ITO inside the pore. Further work is needed to solve this problem.

It is clear that nanotechnology offers a wealth of opportunities for improving the performance and reducing the cost of large-scale production in thin-film solar cells. However, several practical challenges like the quality of the CIS film at nanoscale still remain. Also, a clear theoretical understanding of the heterojunction characteristics at nanoscale is indispensable and needs to be developed further.

ACKNOWLEDGMENTS

The authors thank the Kentucky Science and Engineering foundation for support through Grants KSEF-148-502-02-27 and KSEF-148-502-03-68. The authors also thank Srikalyan Sanagapalli, Vivekanand Jayaraman, Visweshwaran Jayaraman and Saikanth Mahendra for their contributions to the nanostructured semiconductors project.

REFERENCES

[1] Y. Hashimoto, N. Kohara, T. Negami, N. Nishitani and T. Wada, Sol. Energy Mater. Sol. Cells, 50 (1998) 71–77.
[2] F. Kessler and D. Rudmann, Sol. Energy, 77 (2004) 685–695.
[3] J. AbuShama, R. Noufi, S. Johnston, S. Ward and X. Wu, Conference Record of the Thirty-First IEEE Photovoltaic Specialist Conference (IEEE Cat. No. 05CH37608) (2005) 299–302.

[4]   R. Wieting, R. Gay, H. Nguyen, J. Palm, C. Rischmiller, A. Seapan, D. Tamant and D. Willett, Conference Record of the Thirty-first IEEE Photovoltaic Specialist Conference (IEEE Cat. No. 05CH37608) (2005) 177–182.

[5]   A. Gupta and A.D. Compaan, Conference Record of the Thirty-first IEEE Photovoltaic Specialist Conference (IEEE Cat. No. 05CH37608) (2005) 235–238.

[6]   B.E. McCandless and R.W. Birkmire, Conference Record of the Thirty-first IEEE Photovoltaic Specialist Conference (IEEE Cat. No. 05CH37608) (2005) 398–401.

[7]   V.P. Singh and J. McClure, Sol. Energy Mater. Sol. Cells, 76 (2002) 369–385.

[8]   V.P. Singh, D.L. Linam, D.D. Dils, J.C. McClure and G.B. Lush, Sol. Energy Mater. and Sol. Cells, 63 (2000) 445–466.

[9]   S. Wagner, J.L. Shay, P. Migliorato and H.M. Kasper, Appl. Phys. Lett., 25 (1975) 434.

[10] B.M. Basol, V.K. Kapur and A. Halani, Proc. 22nd IEEE PVSC (1991) 893.

[11] B.M. Basol and V.K. Kapur, Proc. 21st IEEE PVSC (1990) 546.

[12] B.M. Basol, V.K. Kapur and R.J. Matson, Proc. 22nd IEEE PVSC (1991) 1179.

[13] B.M. Basol, V.K. Kapur, A. Halani, A. Minnick and C. Leidholm, Proc. 23rd IEEE PVSC (1993) 426.

[14] N. Romeo, A. Bosio, V. Canevari and L. Zanotti, 7th E.C PV Sol. Energy Conf. (1986) 656–661.

[15] L.C. Yang, G. Berry, L.J. Chou, G. Kenshole, A. Rockett, C.A. Mullan and C.J. Kiely, 23rd IEEE PVSC (1993) 505–509.

[16] M.E. Beck and M. Cocivera, Thin Solid Films, 272 (1996) 71–82.

[17] M.S. Sadigov, M. Ozkan, E. Bacaksiz, M. Altunbas and A.I. Kopya, J. Mater. Sci., 34 (1999) 4579–4584.

[18] F.J. Garcia and M.S. Tomar, Proc. 14th Conf. Solid State Devices, Jpn. J. Appl. Phys., (1982) 535–538.

[19] A.P. Alivisatos, J. Phys. Chem., 100 (1996) 13226–13239.

[20] V. Jayaraman, M.S. Thesis University of Kentucky, May 2005.

[21] S.P. Albright, V.P. Singh and J.F. Jordan, Sol. Cells, 24 (1988) 43–56.

[22] T. Aramoto, S. Kumazawa, H. Higuchi, T. Arita, S. Shibutani, T. Nishio, J. Nakajima, M. Tsuji, A. Hanafusa, T. Hibino, K. Omura, H. Ohyama and M. Murozono, Jpn. J. Appl. Phys., Part 1 (Regular Papers, Short Notes & Review Papers), 36 (1997) 6304–6305.

[23] H. Chavez, R. Santiesteban, J.C. McClure and V.P. Singh, J. Mater. Sci.: Mater. Electron., 6 (1995) 21–24.

[24] J.C. McClure, V.P. Singh, G.B. Lush, E. Clark and G.W. Thompson, Sol. Energy Mater. Sol. Cells, 55 (1998) 141–148.

[25] V.P. Singh, J.C. McClure, G.B. Lush, W. Wang, X. Wang, G.W. Thompson and E. Clark, Sol. Energy Mater. Sol. Cells, 59 (1999) 145–161.

[26] D.L. Linam, V.P. Singh, D.W. Dils, J.C. McClure and G.B. Lush, Proc. SPIE–The Int. Soc. Opt. Eng., 3975 (2000) 1258–1261.

[27] H. Chavez, M. Jordan, J.C. McClure, G. Lush and V.P. Singh, J. Mater. Sci.: Mater. Electron, 8 (1997) 151–154.

[28] V.P. Singh, Proc. Eleventh Int. Workshop Phys. Semicond. Devices SPIE 4746 (2002) 65–72.

[29] V.P. Singh, O.M. Erickson and J.H. Chao, J. Appl. Phys., 78 (1995) 4538–4542.

[30] O.M. Erickson, J.C. McClure and V.P. Singh, Proc. 6 Int. Photovoltaic Sci. Eng. Conf., (1992) 97.

[31] D.L. Linam, V.P. Singh, J.C. McClure, G.B. Lush, X. Mathew and P.J. Sebastian, Sol. Energy Mater. Sol. Cells, 70 (2001) 335–344.

[32] V.P. Singh, Proc. 13th IEEE PVSC, Washington, DC (1978) 507–512.

[33] V.P. Singh, US Patent No. 4,404,734, Method of Making a CdS/Cu$_x$S Photovoltaic Cell, Photon Power Inc., El Paso, 1983.

[34] A.M. Barnett, J.A. Bragagnolo, R.B. Hall, J.E. Phillips and J.D. Meakin, Proc. 13th IEEE PVSC, Washington, DC (1978) 419.

[35] R.B. Hall and J.D. Meakin, Thin Solid Films, 63 (1979) 203.

[36] E. Aperathitis, F.J. Bryant and C.G. Scott, Sol. Energy Mater., 20 (1990) 15–28.

[37] J.A. Thornton, D.G. Cornog, W.W. Anderson, R.B. Hall and J.E. Phillips, Conf. Rec. IEEE PVSC (1982) 737–742.

[38] R.B. Hall, R.W. Birkmire, J.E. Phillips and J.D. Meakin, Appl. Phys. Lett., 38 (1981) 925.

[39] A. Goldenblum, G. Popovici, E. Elena, A. Oprea and C. Nae, Thin Solid Films, 141 (1986) 215.

[40] W. Palz, J. Besson, T.D. Nguyen and J. Vedel, Proc. 9th IEEE PVSC (1972) 91.

[41] L.D. Partain, R.A. Schneider, L.F. Donaghey and P.S. McLeod, J. Appl. Phys., 57 (1985) 5056.

[42] A.M. Al-Dhafiri, G.J. Russell and J. Woods, Semicond. Sci. Technol., 7 (1992) 1052–1057.

[43] K.L. Chopra and S.R. Das, Thin Film Solar Cells, 1983 Plenum Press, New York.

[44] K. Ernst, A. Belaidi and R. Konenkamp, 18 (2003) 475–479.

[45] R. Konenkamp , L. Dloczik, K. Ernst and C. Olesch, Physica E, 14 (2002) 219–223.

[46] N. Kouklin, L. Menon, A.Z. Wong, D.W. Thompson, J.A. Woollam, P.F. Williams and S. Bandyopadhyaya, Appl. Phys. Lett., 79 (2001) 4423.

[47] A.W. Zhao, G.W. Mengl, D. Zhang, T. Gao, S.H. Sun and Y.T. Pang, Appl. Phys. A, 76 (2003) 537–539.

[48] Y. Lin, G.S. Wu, X.Y. Yuan, T. Xie and L.D. Zhang, J. Phys.: Condens. Matter 15 (2003) 2917–2922.

[49] A. Boden, D. Braunig, J. Klaer, F.H. Karg, B. Hosslebarth and G. La Roche, 28th IEEE Photovoltaics Specialists Conf., (2000) 1038.

[50] N.G. Dhere and R.G. Dhere, J. Vac. Sci. Technol. A: Vac. Surf. Films, 23 (2005) 1208–1214.

[51] M. Emziane, K. Durose, D.P. Halliday, N. Romeo and A. Bosio, J. Appl. Phys., 97 (2005) 114910-1-6

[52] S.J.C. Irvine, V. Barrioz, A. Stafford and K. Durose, Thin Solid Films, 480 (2005) 76–81.

[53] E. Belas, R. Grill, A.L. Toth, J. Franc, P. Moravec, P. Horodysky, P. Hoschl, T. Wiehert and H. Wolf, IEEE Nucl. Sci. Symp. Conf. Rec., v 7, 2004; IEEE Nucl. Sci. Symp. Conf. Rec., (2004) 4466–4469.

[54] V.P. Singh, R.S. Singh, G.W. Thompson, V. Jayaraman, S. Sanagapalli and V.K. Rangari, Sol. Energy Mater. Sol. Cells, 81 (2004) 293–303.

[55] R.S. Singh, S. Sanagapalli, V. Jayaraman and V.P. Singh, J. Nanosci. Nanotechnol., 4 (2004) 176–182.

[56] E. Arici, H. Hoppe, F. Schaffler, D. Meissner, M.A. Malik and N.S. Sariciftci, Thin Solid Films, 451 (2004) 612–618.

[57] M. Nanu, J. Schoonman and A. Goossens, Nano Letters, 5 (2005) 1716–1719.

[58] R.S. Singh, V.K. Rangari, S. Sanagapalli, V. Jayaraman, S. Mahendra and V.P. Singh, Sol. Energy Mater. Sol. Cells, 82 (2004) 315–330.

[59] G. Liu, T. Schulmeyer, J. Brotz, A. Klein and W. Jaegermann, Thin Solid Films, 431 (2003) 477–482.

[60] S.-Z. Chu, K. Wada, S. Inoue and S.-i. Todoroki, Chem. Mater., 14 (2002) 266–272.

PART III

DYE-SENSITIZED SOLAR CELLS

Nanostructured Materials for Solar Energy Conversion
T. Soga (editor)

Chapter 7

TiO$_2$-Based Dye-Sensitized Solar Cell

Shogo Mori[a] and Shozo Yanagida[b]

[a]Department of Fine Materials Engineering, Faculty of Textile Science and Technology, Shinshu University, Ueda 386-8567, Japan
[b]Center for Advanced Science and Innovation, Osaka University, Suita, Osaka 565-0871, Japan

1. INTRODUCTION

Titanium dioxide (TiO$_2$) is an n-type wide band gap semiconductor, which is transparent for visible light. Dye-sensitized solar cells (DSCs) convert visible light energy to electrical energy through charge separation in sensitizer dyes adsorbed on a wide band gap semiconductor. Energy conversion from dye-sensitized TiO$_2$ electrode immersed in an electrolyte was reported by Vlachopoulos [1] in 1988. Although the charge separation was able to occur at high efficiency, energy conversion efficiency was not high because of low light absorption coefficient of the solar cells. This was because the dyes were adsorbed onto a relatively flat surface of the semiconductor electrode, and light absorption by the monolayer of the dye was limited. In 1991, O'Regan and Grätzel [2] solved the issue by employing nanoporous TiO$_2$ electrode. Since the electrode has huge surface area per projected area, solar cells made from the dye-adsorbed nanoporous TiO$_2$ can drastically increase effective light absorption. By designing proper electrode thickness and sensitization dyes, DSCs are able to absorb most of the visible light. The current highest energy conversion efficiency is over 11% [3], and further increase of the efficiency is possible.

DSC comprises a dye-sensitized nanoporous TiO$_2$ electrode on transparent conductive oxide (TCO) electrode, electrolytes containing I$^-$/I$_3^-$ redox couple filling the pore of the electrode, and a platinum counter electrode placed on the top of the TiO$_2$ electrode. Fig. 1 shows the structure of the DSC, a SEM image of typical nanoporous TiO$_2$ electrodes and the structure of a ruthenium (Ru) complex dye designed for DSCs. The nanoporous electrodes

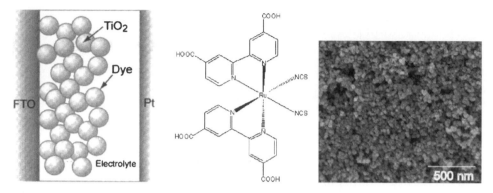

Fig. 1. Structure of DSC, a Ru complex dye (known as N3) and a SEM image of nanoporous TiO$_2$ electrode.

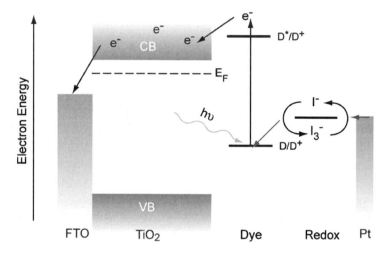

Fig. 2. Schematic of charge transport and interfacial transfer in DSCs.

are commonly prepared by sintering TiO$_2$ nanoparticles, whose diameter is around 20 nm. Examples of porosity, surface area, and average pore diameter of the electrode are about 60%, 50 m^{2}g^{-1} and 20 nm, respectively [4]. In the structure, injected electrons in TiO$_2$ and acceptors, such as dye cation and I$_3^-$ in the electrolyte are located within the scale of nanometer. Such situation likely results in short electron lifetime, decreasing the efficiency of the solar cells, whereas DSCs have significantly long electron lifetime. This is one of the prominent features of the DSCs. Fig. 2 shows a schematic of charge transport and interfacial transfer processes in DSC. Under light irradiation, excited electrons in the adsorbed dyes are injected into the conduction band of TiO_2 and the injected electrons diffuse in the TiO_2 to TiO_2/TCO interface, where electrons are extracted to external load. Resulting dye cations are reduced by I^-, and generated I_3^- ions diffuse to the platinum electrode where the I_3^- is reduced back to I^-. For the purpose of energy conversion, all incoming photons are ideally converted to electrons without losing the energy of the photons. Incident photon-to-current conversion efficiency (IPCE) is determined by (1) light-harvesting efficiency (LHE), (2) charge injection efficiency (CIE), and (3) charge collection efficiency (CCE). LHE is simply the ratio of the amount of absorbed photons to that of incoming photons. Thus, LHE can be improved by increasing the absorption coefficient of dye, the density of adsorbed dye, or the thickness of dye adsorbed nanoporous electrodes. CIE is determined by several factors such as potential difference between the conduction band edge of TiO_2 and lowest unoccupied molecular orbital (LUMO) of the adsorbed dyes, acceptor density in TiO_2, and spatial distance between the surface of TiO_2 and the dye. For CCE, electron diffusion length is a measure of assessment. During diffusion, photogenerated electrons can recombine with acceptors. Thus, to extract the photogenerated electrons for external load, the electrons must reach the TCO faster than the recombination process. Diffusion length can be expressed by $L = (D\tau)^{1/2}$, where D is electron diffusion coefficients and τ electron lifetime in nanoporous TiO_2. To realize high CCE, the thickness of the TiO_2 electrode (w) must be less than L. In view of LHE, w must be thick enough to absorb all incoming photons. In nanoporous TiO_2 electrodes and Ru complex dyes, 10–30 μm thickness is required to achieve high LHE for the wide range of the absorption spectrum of the dyes. Thus, L must be longer than the thickness. The highest IPCE of highly efficient DSC is over 80% [3], suggesting that L is longer than the thickness. This was surprising because TiO_2 had been regarded as an insulator rather than semiconductor.

To achieve high-energy conversion efficiency, high open-circuit voltage (V_{oc}) is also required. Although the origin of photovoltage has not yet been fully understood, V_{oc} is most likely related with the difference between the Fermi level of semiconductor electrode and redox potential [5]. The largest potential difference is limited by the energy level of highest occupied molecular orbital (HOMO) and LUMO of the sensitizer dyes. For example, dyes absorbing up to 800 nm have the potential difference of approximately 1.5 V. To obtain efficient electron injection from the dye to semiconductor and efficient reduction of dye cation by redox couple, certain potential differences, are needed at these interfaces; otherwise, the interfacial electron transfer rates become slow, decreasing the transfer efficiency [6–8]. The Fermi level of TiO$_2$ is related with the density of injected electrons and the density of charge traps in the band gap of TiO$_2$. Taking these into account, maximum V_{oc} of the TiO$_2$-based DSCs seems to be between 0.85 and 0.9 V. Assuming that a dye can absorb light between 400 and 800 nm, and taking the absorption by transparent conductive oxide substrate into account, maximum J_{sc} is approximately 22 mA cm^{-2} [9]. By using fill factor (FF) of 0.75, maximum conversion efficiency is expected to be around 14–15%.

Owing to the high-energy conversion efficiency and the potential of low-production cost in comparison to conventional solar cells, attempts have been made toward commercialization over the world. It remains an unsolved issue since the first report of the DSC is long-term stability. Organic electrolyte has substantial vapor pressure under normal operating condition, and I$^-$/I$_3^-$ redox couple can sublimate through I$_2$. Sealing such volatile electrolyte for tens of years is not trivial. To solve the issues, several approaches, such as employing solid state hole conductor [10–12], gelation of electrolyte [13, 14], using ionic liquid [15, 16], and alternative redox couples [17–21], have been examined. However, at present, the efficiencies of DSCs using these alternative materials have been lower than that of DSCs using organic solvent and I$^-$/I$_3^-$ redox couple. In addition to the sealing problem, chemical decomposition and reactions of electrolytes during usage can also decrease the performance of DSCs [22–24].

In order to increase the energy conversion efficiency and to improve the durability of the DSCs, it is necessary to understand the mechanism of the energy conversion processes. This chapter reviews the current models of electron transport and interfacial electron transfer, and parameters affecting these in the TiO$_2$-based DSCs. The parameters are often interdependent, and it makes the designing DSCs very complex. The chapter also discusses the charge transport/transfer properties in view of designing and optimization of DSCs.

2. ELECTRON TRANSPORT MODEL

In DSC, electrolytes penetrate into nanosized pore of the TiO$_2$ electrodes. Since high concentration of I$^-$/I$_3^-$ redox couple is needed to compensate for the large electron flow in TiO$_2$, the concentration of counter cation is also high. The cations screen the electrons in TiO$_2$ effectively and thus, there is probably no large electric field gradient in the TiO$_2$ electrode [25–27]. Since the size of each TiO$_2$ particle is typically around 20 nm, depletion layer cannot be formed in the nanoporous electrode, and charge separation process

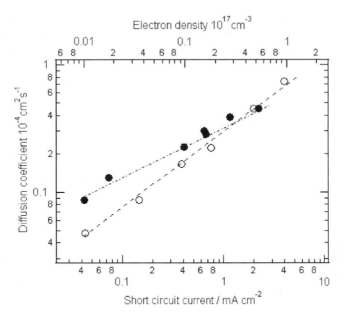

Fig. 3. Typical plot of electron diffusion coefficients in DSCs (open circle, bottom axis) and in dye-adsorbed nanoporous TiO₂ immersed in an electrolyte containing 0.7 M of LiClO₄ in ethanol (closed circle, top axis).

due to band bending at semiconductor/electrolyte interface, which has been used in the bulk semiconductor, cannot be applied. Based on these considerations, charge transport has been modeled with diffusion. Electron diffusion coefficient was measured by Cao et al. [26], and they found that D depended on incident light intensity and the values were significantly lower than that in the bulk TiO₂. The values of D were then measured by several groups with various techniques, and all showed the same tendencies [28–30]. Fig. 3 shows a typical plot of D vs J_{sc} in DSCs [31]. The value of J_{sc} is generally proportional to the intensity of incident light. To explain the observed results of D, trapping model has been applied [32–34]. The model assumes the existence of intraband charge traps in TiO₂, and the electrons in TiO₂ diffuse with the events of trapping and detrapping. Thus, the measured electron transit time, which is used to estimate D, is the sum of the time spent for diffusion and the time spent at the charge traps. Based on the model, measured D is not the D describing the material properties of TiO₂ but apparent D describing the properties of nanoporous electrode including the charge traps. D is also affected by large number of boundaries between nanosized crystals of TiO₂. The light intensity-dependent D was well simulated with an assumption that the energy level of the traps has an exponential distribution [33]. Under low light intensity, that is, low electron density in TiO_2, once electrons are trapped at deep trap sites, it takes long time to detrap by thermal activation and to reach TCO. Under high light intensity, deep traps are filled by electrons; electrons, which contribute currents, experience mostly shallow traps. The origin of charge traps has not yet been understood. It has been speculated with oxygen vacancy, surface amorphous state of titanium oxide, and impurities. Although there was no direct observation of the traps, the density and distribution of the traps have been investigated by infrared spectroscopy [35] and charge extraction method [36].

3. ELECTRON LIFETIME

Electron lifetime in the DSCs has also been measured by various groups [30, 37–39]. Observed lifetime was very long in comparison to conventional solar cells, and had light intensity dependence. In contrast with the model for electron transport, to explain the measured long electron lifetime and light intensity dependence, several models have been proposed. One was based on the two-electron reaction between I_3^- and conduction band electrons [38].

$$I_3^- + 2e^- \rightarrow 3I^- \tag{1}$$

Later, the light intensity dependence was also observed with DSCs using polymer hole transport materials [40] and with DSCs using one electron redox couple of cobalt complexes [19]. Second and third models take intraband charge traps into account, but the role of the traps is different. The second model assumes that the traps are located at the TiO_2 surface and the electrons would transfer directly from the traps to acceptors [37]. In this case, the driving force depends on the energy level of the highest electron-filled traps. With an increase of the potential difference between the traps and redox potential, that is, as filling the trap sites, the driving force increases and the electron lifetime decreases. Third model assumes that electrons would transfer from the conduction band of TiO_2 to acceptors [41, 42]. The rate of transfer depends on the number of electrons at the conduction band. As filling the trap sites, the potential difference between the highest filled traps and the conduction band decreases, increasing excitation frequency from the traps to the conduction band. This increases the density of electrons at the band, increasing the probability of the interfacial electron transfer. Fig. 4 depicts the possible electron transfer processes at the TiO_2/dye/electrolyte interfaces. The difference

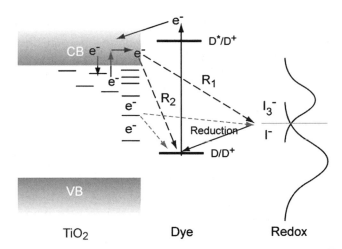

Fig. 4. Schematics of charge transfer at the TiO₂/dye/electrolyte interfaces. There are two recombination paths: R1, TiO₂ to I_3^-, and R2, TiO₂ to dye cation. The electrons can transfer from the conduction band or deep trap sites.

between the two models occurs whether the charge traps act as recombination center or not. If the majority of traps were located either at the surface or in the bulk, these two models could be distinguishable when the electron lifetime increases with the increase of trap density. However, the traps probably exist both at the surface and in the bulk. Moreover, the density of states for oxidized species has Gaussian distribution, which should be taken into account [43]. The observed electron lifetime has also been discussed with Marcus inverted region [44–46]. These complicate the situation further, and the mechanism of the recombination has not yet been fully elucidated. These issues will be discussed later with experimental results.

4. KINETIC COMPETITIONS

In DSCs, electrons in TiO₂ can transfer to dye cation and I_3^-. Ideally all dye cations are reduced by I^-, no electrons transfer to I_3^- at short circuit, and electrons transfer to I_3^- as slow as possible at open circuit. The rate of charge transfer from TiO₂ to dye cations is between 10^{-9} and 10^{-3} s, while dye cation reduction rate can be less than 10^{-5} s [47, 48]. Obviously, the reduction rate depends on the concentration of I^- in the electrolyte solution. In addition, the rate depends on the nature of counter cations of I^-/I_3^-. For example, Li^+ showed faster reduction rate in comparison to tetrabuthylammonium ion (TBA$^+$) [49]. In order to have faster dye cation reduction by I$^-$ than the conduction band electrons, appropriate I$^-$ concentration and species of counter cation should be chosen. In Li$^+$ in acetonitrile, 20 mM of I$^-$ is required [50].

In view of long-term stability, employing I$^-$/I$_3^-$ redox couple itself could cause degradation due to sublimation of I$_2$ and corrosion of, for example, platinum counter electrode. Therefore, alternative redox couples, such as cobalt complexes [17, 19], copper complexes [20], and SeCN [51], have been investigated. However, comparable redox couple with I$^-$/I$_3^-$ redox couple has not yet been found. In addition to the stability, important criterion to design redox couple is the potential difference between the redox couple and HOMO level of dyes. On the basis of the available data, 0.4 V seems to be needed [44]; otherwise, the dye cation reduction rate decreases, increasing the charge recombination rate between the electrons in TiO$_2$ and the dye cation. The required potential difference limits the Voc [52].

5. PARAMETERS INFLUENCING ELECTRON DIFFUSION COEFFICIENT AND LIFETIME

In previous sections, we saw that electron diffusion coefficient and lifetime were models with intraband charge traps. In addition, the lifetime is influenced by the dye cation reduction rate by redox couple. In view of the selection of chemicals and fabrication processes, what are the parameters influencing the electron transport and transfer rates, and consequently the energy conversion efficiency? In this section, we will see various parameters influencing these, and we will correlate these with energy conversion efficiency.

5.1. TiO$_2$ Particle Preparation Method

Nanoporous TiO$_2$ electrodes are typically prepared from colloidal suspension containing nanosized TiO$_2$ particles. The suspension is applied on a TCO with doctor blade techniques, and the resulting film is sintered by annealing between 450°C and 550°C. There are many commercially available nanosized TiO$_2$ particles and many methods to synthesize nanosized TiO$_2$ particles. On the other hand, not many commercial TiO$_2$ particles and methods are suitable for highly efficient DSCs. The differences among the particles can be crystal structure, crystallinity, crystal size, surface structure, and the density and distribution of charge traps. These parameters influence D, τ, and CIE. Table 1 shows electron diffusion coefficient for nanoporous TiO$_2$ electrodes prepared from eight different TiO$_2$ particles [30, 53]. In the TiO$_2$

Table 1
Electron diffusion coefficients (D) in nanoporous TiO_2 electrodes having various characteristics.

	Structure	Shape	Size (nm)[a]	D ($\times 10^{-5}$ cm²s⁻¹)
S1	A	Spheric	19	12
S2	A/Amor	Spheric	12	2.2
S2 TiCl₄[b]	A/Amor	Spheric	12	2.2
A2	A	Cubic	12	0.3
A2 550[c]	A	Cubic	12	2
A3	A	Rod-like	13/34	4
A4	A	Cubic	11	4.1
P25	A/R	Spheric	21	4
P25 large[d]	A/R	Spheric	21	4
R1	R	Spheric	27	0.1
R1 TiCl₄[b]	R	Spheric	27	0.4
R2	R	Rod-like	23/73	0.3

Note: S1 was prepared from hydrolysis of aqueous $TiCl_4$ solution; S2 was prepared from hydrolysis of titanium tetraisopropoxide in the presence of nitric acid.
A = anatase structure, Amor = amorphous phase, R = rutile structure.
[a]Average size.
[b]Treated with aqueous $TiCl_4$ solution.
[c]Annealed at 550°C.
[d]With 20 w% of large TiO_2 (Fluka). Electrodes were prepared by annealing at 450°C otherwise stated. Diffusion coefficients were measured without dye adsorption.

particles having high crystallinity (denoted as A2), that is, containing little amorphous titanium oxide, an increase of annealing temperature from 450°C to 550°C improved the electron diffusion coefficient significantly. The result indicates that in addition to the crystal structure, necking condition between particles is also important. Fig. 5 shows the surface area of nanoporous TiO_2 electrode prepared from the TiO_2 particles, S1, S2 and P25, at different annealing temperatures [30]. S1 and S2 show the decreases of surface area with the increase of annealing temperature, although P25 showed little change. The decrease of the surface area implies that during sintering process, the particles form necks having larger cross-sectional area. Fig. 6 shows electron diffusion coefficients as a function of electron density in the electrodes. The particles consisting only of anatase and showing the decrease of surface area during annealing exhibited the largest values of D. Among the samples prepared at 450°C annealing, the sample giving the large values of D showed the shortest electron lifetime.

Shogo Mori and Shozo Yanagida

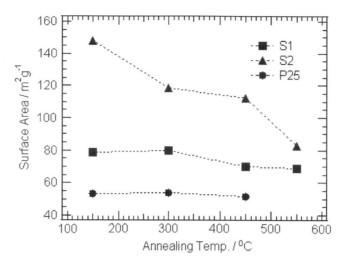

Fig. 5. Surface area of nanoporous TiO_2 electrodes prepared from three different TiO_2 particles. Samples are the same used in Table 1.

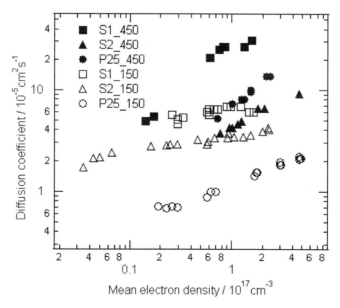

Fig. 6. Electron diffusion coefficients in nanoporous TiO_2 electrodes prepared from three different TiO_2 particles with two different annealing temperatures. S1_450 is the electrode prepared from S1 by annealing at 450°C.

5.2. TiO₂ Annealing Temperature

Annealing temperature is an important parameter for several aspects. To achieve high efficiency, nanoporous electrodes must contain well-connected particles with certain thickness; thus, higher temperature is required. On the other hand, some particles and colloidal suspension shows large differences in heat expansion coefficients between TCO and TiO_2, and high temperature annealing would cause detachment of the nanoporous film from TCO, limiting the process temperature and the electrode thickness.

Annealing temperature is also important in view of applications. Many attempts have been made to fabricate flexible DSCs [54–57]. For the application, nanoporous TiO_2 electrode is to be fabricated on a transparent plastic substrate, which probably cannot stand more than 200°C. Fig. 6 also shows the diffusion coefficients of the samples prepared by 150°C annealing [30]. In low-temperature annealing, TiO_2 particles, which showed the decrease of surface area with the increase of annealing temperature, gave higher electron diffusion coefficients. The results imply that surface amorphous state would act as a kind of glue, resulting better necking even by low-temperature annealing. Fig. 7 shows the electron lifetimes of DSCs prepared with 150°C and 450°C annealing [58]. Low-annealing temperature resulted in shorter electron

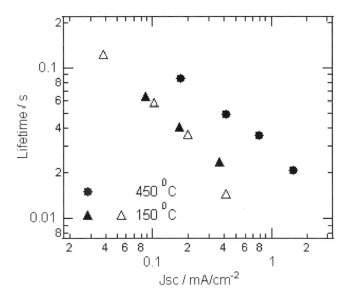

Fig. 7. Electron lifetime at open-circuit conditions in DSCs prepared by two different TiO_2 annealing temperatures.

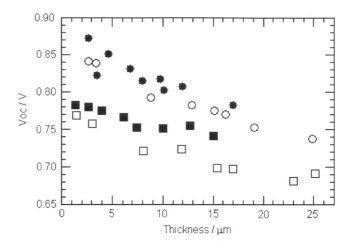

Fig. 8. Open circuit voltage of DSCs vs. TiO₂ electrode thickness. The DSCs were prepared from P25 at two different temperatures: 450°C (open symbol) and 150°C (closed symbol). Electrolytes were 0.1 M of LiI, 0.6 M of DMPImI, 0.5 M of tBP and 0.05 M of I₂ in AN (square) and 0.1 M of LiI, 0.5 M of DMPImI, 0.1 M of THAI, 0.5 M of tBP and 0.05 M of I₂ in AN (circle).

lifetime, and the result was in agreement with the cases for other TiO₂ particles [59]. The observed shorter lifetime could be due to the larger amount of charge traps in the low-temperature annealed films, as speculated by us [30], resulting in higher probability of electron transfer from the traps, or due to less density of traps, as mentioned by Park et al. [59], resulting in higher density of conduction band electrons. Fig. 8 shows the V_{oc} as a function of TiO₂ thickness for the two different annealing temperatures [58]. Low temperature annealing resulted in higher V_{oc}. As it was mentioned earlier, V_{oc} should scale with the potential difference between the Fermi level of the TiO₂ electrode and redox potential. The Fermi level is related with the electron density, trap density, and the conduction band edge potential (E_{cb}) of TiO₂ relative to the redox potential. The E_{cb} is affected by the species of electrolytes. In other words, when electrolytes are the same, E_{cb} is also probably the same. Electron density in the TiO₂ electrode may be expressed with

$$dn/dt = G - n/\tau \tag{2}$$

where n is electron density, G the electron injection rate from dye, and τ electron lifetime. Note that τ depends on n so that Eq. (2) can only provide rough estimation. At open-circuit condition, $dn/dt = 0$. Thus, n is determined by $G\tau$. Since τ is shorter in the low-temperature annealed electrode, electron density should be lower under the same light intensity. Based on these, low-temperature-annealed electrodes probably have less number of charge traps. However, particles are not sintered well among particles; electrons experience more frequent scattering at boundaries between particles, resulting in lower values of D. This reduces the electron diffusion length; thus, the thickness giving maximum J_{sc} was less than $10\,\mu m$, although J_{sc} from the DSC using high-temperature-annealed films increased with the increase of thickness up to $25\,\mu m$ [58].

5.3. TiO₂ Particle Size

Highly efficient DSCs are typically prepared from the particles having diameter between 10 and 20 nm. In view of electron transport, larger value of D is expected with the increase of particle size, due to the lower number of boundary among the particles. If the charge traps were located at the surface, the trap density in the electrode would also be decreased with the increase of particle size, that is, with the decrease of surface area. Fig. 9 shows the electron diffusion coefficients and lifetimes in DSCs prepared from TiO₂ particles that have different particle sizes [60]. As is seen, D increases and τ decreases with the increase of particle size. It seemed that D scaled with the number of boundary between 14 and 19 nm, and with the surface area between 19 and 32 nm. Kopidakis et al. [61] conducted the similar experiments between the average particle sizes of 20.5–41.5 nm, and they showed that the increase of D was well explained with the decrease of surface area. This suggests that, above 20 nm, most charge traps are located at the surface, and they play dominant role on the electron transport.

In view of charge recombination, τ seems to be inversely proportional to the values of D. The correlation implies that most traps do not act as the recombination center. This seems valid in the range of light intensities used for normal solar cell operations, that is, from under sunlight to under room light conditions. When the electron density becomes very low, most electrons are trapped by deep trap sites and the time to escape from the traps to the conduction band becomes longer than the time to transfer to I_3^- or dye cation. The density of state of oxidized species becomes also high at the potential of the deep trap sites in TiO₂. Thus, both models 2 and 3 explained in Section 3 seem correct, depending on the electron density in TiO₂ [43].

For DSCs, what is the optimized TiO₂ particle size? Electron diffusion length seems independent from the preparation method of the nanoporous

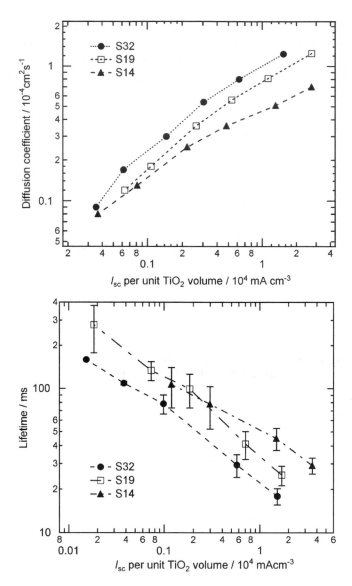

Fig. 9. Electron diffusion coefficients and lifetimes in DSCs prepared from three different TiO$_2$ particles, whose average diameters are 32 nm (S32), 19 nm (S19), and 14 nm (S14). Bottom axis shows the J_{sc} divided by the volume of TiO$_2$ to normalize the different porosity among the samples.

TiO_2 electrode, because of the correlation between D and τ. This would be true if there were no boundaries in the electrode. In reality, the boundaries increase the resistance for electron transport. Cross-sectional area of the boundary also relates with the resistance. As an electrode, the resistance becomes large with the decrease of the particle size due to the increase of the number of boundaries. The resistance also increases with the increase of the particle size due to the decrease of total cross-sectional area. This means that the value of D does not increase monotonically, and there is a particle size giving maximum diffusion length. From the view-point of light absorption, smaller particle is better because of higher roughness factor. However, employing too small particles may result in the lower energy conversion efficiency due to shorter electron diffusion length and lower CIE [62]. Surface binding forms of dyes, which would influence the CIE, are also influenced by TiO_2 particle size [63].

5.4. TiO₂ Surface Treatment

To retard charge transfer from the conduction band to dye cation and I_3^-, one idea is to form insulating layer on the top of the TiO_2 surface [64–68]. Insulating layer could also prevent from charge injection from dye. Thus, the thickness of insulating layer should be so thin that it does not retard the injection rate too much [67]. In Ru complex dyes, the injection rate can be less than picoseconds, which is unnecessarily fast [8]. Thus, the addition of insulating layer having proper thickness works to retard charge recombination without losing charge injection yield.

To reduce the charge recombination, various thin metal oxide layers have been investigated. These oxides act as insulating layers due to higher E_{cb} of these oxides or due to the nature of their amorphous structure. The metal oxides could also shift the E_{cb} positively as well as the retardation of recombination [67]. Thus, species of metal should be chosen to avoid to the positive shift, which reduces V_{oc} of DSCs. In addition, it has been proposed that metal oxide can passivate surface trap states [69].

Another method to form surface insulating layer is electro deposition of insulating materials. Various metal hydroxides have been deposited on the top of the TiO_2 surface [68]. The increase of V_{oc} due to the deposition was well explained with the increased electron density by the retardation of recombination. In comparison to $Al(OH)_3$, $Zn(OH)_2$, and $La(OH)_3$, $Mg(OH)_2$ showed the longest electron lifetime. Because the atomic size of Mg^{2+} was close to that of Ti^{4+}, $Mg(OH)_2$ was probably formed more compactly.

5.5. Redox Couple Concentration

In DSCs, conduction band electrons can transfer to dye cation and I_3^-. Therefore, electron density can be expressed by

$$dn/dt = G - n/\tau_{I_3^-} - n/\tau_{\text{dye}} \qquad (3)$$

where $\tau_{I_3^-}$ and τ_{dye} are the lifetimes determined by the electron transfer to I_3^- and dye cation, respectively. Thus, measurable electron lifetime by means of photovoltage responses is the reciprocal of $(1/\tau_{I_3^-} + 1/\tau_{\text{dye}})$. Since the density of dye cation depends on the concentration of I^-, the lifetime of the photogenerated electrons in TiO_2 depends on both concentrations of I^- and I_3^-. When τ_{dye} becomes very long, that is, the concentration of I^- is so high that the density of dye cation becomes low, Eq. (3) becomes Eq. (2). Fig. 10 shows the electron lifetime as a function of I^- and I_3^- concentrations [50]. The lifetime increases with the increase of I^- and is leveled off above the concentration of 20 mM, which means that above 20 mM of I^-, most dye cations are reduced by I^- faster than by the conduction band electrons, and the lifetime becomes constant due to the constant concentration of I_3^-. When the concentration of

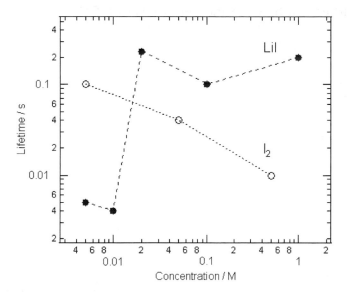

Fig. 10. Electron lifetime at open circuit conditions in DSCs employing various LiI at fixed I_2 concentrations (2 mM, closed symbol, measured under light intensity giving 0.2 mAcm^{-2}) and various I_2 at fixed LiI concentrations (0.7 M, open symbol, measured under light intensity giving 1 mAcm^{-2}).

I_3^- was increased, the electron lifetime decreased as expected. The minimum concentration required for efficient dye cation reduction depends on the electron transfer rate from the redox couple to dye cation. This transfer rate depends on the parameters such as free energy, reorganization energy, and collision frequency. For example, the transfer rate depends on the structure of Co complex redox couples, that is, a Co complex redox couple requires higher concentration for efficient dye reduction in comparison to other Co redox couples. This was probably due to higher reorganization energy of the former Co complex [19]. Collision frequency is related with the viscosity of solvent. In such electrolyte, higher concentration of redox couple is needed for sufficient electron transfer [70].

5.6. Counter Cations

When I^-/I_3^- redox couple is used, counter cations of the redox couple are needed. Widely used electrolyte compositions for DSCs are 0.1 M of LiI, 0.6 M of 1,2-dimethyl-3-propylimidazorium iodide (DMPImI), 0.05 M of I_2 and 0.5 M of 4-*tert*-butylpyridine (tBP). When only Li^+ is employed, it results in high J_{sc} with low V_{oc} [71]. This is because the cation adsorbs on the TiO_2 surface and it shifts the E_{cb} of TiO_2 positively. The shift decreases the potential difference between the Fermi level of TiO_2 and redox potential, decreasing V_{oc}, while it increases the electron injection efficiency from adsorbed dye [6, 72]. To minimize the decrease of V_{oc}, the amount of Li cation should be minimized. On the other hand, electrolyte containing the redox couple must have enough conductivity to compensate the electron current in TiO_2. This determines the amount of the redox couple. The concentration required to achieve more than 10% of conversion efficiency is at least 0.7 M of I^-, and therefore, most of the counter cations should be chosen except from Li^+.

In view of electron diffusion coefficients, cations play important roles. Since the nanoporous electrode is immersed in an electrolyte, to keep charge neutrality, diffusion of electrons should be along with the compensation of cations. This situation induces ambipolar diffusion [73, 74]. Electron diffusion coefficients in the nanoporous TiO_2 is between 10^{-6} and $10^{-4}\,cm^2s^{-1}$ under normal operation conditions, although the diffusion coefficients of cations is in the order of $10^{-6}\,cm^2s^{-1}$. To exploit faster electron diffusion ability in TiO_2, higher cation density than electron density is needed. This is generally realized since the electron density in the solar cells is less than $10^{18}\,cm^{-3}$, as the cation density is much more than that. In addition, the cation concentration at the surface of TiO_2, in the electric double layer is higher than the bulk concentration of electrolytes. Thus, the amount of cation usually does not limit the electron diffusion [50]. Concerning the influence of species of cations, those, which adsorb at the TiO$_2$ surface, such as Li$^+$ and DMPIm$^+$, increase the value of D [75]. In Li$^+$, it can intercalate into TiO$_2$. When the intercalation occurs, Li$^+$ act as electron traps, and the values of D decrease with the increase of electron lifetime [76].

    Species of cation affects not only the values of D, but also electron life-time. Fig. 11 shows the electron lifetime in DSCs using various cations as the counter charges of I$^-$/I$_3^-$ [50, 58]. The sensitizer dye was *cis*-[Ru^{II}-(4,4'-dicarboxylate-*H*-2,2'-bipyridine)$_2$(NCS)$_2$](TBA)$_2$, (known as N719). As is seen, Li$^+$ and DMPIm$^+$ cations gave comparable electron lifetime, while quaternary ammonium cation resulted in longer electron lifetime. Further increase of the lifetime was observed with the ammonium cation having longer alkyl chain. The results can be interpreted with the condition of electric double layer formed at the TiO$_2$ surface [50]. Under light irradiation, the TiO$_2$ electrode is negatively charged by the injected electrons. Then, electric double layer is formed by cations to screen the electrode. When the size of cations is large, the cations would not penetrate into dye molecule layer and the double layer would be formed on the top of the dye layer. For anions such as I$_3^-$, in order to approach the TiO$_2$ surface through penetrating the dye layer, the anions feel a repulsive

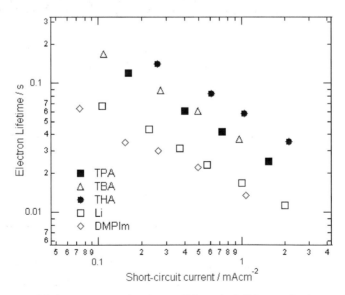

Fig. 11. Electron lifetime at open circuit conditions in DSCs employing various counter cations for I$^-$/I$_3^-$ redox couples. Redox concentrations were 0.65 M I$^-$ and 0.05 M of I$_2$.

force. This reduces the concentration of I_3^- in the vicinity of the TiO_2 surface, decreasing the probability of charge transfer from TiO_2 to I_3^-. When the amount of the dye was reduced so that the bulky cations could approach the TiO_2 surface, observed electron lifetimes were comparable regardless of the examined cations, supporting the model using electric double layer. The effect of quaternary ammonium cation was not observed with N3 and Z907 [77]. This suggests that the effect depends also on the structure of dye molecules, that is, the concentration and orientation of adsorbed dyes should be taken into account.

Although longer electron lifetime was achieved by the quaternary ammonium cation, using only the cation might not be enough to increase the energy conversion efficiency largely. If the size of the cation is too large to diffuse in the nanopore of the electrode, it could decrease FF. In addition, quaternary ammonium cation has disadvantages in term of CIE from dye to TiO_2 and the dye cation reduction rate by I_3^-. Therefore, a mixture of electrolytes is needed to add desired functions onto the electrolyte. Through optimization, it was found that replacing 0.1 M of DMPImI in the conventional electrolyte with 0.1 M of tetrahexylammonium iodides (THAI) gave longer electron lifetime without disturbing the CIE. The electrolyte was especially effective for the DSCs prepared with low-temperature-annealed TiO_2 electrodes. Since such DSCs are affected by short electron lifetime, the electrolyte is expected to improve the efficiency by increasing electron diffusion length by the increase of electron lifetime. Fig. 12 shows the efficiency with various DSCs having different TiO_2 thickness [58]. With the electrolyte containing THAI, the thickness giving the highest efficiency was increased due to the increased diffusion length. The V_{oc} was also increased due to higher electron density realized by the increase of the electron lifetime (Fig. 8).

5.7. Additives

In electrolytes, various additives have been employed [3, 78, 79]. Most famous additive is tBP. The additive is known to increase V_{oc} with a small decrease of J_{sc}. The effect was initially interpreted with the retardation of charge recombination due to the adsorption of tBP at the TiO_2 surface [80]. The interpretation was derived by analyzing $I–V$ characteristics of the solar cells with an equation of

$$V_{oc} = \frac{kT}{qu\alpha} \ln\left(\frac{AI_0}{n_0^{u\alpha} k_{et} c_{ox}^m} \right) \qquad (4)$$

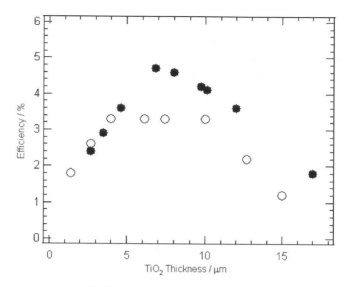

Fig. 12. Energy conversion efficiency of DSCs prepared by annealing temperature at 150°C. Electrolytes were standard (open) and one containing THAI (closed).

where k_{et} is the rate constant for charge recombination, I_0 the incident photon flux, c_{ox} the concentration of I_3^-, n_0 the electron density in the dark, and m, u and α are constants. When different electrolytes are employed, the E_{cb} may differ depending on the adsorbed species. Eq. (4) is not taken into account of the E_{cb}, and thus, from Eq. (4), the effects of recombination and of potential shift cannot be ruled out. Later, the influence of tBP was investigated by measuring electron lifetime and it was not significant, suggesting that the V_{oc} was improved by the change of E_{cb} [37, 50]. The effect of tBP on V_{oc} is significant when Li^+ was dissolved in the electrolyte. Higher V_{oc} observed with tBP seems to be caused by the suppression of the specific adsorption of Li^+ through the formation of complex with tBP, retarding the positive shift of the conduction band edge [50]. In addition, basicity of tBP itself induces the negative shift of the conduction band edge potential, increasing V_{oc}.

5.8. Spatial Distance

In DSCs, charge recombination occurs through interfacial electron transfer from TiO_2 to HOMO of the dye and acceptors in the electrolyte. The probability of transfer is related not only with the free energy and reorganization energy differences but also with the special distance between them. As the distance increases, tunneling probability also decreases exponentially.

When quaternary ammonium cations were used as electrolyte, resulting longer electron lifetime was interpreted with the lower concentration of I_3^- at the TiO_2 surface [50]. In other words, the distance between the TiO_2 surface and location of I_3^- was increased, decreasing the transfer probability. The idea of the distance can also be applied to design dye molecule. If HOMO is located far from the TiO_2 surface, the transfer rate is decreased [46]. The lifetime of dye cation can be monitored by the transient absorption of dye cations. When HOMO is located far from the TiO_2 surface, the transients follow an exponential decay, while others show stretched exponential decay. When the rate of transfer from the conduction band to dye cation is fast, electron lifetime is limited by the rate of transport in the TiO_2 surface or more precisely, limited by the rate of detrapping in TiO_2. In this case, stretched exponential decay occurs due to the change of detrapping rate as the electron density decreases. When the interfacial transfer becomes the limiting process, the decay becomes normal exponential function [46].

5.9. Solvent

Solvent is chosen in terms of solubility, viscosity, and boiling point. Lower viscosity is required for conductivity. Viscosity is generally related with boiling point, that is, solvents having high-boiling point are typically found with high viscosity. Low viscous solvent increases a difficulty of sealing of the solvent against evaporation. For long-term stability, highly viscous electrolyte would be preferred. When the viscosity of electrolyte increases, mobility of ions also decreases, inducing the decrease of both FF and dye cation reduction rate. On the other hand, charge transfer rate between the conduction band electrons and I_3^- should be decreased by the lower collision frequency between them. Fig. 13 shows the electron lifetime for the DSCs using two different solvents and two different cations as counter charges of I^-/I_3^- [70]. When there were enough I^-, that is, dye cations were effectively reduced, electron lifetime was longer in the DSCs using highly viscose solvent. Under this condition, the efficiency of the DSCs using highly viscous electrolyte became comparable.

6. SENSITIZING DYES

6.1. Electron Injection Dynamics

The sensitizing dyes giving high efficiency at this moment are a series of Ru bipyridine complexes. The complexes are adsorbed through ester-like linkage, bridge, or chelating of carboxyl group to the surface of TiO_2 [63, 81, 82].

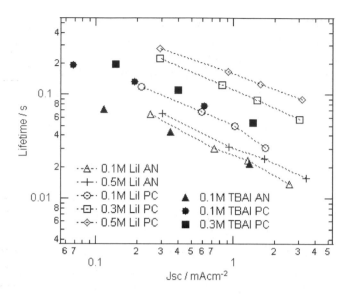

Fig. 13. Electron lifetimes in DSC using two different solvents: propylene carbonate (PC) and acetonitrile (AN) and two different counter cations (Li$^+$ and TBA$^+$) for I$^-$/ I$_3^-$ redox couple.

Visible light absorption of the Ru dyes occurs by metal center to ligand (bipyridyl) charge transfer. Then the electrons are injected through the carboxyl groups attached to the bipyridine. The injection occurs between less than 10^{-12} and 10^{-9} [83–85], depending on the relative potential of LUMO to the TiO$_2$ E_{cb}. The relative potential can be controlled by charged species and their concentrations in electrolytes. When adsorptive cations, such as Li$^+$ and Na$^+$, are employed, the cations shift the TiO$_2$ E_{cb} to positive, while the potential of the LUMO stays the same level. Since the density of states at the conduction band is high, the positive shift makes the electron transfer easy, increasing the injection rate [85].

6.2. Metal Complex Sensitizing Dyes

Most well-known Ru complexes for DSCs are N3 [78], N719 [86], N749 [87] and Z907 [88]. N3 has two bipyridine and two NCS ligands. Absorption spectrum is up to 800 nm due to the loosely attached NCS groups. Although the dye can provide high J_{sc}, it could not give high V_{oc}. N719 dye has the same structure with N3 dye but having TBA$^+$ instead of H$^+$ at two carboxyl groups over four of them. The replacement of the cations increases V_{oc} in comparison to N3 dye. The different V_{oc} between them can be rationalized with the difference of proton concentrations at the surfaces [86]. Since N3 dye can provide at most four protons per dye, the protons can adsorb the basic sites of the TiO$_2$ surface and can shift the E_{cb} to positive.

These two dyes have absorption spectrum up to 800 nm and this limits the maximum J_{sc}. N749 dye, which is called black dye, has achieved the maximum absorption up to 860 nm, and has a potential overwhelming the N3 and N719 dyes. However, the absorption coefficient of the N749 is lower than N3 and N719 dyes. These low-absorption coefficients require thicker TiO$_2$ electrodes to adsorb more dyes. The increase of the thickness has disadvantages in view of electron transport and open circuit voltage, that is, J_{sc} and V_{oc} could decrease. V_{oc} of the DSC using N749 is somehow lower than N719, which cannot be explained by proton effect. Based on these, N749 has not shown higher efficiency than N719.

Z907 has different features than others. DSCs using N3 and N719 have shown degradation during long-term operation. This is because water molecules penetrate into the electrolyte and they desorb the dyes from the TiO$_2$ surface. Z907 has hydrophobic alkyl chains attached to one of the bipyridine ligands and it keeps water molecules away from the chemical bonds between the dye and TiO$_2$. Durability tests showed stable performance of the DSCs with Z907 for 1000 h under light irradiation [89].

When the dye was applied with polymer hole conductor, electron lifetime was increased. The degree of the increase was correlated with the length of the alkyl chain, suggesting that the alkyl chain increased the special distance from the TiO$_2$ surface to the hole conductors [90]. In order to increase the electron lifetime, another approach is to attach additional doner ligand to Ru complex dye so that the photogenerated holes are rapidly transferred to the ligand, which is located far from the surface of TiO$_2$ [91].

Instead of Ru, several attempts were made to replace Ru with other metals, such as Fe [92] and Os [93, 94]. In Fe, Fe^{II}(2,2'-bipyridine-4, 4'-dicarboxylic acid)$_2$(CN)$_2$ showed photosensitization with very low photocurrent. The authors interpreted the results with antibonding e_g orbitals, which are located lower in energy than the π^* orbitals. Since the e_g oribitals do not overlap with the oribitals on the TiO$_2$ surface there is no injection from the orbitals. Thus, only the electron excited to metal to ligand charge transfer state can inject to TiO$_2$. In addition, MLCT state has probably picoseconds time scale lifetime, decreasing also the charge injection yield. In Os, several Os complexes showed comparable incident photon to current conversion efficiency and charge injection dynamics with Ru complexes.

6.3. Organic Dyes

Ru is a rare and expensive metal. When mass production of DSC becomes feasible, the Ru complex dyes would not be preferred. Thus, large efforts were made to synthesize organic dyes for DSCs. It has been reported that benzothiazol merocianines dyes [95], coumarin dyes [7, 96], polyene dyes [97], and indoline dyes [98] showed conversion efficiency of 4–8% with respectable IPCE. One strategy to improve the efficiency with organic dyes is to increase the range of absorption spectrum by expanding π-conjugation. Hara et al., systematically increased π-conjugation by introducing methine unit ($-CH=CH-$) into coumarin dyes, and found that absorption spectra and energy conversion efficiency were increased with the increase of the methine group length [7]. They also reported that the addition of cyano group along with carboxyl group caused red-shift in the absorption spectrum due to the electron-withdrawing ability of the group. Too much increase of the methine unit would cause instability of the dyes of gradual extent. To solve the problem, the same group introduced thiophene group similar to the methine chain and successfully expanded the absorption spectrum further [96]. The resulting efficiency was 7.7%. Fig. 14 shows the structure of these organic dyes. Charge injection dynamics of organic dyes, which include carboxyl and cyano groups attached to methine groups, have been measured, showing femtosecond scale injection [99]. It was comparable to the time scale observed in the Ru complex dyes. On the other hand, DSCs using organic dyes tend to have shorter electron lifetime in comparison to the conventional Ru dyes [100].

6.4. Coadsorbent

When sensitizer dyes are adsorbed on the TiO_2 surface, coadsorbents like cholic acid derivative are also commonly employed. Some dyes have a tendency to aggregate when they are adsorbed. Aggregation is preferred if the absorption spectrum is widened and if it contributes to an increase of J_{sc}. On the other hand, most aggregation through $\pi-\pi$ stacking could result in the decrease of electron injection yield, decreasing J_{sc}. To avoid the negative effects, various coadsorbents, such as cholic acid [101], deoxycholic acid [102], TBA chenodeoxycholate [103], hexadecymalonic acid [104], have been used to increase energy conversion efficiency. When these coadsorbents are employed, the amount of adsorbed dyes is decreased. However, the decrease does not influence significantly on J_{sc}. The small changes of J_{sc} relative to the large decrease of the amount of loaded dyes may be due to the result of the increase of charge injection yield. Increases of V_{oc} in the DSCs using coadsorbents are more prominent. The increase could be caused by the

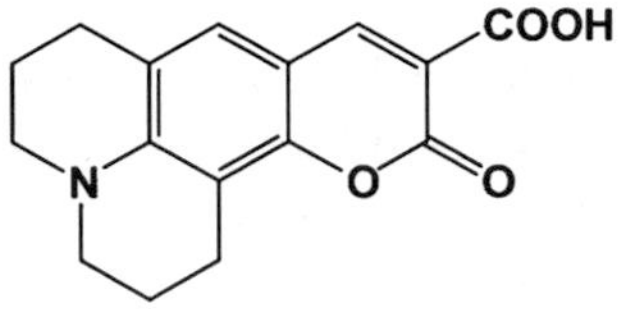

Fig. 14. Structures of organic dyes. Wavelength edge of IPCE and energy conversion efficiency of DSCs using these dyes are 600 nm and 0.9% for C343, 750 nm and 6.0% for NKX2311, and 900 nm and 7.7% for NKX2677.

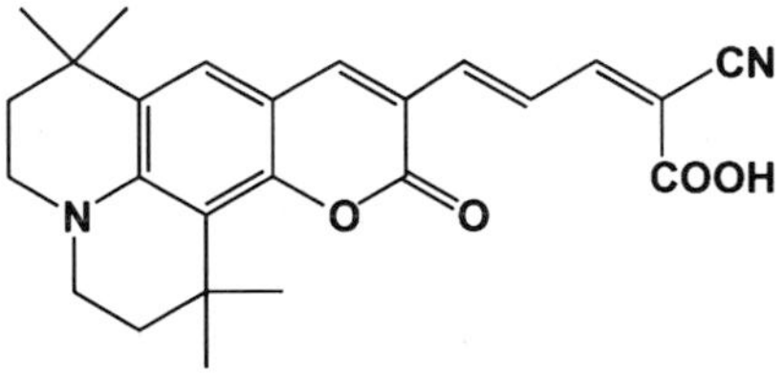

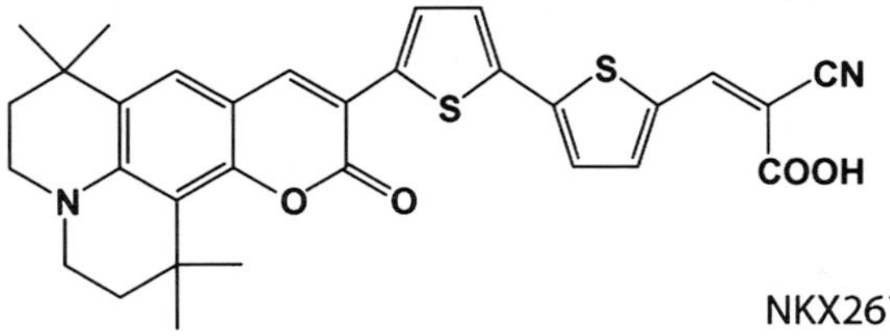

negative shift of TiO_2 E_{cb} and/or by the suppression of charge recombination. In the case of TBA chenodeoxycholate, the coadsorbent causes negative shift of TiO_2 conduction band edge with simultaneous accelerations of charge recombination, which is opposite to the speculations that the coadsorbent suppresses recombination [103]. In view of covering the exposed TiO_2 surface between adsorbed dyes by coadsorbent, adsorption of smaller molecules after dye adsorption would also work. Based on the strategy, carboxylic acid has been used to cover the vacant surface of the TiO_2 electrodes [105].

7. HOLE TRANSPORT LAYER

In order to gain long-term stability, large efforts have been made to replace highly volatile organic electrolyte with quasi-solid electrolytes or solid state hole conductors. The approaches are gelation of liquid electrolyte, using molten salt electrolytes, combination of gelator and ionic liquids, employing organic hole conducting materials, including p-type semiconducting polymer, and using p-type solid state semiconductor. When electrolytes are solidified, it is expected to reduce largely the diffusion coefficients of redox species, decreasing the conductivity. This is not always the case, that is, the conductivity is not decreased as the increase of viscosity [13]. This can be interpreted with Grotthuss mechanism, which describes charge transport through the exchange of I^- and I_3^-. However, to add sufficient conductivity to the solid electrolytes, larger amount of redox couples than that in conventional liquid electrolyte may be needed. This approach has side effects, which are the decrease of the light-harvesting efficiency due to light absorption by poly-iodide and the decrease of electron lifetime due to the increased amount of I_3^- [15]. Remedy of the issues would be decreasing the viscosity of solvents and/or increasing the conductivity by enhancing the ion exchange ability. Former approach can be seen in employing lower viscous ionic liquid electrolyte [16], and latter was realized by using nanoparticles as geletor [14, 106] or using liquid crystal as solvent so that redox couples ions exist with arranged manner [107].

In the organic hole conducting materials, one obstacle is how to fill the nanosized pore on the TiO_2 electrode. Methods will be applied to small p-type conducting molecules to polymerize the monomer photoelectrochemically in the pore [11, 108]. The efficiency of DSCs with these methods has been low but is gradually improving.

8. COUNTER ELECTRODES

The DSCs using I^-/I_3^- redox couple are mostly fabricated with platinum counter electrode. As Ru may not be preferred for mass production, platinum would be replaced with others. Graphite works as catalyst for I^-/I_3^- redox couple but the performance is not as good as platinum. Among alternative materials, poly (3,4-ethylenedioxythiophene (PEDOT)) has shown equivalent ability with platinum [109].

Instead of using catalytic materials, using p-type semiconductor is an interesting approach. By using the p-type semiconductor as photocathode, the DSC becomes a tandem cell. In this case, V_{oc} is the difference of quasi-Fermi level of the TiO_2 and p-type semiconductor. At the moment, tandem DSCs using dye-sensitized TiO_2 and dye-sensitized NiO has been reported [110, 111]. Hole injection from the dye to NiO was measured [112], showing the feasibility of the idea. Although the sum of V_{oc} from the two electrodes of the tandem DSC was observed as expected, the efficiency was less than 1%. This was due to low energy conversion efficiency of the dye-sensitized NiO electrode. However, in order to achieve energy conversion efficiency beyond 15%, tandem structure will be probably inevitable.

9. CONCLUSION

In the TiO_2-based DSCs, photo-induced charge separation occurs by electron transferring at the TiO_2/dye/electrolyte interfaces, and the charges travel in TiO_2 and an electrolyte separately. In order to achieve high efficiency of the solar cells, these charge transfer and transport rates should be controlled so that all electrons are extracted for external load without charge recombination. The transport and transfer rates are interdependent, and thus simultaneous control of the rates is important to achieve high efficiency. Challenging of the DSCs is that the development of the solar cell covers the wide range of chemistry and physics, and all components including organic and inorganic materials should be designed properly in view of kinetics of electron transport and transfer, whose mechanism is governed by nanosized structure. It is also an interesting aspect where collaboration among different fields of scientists is important to develop the solar cells and to elucidate the mechanism of solar cells' working principle. Further increase of the efficiency is gained by the interdisciplinary research.

APPENDIX: METHODS TO MEASURE ELECTRON DIFFUSION COEFFICIENTS AND ELECTRON LIFETIME

The values of D and τ can be measured by analyzing photocurrent or photovoltage responses against the perturbation of light [26, 31, 33, 38, 70, 113, 114] or applied bias potential [69, 115]. First step is to choose the method of perturbation and solve a differential equation describing the density of electrons in TiO_2. By assuming that the electron transport occurs by diffusion, and D and τ are independent of electron density, time-dependent electron density $n(x,t)$ can be written by

$$\frac{\partial n(x,t)}{\partial t} = D\frac{\partial^2 n(x,t)}{\partial x^2} + G(x,t) - R(x,t) \tag{A1}$$

where x is the distance from TCO along the thickness of the TiO_2 electrode, t the time, G the charge injection rate, and R the recombination rate. Light perturbation can be applied by sinusoidal modulated light source [26, 37, 38], pulsed laser [31, 33, 114], or step-wise change of light intensity [70]. In order to solve the differential equation easily, using pulsed laser would be the choice. When the pulse length is much shorter than the transient time, the term of $G(x,t)$ can be neglected. If the wavelength of the pulsed laser is short, electrons are mostly generated at the surface of the TiO_2 electrode and the initial condition of electron density can be approximated by a delta function [114]. Next, by choosing the TiO_2 thickness, which allows electrons to travel to TCO before meeting charge acceptors, the term of $R(x,t)$ can also be dropped, and the equation becomes very simple [31].

In the electron lifetime measurements, it is not necessary to have the photoelectrons flow. By selecting wavelength where the sensitizer dyes barely absorb, homogeneous electron generation can be expected. Under the condition, the differential equation becomes independent of the position, and diffusion term in Eq. (A1) can be dropped. Then, Eq. (A1) becomes Eq. (2). V_{oc} scales with the Fermi level, which is related the density of electrons in TiO_2 (n) through $\log(n/n_0)$. By measuring the V_{oc} change against the light perturbation, the time-course change of n can be also monitored. In reality, both D and τ depend on n. Solving Eq. (A1) including the dependency is not facile. To avoid the situation, perturbation of very small fraction of applied light intensity is effective. Under this condition, the DSCs can be approximated as a steady condition, and thus D and τ can be dealt with a constant value. Drawback of the method is that one has to measure very small changes of the V_{oc} and J_{sc}, e.g., less than 1 mV and 100 nA on large background signals. For the measurements, appropriate amplifiers and/or high-resolution multimeters are needed.

Since D and τ depend on n, measured values should be compared under the same electron density. Electron density can be measured by integration of current transients [114], charge extraction methods [116], and monitoring light absorption at infrared region [103]. For qualitative analysis, short-circuit current may be used as a measure of electron density. Electron lifetime measured at open circuit conditions has been plotted as a function of J_{sc}, which is measured under the same light intensity used for the lifetime measurements. When the results are replotted as a function of electron density, the tendency of the results is the same but the degree of the difference becomes large [50].

ACKNOWLEDGMENTS

We would like to thank Dr. W. Kubo, Dr. S. Kambe, Dr. Y. Saito, Dr. N. Masaki, Dr. T. Kitamura, Dr. Y. Wada, Miss M. Matsuda and Mr. T. Kanzaki for their valuable discussions.

REFERENCES

[1]   N. Vlachopoulos, P. Liska, J. Augustynski and M. Grätzel, J. Am. Chem. Soc., 110 (1988) 1216–1220.

[2]   B. O'Regan and M. Grätzel, Nature, 353 (1991) 737.

[3]   M. Gratzel, J. Photochem. Photobiol. A 164 (2004) 3.

[4]   Y. Saito, S. Kambe, T. Kitamura, Y. Wada and S. Yanagida, Sol. Energy Mater. Sol. Cells, 83 (2004) 1–13.

[5]   F. Pichot and B.A. Gregg, J. Phys. Chem. B 104 (2000) 6–10.

[6]   C.A. Kelly, F. Farzad, D.W. Thompson, J.M. Stipkala and G.J. Meyer, Langmuir, 15 (1999) 7074.

[7]   K. Hara, T. Sato, R. Katoh, A. Furube, Y. Ohga, A. Shinpo, S. Suga, K. Sayama, H. Sugihara and H. Arakawa, J. Phys. Chem. B 107 (2003) 597–606.

[8]   S.A. Haque, E. Palomares, B.M. Cho, A.N.M. Green, N. Hirata, D.R. Klug and J.R. Durrant, J. Am. Chem. Soc., 127 (2005) 3456–3462.

[9]   Y. Tachibana, K. Hara, K. Sayama and H. Arakawa, Chem. Mater., 14 (2002) 2527–2535.

[10] G.R.A. Kumara, S. Kaneko, M. Okuya and K. Tennakone, Langmuir, 18 (2002) 10493–10495.

[11] K. Murakoshi, R. Kogure, Y. Wada and S. Yanagida, Chem. Lett., 26 (1997) 471–472.

[12] U. Bach, D. Lupo, P. Comte, J.E. Moser, F. Weissörtcl, J. Salbeck, H. Spreitzer and M. Grätzel, Nature, 395 (1998) 583–585.

[13] W. Kubo, K. Murakoshi, T. Kitamura, S. Yoshida, M. Haruki, K. Hanabusa, H. Shirai, Y. Wada and S. Yanagida, J. Phys. Chem. B 105 (2001) 12809–12815.

[14] H. Usui, H. Matsui, N. Tanabe and S. Yanagida, J. Photochem. Photobiol. A 164 (2004) 97–101.

[15] W. Kubo, S. Kambe, S. Nakade, T. Kitamura, K. Hanabusa, Y. Wada and S. Yanagida, J. Phys. Chem. B 107 (2003) 4374–4381.

[16] P. Wang, S.M. Zakeeruddin, J.E. Moser and M. Grätzel, J. Phys. Chem. B 107 (2003) 13280–13285.

[17] S.A. Sapp, C.M. Elliott, C. Contado, S. Caramori and C.A. Bignozzi, J. Am. Chem. Soc., 124 (2002) 11215–11222.

[18] H. Nusbaumer, J.E. Moser, S.M. Zakeeruddin, M.K. Nazeeruddin and M. Grätzel, J. Phys. Chem. B 105 (2001) 10461–10464.

[19] S. Nakade, Y. Makimoto, W. Kubo, T. Kitamura, Y. Wada and S. Yanagida, J. Phys. Chem. B 109 (2005) 3488–3493.

[20] S. Hattori, Y. Wada, S. Yanagida and S. Fukuzumi, J. Am. Chem. Soc., 127 (2005) 9648–9654.

[21] P. Wang, S.M. Zakeeruddin, J.E. Moser, R. Humphry-Baker and M. Grätzel, J. Am. Chem. Soc., 126 (2004) 7164–7165.

[22] A. Hinsch, J.M. Kroon, R. Kern, I. Uhlendorf, J. Holzbock, A. Meyer and J. Ferber, Prog. Photovoltaics, 9 (2001) 425–438.

[23] P.M. Sommeling, M. Späth, H.J.P. Smit, N.J. Bakker and J.M. Kroon, J. Photochem. Photobiol. A 164 (2004) 137–144.

[24] S. Nakade, T. Kanzaki, S. Kambe, Y. Wada and S. Yanagida, Langmuir, 21 (2005) 11414–11417.

[25] S. Södergren, A. Hagfeldt, J. Olsson and S.E. Lindquist, J. Phys. Chem., 98 (1994) 5552–5556.

[26] F. Cao, G. Oskam, G.J. Meyer and P.C. Searson, J. Phys. Chem., 100 (1996) 17021–17027.

[27] J. van de Lagemaat, N.G. Park and A.J. Frank, J. Phys. Chem. B 104 (2000) 2044–2052.

[28] L. Dloczik, O. Ileperuma, I. Lauermann, L.M. Peter, E.A. Ponomarev, G. Redmond, N.J. Shaw and I. Uhlendorf, J. Phys. Chem. B 101 (1997) 10281–10289.

[29] N.G. Park, J. van de Lagemaat and A.J. Frank, J. Phys. Chem. B 104 (2000) 8989–8994.

[30] S. Nakade, M. Matsuda, S. Kambe, Y. Saito, T. Kitamura, T. Sakata, Y. Wada, H. Mori and S. Yanagida, J. Phys. Chem. B 106 (2002) 10004–10010.

[31] S. Nakade, W. Kubo, Y. Saito, T. Kanzaki, T. Kitamura, Y. Wada and S. Yanagida, J. Phys. Chem. B 107 (2003) 14244–14248.

[32] J. Nelson, Phys. Rev. B 59 (1999) 15374–15380.

[33] J. van de Lagemaat and A.J. Frank, J. Phys. Chem. B 105 (2001) 11194–11205.

[34] J. Bisquert, J. Phys. Chem. B 108 (2004) 2323–2332.

[35] K. Takeshita, Y. Sasaki, M. Kobashi, Y. Tanaka, S. Maeda, A. Yamakata, T.A. Ishibashi and H. Onishi, J. Phys. Chem. B 108 (2004) 2963–2969.

[36] M. Bailes, P.J. Cameron, K. Lobato and L.M. Peter, J. Phys. Chem. B 109 (2005) 15429–15435.

[37] G. Schlichthörl, S.Y. Huang, J. Sprague and A.J. Frank, J. Phys. Chem. B 101 (1997) 8141–8155.

[38] A.C. Fisher, L.M. Peter, E.A. Ponomarev, A.B. Walker and K.G.U. Wijayantha, J. Phys. Chem. B 104 (2000) 949–958.

[39] A. Zaban, M. Greenshtein and J. Bisquert, Chem. Phys. Chem., 4 (2003) 859–864.

[40] J. Krüger, R. Plass, M. Grätzel, P.J. Cameron and L.M. Peter, J. Phys. Chem. B 107 (2003) 7536–7539.

[41] J. Nelson, S.A. Haque, D.R. Klug and J.R. Durrant, Phys. Rev. B 63 (2001) 2053211–2053219.

[42] J. Bisquert and V.S. Vikhrenko, J. Phys. Chem. B 108 (2004) 2313–2322.

[43] J. Bisquert, A. Zaban, M. Greenshtein and I. Mora-Sero, J. Am. Chem. Soc., 126 (2004) 13550–13559.

[44] A. Hagfeldt and M. Grätzel, Chem. Rev., 95 (1995) 49–68.

[45] D. Kuciauskas, M.S. Freund, H.B. Gray, J.R. Winkler and N.S. Lewis, J. Phys. Chem. B 105 (2001) 392–403.

[46] J.N. Clifford, E. Palomares, M.K. Nazeeruddin, M. Grätzel, J. Nelson, X. Li, N.J. Long and J.R. Durrant, J. Am. Chem. Soc., 126 (2004) 5225–5233.

[47] S.A. Haque, Y. Tachibana, R.L. Willis, J.E. Moser, M. Grätzel, D.R. Klug and J.R. Durrant, J. Phys. Chem. B 104 (2000) 538–547.

[48] I. Montanari, J. Nelson and J.R. Durrant, J. Phys. Chem. B 106 (2002) 12203–12210.

[49] S. Pelet, J.E. Moser and M. Grätzel, J. Phys. Chem. B 104 (2000) 1791–1795.

[50] S. Nakade, T. Kanzaki, W. Kubo, T. Kitamura, Y. Wada and S. Yanagida, J. Phys. Chem. B 109 (2005) 3480–3487.

[51] G. Oskam, B.V. Bergeron, G.J. Meyer and P.C. Searson, J. Phys. Chem. B 105 (2001) 6867–6873.

[52] A. Hagfeld and M. Grätzel, Chem. Rev., 95 (1995) 49–68.

[53] S. Kambe, S. Nakade, Y. Wada, T. Kitamura and S. Yanagida, J. Mat. Chem., 12 (2002) 723–728.

[54] H. Lindström, A. Holmberg, E. Magnusson, S.E. Lindquist, L. Malmqvist and A. Hagfeldt, Nano Letters, 1 (2001) 97–100.

[55] T. Miyasaka, Y. Kijitori, T.N. Murakami, N. Kawashima, Proc. SPIE Int. Soc. Opt. Eng., 5215 (2004) 219–225.

[56] T. Kado, M. Yamaguchi, Y. Yamada and S. Hayase, Chem. Lett., 32 (2003) 1056–1057.

[57] D. Zhang, T. Yoshida, K. Furuta and H. Minoura, J. Photochem. Photobiol. A 164 (2004) 159–166.

[58] T. Kanzaki, S. Nakade, Y. Wada and S. Yanagida, Photochem. Photobiol. Sci., 5 (2006) 389–394.

[59] N.G. Park, G. Schlichthörl, J. van de Lagemaat, H.M. Cheong, A. Mascarenhas and A.J. Frank, J. Phys. Chem. B 103 (1999) 3308–3314.

[60] S. Nakade, Y. Saito, W. Kubo, T. Kitamura, Y. Wada and S. Yanagida, J. Phys. Chem. B 107 (2003) 8607–8611.

[61] N. Kopidakis, N.R. Neale, K. Zhu, J. van de Lagemaat and A.J. Frank, Appl. Phys. Lett., 87 (2005) 1–3.

[62] H. Lindström, H. Rensmo, S.E. Lindquist, A. Hagfeldt, A. Henningsson, S. Södergren and H. Siegbahn, Thin Solid Films, 323 (1998) 141–145.

[63] Q.L. Zhang, L.C. Du, Y.X. Weng, L. Wang, H.Y. Chen and J.Q. Li, J. Phys. Chem. B 108 (2004) 15077–15083.

[64] B.A. Gregg, F. Pichot, S. Ferrere and C.L. Fields, J. Phys. Chem. B 105 (2001) 1422–1429.

[65] S.G. Chen, S. Chappel, Y. Diamant and A. Zaban, Chem. Mater., 13 (2001) 4629–4634.

[66] A. Kay and M. Grätzel, Chem. Mater., 14 (2002) 2930–2935.

[67] E. Palomares, J.N. Clifford, S.A. Haque, T. Lutz and J.R. Durrant, J. Am. Chem. Soc., 125 (2003) 475–482.

[68] J.-H. Yum, S. Nakade, D.-Y. Kim and S. Yanagida, J. Phys. Chem. B 110 (2006) 3215–3219.

[69] F. Fabregat-Santiago, J. Garcia-Canadas, E. Palomares, J.N. Clifford, S.A. Haque, J.R. Durrant, G. Garcia-Belmonte and J. Bisquert, J. Appl. Phys., 96 (2004) 6903–6907.

[70] S. Nakade, T. Kanzaki, Y. Wada and S. Yanagida, Langmuir, 21 (2005) 10803–10807.

[71] Y. Liu, A. Hagfeldt, X.R. Xiao and S.E. Lindquist, Sol. Energy Mater. Sol. Cells, 55 (1998) 267–281.

[72] A. Hagfeldt, S.E. Lindquist, Y. Liu and X.R. Xiao, Sol. Energy Mater. Sol. Cells, 55 (1998) 267–281.

[73] N. Kopidakis, E.A. Schiff, N.G. Park, J. van de Lagemaat and A.J. Frank, J. Phys. Chem. B 104 (2000) 3930–3936.

[74] S. Nakade, S. Kambe, T. Kitamura, Y. Wada and S. Yanagida, J. Phys. Chem. B 105 (2001) 9150–9152.

[75] S. Kambe, S. Nakade, T. Kitamura, Y. Wada and S. Yanagida, J. Phys. Chem. B 106 (2002) 2967–2972.

[76] N. Kopidakis, K.D. Benkstein, J. van de Lagemaat and A.J. Frank, J. Phys. Chem. B 107 (2003) 11307–11315.

[77] S. Nakade, W. Kubo, T. Kanzaki, N. Masaki, Y. Wada, S. Yanagida, Denki Kagaku Kai Fall Meeting Abstract 1E06 (2005).

[78] M.K. Nazeeruddin, A. Kay, I. Rodicio, R. Humphry-Baker, E. Müller, P. Liska, N. Vlachopoulos and M. Grätzel, J. Am. Chem. Soc., 115 (1993) 6382–6390.

[79] H. Kusama, M. Kurashige and H. Arakawa, J. Photochem. Photobiol. A 169 (2005) 169–176.

[80] S.Y. Huang, G. Schlichthörl, A.J. Nozik, M. Grätzel and A.J. Frank, J. Phys. Chem. B 101 (1997) 2576–2582.

[81] K. Murakoshi, G. Kano, Y. Wada, S. Yanagida, H. Miyazaki, M. Matsumoto and S. Murasawa, J. Electroanal. Chem., 396 (1995) 27–34.

[82] M.K. Nazeeruddin, R. Humphry-Baker, P. Liska and M. Grätzel, J. Phys. Chem. B 107 (2003) 8981–8987.

[83] Y. Tachibana, S.A. Haque, I.P. Mercer, J.E. Moser, D.R. Klug and J.R. Durrant, J. Phys. Chem. B 105 (2001) 7424–7431.

[84] J.B. Asbury, E. Hao, Y. Wang, H.N. Ghosh and T. Lian, J. Phys. Chem. B 105 (2001) 4545–4557.

[85] A. Furube, R. Katoh, K. Hara, T. Sato, S. Murata, H. Arakawa and M. Tachiya, J. Phys. Chem. B 109 (2005) 16406–16414.

[86] M.K. Nazeeruddin, S.M. Zakeeruddin, R. Humphry-Baker, M. Jirousek, P. Liska, N. Vlachopoulos, V. Shklover, C.H. Fischer and M. Grätzel, Inorg. Chem., 38 (1999) 6298–6305.

[87] M.K. Nazeeruddin, P. Péchy, T. Renouard, S.M. Zakeeruddin, R. Humphry-Baker, P. Cointe, P. Liska, L. Cevey, E. Costa, V. Shklover, L. Spiccia, G.B. Deacon, C.A. Bignozzi and M. Grätzel, J. Am. Chem. Soc., 123 (2001) 1613–1624.

[88] S.M. Zakeeruddin, M.K. Nazeeruddin, R. Humphry-Baker, P. Péchy, P. Quagliotto, C. Barolo, G. Viscardi and M. Grätzel, Langmuir, 18 (2002) 952–954.

[89] P. Wang, S.M. Zakeeruddin, J.E. Moser, M.K. Nazeeruddin, M. Grätzel and T. Sekiguchi, Nature Mater., 2 (2003) 402–407.

[90] L. Schmidt-Mende, J.E. Kroeze, J.R. Durrant, M.K. Nazeeruddin and M. Grätzel, Nano Letters, 5 (2005) 1315–1320.

[91] N. Hirata, J.J. Lagref, E.J. Palomares, J.R. Durrant, M.K. Nazeeruddin, M. Grätzel and D. Di Censo, Chem. Eur. J., 10 (2004) 595–602.

[92] S. Ferrere, Chem. Mater., 12 (2000) 1083–1089.

[93] G. Sauvè, M.E. Cass, S.J. Dolg, I. Lauermann, K. Pomykal and N.S. Lewis, J. Phys. Chem. B 104 (2000) 3488–3491.

[94] G. Sauvè, M.E. Cass, G. Coia, S.J. Doig, I. Lauermann, K.E. Pomykal and N.S. Lewis, J. Phys. Chem. B 104 (2000) 6821–6836.

[95] K. Sayama, S. Tsukagoshi, K. Hara, Y. Ohga, A. Shinpou, Y. Abe, S. Suga and H. Arakawa, J. Phys. Chem. B 106 (2002) 1363–1371.

[96]   K. Hara, M. Kurashige, Y. Dan-Oh, C. Kasada, A. Shinpo, S. Suga, K. Sayama and
       H. Arakawa, New J. Chem., 27 (2003) 783–785.
[97]   K. Hara, M. Kurashige, S. Ito, A. Shinpo, S. Suga, K. Sayama and H. Arakawa,
       Chem. Commun., (2003) 252–253.
[98]   T. Horiuchi, H. Miura, K. Sumioka and S. Uchida, J. Am. Chem. Soc., 126 (2004)
       12218–12219.
[99]   T. Kitamura, M. Ikeda, K. Shigaki, T. Inoue, N.A. Anderson, X. Ai, T. Lian and
       S. Yanagida, Chem. Mater., 16 (2004) 1806–1812.
[100] K. Hara, K. Miyamoto, Y. Abe and M. Yanagida, J. Phys. Chem. B 109 (2005)
       23776–23778.
[101] A. Kay and M. Grätzel, J. Phys. Chem., 97 (1993) 6272–6277.
[102] K. Hara, Y. Dan-Oh, C. Kasada, Y. Ohga, A. Shinpo, S. Suga, K. Sayama and
       H. Arakawa, Langmuir, 20 (2004) 4205–4210.
[103] N.R. Neale, N. Kopidakis, J. van de Lagemaat, M. Grätzel and A.J. Frank, J. Phys.
       Chem. B 109 (2005) 23183–23189.
[104] P. Wang, S.M. Zakeeruddin, P. Comte, R. Charvet, R. Humphry-Baker and
       M. Grätzel, J. Phys. Chem. B 107 (2003) 14336–14341.
[105] S. Sakaguchi, H. Ueki, T. Kato, T. Kado, R. Shiratuchi, W. Takashima, K. Kaneto
       and S. Hayase, J. Photochem. Photobiol. A 164 (2004) 117–122.
[106] P. Wang, S.M. Zakeeruddin, P. Comte, I. Exnar and M. Grätzel, J. Am. Chem.
       Soc., 125 (2003) 1166–1167.
[107] N. Yamanaka, R. Kawano, W. Kubo, T. Kitamura, Y. Wada, M. Watanabe and
       S. Yanagida, Chem. Commun., (2005) 740–742.
[108] Y. Saito, N. Fukuri, R. Senadeera, T. Kitamura, Y. Wada and S. Yanagida,
       Electrochem. Commun., 6 (2004) 71–74.
[109] Y. Saito, W. Kubo, T. Kitamura, Y. Wada and S. Yanagida, J. Photochem.
       Photobiol., A, 164 (2004) 153–157.
[110] J. He, H. Lindström, A. Hagfeldt and S.E. Lindquist, Sol. Energy Mater. Sol.
       Cells, 62 (2000) 265–273.
[111] A. Nakasa, H. Usami, S. Sumikura, S. Hasegawa, T. Koyama and E. Suzuki,
       Chem. Lett., 34 (2005) 500–501.
[112] A. Morandeira, G. Boschloo, A. Hagfeldt and L. Hammarström, J. Phys. Chem. B
       109 (2005) 19403–19410.
[113] S. Nakade, Y. Saito, W. Kubo, T. Kanzaki, T. Kitamura, Y. Wada and S. Yanagida,
       J. Phys. Chem. B 108 (2004) 1628–1633.
[114] A. Solbrand, H. Lindström, H. Rensmo, A. Hagfeldt, S.E. Lindquist and
       S. Södergren, J. Phys. Chem. B 101 (1997) 2514–2518.
[115] F. Fabregat-Santiago, J. Bisquert, G. Garcia-Belmonte, G. Boschloo and
       A. Hagfeldt, Sol. Energy Mater. Sol. Cells, 87 (2005) 117–131.
[116] N.W. Duffy, L.M. Peter, R.M.G. Rajapakse and K.G.U. Wijayantha, Electrochem.
       Commun., 2 (2000) 658–662.

Nanostructured Materials for Solar Energy Conversion
T. Soga (editor)

Chapter 8

Dye-Sensitized Nanostructured ZnO Electrodes for Solar Cell Applications

Gerrit Boschloo, Tomas Edvinsson and Anders Hagfeldt

Center of Molecular Devices, Department of Chemistry and Physical Chemistry, Royal Institute of Technology (KTH), Teknikringen 30, 100 44 Stockholm, Sweden

1. INTRODUCTION

Dye-sensitized nanostructured solar cells (DNSCs) based on nanostructured metal oxide films have attracted much attention in recent years [1–4]. They offer the prospect of low-cost photovoltaic energy conversion. Promising solar-to-electrical energy conversion efficiencies of more than 10% has been achieved [5, 6] and good progress has been made on long-term stability [7]. The working mechanism of dye-sensitized solar cells differs completely from conventional p–n junction solar cells, but is, after more than 15 years of research, still not completely resolved. A schematic representation of the DNSC is given in Fig. 1. Research has largely focused on nanostructured TiO_2 (anatase) as the metal oxide to which the dye is bound. Good results have, however, also been obtained using other n-type metal oxides, such as ZnO, Nb_2O_5 and SnO_2. In this review, we will focus on ZnO as a material for DNSC.

ZnO is an attractive material for nanoscale optoelectronic devices, as it is a wide-band gap semiconductor with good carrier mobility and can be doped both n and p-type [8]. The electron mobility is much higher in ZnO than in TiO_2, while the conduction band edge of both materials is located at approximately the same level. The material properties will be discussed in more detail in Section 2.

A large range of fabrication procedures is available for ZnO nanostructures, such as sol–gel processes, chemical bath deposition, electrodeposition and vapor-phase processes. Different morphologies such as spherical particles, rods, wires and hollow tubes can be prepared with relative ease.

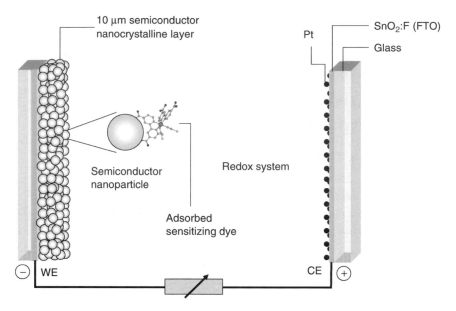

Fig. 1. Schematic illustration of the working principle of the DNSC. Incoming light excites the sensitizing dye, which induces an electron injection from the dye into the conduction band of the semiconductor nanoparticle. The electron is transported through the nanoporous structure to the back contact of the working electrode (WE). The electrons can then be utilized in work in an external circuit. The circuit is connected to the counter electrode (CE) where the redox system is effectively reduced with the help of a catalyst (nanoparticulate Pt in the case of I^-/I_3^-). The circuit is closed when the redox system reduces the sensitizing dye.

ZnO shows more flexibility in synthesis and morphology than TiO_2. The chemical stability of ZnO is, however, less than that of TiO_2, which can lead to problems in the dye adsorption procedure [9]. An overview of the preparation and dye-sensitization of nanostructured ZnO is presented in Section 3.

The performance of dye-sensitized ZnO solar cells in terms of solar-to-electrical energy conversion efficiencies is so far significantly lower than that of TiO_2, reaching currently about 4–5%. An analysis of the energetics and kinetics of ZnO-based DNSCs suggests that this is mainly due to the lesser degree of optimization in case of ZnO compared to TiO_2-based DNSCs. The main problem in dye-sensitized ZnO solar cells appears to be the troublesome dye-adsorption process. A summary of the best results obtained with dye-sensitized ZnO solar cells is given in Section 4, along with a discussion about the energetics and the kinetics in these cells.

Finally, in Section 5, we will present a short outlook for ZnO as a solar cell material for the future.

2. MATERIAL PROPERTIES

Zinc oxide (ZnO) has a large application potential owing to the diverse physical properties and the fine-tuning in the preparation process. Historically, ZnO has been used for a long time as pigment and protective coatings on metal surfaces. Its wide band gap of 3.2 eV at room temperature has rendered the use as protective UV-absorbing additive in everything from skin cream to advanced plastic and rubber composites. In the new era of nanotechnology, the potential ranges are spanning the vast fields of nano-electronics and acousto-optics. For example, ZnO has been used in photocopy machines, transparent conducting layers, varistors, as optical wave guides, surface acoustic wave transducers and thin film transistors [10]. The wide band gap has also made it suitable for short-wavelength optoelectronic devices, including UV detectors, photocatalysts, laser diodes and light-emitting diodes (LEDs) [8]. The high exciton binding energy of ZnO, 60 meV, gives a higher resistance for thermally induced exciton recombination compared to other semiconductors. For comparison, the thermal energy is 26 meV at room temperature and the exciton-binding energies of e.g. ZnSe and GaN are 22 and 25 meV, respectively. This makes ZnO a promising candidate for room temperature exciton lasers in UV or near UV. There are also some promising features for tuning of the emission via state confinement in quantum dots and organic end-capped ZnO nanoparticles [11–13]. Furthermore, the piezoelectric, ferroelectric and ferromagnetic effects in ZnO with different dopants show promise in different multifunctional applications using both the semi conductance and the spin configuration.

Here we will describe the most relevant photochemical and physical properties of zinc oxide (ZnO) for the application in nanostructured solar cells. For a comprehensive review of ZnO materials and devices, we refer to the review of Ozgur et al. [10]. Relevant comparisons will be made with TiO$_2$ (anatase), the most commonly used semiconductor in DNSC.

2.1. Bulk Properties of ZnO

In Table 1, we have summarized selected material properties for ZnO (wurtzite) and TiO$_2$ (anatase). Zinc oxide in the DNSCs has normally a hexagonal (wurtzite) structure with the lattice parameters a, $b = 3.25$ and $c = 5.12$ Å. The structure is shown in Fig. 2. ZnO is a direct band gap semiconductor with an optical band gap of 3.2 eV at room temperature. TiO$_2$ anatase has a similar band gap, but its lowest energy optical transition is indirect (see Fig. 3).

Table 1
Selected material properties of single-crystalline ZnO and TiO$_2$

		ZnO	TiO$_2$	Unit	References
Crystal structure		Wurtzite	Anatase		
Lattice constant, a		3.25	3.78	Å	[110, 111]
Lattice constant, c		5.12	9.51	Å	[110, 111]
Density		5.6	3.79	g cm^{-3}	[30, 110]
Static dielectric constant	ε_s	7.9	31		[112, 113]
Optical dielectric constant[a]	ε_∞	3.7	6.25		[30, 113]
Optical band gap	E_{obg}	3.2	3.2	eV	[22, 114]
Flat band potential[b]	E_{fb}	−0.5	−0.5	V vs. SCE	[20, 22]
Effective electron mass	M^*	0.24–0.3 m_e	1.0 m_e	$m_e = 9.11 \times 10^{-31}$ kg	[37, 114–116]
Effective hole mass	m_h^*	0.45–0.6 m_e	0.8 m_e	$m_e = 9.11 \times 10^{-31}$ kg	[37, 115, 116]
Electron mobility[c]	μ_e	200	30	cm^{2}V^{-1}s^{-1}	[117–119]
Point of zero charge	Pzc	8–9	5.5–6.5	pH	[24, 25]

[a]ε_∞ for anatase was calculated from n^2 from Ref. [30].
[b]The value from Ref. [20] is recalculated to pH = 5 with a Nernstian shift of −59mV/pH.
[c]Hall mobility for the n-type materials at 300 K.

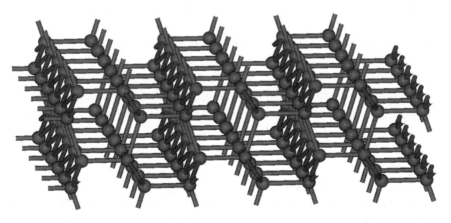

Fig. 2. The crystal structure of wurtzite ZnO in a 6 × 2 × 3 unit cell lattice.

The electrical conductivity of ZnO is determined by defects in the material that are present intrinsically or incorporated on purpose. ZnO is intrinsically doped via oxygen vacancies and/or zinc interstitials which act as n-type donors [14, 15]. Recently it has been shown by experiments [16–18] and

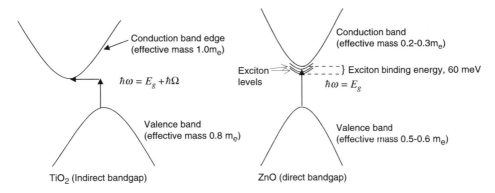

Fig. 3. A potential energy representation of the conduction and valence band profiles of the indirect band gap of TiO$_2$ (anatase) and the direct band gap of ZnO (wurtzite).

first-principle calculations [19] that the incorporation of hydrogen produces shallow states 30–40 meV below the conduction band edge. Furthermore, ZnO is frequently doped with aluminum or fluorine to obtain highly n-type conducting films. The electron mobility in single-crystalline ZnO is much higher than that of TiO$_2$ anatase. Mobilities tend to decrease upon doping due to scattering of the electrons at the impurities, and more substantial decrease when going from single crystals to polycrystalline and nanocrystalline materials, due to scattering at grain boundaries and energy barriers at these boundaries.

The conduction band edge positions in ZnO and TiO$_2$ are very similar, about -0.5 V vs. SCE (Saturated Calomel Electrodes) at pH 5 [20–23], and shows a Nernstian shift of -59 mV/pH. The point of zero charge (pzc) for ZnO is determined to pH $= 8$–9 depending on the preparation method and the experimental conditions [24], whereas credible values of the pzc for titanium dioxide reported in literature range from 5.5 to 6.5 [25, 26].

There is a large difference in the static dielectric constants of ZnO (7.9) and TiO$_2$ anatase (31–58) [22, 27]. The static dielectric constant for anatase may depend on preparation conditions, as values ranging from 35 to 108 have been reported, also within the same method of preparation [28, 29]. TiO$_2$ in the rutile form has a static dielectric constant of 90 perpendicular to the c-axis and 170 along the c-axis [30].

For relatively large ZnO particles (10–150 nm), substantially larger than the Bohr radius (as defined below), the similar optical properties and the density of states are very similar for the particles and the single-crystalline phase [21, 31]. The band edges of semiconductor nanoparticles are known to shift with the type and amount of cations present at the interface [32].

In a bulk n-type semiconductor in contact with a material with a different Fermi level (a metal, p-type semiconductor or an electrolyte); a space charge layer will be formed resulting in a band bending. We can anticipate a band bending for relatively large particles (150 nm), whereas for particles in the range of 10–20 nm, the estimated potential difference between the surface and the bulk using the Poisson–Boltzmann equation with an ionic surrounding, is negligible and thus indicating the absence of band bending [3, 33, 34].

2.2. Nanoscale Particles–Quantum Size Effects

In order to discuss the change in physical properties when going from bulk to nanosystems, we will briefly touch upon the effects of nano-confinement. In a nanoscale semiconductor, the crystal dimensions are too small for development of band bending and there will be an energy band sensitive to the surface conditions [3, 23]. If the size is comparable or smaller than the localization range of an electron, as quantified by the effective electron mass, one will enter the regime of quantum confinement.

Quantum confinement will affect the electronic spectra if the radius of the semiconductor particle is commensurable or smaller than the Bohr radius. The Bohr radius for the hydrogen in vacuum, a_0, is determined by $a_0 = \hbar^2 \varepsilon_0 / q^2 m_e = 0.53$ Å where $\hbar$ is the Planck's constant divided by 2π, $\varepsilon_0 e$ the dielectric constant in vacuum, q the charge and m_e the electron rest mass. The electronic structure of donors in bulk semiconductors and the range where quantum confinement starts to interfere with the electronic structure are successfully described by the effective mass approximation, which uses a formal analogy with the hydrogen atom. Within the effective mass approximation, the Bohr radius can be determined via [35]

$$a^* = \frac{\hbar^2 \varepsilon_r}{q^2 m^*} = \frac{\varepsilon_\infty}{m^* / m_e} a_0 \tag{1}$$

where m^* is the effective electron mass and ε_∞ the optical (high frequency) dielectric constant. Using $\varepsilon_\infty = 3.7$ and the effective mass $0.3\, m_e$ for ZnO, we obtain $a^* = 6.5$ nm. For TiO_2 (anatase), with $\varepsilon_\infty = 6.25$ and the effective mass $1.0\, m_e$ we arrive at $a^* = 3.3$ nm. The effect of quantum-sized confinement can most directly be seen for ZnO semiconductor particles with size smaller than the Bohr radius of the material. For these particles (< 4–5 nm diameter) an increase in band gap energies is seen arising from the confinement of the electronic states [31, 36, 37]. For anatase, the quantum effect can only be seen for particles smaller than 2–3 nm. In practice, spectroscopy on these small quantum dots can be complicated. The onset of absorption may not be very sharp due to band-tailing and polydispersity of the dots. In addition, one should also be aware of the Burstein–Moss shift, where filling of band states leads to an apparent increase in the band gap energy.

For particles with dimensions approaching the Bohr radius, also the model has limitations. The electronic response in an electromagnetic field cannot rigorously be analyzed with the same molecular parameters as the large single-crystalline phase. At these circumstances the carriers arc confined within the nanoparticle and the boundaries will to a larger extent affect the molecular parameters and pose a limit to e.g. the effective mass approximation [38]. In practice, however, analysis based on the effective mass approximation has proven to predict many properties of confinement effects in quantum dots [31, 36, 37].

The interest of going from spherical particles to nanorods is to obtain a more directional transport to the conductive substrate. Here one would not expect any grain boundary barriers, as been reported for spherical non-epitaxially oriented ZnO [39–42]. Another advantage is the easier pore filling with a hole-conducting medium. Quantum confinement effects in nanorods are in many aspects similar to the situation for the spherical nanoparticles, with the difference that the nanorods are confined only in two dimensions compared to three dimensions. For spherical nanoparticles, it is sufficient to consider the average effective mass, $m^* = (m^*_1 \cdot m^*_{t1} \cdot m^*_{t2})^{1/3}$ where m_1 is the longitudinal and m_{t1} and m_{t2} are the two transverse effective masses. The subtleties of the different effective masses in different crystal directions for the particles in these systems are obscured by differences in orientation in the nanostructured film. For single crystal ZnO nanowires, the crystal lattice orientation is not randomized but epitaxially grown and the different effective masses should be considered.

3. PREPARATION OF ZnO FOR DNSC APPLICATION

3.1. Synthesis of ZnO Nanoparticles

Nanostructured ZnO electrodes can be either synthesized from ZnO nanoparticles that are prepared in a separate procedure or produced in a single synthetic step. An overview of different synthetic procedures and resulting morphologies is given in Table 1. We will first discuss several synthetic routes for ZnO nanoparticles.

Crystalline ZnO nanoparticles can be prepared at room temperature in non-aqueous solutions [43–47]. In a typical preparation, a solution of zinc

Table 2
Preparation methods for ZnO nanostructures

Preparation method	Morphology	References
Non-aqueous solution chemistry	Spherical nanoparticles	[43–47]
	Nanorods	[48]
Controlled precipitation (aqueous solution)	Aggregated particles	[49, 50]
Hydrothermal synthesis	Nanorods	[53]
Electrodeposition	Nanocolumns on substrate	[58, 59]
	Nanostructured film on substrate	[60, 61]
Chemical bath deposition	Nanorods/wire on substrate	[54–56]
	Nanotube on substrate	
Vapor–liquid–solid deposition	Nanorods/wire on substrate	[64, 65]

acetate in alcohol is mixed with an equimolar amount of hydroxide. The dehydrating properties of the solvent prevent the formation of zinc hydroxide and promote formation of crystalline ZnO. Transparent ZnO colloidal solutions are easily prepared. The particles tend to be approximately spherical and their size depends strongly on the preparation temperature. Quantum-sized particles are obtained at low temperatures. Prolonged reflux of the colloidal solution results in formation of ZnO nanorods [48].

In aqueous preparations, zinc hydroxide rather than zinc oxide tends to be formed. Zinc hydroxide decomposes at 125°C to form zinc oxide. Several preparation methods have been developed (see Table 2), in which crystalline ZnO is directly formed in aqueous solution by controlled precipitation at temperatures below 125°C [49, 50]. In these methods elevated temperatures are used and chelating agents are used to prevent formation of zinc hydroxide. Alternatively, zinc hydroxide or carbonate nanoparticles can be precipitated from aqueous solution and converted to ZnO by heat treatment [51, 52]. Hydrothermal treatment at 180°C has been used to form ZnO nanorods with a diameter of about 50 nm [53]. Other methods to prepare ZnO nanoscale powders are spray pyrolysis and other gas-phase methods.

3.2. Synthesis of Nanostructured ZnO Films

3.2.1. *Preparation of Nanostructured ZnO Films from Nanoparticle Suspension*

Nanostructured electrodes can be prepared by application of a nanoparticle dispersion onto a conducting substrate, followed by heat treatment. The conducting substrate is usually conducting glass, consisting of glass coated with a thin layer of a transparent conducting oxide (TCO), such as F-doped SnO_2 or Sn-doped In_2O_3 (ITO). Typical application methods for the dispersion are doctor blading, screen printing, spin coating, dip coating and spraying.

3.2.2. Direct Synthesis of Nano-structured ZnO Films onto Substrates

Chemical Bath Deposition. Chemical bath deposition (CBD) is a low temperature method ($<150°C$) in which a substrate is immersed in a semi-stable precursor solution [54–56]. For ZnO preparations the solution contains Zn^{2+} ions and a chelating agent, such as methenamine. Instead of a chelator, an excess of hydroxide can be used so that the $Zn(OH)_4^{2-}$ complex is formed [57]. Reactions will occur preferentially at the substrate where crystalline ZnO can be formed, but may also occur to some extent in solution. The deposition rate can be controlled by adjusting the temperature, the pH and the relative concentration of the reactants in the solution. The morphology and crystallinity of the deposit depends strongly on the substrate and seeding layers that may be present. Vayssieres et al. [55, 56] described formation of ordered crystalline ZnO microstructures (rods and tubes) and nanorods on several types of substrate. Peterson et al. [57] prepared uniform arrays of nanorods grown epitaxially onto a sputtered ZnO film. Fig. 4 displays SEM pictures of ZnO nanorods grown with similar methods in our laboratory.

Electrodeposition. Electrodeposition is a low-temperature method, where the ZnO is formed directly on the substrate. Film morphology and thickness

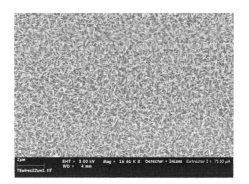 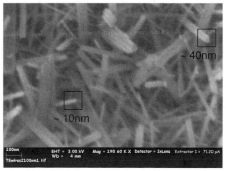

Fig. 4. SEM micrographs of epitaxially grown ZnO nanowires. The nanorods were grown from a nucleation matrix directly on a conducting surface. The nucleation matrix was created with spray pyrolysis using a supersaturated Zn^{2+} solution and the growth was performed by immersing the substrate in a second, slightly altered supersaturated solution of Zn^{2+} at 80°C. The scale bars in the left and right picture corresponds to 2 μm and 100 nm, respectively.

can be controlled by applied potential, current density, reactant concentrations, temperature and deposition time. Nano-columns of ZnO were prepared by electrodeposition on several substrates [58, 59]. Using non-aqueous conditions, deposition of nanostructured ZnO films was achieved [60, 61]. In addition, successful electrodeposition of ZnO in the presence of dye was reported, [62] which will be discussed further in Section 3.3.2.

Controlled Oxidation of Zn in Solution. In this method, similar to CBD, but also electrochemical in nature, a Zn substrate is immersed in an aqueous formamide solution at slightly elevated temperatures [63]. Zn is oxidized by oxygen to form soluble Zn^{2+}-formamide complexes, which in turn can deposit and react to ZnO on the Zn substrate. Aligned ZnO nanorods are formed. Deposition can also take place on other substrates in close vicinity of the Zn plate.

Gas-phase Methods. Nanorods and wires have been produced by vapor–liquid–solid deposition [64]. The substrate is first coated with a thin layer of a catalyst such as gold. A ZnO/graphite mix is placed close to the substrates in a furnace tube and heated about 900°C under an argon flow. Zn and CO vapor is formed, and the Zn forms an alloy with the Au particles. ZnO starts to grow when the alloy becomes saturated with Zn. Epitaxial growth of oriented nanowires on silicon substrates has been reported [65].

3.3. Dye-Sensitization

The dye-sensitization process is an important step in the preparation of DNSC. For TiO_2 films dye-sensitization is achieved by immersing the nanostructured electrode into typically a 0.5 mM ethanolic dye solution. Carboxyl groups are commonly used anchoring groups for chemisorption of dye molecules onto metal oxide surfaces [2]. For TiO_2, the strong adsorption of the carboxyl groups to the surface favors monolayer growth. For ZnO, on the other hand, the sensitization process is more complex as described below.

3.3.1. Conventional Dye Adsorption from Solution

For the frequently used dyes, such as *cis*-bis(isothiocyanato)bis(2, 2_-bipyridyl-4,4_-dicarboxylato)-ruthenium(II) (or $Ru(dcbpyH_2)_2(NCS)_2$, "N3") and *cis*-bis(isothiocyanato)bis(2,2-bipyridyl-4,4-dicarboxylato)-ruthenium(II) bis-tetrabutylammonium (or $Ru[dcbpy(TBA)_2]_2(NCS)_2$, "N719"), sensitization of ZnO may be described by the following processes, see Fig. 5: diffusion of the dye into the ZnO nanostructure, adsorption of the dye to the ZnO surface, dissolution of Zn surface atoms from ZnO and formation of Zn^{2+}/dye complexes in the pores of the ZnO film [9]. Depending

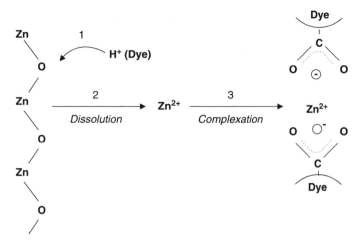

Fig. 5. A schematic drawing of the problems associated with the dye-sensitization process of protonated anchoring groups on ZnO surfaces.

on the rate of these different processes, the outer part of the electrode may be in the process of forming Zn^{2+}/dye complexes in the pores when dye molecules reach the interface between the back contact and the ZnO film. Thus, the latter process is on the way to increasing the efficiency, whereas the former is in the phase of decreasing the efficiency. The results indicated that protons from dyes cause dissolution of Zn surface atoms and the formation of Zn^{2+}/dye complexes in the pores of the nanostructured film, which gives rise to a filter effect (inactive dye molecules). Consequently, the net yield for charge carrier injection is decreased whereas the light harvesting efficiency is increased due to the large number of dye molecules in the film, as discussed below. A schematic drawing of the Zn^{2+} dissolution and Zn^{2+}/dye complexation processes can be seen in Fig. 5. The complex formation between Zn^{2+}-ions originating from ZnO surface and different sensitizers has been reported in the literature [66, 67]. The formation of metal-ion/dye complexes does not take place in the case of TiO_2 electrodes but is apparently extremely important in case of ZnO.

Based on these facts, careful control of the dye composition, concentration, pH and sensitization time is necessary in order to avoid the Zn^{2+}/dye complex formation and to achieve high-efficiency solar cells based on ZnO. One possible reason why protons from the carboxyl groups of the dye initiate a dissolution process of the ZnO and a formation of Zn^{2+}/dye complexes is related to the surface properties of the oxide. The point of zero charge (pzc) of metal oxides is defined as the pH where the concentrations of protonated and deprotonated surface groups are equal. It means that the surface is predominantly positively charged at pH below pzc and negatively charged above this value. The pH for the dye-sensitization process (pH = 5) is much lower than the pzc of ZnO (pzc ≈ 9). Bahnemann [68] has shown that dissolution of ZnO colloids occurs below pH 7.4. Thus, the protons adsorbed on the ZnO surface will dissolve the ZnO. In the case of TiO_2 the pH of the dye solution is similar to the pzc of the oxide. Another difference between TiO_2 and ZnO relates to the fact that Ti atoms are six-fold coordinated in the bulk whereas the Zn atoms are four-fold coordinated. The typical surface coordination of the metal ion is 5 and 3 for TiO_2 and ZnO, respectively, resulting in a relatively larger loss of coordination for ZnO. Moreover, theoretical investigations indicate that the bond length between Zn and O atoms on the ZnO (1 0 1 0) surface increases upon the adsorption of formic acid, making the Zn–O bond weaker and prone to a Zn atom dissolution process [69, 70].

3.3.2. Electrodeposition of ZnO in the Presence of Dye

An interesting alternative to the conventional consolidation of the nanostructured oxide film by doctor blading/screen printing and sintering comes from the work on the electrodeposition of nanostructured ZnO electrodes by Yoshida and co-workers [62, 71–75]. Cathodic electrodeposition of ZnO from aqueous zinc salt solutions in the presence of oxidants and water-soluble dye molecules provides a nanostructured photoanode to be used in dye-sensitized solar cells. Dye molecules with attachment groups such as $-COO^-$ and $-SO_3^-$ adsorb to the growing oxide film and besides being used as sensitizers the dyes are also influential in formulating the structure and morphology of the nanostructured film. Highly porous ZnO has been obtained especially in the presence of O_2 and the red xanthene dye eosin Y.

As for our own studies described above in Section 3.3.1, the solar cell efficiencies are hampered by dye aggregation in the pores of the as-deposited films originating from the use of acidic anchoring groups. Thus, quenching of the photogenerated electrons resulting in low-electron injection efficiency has been observed. This problem has to a large extent been overcome by desorption and readsorption of the dye, leading to the formation of a dye monolayer on the ZnO surface and a significant improvement of the electron injection efficiency being as high as 90% [72]. Although the eosin Y dye has a relatively narrow absorption band, an overall solar-to-electrical energy conversion efficiencies of 2.3% was achieved. There is certainly potential for improving the solar cell efficiency of these materials, in the first place by combining the electrodeposited film with broader absorbing dyes suited for ZnO. Another aspect of these materials is that the preparation process is compatible with plastic film substrates, thus leading to the future realization of flexible DNSCs.

3.3.3. Quantum Dot Sensitization and Solid-State Devices

As an alternative to dye-sensitization of large band gap oxide semiconductors, there have been considerable efforts to replace the dyes with inorganic quantum-sized semiconductor particles [76]. Visible light can be absorbed by the Q-particles which, consequently, transfer electrons into the nanostructured oxide material. In some of the early work using quantum size CdS on ZnO, Hotchandani and Kamat [77] measured quantum yields of up to 15%, whereas very high internal quantum yields of up to 80% were achieved by Vogel et al. [76] for the same type of materials.

For research on complete solid-state devices based on a nanostructured large band gap semiconductor electrode, the concept of the extremely thin absorber (ETA) solar cell has been recently proposed [78]. The basic structure of an ETA solar cell is a nano- or microstructured layer of a wide-band gap material ($E_g > 3$ eV) which also serves as an n-type window layer to the cell, an absorber with $1.1 < E_g < 1.8$ eV conformally coated on the window layer, and a void-filling p-type wide-band gap material ($E_g > 3$ eV). At present there are only few published results of complete ETA solar cells. For ZnO, a very recent and interesting paper describes a complete ETA cell using CdSe as absorber and CuSCN as p-type hole conductor [79]. The electrochemically deposited ZnO consisted of free-standing single-crystal nanowires oriented along the c-axis. The nanowires were ~100–150 nm across and had a length within the range of 1–2 μm. The best solar cell efficiency was measured to 2.3%. The electron lifetime was comparable to the electron transport time (microsecond-regime), implying that recombination was limiting charge collection in these devices. Thus, a reduction of recombination by, for example, improved interfacial properties would give significant increases in photocurrent. A similar type of solid-state device based on nanostructured ZnO films was developed by O'Regan: ruthenium complexes sandwiched between ZnO and CuSCN served as the absorber and an efficiency of 1.5% was achieved [60].

Alternatively n-type ZnO nanoparticles can be combined with a p-type conjugated polymer. Such a device can be seen as a hybrid between DSC and organic bulk heterojunction solar cells, in which ZnO replaces, for example, fullerenes as electron acceptors in the latter technology. The advantage with a hybrid polymer-inorganic nanocomposite could be the combination of a solution processing of polymer semiconductors in, for example, a roll-to-roll process, and the high mobility of inorganic semiconductors. A recent thorough study of a hybrid device using nanoparticulate ZnO films, or ZnO nanowires, blended with the conjugated polymer poly[2-methoxy-5-(3',7'-dimethyloctyloxy)-1,4-phenylenevinylene] (MDMO-PPV) as electron donor [80a]. The best performance was found for a blend with 30 vol.% ZnO (nanoparticles) and thickness of ~100 nm, sandwiched between a transparent (PEDOT:PSS on ITO) and a metal (aluminum) electrode. For such a device, an efficiency of about 1.6% could be estimated under AM (Air Mass) 1.5 conditions. A better control over the complex morphology of the nanoparticulate ZnO: conjugated polymer blends is then the main focus point for improving the photovoltaic effect of these hybrid devices.

4. THE DYE-SENSITIZED NANOSTRUCTURED ZnO SOLAR CELL

DNSC technology gives a large flexibility in the choice of the components allowing the ability to fine-tune the properties of the solar cell. The choice of metal oxide for the nanostructured electrode will affect the range of band gap, band edge positions, the density of states and the electron transport and loss properties. The choice will also affect the possibility to obtain different shapes and doping densities of the nanoparticles. For each choice of oxide material the particle size, the surface structure and the porosity and film thickness are also important parameters to optimize for each type of sensitizing dye and redox system.

Systems based on TiO_2 sensitized with ruthenium complexes with bis- or tris-pyridines and thiocyanate ligands and carboxyl acids as attachment groups are the most extensively studied. These systems has so far also shown to be the most efficient for dye-sensitized solar cells, reaching over 10% overall conversion efficiency [5, 6]. ZnO has also been extensively studied in this context. It started in the 1960s where a lot of interest was steered toward increase in photoconductivity in dye-sensitized ZnO. The earliest work traces back to the 1950s, see e.g. the references in the review by Meier on dye-sensitization from 1965 [80b]. As a photovoltaic device, it started with the work of Gerischer and Tributsch [81–83] in the late 1960s when monocrystalline ZnO was sensitized with ruthenium pyridinium complexes. The use of a porous ZnO electrode with an enhanced surface area was first reported in 1976 by Tsubomura et al. [84].

Table 3
Compilation of the overall efficiencies reached in sensitized nanostructured ZnO solar cells

Size/film thickness	Sensitizer	Intensity (W m^{-2})	Efficiency (%)	Reference
150 nm	N719	100	5	[85]
11 nm	N719	1000	4.1	[120]
10–20 nm	Mercurochrome	1000	2.5	[121]
5 nm	MDMO-PPV[a]	710	1.6	[80]
Porous crystal	Eosin Y	1000	2.3	[72]
Wires 100–150 nm	CdSe	360	2.3	[79]
Wires 130–200 nm	N719	1000	1.5	[122]
ZnO sheets	N719	1000	3.9	[123]

[a]MDMO-PPV = poly[2-methoxy-5-(3′,7′-dimethyloctyloxy)-1,4-phenylenevinylene].

Solar cell efficiencies of dye-sensitized ZnO have reached 5% at 100 W cm^{-2} [85] and 4.6% at 1000 W cm^{-2} [86] showing promise for the future. In Table 3, we summarize overall efficiencies reached with the dye-sensitized nanostructured ZnO solar cell. DNSCs based on ZnO nanowires have not yet been able to give better performance than those based on spherical particles (see Table 3). Although the electron transport is expected to be improved in the nanowires, the light absorption of the film is insufficient so far. Further fine-tuning of the ZnO nanostructure is necessary to obtain optimized DNSCs with respect to charge transport as well as dye adsorption and minimized recombination.

The use of zinc oxide as an outer shell on nanoparticulate SnO$_2$ electrodes has given promising results with overall solar cell efficiencies ranging from 5.2 to 6.3% using the N719 dye [87, 88]. The reported 9% efficiency [89] should be treated with care, as an anomalously high short circuit photocurrent was recorded, indicating a spectral or intensity mismatch in the measurement equipment. Nevertheless, the comparison with TiO$_2$ in the same study showed great potential for the ZnO/SnO$_2$ system. In the following, we describe the current understanding of the energetics and kinetics of the ZnO DNSC device.

4.1. Energetics

The energy level matching between dye, semiconductor and redox electrolyte is crucial for the performance of the DNSC. The dye should absorb a large fraction of the incoming sunlight. The energy difference between the highest occupied molecular orbitals (HOMO) and the lowest unoccupied

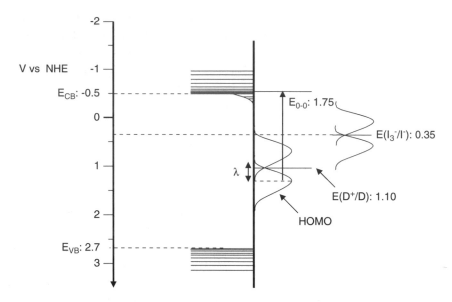

Fig. 6. Schematic diagram of the energetics in a solar cell with ZnO, N719 and I^-/I_3^- electrolyte. The energy-level distribution of the energy states in solution are illustrated as Gaussian distributions in the diagram. The electronic excitation of the dye is most likely from the average energy level in the ground state distribution. The HOMO is estimated via the oxidation potential of the dye and the reorganization energy, λ. The zeroth–zeroth transition (E_{0-0}) is determined via the normalized intercept of the absorption and emission spectra.

molecular orbitals (LUMO) of the dye is important in this respect. A smaller gap would result in the possibility to absorb more of the solar spectrum, but at the cost of a lower voltage output. In order to obtain good injection efficiency of excited electrons from dye molecules into the conduction band of the semiconductor, a good overlap between the excited state levels of the dye molecule with conduction band levels is required. Recombination of injected electrons with the oxidized dye should be prevented. Fast interception of the oxidized dye by the reductant in the redox electrolyte is therefore essential. There should be sufficient driving force for electron transfer between the reductant and the oxidized dye. The maximum voltage that can be obtained from the DNSC is given by the potential difference between the conduction band edge and the redox potential. An overview of the energetics of the dye-sensitized nanostructured ZnO solar cell is shown in Fig. 6.

Experimental methods to determine the energetics in DNSCs include electrochemistry and photoelectron spectroscopy (PES). PES has been used

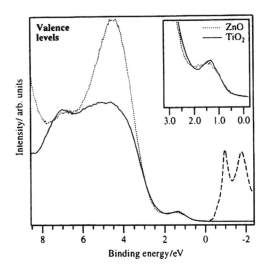

Fig. 7. Photoelectron spectroscopy: valence level spectra of N3-sensitized nanostructured ZnO and TiO$_2$. The striped line (- -) between 0 and −2 shows the calculated LUMO level of N3 and the inset shows a close up of the HOMO region.

to investigate the energy level matching for Ru(dcbpyH$_2$)$_2$(NCS)$_2$ adsorbed on nanostructured ZnO compared with that of the same complex adsorbed on TiO$_2$ [90]. In Fig. 7, the valence PES spectra of this ruthenium complex adsorbed on TiO$_2$ and ZnO are shown. The dye HOMO level is clearly distinguished above the valence band edge for both semiconductors. The energy position of this peak with respect to the valence band of the semiconductor is similar in both cases. Since the band gap is 3.2 eV for both zinc oxide and titanium dioxide, this result indicates that the energy matching between the dye and the metal oxide is similar for ZnO and TiO$_2$.

4.2. Electron Transfer Kinetics

The functioning of dye-sensitized solar cells based on nanostructured ZnO depends on favorable kinetics for the forward reactions with respect to the recombination reactions. Current is generated as follows: light is absorbed by the dye and excited dye molecules inject electrons into the conduction band of ZnO. These electrons are collected at the substrate and extracted as a current. These reactions compete with several recombination reactions. The incident photon-to-current conversion efficiency (IPCE) is given by

$$\text{IPCE}(\lambda) = \text{LHE}(\lambda) \times \phi_{\text{inj}}(\lambda) \times \varphi_{\text{red}} \times \eta_{\text{c}}(\lambda) \qquad (2)$$

where LHE is the light-harvesting efficiency (the fraction of light that is absorbed at a specific wavelength), ϕ_{inj} the quantum yield of charge injection, φ_{red} the quantum yield of the reduction of the oxidized dye and η_c the efficiency of collecting the injected charge at the back contact.

4.2.1. Photoinduced Electron Injection

The quantum yield for electron injection is given by the following relation:

$$\varphi_{inj} = \frac{k_{inj}}{k_{inj} + k_{decay}} \tag{3}$$

where k_{inj} is the rate for electron injection and k_{decay} the rate for radiative and non-radiative decay of the excited state of the dye to the ground state. As the energy level from which the electron in the excited dye is injected may depend on the wavelength of the absorbed photon, there can be a wavelength dependence in k_{inj}.

The number of studies on photoinduced electron injection in dye-sensitized ZnO are relatively few, compared to that of the dye-sensitized TiO$_2$ system. The main reason is that solar cells based on ZnO show lower efficiencies compared to those based on TiO$_2$. Nevertheless, for scientific understanding of ultrafast electron transfer reactions the comparison of the ZnO and TiO$_2$ systems can provide very useful information. The energetic position of the conduction band of ZnO is similar to that of TiO$_2$ (anatase). The density of states is, however, expected to be about one order of magnitude lower in ZnO. Also the structure of the conduction bands differs. In ZnO it is mainly composed of Zn 4s and 4p orbitals, while in TiO$_2$ it is mainly composed of Ti 3d orbitals. As electron injection rates will depend on both density of acceptor states and electronic coupling, significant differences between ZnO and TiO$_2$ as acceptor may be expected.

Lian and coworkers reported multiple-exponential electron injection from Ru-polypyridyl complexes into ZnO. About 20% occurred very fast, <1 ps, while the rest took place on a 20–200 ps timescale [91, 92]. The difference with dye-sensitized TiO$_2$, where significantly faster electron injection is found, was explained by differences in electronic coupling and density of acceptor levels. Bauer et al. [93], however, reported that injection in the N719–ZnO system occurred in the sub-picosecond range only. As only one probe wavelength was used, their results are considered to be not fully conclusive.

Tashiya and coworkers investigated the N3–ZnO system and found evidence for stepwise electron injection [94]. First, formation of an intermediate state was observed (100 fs), followed by electron injection on a picosecond to nanosecond timescale. The same group suggested a similar mechanism for the coumarin derivate NKX2311–ZnO system that they studied, but found also evidence for ~40% direct electron injection (500 fs) [95]. The intermediate state was proposed to be an exciplex formed by the excited dye and a ZnO-based acceptor surface state [94, 95].

As mentioned before, aggregation of dyes is a problem in case of dye-sensitized ZnO. Solar cell efficiency was found to depend strongly on both ZnO preparation conditions and dye adsorption conditions. Dye aggregation is expected to affect electron injection kinetics in ZnO. Differences in the ultra-fast spectroscopy studies mentioned above can be at least in part attributed to the degree of dye aggregation that was present. Under conditions where aggregation is avoided in the N719 (N3)–ZnO system, quantitative electron injection was reported based on measured photocurrent quantum yields [93] or electron injection yields from transient absorption [96]. Horiuchi et al. [97] reported that N3–Zn^{2+} aggregates are not effective in electron injection. Anderson et al. [92] found that aggregates did not contribute to electron injection within 1 ns, as they observed similar kinetics for different preparation conditions, with and without aggregates.

A study that clearly demonstrates sub-picosecond photoinduced electron injection in dye-sensitized ZnO is that of Furube et al. [95] on a coumarin–ZnO system. They mentioned reasonable photocurrent conversion values for this system. A quantitative estimate of the injection efficiency was 0.8 [96]. The fact the sub-picosecond injection occurs in ZnO as well as TiO_2 shows that the lower density of conduction band states in ZnO does not prevent ultrafast injection kinetics.

The efficiency of photoinduced electron injection in dye-sensitized ZnO has been studied using nanosecond-spectroscopy by Katoh, Yoshihara et al. [97, 98]. A series of dyes was tested and efficient injection was observed when the potential of the relaxed excited state (E_{D0-0}) was more than 0.2 V negative of the ZnO conduction band edge. E_{D0-0} was estimated by adding the energy of the relaxed excited state, equal to the HOMO–LUMO difference, to the reversible oxidation potential of the dye $E^0(D^+/D)$.

Ultrafast spectroscopic measurements have also been performed on hybrid ZnO–polymer solar cells [80]. These cells consisted of a blend of a photoactive polymer and ZnO colloids. Electron injection was found to occur on a sub-picosecond timescale in these systems, while charge carriers were detectable up to milliseconds after excitation. This further confirms the possibility of ultrafast kinetics in dye-sensitized ZnO. As (radiative) decay times of excited dye molecules are generally larger than 1 ns, high injection yield may be expected for optimized dye-sensitized ZnO systems.

4.2.2. Recombination of Electrons in ZnO with Oxidized Dye Molecules

After electron injection, rapid regeneration of the oxidized dye should take place by the reduced part of the redox couple in the electrolyte, which is usually iodide in practical systems. Little experimental data is available on this reaction for dyes attached to ZnO. In 1983, Matsumura et al. [99] used a potential modulation technique to study this reaction in a ZnO-sintered disc electrode sensitized by rose bengal in aqueous electrolyte. They found a rate constant of about $10^4 \, M^{-1} s^{-1}$ using low concentrations of potassium iodide. Reduction of the oxidized N3 or N719 dye on TiO_2 was reported to take place on a $0.1–10 \, \mu s$ timescale [100, 101]. Similar values may be expected for optimized dye-sensitized ZnO systems, i.e. the reduction rate constant k_{red} is expected to be about $10^6 \, s^{-1}$. Competing with this dye-reduction reaction is the recombination of oxidized dye molecules with electrons in the ZnO, with a rate constant k_{rec}. A quantum yield may be defined for reduction of the oxidized dye by the redox electrolyte:

$$\phi_{red} = \frac{k_{red}}{k_{red} + k_{rec}} \tag{4}$$

The back electron transfer of injected electrons in dye-sensitized ZnO to oxidized dye molecules appears to be similar to dye-sensitized TiO_2 [93, 102]. Bauer et al. [93] found that recombination kinetics of both N719–ZnO and N719–TiO_2 systems under identical conditions could be fitted to a biexponential function with time constants of about 0.3 (50%) and $300 \, \mu s$ (50%). Using the same experimental technique, nanosecond flash photolysis, but with lower laser pulse intensity, Willis et al. reported stretched exponential decay kinetics on a $10^{-6}–10^{-1}$ second timescale for a similar Ru(HPterpy) $(Me_2 bpy)$NCS–ZnO system [102]. This behavior was similar to that of dye-sensitized TiO_2 in their system. They also investigated the effect of applied potential on recombination kinetics. Here a significant difference was found between ZnO and TiO_2: while recombination half times decrease with more negative applied potentials for both nanostructured semiconductors, this trend started at a 400 mV more positive potential for ZnO. They demonstrated that a numerical model based on random walk of electrons between localized sub-band gap states (traps) can account for the observed stretched exponential recombination kinetics in the nanostructured metal oxides. The difference between ZnO and TiO$_2$ was attributed mainly to a higher density of traps in ZnO [102].

Although experimental data of the dye-sensitized ZnO solar cell is still rather limited, it appears that ϕ_{red} will be close to one in optimized systems under typical operation conditions.

4.2.3. Recombination of Electrons in ZnO with Redox Species in the Electrolyte

Injected electrons in the nanostructured ZnO electrode have to travel a considerable distance and time before they are collected at the substrate. During this time they are always close to the semiconductor–electrolyte interface and thus at risk to recombine with the oxidized form of the redox couple, in most cases triiodide. The charge collection efficiency can be calculated as follows: [103]

$$\eta_C = 1 - \frac{\tau_{IMPS}}{\tau_n} \tag{5}$$

where τ_{IMPS} is the time constant obtained from intensity-modulated photocurrent spectroscopy (IMPS or a comparable technique), corresponding approximately to the average electron transport time, and τ_n the electron lifetime, i.e. the inverse of the (pseudo) first-order rate constant for the reaction of electrons in the ZnO with the redox electrolyte.

Very few studies deal with the electron transfer of electrons in dye sensitized ZnO to the oxidized part of the redox couple (i.e. triiodide). This reaction can be studied in complete nanostructured solar cells using intensity-modulated photovoltage spectroscopy (IMVS). The electron lifetime in ZnO associated with this process decreases with increasing light intensity, as is the case for nanostructured TiO$_2$ solar cells. Oekermann et al. [104] measured the electron lifetime in eosin Y-sensitized ZnO at open circuit conditions. In electrodeposited nanostructures, the electron lifetime was approximately four times smaller than in nanostructured films prepared from 30 nm-sized particles at the same light intensity. The difference may be attributed to the smaller size of the ZnO nanostructure in the electrodeposited electrodes, resulting in a closer distance between electrons in the semiconductor and triiodide in the electrolyte. Recent experiments in our laboratory suggest that the electron lifetime in ZnO is significantly larger than in TiO$_2$ in

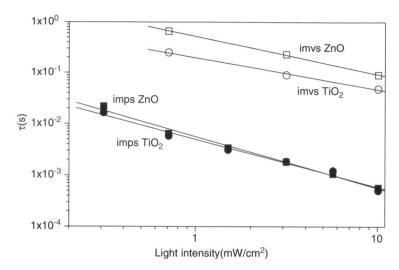

Fig. 8. Electron transport characterized with IMPS for ZnO and TiO$_2$ DNSCs. IMVS is here used to measure the electron lifetime in ZnO under open circuit conditions. The films were made with similar nanocrystal size (15 nm) and film thickness (4 μm). The transport in ZnO and TiO$_2$ are very similar under these conditions, whereas the electron lifetime is notably higher for ZnO.

nanostructured solar cells with similar nanocrystal size, see Fig. 8 [105]. Using a charge extraction method, Duffy et al. [106] found that the back reaction between electrons and triiodide was second order in electron concentration for both ZnO and TiO$_2$-based dye-sensitized solar cells.

4.2.4. *Electron Transport in Nanostructured ZnO*

Electron transport in electrolyte-permeated nanostructured ZnO films has been studied using different techniques such as photocurrent transients, IMPS and conductivity measurements. Hoyer and Weller used photocurrent transients to study electron transport in bare nanostructured ZnO electrodes. The photocurrent response became faster with more negative applied potentials, and electron transport times were found to decrease with increasing laser pulse intensity. Similar effects were observed by Solbrand et al. [107], who additionally showed that transport time was dependent on electrolyte concentration, becoming faster with higher concentrations. In contrast, de Jongh et al. [108], did not find an intensity dependence in an IMPS study, but this was attributed to a slow tunneling process of electrons from ZnO nanoparticles to the substrate. Using conductivity measurements, Meulenkamp [109] found that the mobility of electrons in nanostructured ZnO depended on the electron density in the film and was about four orders of magnitude lower than the mobility in ideal single crystals.

Electron transport in actual dye-sensitized ZnO solar cells becomes faster with increasing light intensity, as is the case for nanostructured TiO_2 solar cells. In electrodeposited eosin Y-sensitized ZnO nanostructures the transport time was found to be two orders of magnitude faster than in films based on 30 nm-sized ZnO particles having a similar dye loading (the thickness of the latter was about seven times more). It was concluded that the high crystallinity and the lower number of grain boundaries in the electrodeposited film was responsible for the faster transport [104]. Moreover, a recent study suggests that the porosity of the electrodeposited films increases with film thickness and the films therefore consist of two parts; a less porous part deposited in the first few minutes that exhibit field-driven electron transport and a more porous outer part where electron transport is governed by diffusion [71]. The electron transport times were much smaller than electron lifetimes, leading to a very efficient electron collection. In a comparison of dye-sensitized ZnO and TiO_2 nanostructured film with similar crystallite size, it was found that transport times are very similar, see Fig. 8 [105]. In single crystals, however, the electron mobility in ZnO is much higher than in TiO_2 (anatase), see Table 1. This suggests that the slow transport in both nanostructured ZnO and TiO_2 is mainly caused by the grain boundaries and/or trap states induced by the nanocrystallinity. Additionally, the fact that electrons are charge compensated by ions in the electrolyte may play an important role in the transport.

5. FUTURE OUTLOOK

DNSCs are devices whose essential characteristic behavior at the molecular level is collective. Conventional approaches are inadequate to describe such complex molecular systems where new and unexpected behavior emerge in response to the strong and competing interactions among the elementary constituents. For DNSC devices, based on nanostructured TiO_2 electrodes, top efficiencies of 11% on small laboratory samples have been reported. There is a potential for boosting this performance substantially by imaginative approaches that exploit nanoscience and an improved understanding of complex molecular devices. This will be made by developing new functional material components which are tailor-made for the systems. Since the basic knowledge of how the DNSC molecular system operates still remains unclear, the specifications on how to tailor-make the components are incomplete and there is still a lot of fundamental and exploratory work to be done. In this context, nanostructured ZnO electrodes stand out as one of the most exciting materials. A large range of fabrication procedures is available to create ZnO nanostructures. Different morphologies such as spherical particles, rods, wires and hollow tubes can be prepared with relative ease. Transparent conducting oxide substrates based on aluminum doped ZnO are reported to combine high conductivity and transparency. A plethora of ZnO materials with different particle sizes, morphologies and compositions are at present being reported and there are so far relatively few studies on the use of these materials for DNSC applications.

The performance of dye-sensitized ZnO solar cells in terms of solar-to-electrical energy conversion efficiencies is so far lower than that of TiO_2, reaching currently about 4–5%. This can, at least partly, be attributed to the much lower research activity on the use of ZnO. The main problem with ZnO at present is related to a complex dye-sensitization process. Acidic carboxyl groups are commonly used as anchoring groups and protons from these groups cause dissolution of Zn surface atoms with subsequent formation of Zn^{2+}/dye complexes in the pores of the nanostructured films. So far, virtually no approaches have been reported to circumvent this problem. Thus, a large potential exists to improve on the performance of dye-sensitized ZnO solar cells by learning how to use new types of anchoring groups and controlling the chemistry at the oxide/dye/electrolyte interface. This, in combination with the possibilities to tailor-make ZnO materials, manifests the opportunities for future research and development of these devices.

ACKNOWLEDGMENTS

We would like to thank BASF AG and the Swedish Energy Agency for their financial support.

REFERENCES

[1]   B. O'Regan and M. Grätzel, Nature (London), 353 (1991) 737–740.
[2]   M.K. Nazeeruddin, A. Kay, I. Rodicio, R. Humphry-Baker, E. Müller, P. Liska, N. Vlachopoulos and M. Grätzel, J. Am. Chem. Soc., 115 (1993) 6382–6390.
[3]   A. Hagfeldt and M. Grätzel, Chem. Rev., 95 (1995) 49–68.
[4]   A. Hagfeldt and M. Grätzel, Acc. Chem. Res., 33 (2000) 269–277.
[5]   M.K. Nazeeruddin, P. Péchy, T. Renouard, S.M. Zakeeruddin, R. Humphry-Baker, P. Comte, P. Liska, L. Cevey, E. Costa, V. Shklover, L. Spiccia, G.B. Deacon, C.A. Bignozzi and M. Grätzel, J. Am. Chem. Soc., 123 (2001) 1613–1624.

[6] Z.S. Wang, H. Kawauchi, T. Kashima and H. Arakawa, Coord. Chem. Rev. 248 (2004) 1381–1389.

[7] P. Wang, S.M. Zakeeruddin, J.E. Moser, M.K. Nazeeruddin, T. Sekiguchi and M. Grätzel, Nat. Mater, 2 (2003) 402–407.

[8] A. Tsukazaki, A. Ohtomo, T. Onuma, M. Ohtani, T. Makino, M. Sumiya, K. Ohtani, S.F. Chichibu, S. Fuke, Y. Segawa, H. Ohno, H. Koinuma and M. Kawasaki, Nat. Mater, 4 (2005) 42–46.

[9] K. Keis, J. Lindgren, S.-E. Lindquist and A. Hagfeldt, Langmuir, 16 (2000) 4688–4694.

[10] U. Ozgur, Y.I. Alivov, C. Liu, A. Teke, M.A. Reshchikov, S. Dogan, V. Avrutin, S.J. Cho and H. Morkoc, J. Appl. Phys., 98 (2005) 041301.

[11] H.-M. Xiong, D.-P. Liu, Y.-Y. Xia and J.-S. Chen, Chem. Mater, 17 (2005) 3062–3064.

[12] M. Shim and P. Guyot-Sionnest, J. Am. Chem. Soc., 123 (2001) 11651–11654.

[13] M. Shim and P. Guyot-Sionnest, Nature, 407 (2000) 981–983.

[14] S.B. Zhang, S.-H. Wei and A. Zunger, Phys. Rev. B, 63 (2001) 075205.

[15] D.C. Look, G.C. Farlow, P. Reunchan, S. Limpijumnong, S.B. Zhang and K. Nordlund, Phys. Rev. Lett., 95 (2005) 225502.

[16] D.M. Hofmann, A. Hofstaetter, F. Leiter, H. Zhou, F. Henecker and B.K. Meyer, Phys. Rev. Lett., 88 (2002) 045504-045501–045504-045504.

[17] S.J. Jokela and M.D. McCluskey, Phys. Rev. B, 72 (2005) 113201.

[18] M.D. McCluskey, S.J. Jokela, K.K. Zhuravlev, P.J. Simpson and K.G. Lynn, Appl. Phys. Lett., 81 (2002) 3807–3809.

[19] C.G. Van de Walle, Phys. Rev. Lett., 85 (2000) 1012–1015.

[20] M.A. Butler and D.S. Ginley, J. Electrochem. Soc., 125 (1978) 228–232.

[21] G. Redmond, A. O'Keeffe, C. Burgess, C. MacHale and D. Fitzmaurize, J. Phys. Chem., 97 (1993) 11081–11086.

[22] L. Kavan, M. Grätzel, S.E. Gilbert, C. Klemenz and H.J. Scheel, J. Am. Chem. Soc., 118 (1996) 6716–6723.

[23] B. O'Regan, J. Moser, M. Anderson and M. Grätzel, J. Phys. Chem., 94 (1990) 8720–8726.

[24] L. Blok and P.L.D. Bruyn, J. Coll. Int. Sci., 32 (1970) 518–526.

[25] M. Kosmulski, Adv. Coll. Int. Sci., 99 (2002) 255–264.

[26] N. Kallay, D. Babic and E. Matijevic, Colloids Surf., 19 (1986) 375–386.

[27] R. van de Krol, A. Goossens and J. Schoonman, J. Electrochem. Soc., 144 (1997) 1723–1727.

[28] K. Fukushima and I. Yamada, J. Appl. Phys., 65 (1989) 619–623.

[29] K.-T. Lin and J.-M. Wu, Jpn. J. Appl. Phys., 43 (2004) 232–236.

[30] D.R. Lide, Handbook of Chemistry and Physics, CRC Press, Boca Raton, 2004–2005.

[31] C.L. Dong, C. Persson, L. Vayssieres, A. Augustsson, T. Schmitt, M. Mattesini, R. Ahuja, C.L. Chang and J.-H. Guol, Phys. Rev. B, 70 (2004) 195321–195325.

[32] Y. Liu, A. Hagfeldt, X.-R. Xiao and S.-E. Lindquist, Sol. Energy Mater. Sol. Cells, 55 (1998) 267–281.

[33] W.J. Albery and P.N. Bartlett, J. Electrochem. Soc., 131 (1984) 315.

[34] A. Goossens, J. Electrochem. Soc., 143 (1996) L131–L133.

[35] J.I. Pankove, Optical Processes in Semiconductors, Dover Publications, Inc., New York, 1975.

[36] L. Brus and J. Phys. Chem., 90 (1986) 2555–2560.

[37] L. Brus, J. Phys. Chem., 80 (1984) 4403–4409.

[38] S.B. Orlinskii, J. Schmidt, E.J.J. Groenen, P.G. Baranov, C.d.M. Donegá and A. Meijerink, Phys. Rev. Lett., 94 (2005) 097602.

[39] J.P. Gambino, W.D. Kingery, G.E. Pike, H.R. Philipp and L.M. Levinson, J. Appl. Phys., 61 (1987) 2571–2574.

[40] F.M. Hossain, J. Nishii, S. Takagi, A. Ohtomo, T. Fukumura, H. Fujioka, H. Ohno, H. Koinuma and M. Kawasaki, J. Appl. Phys., 94 (2003) 7768–7777.

[41] S.D. Helder and P.D. Bristowe, Comp. Mater. Sci., 22 (2001) 38–43.

[42] F. Oba, S.R. Nishitani, H. Adachi, I. Tanaka, M. Kohyama and S. Tanaka, Phys. Rev. B, 63 (2001) 045410.

[43] U. Koch, A. Fojtik, H. Weller and A. Henglein, Chem. Phys. Lett., 122 (1985) 507–510.

[44] D.W. Bahnemann, C. Kormann and M.R. Hoffmann, J. Phys. Chem., 91 (1987) 3789–3798.

[45] M. Haase, H. Weller and A. Henglein, J. Phys. Chem., 92 (1988) 482–487.

[46] L. Spanhel and M.A. Anderson, J. Am. Chem. Soc., 113 (1991) 2826–2833.

[47] M. Hilgendorff, L. Spanhel, C. Rothenhausler and G. Muller, J. Electrochem. Soc., 145 (1998) 3632–3637.

[48] C. Pacholski, A. Kornowski and H. Weller, Angew. Chem. Int. Ed., 41 (2002) 1188.

[49] T. Trindade, J.D. Pedrosa de Jesus and P.O'Brien, J. Mater. Chem., 4 (1994) 1611–1617.

[50] K. Keis, L. Vayssieres, S.E. Lindquist and A. Hagfeldt, Nanostruct. Mater, 12 (1999) 487–490.

[51] M. Castellanot and E. Matijevic, Chem. Mater, 1 (1989) 78–82.

[52] H. Rensmo, K. Keis, H. Lindström, S. Södergren, A. Solbrand, A. Hagfeldt, S. E. Lindquist, L.N. Wang and M. Muhammed, J. Phys. Chem. B, 101 (1997) 2598–2601.

[53] B. Liu and H.C. Zeng, J. Am. Chem. Soc., 125 (2002) 4430–4431.

[54] P. O'Brien, T. Saeeda and J. Knowles, J. Mater. Chem., 6 (1996) 1135–1139.

[55] L. Vayssieres, K. Keis, A. Hagfeldt and S.E. Lindquist, Chem. Mater, 13 (2001) 4395.

[56] L. Vayssieres, Adv. Mater, 15 (2003) 464–466.

[57] R.B. Peterson, C.L. Fields and B.A. Gregg, Langmuir, 20 (2004) 5114–5118.

[58] T. Pauporte and D. Lincot, Appl. Phys. Lett., 75 (1999) 3817–3819.

[59] R. Könenkamp, K. Boedecker, M.C. Lux-Steiner, M. Poschenrieder, F. Zenia, C. Levy-Clement and S. Wagner, Appl. Phys. Lett., 77 (2000) 2575–2577.

[60] B. O'Regan, D.T. Schwartz, S.M. Zakeeruddin and M. Grätzel, Adv. Mater, 12 (2000) 1263–1267.

[61] B. O'Regan, V. Sklover and M. Grätzel, J. Electrochem. Soc., 148 (2001) C498–C505.

[62] T. Yoshida, M. Tochimoto, D. Schlettwein, D. Wöhrle, T. Sugiura and H. Minoura, Chem. Mater, 11 (1999) 2657–2667.

[63] Z. Zhang, H. Yu, X. Shao and M. Han, Chem. Eur. J., 11 (2005) 3149–3154.

[64] M.H. Huang, Y. Wu, Henning Feick, N. Tran, E. Weber and P. Yang, Adv. Mater, 13 (2001) 113–116.

[65] M.H. Huang, S. Mao, H. Feick, H. Yan, Y. Wu, H. Kind, E. Weber, R. Russo and P. Yang, Sci., 292 (2001) 1897–1899.

[66] R. Baumeler, P. Rys and H. Zollingen, Helv. Chim. Acta., 56 (1973) 2451.

[67]   R. Brändli, P. Rys, H. Zollingen, H. Oswald and F. Schweizer, Helv. Chim. Acta., 53 (1970) 1133.

[68]   D. Bahnemann, Isr. J. Chem., 33 (1993) 115–136.

[69]   P. Persson and L. Ojamäe, Chem. Phys. Lett., 321 (2000) 302–308.

[70]   P. Persson, S. Lunell and L. Ojamäe, Int. J. Quant. Chem., 89 (2002) 172–180.

[71]   T. Oekermann, T. Yoshida, C. Boeckler, J. Caro and H. Minoura, J. Phys. Chem. B, 109 (2005) 12560–12566.

[72]   T. Yoshida, M. Iwaya, H. Ando, T. Oekermann, K. Nonomura, D. Schlettwein, D. Wohrle and H. Minoura, Chem. Com., (2004) 400–401.

[73]   T. Oekermann, S. Karuppuchamy, T. Yoshida, D. Schlettwein, D. Wohrle and H. Minoura, J. Electrochem. Soc., 151 (2004) C62–C68.

[74]   T. Yoshida, K. Terada, D. Schlettwein, T. Oekermann, T. Sugiura and H. Minoura, Adv. Mater, 12 (2000) 1214–1217.

[75]   T. Yoshida and H. Minoura, Adv. Mater, 12 (2000) 1219–1222.

[76]   R. Vogel, P. Hoyer and H. Weller, J. Phys. Chem., 98 (1994) 3183–3188.

[77]   S. Hotchandani and P.V. Kamat, J. Phys. Chem., 96 (1992) 6834–6839.

[78]   K. Ernst, A. Belaidi and R. Konenkamp, Semicond. Sci. Techn., 18 (2003) 475–479.

[79]   C. Levy-Clement, R. Tena-Zaera, M.A. Ryan, A. Katty and G. Hodes, Adv. Mater, 17 (2005) 1512–1515.

[80a] W.J.E. Beek, M.M. Wienk, M. Kemerink, X.N. Yang and R.A.J. Janssen, J. Phys. Chem. B, 109 (2005) 9505–9516.

[80b] H. Meier, J. Phys. Chem., 69, 3 (1965) 719 729.

[81]   H. Gerischer and H. Tributsch, Ber. Bunsenges. Phys. Chem., 72 (1968) 437–445.

[82]   H. Tributsch and H. Gerischer, Ber. Bunsenges. Phys. Chem., 73 (1969) 850–854.

[83]   H. Tributsch and H. Gerischer, Ber. Bunsenges. Phys. Chem., 73 (1969) 251–260.

[84]   H. Tsubomura, M. Matsumura, Y. Nomura and T. Amamiya, Nature, 261 (1976) 402–403.

[85]   K. Keis, E. Magnusson, H. Lindström, S.-E. Lindquist and A. Hagfeldt, Sol. Energy Mater. Sol. Cells, 73 (2002) 51–58.

[86]   T. Yoshida, Private Communication Gifu University, Japan, 2005.

[87]   D. Niinobe, Y. Makari, T. Kitamura, Y. Wada and S. Yanagida, J. Phys. Chem. B, 109 (2005) 17892–17900.

[88]   A. Kay and M. Grätzel, Chem. Mater, 14 (2002) 2930–2935.

[89]   K. Tennakone, G.K.R. Senadeera, V.P.S. Perera, I.R.M. Kottegoda and L.A.A. De Silva, Chem. Mater, 11 (1999) 2474–2477.

[90]   K. Westermark, H. Rensmo, H. Siegbahn, K. Keis, A. Hagfeldt, L. Ojamae and P. Persson, J. Phys. Chem. B, 106 (2002) 10102–10107.

[91]   J.B. Asbury, Y. Wang and T.Q. Lian, J. Phys. Chem. B, 103 (1999) 6643–6647.

[92]   N.A. Anderson, X. Ai and T.Q. Lian, J. Phys. Chem. B, 107 (2003) 14414–14421.

[93]   C. Bauer, G. Boschloo, E. Mukhtar and A. Hagfeldt, J. Phys. Chem. B, 105 (2001) 5585–5588.

[94]   A. Furube, R. Katoh, K. Hara, S. Murata, H. Arakawa and M. Tachiya, J. Phys. Chem. B, 107 (2003) 4162–4166.

[95]   A. Furube, R. Katoh, T. Yoshihara, K. Hara, S. Murata, H. Arakawa and M. Tachiya, J. Phys. Chem. B, 108 (2004) 12583–12592.

[96]   T. Yoshihara, R. Katoh, A. Furube, M. Murai, Y. Tamaki, K. Hara, S. Murata, H. Arakawa and M. Tachiya, J. Phys. Chem. B, 108 (2004) 2643–2647.

[97]   H. Horiuchi, R. Katoh, K. Hara, M. Yanagida, S. Murata, H. Arakawa and M. Tachiya, J. Phys. Chem. B, 107 (2003) 2570–2574.

[98]   R. Katoh, A. Furube, K. Hara, S. Murata, H. Sugihara, H. Arakawa and M. Tachiya, J. Phys. Chem. B, 106 (2002) 12957–12964.

[99]   M. Matsumura, K. Mitsuda and H. Tsubomura, J. Phys. Chem., 87 (1983) 5248–5251.

[100]   S.A. Haque, Y. Tachibana, D.R. Klug and J.R. Durrant, J. Phys. Chem. B, 102 (1998) 1745–1749.

[101]   S. Pelet, J.-E. Moser and M. Grätzel, J. Phys. Chem. B, 104 (2000) 1791–1795.

[102]   R.L. Willis, C. Olson, B. O'Regan, T. Lutz, J. Nelson and J.R. Durrant, J. Phys. Chem. B, 106 (2002) 7650–7613.

[103]   J. van de Lagemaat, N.-G. Park and A.J. Frank, J. Phys. Chem. B, 104 (2000) 2044–2052.

[104]   T. Oekermann, T. Yoshida, H. Minoura, K.G.U. Wijayantha and L.M. Peter, J. Phys. Chem. B, 108 (2004) 8364–8370.

[105]   M. Quintana, G. Boschloo and A. Hagfeldt, J. Phys. Chem. B, (2005).

[106]   N.W. Duffy, L.M. Peter, R.M.G. Rajapakse and K.G.U. Wijayantha, Electrochem. Commun., 2 (2000) 658–662.

[107]   A. Solbrand, K. Keis, S. Södergren, H. Lindström, S.-E. Lindquist and A. Hagfeldt, Sol. Energy. Mater. Sol. Cells, 60 (2000) 181–193.

[108]   P.E.d. Jongh, E.A. Meulenkamp, D. Vanmaekelbergh and J.J. Kelly, J. Phys. Chem. B, 104 (2000) 7686–7693.

[109]   E. Meulenkamp, J. Phys. Chem. B, 103 (1999) 7831–7838.

[110]   D.T. Cromer and K. Herrington, J. Am. Chem. Soc., 77 (1955) 4708–4709.

[111]   J.K. Burdett, T. Hughbanks, G.J. Miller, J. James, W. Richardson and J.V. Smith, J. Am. Chem. Soc., 109 (1987) 3639–3646.

[112]   S. Roberts, Phys. Rev., 76 (1949) 1215–1220.

[113]   H. Yoshikawa and S. Adachi, Jpn. J. Appl. Phys., 36 (1997) 6237–6243.

[114]   H. Tang, K. Prasad, R. Sanjinbs, P.E. Schmid and F. Lévy, J. Appl. Phys., 75 (1994) 2042–2047.

[115]   N.N. Syrbu, I.M. Tiginyanu, V.V. Zalamai, V.V. Ursaki and E.V. Rusu, Phys. B, 353 (2004) 111–115.

[116]   B. Enright and D. Fitzmaurice, J. Phys. Chem., 100 (1996) 1027–1035.

[117]   D.C. Look, D.C. Reynolds, J.R. Sizelove, R.L. Jones, C.W. Litton, G. Cantwe and W.C. Harsch, Sol. Stat. Com., 105 (1998) 399–401.

[118]   C.H. Seager and S.M. Myers, J. Appl. Phys., 94 (2003) 2888–2894.

[119]   L. Forro, O. Chauvet, D. Emin, L. Zuppiroli, H. Berger and F. Lévy, J. Appl. Phys., 75 (1994) 633–635.

[120]   K. Kakiuchi, E. Hosono and S. Fujihara, J. Photochem. Photobiol. A. Chem., 179 (2006) 81–86.

[121]   K. Hara, T. Horiguchi, T. Kinoshita, K. Sayama, H. Sugihara and H. Arakawa, Chem. Lett., (2000) 316–317.

[122]   M. Law, L.E. Greene, J.C. Johnson, R. Saykally and P.D. Yang, Nat. Mater, 4 (2005) 455–459.

[123]   E. Hosono, S. Fujihara, I. Honna and H.S. Zhou, Adv. Mater, 17 (2005) 2091.

Nanostructured Materials for Solar Energy Conversion
T. Soga (editor)

Chapter 9

Solid-State Dye-Sensitized Solar Cells

Akira Fujishima and Xin-Tong Zhang

Kanagawa Academy of Science and Technology, KSP Bldg. West 614, 3-2-1 Sakado, Takatsu-ku, Kawasaki, Kanagawa 213-0012, Japan

1. INTRODUCTION

The all solid-state dye-sensitized solar cell (SSDSSC) has been a subject of research in the last 10 years [1–4]. This work has appeared as an offshoot of that on dye-sensitized photoelectrochemical cells (DSPECs); thus, the solid-state cell always exhibits a structure similar to the latter except for the replacement of the liquid electrolyte with a p-type semiconductor or p-type organic materials. This type of cell inherits the advantage of the DSPEC in terms of separating charge generation from charge transport. Therefore, the quality of the electrode material (purity and crystallinity), for either n-type or p-type semiconductors, is not as critical as that for classical photovoltaic cells, which have to provide good photoresponse as well as good charge mobility within the same compound. Moreover, being a kind of solid-state device, these cells are expected to be manufactured with less expensive technology than that for PECs because problems such as leakage, packaging, and corrosion, which exist for liquid electrolytes, could be avoided.

This chapter deals with the background and recent efforts in the area of SSDSSCs. Section 2 briefly introduces the general aspects of SSDSSCs. Section 3 describes the fabrication methods used for SSDSSCs, with the TiO$_2$/Ru dye/CuI cell as an example. Section 4 discusses the performance of SSDSSCs in terms of solar-energy conversion efficiency. Some problems of solid-state cells are also addressed in this section. Section 5 focuses on studies that have employed interfacial blocking layers to improve cell performance. Section 6 discusses the operational stability of SSDSSCs. The future outlook of these cells is covered in Section 7.

2. GENERAL ASPECTS

The concept of the SSDSSC was described by Tennakone and his research group in 1995 [1]. In their pioneering work, they assembled the solid-state solar cell from a flower pigment (cyanidin)-sensitized titania (TiO$_2$) nano-porous electrode and a p-type copper iodide (CuI) semiconductor. CuI was filled into the voids of the TiO$_2$ film by solution deposition. The operational principle (Fig. 1) of the solar cell was explained as follows: the cyanidin molecule injects an electron into the TiO$_2$ electrode conduction band under light excitation, the reduced state of the oxidized cyanidin molecule is regenerated by injecting a hole into p-type CuI, and the injected electron and hole are transported through the TiO$_2$ electrode and the CuI electrode, respectively, to the external circuit. The energy-conversion efficiency of the cell was estimated to be about 0.8% in direct sunlight, a rather low value. However, this was an exciting result, because it demonstrated the possibility of fabricating all solid-state dye-sensitized photovoltaic devices from inexpensive, low crystalline quality materials.

The advantages of a solid-state dye-sensitized photovoltaic device are apparent in manufacturing and operational stability, compared with dye-sensitized photoelectrochemical cells. In the solid-state device, the p-type semiconductor and the dye are required to have special properties. As summarized by Tennakone [1]: (1) the p-type material must be transparent in the visible spectrum, where the dye absorbs light; in other words, the p-type material must have a wide band gap; (2) a method must be available for

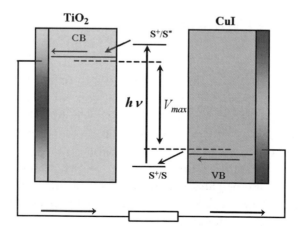

Fig. 1. Principle of operation of the solid-state dye-sensitized solar cell (SSDSSC).

depositing the p-type material without dissolving or degrading the monolayer of dye on the TiO_2 nanocrystallites; (3) the dye must be such that its lowest unoccupied molecular orbital (LUMO) level is located above the bottom of the conduction band of TiO_2 and its highest occupied molecular orbital (HOMO) level should be located below the upper edge of the valence band of the p-type material. CuI is a good candidate for the SSDSSC, because it has a wide band gap (3.1 eV) and can be easily dissolved in acetonitrile for deposition. The valence band edge of CuI (−5.3 V vs. the vacuum level) also matches the HOMO level of the ruthenium (Ru) bipyridyl dye used in the Grätzel cell. In a later report, Tennakone et al. [5] reported an energy conversion efficiency >3% by assembling the cell with a Ru dye-sensitized TiO_2 porous film and CuI.

Many p-type semiconducting materials, besides CuI, have been applied to SSDSSCs. One such material is copper thiocyanate (CuSCN), a stable Cu(I) p-type semiconductor [3, 6–9]. This material has a band gap of 3.6 eV and a valence band edge of −5.1 V with respect to the vacuum scale, which fits the requirements of solid-state dye-sensitized devices. The only shortcoming of CuSCN is its slight solubility in organic solvents. So far, the best solvent for CuSCN is dipropyl sulfide, a malodorous, toxic compound.

Spirobisfluorene-connected arylamine (*spiro*-OMeTAD) is another important p-type material for SSDSSCs [2, 10–11]. This material was first used in electroluminescent devices as a hole transmitter. The Grätzel group reported the application of this material in SSDSSCs by a very simple spin-coating method [2]. The material is amorphous, so that it can contact well with the dye monolayer. Transient spectroscopic data showed that *spiro*-OMeTAD reduced the oxidized Ru dye within times on the ns scale [10].

3. FABRICATION PROCEDURES

SSDSSCs are based on the same concept as that of DSPECs, that is, the use of a nanoporous TiO_2 film to enhance the light absorption of the dye monolayer. However, the fabrication procedures for the former are quite different from those for the latter. Fig. 2 illustrates the profile structure for a TiO_2/dye/CuI solid-state cell. The cell is made up, from left to right, of a conductive glass electrode, a compact TiO_2 thin film, a nanoporous TiO_2 film and dye monolayer, a CuI layer interpenetrated with the TiO_2 porous film, and a gold-coated conducting glass electrode. The 100-nm thick compact TiO_2 thin layer is necessary for SSDSSCs to avoid direct contact between the hole conductor and the conducting glass cathode, which would short-circuit the cell. The compact TiO_2 layer, however, is not needed for DSPECs. Gold

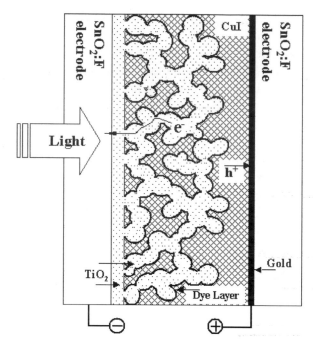

Fig. 2. Schematic of a cross-section of a SSDSSC [23].

film has always been used as the anode for SSDSSCs, but other conducting materials, such as platinum and graphite, also work well. DSPECs, however, can only use platinum as the counter electrode for the fast regeneration of the iodide electrolyte. The structure shown in Fig. 2 is a common one for all types of SSDSSCs, no matter what type of hole conductor is used.

The fabrication procedures begin with the compact, thin TiO_2 barrier layer on conducting glass. As mentioned earlier, this layer is very important to avoid the inner short circuit of the cell, and thus it must be pinhole-free. This layer also must be very thin (<100 nm) to not increase the internal resistance of the cell. The most common method to prepare such a compact, pinhole-free TiO_2 thin layer is spray pyrolysis, in which a titanium precursor solution, for example, $0.05\,mol\,L^{-1}$ titanium oxyactylactonate solution in ethanol, is sprayed onto heated conductive glass (over 450°C) through a nozzle. The film thickness can easily be controlled by appropriate choice of the volume of precursor solution. Electrochemical studies showed that the TiO_2-coated conductive glass became inactive for the oxidation of $Fe(CN)_6^{4-}$ ions. The sputtering and electron beam evaporation methods can be used to prepare pinhole-free TiO_2 layers of even higher quality, but the disadvantage is their high cost.

The sol-gel method is inexpensive, but it is difficult to prepare pinhole-free TiO_2 thin films via this method.

In the second step, a nanoporous TiO_2 film is prepared on the compact TiO_2 layer-coated conductive glass. The nanoporous TiO_2 film can be of the same type as that used in DSPECs, except for the smaller thickness that is used due to the pore filling and interfacial recombination problems of SSDSSCs, as we will discuss later. Studies of Krüger et al. [11] suggested that a 2-μm thick TiO_2 film was optimal for cells employing *spiro*-OMeTAD as hole conductors, while DSPECs can make use of similar types of nanoporous films as thick as 20 μm.

Tennakone et al. [5] developed another type of TiO_2 nanoporous film for SSDSSCs. This type of film is prepared by repeated deposition of a colloidal TiO_2 solution on a heated conductive glass electrode. Large pores from the surface down to the conductive glass substrate can be formed in the film due to the vigorous evaporation of solvent during preparation. This type was shown to be suitable for the deposition of hole conductors from solution, since a thick TiO_2 film ($\sim$10 μm) was often used in their studies. The disadvantage of this film is the time-consuming preparation process. In the following section, if not specifically mentioned otherwise, all of the nanoporous TiO_2 films were prepared by the Tennakone method.

In the third step, a dye monolayer is loaded onto the TiO_2 nanoporous film by the same method used for DSPECs, that is, by immersing the nanoporous electrode into a dye solution for several hours. Various types of dye molecules include cyanidin dyes [1], Ru bipyridyl dyes [5, 8–12], zinc porphyrins [13], coumarins [14], and indolines [15], have been employing in SSDSSCs, with the Ru bipyridyl compound $RuL_2(SCN)_2$, where $L - 4,4'$-dicarboxy- $2,2'$-bipyridine (N3), being the most popular choice. There are many advantages of Ru bipyridyl dyes for SSDSSCs. For example, they are very stable, and they can effectively sensitize TiO_2 over most of the visible spectrum. In particular, when CuI or CuSCN is used as hole conductors, the thiocyanate ligand of the Ru dye can form a firm bond with a surface copper atom of the hole conductor, and this facilitates the fast regeneration of oxidized dye molecules [16].

Now, let us turn to the fourth step, which is the deposition of the hole conductor into the dye-loaded TiO_2 porous film. Hole conductors are often deposited from their solutions. Organic hole conductors can be dissolved in organic solvents easily, and the deposition is carried out by a spin-coating process. For CuI, the best solvent is acetonitrile, with a solubility of over $30\,g\,L^{-1}$. CuSCN is difficult to dissolve in any solvent, except for dipropyl sulfide. Repeated pipetting of CuI or CuSCN solution onto the porous electrode results in pore filling and the formation of a continuous hole conductor layer on the surface of the electrode [1, 8]. To facilitate pore filling and to avoid water adsorption, the porous electrode should be heated to a temperature exceeding 80°C.

O'Regan et al. [17, 18] developed an automated filling apparatus for CuSCN deposition. The same idea was also adopted by us for CuI deposition. The auto-filling apparatus consists of a hot stage onto which the porous electrode is placed. CuI solution is dispensed onto the surface of the porous electrode from a movable horizontal stainless needle, closed at the end and having a series of small holes along the top. The solution is pumped through the needle at 40 μL min^{-1} with a standard syringe pump. Acetonitrile is not a good solvent for precise control of the filling procedure because of its low boiling point. Butyronitrile has proven to be a good solvent, even though its solubility for CuI is only half of that for acetonitrile. The deposition of CuI was examined with a scanning electronic microscope (SEM). As shown in Fig. 3, the pore filling and surface deposition of CuI are two simultaneous processes. Pore filling will be terminated after formation of a continuous CuI layer on the surface. Therefore, incomplete pore filling is always an unavoidable problem in the fabrication of SSDSSCs.

In the last step, the hole-conductor-filled electrode is contacted to the counter electrode, followed by sealing. The counter electrode can be prepared by direct deposition of a gold layer via vacuum evaporation [2], or by pressing a layer of graphite powder [17]. For the CuI-filled electrode, it can be simply pressed together with another gold- or platinum-coated conductive glass, followed by sealing with epoxy resin [5]. Irrespective of whether gold, platinum, or graphite powder is used, the cell exhibits very similar current-voltage curves.

4. SOLAR ENERGY CONVERSION EFFICIENCY

Like the DSPEC, the SSDSSC is also a photovoltaic device that separates charge generation from charge transport. Under light excitation, dye molecules inject electrons into the TiO$_2$ nanoporous film, and the oxidized dye molecules are subsequently re-reduced by the hole-conductor layer, generating holes within it. This charge generation is an interfacial process of nearly unit quantum efficiency. The generated electrons and holes are confined to discrete phases: electrons diffusing in the n-type TiO$_2$, and holes diffusing in the p-type material. Because both of the charge carriers are transported as majority carriers, it is possible to use inexpensive, low crystalline quality materials in the fabrication of

Fig. 3. Electronic micrograph of TiO₂ porous electrode in different stages of pore filling with CuI: (a) no pore filling; (b) one-sixth pore filling; (c) one-third pore filling; and (d) complete pore filling.

SSDSSCs. In contrast, classical p–n junction solar cells, driven by minority carriers, must use very high purity semiconductor-grade materials of high crystalline quality, requiring the use of high-cost manufacturing technologies.

The output of SSDSSCs is related to many factors, such as the light-absorption capability, the charge injection efficiency, the charge transport rate, the interfacial recombination rate, and the inner resistance, etc. High roughness TiO_2 porous film is needed to ensure sufficient light absorption because monolayer dye molecules are employed to absorb light. The electron injection from dye monolayer to TiO_2 film could be highly efficient, for example, unit quantum efficiency for Ru dye N3, because the molecules are anchored onto the TiO_2 film. However, the re-reduction of the oxidized dye, in other words, the injection of a hole into the p-type material, is dependent on whether an intimate contact between the p-type material and the dye molecule is formed or not. Certainly, this is not a problem for DSPECs, as

Table 1
Solar energy conversion efficiency of several types of SSDSSCs (for reference, data for a DSPEC is also listed)

Cell type	Dye molecule	Optimal film thickness	Overall efficiency (%)
CuI	Ruthenium dye [20]	~10 μm[a]	~3
CuSCN	Ruthenium dye [21]	~4 μm[b]	~2.3
spiro-OMeTAD	Ruthenium dye + silver ion [27]	~2 μm[b]	~3.2
	Hydrophobic ruthenium dye [12]	~2 μm[b]	~4
	Indonium dye [15]	~1.6 μm[b]	~4
Electrolyte DSPEC	Ruthenium dye [19]	~12 μm[b]	~10.6

[a]Film prepared by the Tennakone method.
[b]Film prepared by the Grätzel method.

the liquid electrolyte can penetrate into the TiO_2 porous film completely and thus can access all of the dye molecules [19]. However, for SSDSSCs, one might imagine that it could be difficult for the deposited p-type materials to achieve intimate contact with the dye monolayer covering the porous TiO_2 electrode. The p-type semiconductors CuI and CuSCN tend to crystallize inside the mesoporous TiO_2 film, which destroys their contact to the dye molecules. However, the results are rather encouraging for the Ru dye-based SSDSSCs. As shown in Table 1, the record efficiencies are more than 3% for CuI-based cells [20], more than 2% for CuSCN-based cells [21], and ~4% for *spiro*-OMeTAD-based cells [12]. Possibly, it is because of the firm bond between the thiocyanate group of the Ru dye and the cuprous ion, which may affect the nucleation and growth of CuI or CuSCN microcrystals inside the porous film. The organic hole conductor, *spiro*-OMeTAD, can contact the dye monolayer intimately, since it forms an amorphous solid inside the TiO_2 porous film owing to its special molecular structure.

However, the incomplete pore filling of p-type materials is a real problem for the efficient re-reduction of oxidized dye molecules; this restricts the thickness, and thus the roughness factor of porous films used in SSDSSCs. The thicknesses used for the record-efficiency solid-state cells in Table 1 are almost several times lower than those common for DSPECs. A solid-state cell prepared from a thin Grätzel-type mesoporous TiO_2 film (<2 μm) was able to produce an output comparable to that of a cell that used liquid electrolyte, while a thicker film worked poorly for the solid-state solar cell [17]. One of the reasons is that the filling of the p-type material into the void volume of the nanoporous film becomes worse, as a thicker film is used. As studied by O'Regan et al. [17], the pore filling by a CuSCN deposit was 100% for a 1.7-μm thick nanoporous film, 76% for a 3-μm thick film, and only 64% for a 5.8-μm thick film. The Tennakone-type film allows better pore filling by the solution-deposition method, and thus a thick film ($\sim$10 μm) was often used for the fabrication of solar cells [5, 20]. However, the method is rather time-consuming and is unsuitable for large-scale manufacturing. Moreover, the roughness factor of a 10-μm thick Tennakone-type film was $\sim$300, which was not large enough for the effective absorption of incident light, and this restricted the energy conversion efficiency of the SSDSSC. In contrast, the mesoporous TiO$_2$ film used in the DSPEC always exhibits a roughness factor $>$1000.

Another factor that restricts the conversion efficiency of the SSDSSC is the interfacial charge recombination, especially when CuI or CuSCN is used as the p-type material. To understand this problem, we must look back at the structure and the operational principle of SSDSSCs. This type of cell is better described as a dye-sensitized heterojunction; such a description can lead to a better understanding of the serious recombination problem. This heterojunction has two features: an extremely large interface and a very weak interfacial electric field, because of its mesostructured interpenetrating nature. As we know, a substantial built-in electric field at the interface is a necessary condition for a high-quality heterojunction. Moreover, the physical heterojunction device always adopts a planar structure, instead of the interpenetrating structure with extremely large interface, to avoid the production of electronic states in the band gap of the semiconductor. Thus, the interfacial recombination between the electrons in the TiO$_2$ phase and the holes in the hole conductor is unavoidable for SSDSSCs, after the initial interfacial charge generation. The DSPEC does not meet with such a serious recombination problem, since it is basically a molecular device, and the iodide/triiodide redox couple is used as charge mediator. The iodide/triiodide couple exhibits irreversible charge transfer kinetics. The reduction of triiodide to iodide on the surface of TiO$_2$ and conducting glass electrode is very sluggish.

The serious interfacial recombination is equivalent to an internal short circuit for the SSDSSC. One may be surprised at the reported considerable output of the cell. However, the truth is that the SSDSSC works well, despite the disastrous recombination that always threatens to ruin the cell performance. The answer to this dilemma is the existence of a dye monolayer at the interface. This monolayer, with a thickness of $\sim$1 nm in the case of a Ru dye, behaves as a physical barrier layer at the interface and blocks the interfacial recombination [5]. Supposing that the monolayer were ideally compact, it would be possible to completely suppress the interfacial recombination between the TiO$_2$ and the hole conductor. However, such a compact dye monolayer is impossible to form, because of the large size of the dye molecule and the electrostatic repulsion between the molecules. Thus, the direct contact between TiO_2 and the hole conductor is nearly unavoidable for SSDSSCs.

There is another kind of interfacial recombination that is between the oxidized dye molecule and the electron injected into the TiO_2. However, this type of recombination can be neglected in the SSDSSC, because the oxidized dye molecule is re-reduced by the hole conductor at a rate much higher than that for the recapture of an electron from TiO_2. The recapture of the electron may occur on the time scale of microseconds to milliseconds; however, the reduction of oxidized dye molecule with *spiro*-OMeTAD has been reported to be on the timescale of nanoseconds [10]. No kinetic data for oxidized dye reduction by CuI or CuSCN have been reported, but this charge transfer should occur at a very high rate, considering the firm bond between the cuprous ion and the NCS group of the dye molecule, whose orbitals mainly contribute to the HOMO of the Ru dye molecule.

O'Regan et al. [18] have measured the recombination rate in an SSDSSC using CuSCN as the hole conductor. They found that the recombination in the SSDSSC was ten times faster than in the DSPEC at the open-circuit potential V_{oc} ($t_{1/2}{\sim}150\,\mu s$), and 100 times faster at short circuit ($t_{1/2}{\sim}450\,\mu s$), although both types of cells exhibited a similar charge transport rate ($t_{1/2}{\sim}200\,\mu s$). The similarity of the charge transport and recombination rates in the SSDSSC results in a low fill factor (FF) and photocurrent losses, both important limiting factors of the efficiency.

Krüger et al. [22] measured the electron diffusion length in the Grätzel-type TiO_2 porous film by means of intensity-modulated photocurrent spectroscopy (IMPS) and intensity-modulated photovoltage spectroscopy (IMVS), when *spiro*-OMeTAD was used as hole conductor. The electron diffusion length was determined as $4.4\,\mu m$, which is a factor of 4 and lower than in the DSPEC, because of the greater interfacial recombination in the former case. Thus, even without the problem of incomplete pore filling, efficient SSDSSCs can only use nanoporous films up to ${\sim}4\,\mu m$ for their system if interfacial recombination is not suppressed.

No recombination data for CuI-based SSDSSCs have been available up to now. However, the fact that CuI-based cells have always exhibited V_{oc} values of 400–500 mV, much lower than the theoretical V_{oc} (${\sim}1\,V$) estimated from the flat-band potentials for CuI and TiO_2, should provide evidence for the existence of serious interfacial recombination. Thus, it is apparent that the key for high-efficiency SSDSSCs is to suppress interfacial recombination.

5. INTERFACIAL BLOCKING LAYERS

We have proposed the idea of employing an interfacial blocking layer to suppress the interfacial recombination in the CuI-based SSDSSC [23–25]. This involves inserting an ultrathin insulating layer at the TiO_2/CuI interface. The insulating layer acts as a physical barrier to avoid the direct contact between TiO_2 and CuI. This layer must be very thin (<1 nm) in order to maintain the tunneling efficiency of electrons from the dye molecules to TiO_2. This idea seems reasonable for suppressing the interfacial recombination. The next question is how to prepare such a blocking layer at the interface. One method is to prepare a film from insulator-coated particles (Fig. 4a). However, it is more interesting to prepare an insulator layer on the porous film by means of a surface reaction (Fig. 4b), because the insulating layer prepared by this method will not increase the inner resistance of the film for electron transport.

We coated the TiO_2 nanoporous film with an ultrathin Al_2O_3 layer by the surface sol-gel method and assembled solar cells [23, 25]. In brief, Al_2O_3, an ultrathin layer, was coated onto the surface of TiO_2 film by stepwise adsorption of aluminum alkoxide with subsequent sintering. The thickness of the Al_2O_3 layer was controlled by the number of adsorption cycles. We investigated the effect of the Al_2O_3 layer on the interfacial recombination

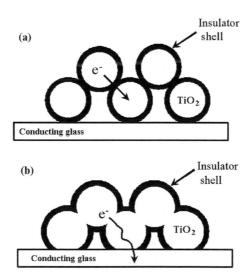

Fig. 4(a,b). Two configurations for the insulating layer-coated TiO_2 porous film electrode.

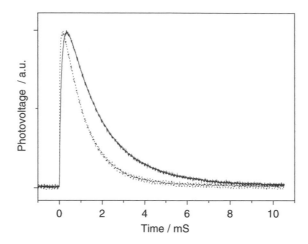

Fig. 5. Photovoltage transients of SSDSSCs prepared from TiO_2 (dotted line) and Al_2O_3-coated TiO_2 (straight line) films [25].

by means of transient photovoltage measurements [25]. This measurement provides a direct investigation on the interfacial recombination, since all the charges generated in the cell by the light pulse will recombine through the interface under open-circuit conditions. The result was promising. A 0.19-nm thick Al_2O_3 layer decreased the decay rate of the photovoltage transient (Fig. 5). The lifetimes for the decay were calculated to be 1.19 ms for a conventional cell and 1.78 ms for the cell containing the Al_2O_3 layer. Thus, we can conclude that a blocking function of the Al_2O_3 layer exists.

The improvement in the cell performance because of the Al_2O_3 layer was very apparent, as shown in Table 2 [25]. The blocking layer improved all of the cell parameters, including the open-circuit voltage (V_{oc}), the short-circuit current (J_{sc}), the FF, and the conversion efficiency. The improvements in V_{oc} and FF are direct evidence for the suppression of interfacial recombination. The two cells exhibited the same J_{sc} under very weak illumination (4 mWcm^{-2}); however, the cell containing the Al_2O_3 layer exhibited a larger value under higher light intensity. This phenomenon also indicated that the charge collection of cell was improved by the blocking function of the Al_2O_3 layer. The J_{sc} values of both cells increased linearly with illumination power in the range of 4.1–31.4 mWcm^{-2}, yet both deviated from the linear function under higher light intensity. Thus, it indicated that the quality of the 0.19-nm thick Al_2O_3 layer was not high enough to resolve the recombination problem.

Table 2
Comparison of the performance parameters of SSDSSCs based on TiO_2 and Al_2O_3-coated TiO_2 films under various intensities of simulated sunlight [25]

	Light intensity (mWcm^{-2})	V_{oc} (V)	J_{sc} (mAcm^{-2})	FF	Efficiency (%)
TiO_2	4.1	0.29	0.40	0.51	1.45
	17.6	0.35	1.97	0.58	2.27
	31.4	0.37	3.62	0.56	2.38
	89.0	0.40	9.10	0.48	1.94
TiO_2–Al_2O_3	4.1	0.35	0.40	0.56	1.92
	17.6	0.42	2.08	0.57	2.86
	31.4	0.45	3.77	0.57	3.03
	89.0	0.47	9.46	0.52	2.59

MgO also did well as a blocking layer in SSDSSCs [24]. Other insulating materials, for example, SiO_2 and Y_2O_3, might also work well as blocking layers for recombination. The blocking function should depend on the thickness and density of the layer, instead of depending on the materials used. Interestingly, some wide band-gap semiconductors, such as ZnO, also showed a blocking function similar to that of an insulating layer [26]. One possible reason is dye aggregation on the film induced by dissolved zinc ions, similar to the silver-ion effect observed by Grätzel et al. [27].

It is pertinent to discuss the effect of insulating layer on the photocurrent in greater detail. As shown in Fig. 6, a cell containing a 0.33-nm thick Al_2O_3 layer exhibited a higher FF and V_{oc} than a 0.19-nm thick Al_2O_3 layer, but its J_{sc} became lower, because the electron injection was exponentially dependent on the thickness of the insulating layer [23]. As an extreme condition, a very thick MgO layer improved the V_{oc} of a CuI-based cell up to 800 mV, with negligible J_{sc} observed. The foregoing discussion shows that the insulating layer in principle can increase the V_{oc} of the cell greatly through suppressing recombination, but the problem is how to keep the J_{sc} from decreasing at the same time. We have to point out that the present preparation method—the surface sol-gel method—is not an optimal one for preparing insulating blocking layers. It depends on the hydrolysis and condensation of metal alkoxide on the surface of the TiO_2 film, through the reaction with adsorbed water. It is possible to prepare an ultrathin insulating layer on the TiO_2 film by this method, but it is impossible to prepare a dense insulating layer, because the cleavage of bulky alkoxy groups will leave pores in the layer. As a result, the insulating layer cannot cover the surface of TiO_2 completely.

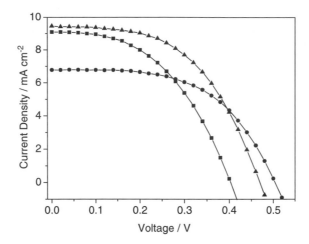

Fig. 6. Current–voltage curves of CuI-based solid-state dye-sensitized solar cells prepared from TiO$_2$ film (squares), 0.19-nm thick Al$_2$O$_3$-coated TiO$_2$ film (triangles), and 0.33-nm thick Al$_2$O$_3$-coated TiO$_2$ film (circles).

If a better method is found for preparing dense insulating layers on the TiO$_2$ surface, it will be possible to resolve the interfacial recombination problem discussed earlier. A physical method, for example, atomic layer chemical-vapor deposition, might be a good choice for the preparation [28]. In addition, post-treating a dye-deposited TiO$_2$ nanoporous film with the precursor of the insulating material is also a promising method to resolve the recombination problem. Because the dye molecules are adsorbed on the TiO$_2$ film, the insulating material will not hinder the electron injection.

6. OPERATIONAL STABILITY

The operational stability is a controversial issue for SSDSSCs. Stability data for the CuSCN- and *spiro*-OMeTAD-based solar cells have not been available. Previous studies suggested that the CuI-based solar cell was not stable for long-term operation or even for long-term storage, possibly because of the growth of CuI microcrystals inside the cell [1]. Recent studies showed that the growth of CuI microcrystals, however, could be suppressed by adding a small amount of imidazolium thiocyanate, a crystal growth inhibitor for CuI [20]. This compound stabilized the surfaces of CuI microcrystals by the strong interaction of the thiocyanate anion and the surface cuprous cation. Other thiocyanate salts also exhibited a similar effect [29].

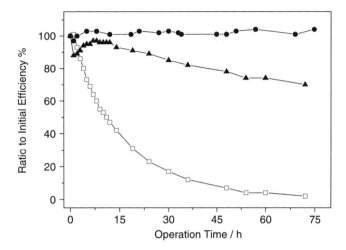

Fig. 7. Operational stability of TiO_2/dye/CuI solid-state solar cells under continuous illumination of simulated sunlight. Open squares, no UV cutoff filter was used; solid circles,
UV cutoff filter (cutoff wavelength 435 nm) was used; solid triangles, an MgO layer was
coated on the TiO_2 surface, and no UV cutoff filter was used. The intensity of simulated
sunlight was $40\,mWcm^{-2}$. The cells were cooled with a fan under test.

Table 3
Operational stability of several solar cells under simulated sunlight

Cell type	Operation time	UV cutoff filter	Percentage of initial value			
			V_{oc}	I_{sc}	FF	Efficiency
TiO_2	72	N	16.1	30.0	51.0	2.5
TiO_2	75	Y	103.0	106.0	95.0	103.7
TiO_2–MgO	72	N	67.8	138.7	74.7	70.2

Operational stability is one of the most important factors for a photovoltaic device. It should be measured at the maximum output of the device.
Previous studies on DSPECs always emphasized the stability under open-
circuit conditions, which could not provide a meaningful and accurate evaluation of the operational stability of the device [19]. Here, we will discuss
the operational stability of the CuI-based solar cell under maximum output.

The CuI-based cell exhibited very fast degradation under simulated sunlight. The output became almost zero after 72 h of continuous illumination,
as shown in Fig. 7 and Table 3 [24]. The illuminated area turned from red (the
color of the Ru dye) to black after the stability test, which was an evidence

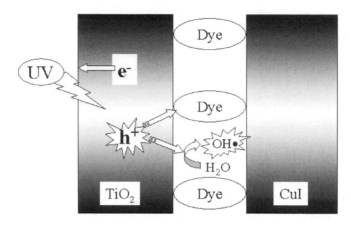

Fig. 8. Schematic illustration of the photodegradation mechanism for SSDSSCs. The highly reactive hydroxyl radicals, generated by the photocatalytic reactions on TiO_2, attack both dye molecules and CuI, which causes the degradation of the SSDSSC.

for the existence of cupric oxide. However, when we carried out the stability test under ultraviolet (UV)-free illumination ($\lambda > 435$ nm), the cell exhibited good operational stability. The cell retained more than 90% of the initial conversion efficiency after continuous illumination for more than 500 h.

X-ray photon-electron spectroscopy (XPS) studies confirmed the oxidation of CuI to cupric compounds under UV light by TiO_2. The oxidation may involve TiO_2, UV light, and residual water in the cell (Fig. 8). Water was oxidized to form OH˙ radicals, which reacted with CuI to form the cupric compounds. Because the reaction happened at the interface, the oxidation of CuI might degrade the heterojunction directly and then finally degrade the cell. This mechanism suggests that the photocatalytic activity of TiO_2 should be purposefully suppressed, and the residual water should be kept to a very low level to fabricate highly stable CuI-based DSSCs. Interestingly, the insulating blocking layer was observed to improve the operational stability of the cell under simulated sunlight [24]. Even without filtering UV light, a cell constructed with an MgO-coated film remained 70.2% efficient after 72 h of continuous illumination. This is an exciting result, since a cell constructed with a bare TiO_2 film lost almost all output under the same conditions. A possible reason is that the MgO layer blocks the photo-oxidative function of TiO_2 film and thus stabilizes the dye-sensitized heterojuction, even under UV excitation.

The results of the operational stability of the CuI-based DSSC were rather promising. It can be expected that the CuSCN-based DSSC should exhibit even better operational stability, since CuSCN itself is more stable than CuI against oxidation.

7. FUTURE OUTLOOK

The development of SSDSSCs is encountering two obstacles: insufficient light absorption because of the limited inner surface area of the TiO_2 film electrode; and the serious interfacial recombination because of the lack of a substantial interfacial electric field. These two problems restrict the energy-conversion efficiency of the SSDSSC to below 4%, even though recent studies have shown that a 4% cell could be fabricated. The operational stability of the CuI-based solar cell was, however, unexpectedly good for a 500 h continuous test.

The interfacial recombination in the SSDSSC can be suppressed by a thin layer of insulating material. However, the present preparation methods for this insulating layer are not satisfactory. A recent study on the ETA solar cell has suggested that the atomic layer–chemical vapor deposition (Al-CVD) method is a possible candidate for the preparation of a thin, dense, insulating layer on a porous structure [28]. It can be predicted that a breakthrough in the energy-conversion efficiency will occur for the SSDSSC when a preparation method for a high-quality insulating blocking layer is developed. Furthermore, the high-quality insulating layer will improve the operation stability of the SSDSSC.

The problem of insufficient light absorption may be overcome by replacing the Ru dye with another organic dye having a greater extinction coefficient. Schmidt-Mende et al. [15] have reported an efficiency of more than 4% for a SSDSSC using indoline dye as sensitizer. The indonium dye has a very strong extinction coefficient ($55800 \, L \, mol^{-1} \, cm^{-1}$ at 491 nm), which is four times stronger than the Ru dye N3 ($13900 \, L \, mol^{-1} \, cm^{-1}$). To obtain a broad optical absorption extending throughout the visible and the near-IR region, one can use a combination of two dyes that complement each other in their spectral features [30]. The semiconductor quantum dot is also a possible option for panchromatic sensitizers. The absorption spectra of the quantum dots, always II–VI and III–V semiconductor nanoparticles, can be adjusted to cover the visible and near-IR region by changing the particle size. Optical design can also improve the light collection in the SSDSSC and thus improve its solar-energy-conversion efficiency.

The morphology of the porous film electrode is also very important for the SSDSSC. A desirable morphology for the porous film used in the

SSDSSC would have the mesoporous channels or nanorods aligned in parallel to each other, and vertically with respect to the conducting glass substrate. Such a morphology would facilitate the pore-filling with p-type materials, allow the formation of a higher quality heterojunction, and exhibit faster charge transport in the film, compared to the random mesoporous morphology. Future studies should also be devoted to developing the fabrication technology for the SSDSSC. Recent progress in solution deposition is encouraging [17, 18]. However, more efforts are required for the reproducible production of the SSDSSC.

ACKNOWLEDGMENTS

Recognition is due to the members having joined in our all solid-state dye-sensitized solar cell project, some of whose work is referenced. We thank Dr. Donald A. Tryk for reading the manuscript and offering us helpful comments. This work is supported by a Grant-in Aid for Scientific Research on Priority Areas (417) from the Ministry of Education, Culture, Sports, Science and Technology (MEXT) of the Japanese Government.

REFERENCES

[1] K. Tennakone, G.R.R.A. Kumara, A.R. Kumarasinghe, K.G.U. Wijayantha and P.M. Simimanne, Semicond. Sci. Technol., 10 (1995) 1689–1693.
[2] U. Bach, D. Lupo, P. Comte, J.E. Moser, F. Weissortel, J. Salbeck, H. Spreitzer and M. Grätzel, Nature, 395 (1998) 583–585.
[3] B. O'Regan and D.T. Schwartz, Chem. Mater., 10 (1998) 1501–1509.
[4] A. Fujishima and X.-T. Zhang, Proc. Jpn. Acad., B 81 (2005) 33–42.
[5] K. Tennakone, G.R.R.A. Kumara, I.R.M. Kottegoda, K.G.U. Wijayantha and V.P.S. Perera, J. Phys. D: Appl. Phys., 31 (1998) 1492–1496.
[6] B. O'Regan and D.T. Schwartz, Chem. Mater., 7 (1995) 1349.
[7] B. O'Regan and D.T. Schwartz, J. Appl. Phys., 80 (1996) 4749.
[8] G.R.R.A. Kumara, A. Konno, G.K.R. Senadeera, P.V.V. Jayaweera, D.B.R.A. De Silva and K. Tennakone, Sol. Energy Mater. Sol. Cells, 69 (2001) 195–199.
[9] B. O'Regan, D.T. Schwartz, S.M. Zakeeruddin and M. Grätzel, Adv. Mater., 12 (2000) 1263–1267.
[10] U. Bach, Y. Tachibana, J.-E. Moser, S.A. Haque, J.R. Durrant, M. Grätzel and D.R. Klug, J. Am. Chem. Soc., 121 (1999) 7445–7446.
[11] J. Krüger, R. Plass, L. Cevey, M. Piccirelli, M. Grätzel and U. Bach, Appl. Phys. Lett., 79 (2001) 2085–2087.
[12] L. Schmidt-Mende, S.M. Zakeeruddin and M. Grätzel, Appl. Phys. Lett., 86 (2005) 013504.

[13] L. Schmidt-Mende, W.M. Campbell, Q. Wang, K.W. Jolley, D.L. Officer, M.K. Nazeeruddin and M. Grätzel, Chem. Phys. Chem., 6 (2005) 1253.

[14] A. Konno and G.R.A. Kumara, The 56th Annual Meeting of the International Society of Electrochemistry, Busan, South Korea, 2005.

[15] L. Schmidt-Mende, U. Bach, R. Humphry-Baker, T. Horiuchi, H. Miura, S. Ito, S. Uchida and M. Grätzel, Adv. Mater., 17 (2005) 813.

[16] B. Mahrov, G. Boschloo, A. Hagfeldt, H. Siegbahn and H. Rensmo, J. Phys. Chem., B, 108, (2004) 11604–11610.

[17] B. O'Regan, F. Lenzmann, R. Muis and J. Wienke, Chem. Mater., 14 (2002) 5023.

[18] B. O'Regan and F. Lenzmann, J. Phys. Chem., B, 108 (2004) 4342.

[19] M. Grätzel, J. Photochem. Photobiol. C: Photochem. Rev., 4 (2003) 145.

[20] G.R.A. Kumara, A. Konno, K. Shiratsuchi, J. Tsukahara, and K. Tennakone, Chem. Mater., 14 (2002) 954.

[21] B.C. O'Regan, S. Scully, A.C. Mayer, E. Palomares and J. Durrant, J. Phys. Chem., 109 (2005) 4616.

[22] J. Krüger, R. Plass, M. Grätzel, P.J. Cameron and L.M. Peter, J. Phys. Chem., B, 107 (2003) 7536.

[23] X.-T. Zhang, I. Sutanto, T. Taguchi, K. Tokuhiro, Q.-B. Meng, T.N. Rao, A. Fujishima, H. Watanabe, T. Nakamori and M. Uragami, Sol. Energy Mater. Sol. Cells, 80 (2003) 315.

[24] T. Taguchi, X.-T. Zhang, I. Sutanto, K. Tokuhiro, T.N. Rao, H. Watanabe, T. Nakamori, M. Uragami and A. Fujishima, Chem. Commun., 19 (2003) 2480.

[25] X.-T. Zhang, H.-W. Liu, T. Taguchi, Q.-B. Meng, O. Sato and A. Fujishima, Sol. Energy Mater. Sol. Cells, 81 (2004) 197.

[26] X.-T. Zhang, I. Sutanto, T. Taguchi, Q.-B. Meng, T.N. Rao, O. Sato and A. Fujishima, Annual Meeting on Photochemistry in Japan, Kyoto, 2002.

[27] J. Krüger, R. Plass, M. Grätzel and H.-J. Matthieu, Appl. Phys. Lett., 81 (2002) 367.

[28] M. Nanu, J. Schoonman and A. Goossens, Adv. Mater., 16 (2004) 453.

[29] G.R.A. Kumara, S. Kaneko, M. Okuya and K. Tennakone, Langmuir, 18 (2002) 10493.

[30] V.P.S. Perera, P.K.D.D.P. Pitigala, P.V.V. Jayaweera, K.M.P. Bandaranayake and K. Tennakone, J. Phys. Chem. B, 107 (2003) 13758.

PART IV

ORGANIC- AND CARBON-BASED SOLAR CELLS

Nanostructured Materials for Solar Energy Conversion
T. Soga (editor)

Chapter 10

Nanostructure and Nanomorphology Engineering in Polymer Solar Cells

H. Hoppe[1] and N.S. Sariciftci[2]

[1]Institut für Physik, Technische Universität Ilmenau, Weimarer Str. 32, D-98693 Ilmenau, Germany
[2]Linz Institute for Organic Solar Cells (LIOS), Physical Chemistry, Johannes Kepler University at Linz, Altenbergerstr. 69, A-4040 Linz, Austria

ABSTRACT

Solar cells based on conjugated polymers have been a rapidly developing area of research during the last decade. Since photoexcitations in conjugated polymers show diffusion lengths of only around 5–20 nm, the structure of the polymeric nanophase within the photoactive layer has a large influence on the device properties and the solar power conversion efficiency. In this chapter, we will address different architectures of polymer solar cells and the influence of their design on solar cell properties. Thereafter we will concentrate on polymer–fullerene bulk heterojunction solar cells and the engineering of their nanostructure toward improved power conversion efficiencies. Bulk heterojunctions constitute intimate blends of organic donor and acceptor materials that allow for efficient charge separation throughout the photoactive layer and provide independent pathways to transport the charge carriers to the contacts. Here domains of donor and acceptor materials serve as hole and electron conducting nanophases, respectively. Finally, a viewpoint is presented with the focus on the engineering of ordered bulk heterojunctions based on conjugated polymers in combination with inorganic scaffolds or diblock copolymers.

1. INTRODUCTION

Conjugated polymers constitute a material's class with an exciting range of properties and possibilities [1, 2]. Although the early research focused on their metallic-like behavior of charge conduction [3], the opportunity for semiconducting properties later resulted in the development of organic light-emitting diodes (OLEDs) [4], organic field effect transistors (OFETs) [5] and organic solar cells (OSCs) [6–8, 11–14]. As an advantage over the traditional inorganic semiconductors, conjugated polymers allow processing from solution at room temperature by application of spin coating or even conventional printing techniques. These features pave the way for low-cost large-scale production of all kinds of different organic optoelectronic devices – even integrated on a single substrate. Since the polymer layers are very thin films (~100 nm), the construction of lightweight and flexible devices drives today's economic interest [15].

1.1. Materials

The attractiveness of conjugated semiconducting polymers lies in the possibility to process them from solution, applying spin coating, doctor blading, or even conventional printing techniques for a controlled film formation. The solubility of these polymers in common organic solvents is usually governed through their side chains. Some examples of conjugated polymers and a fullerene applied in organic solar cells are presented in Fig. 1. Three important

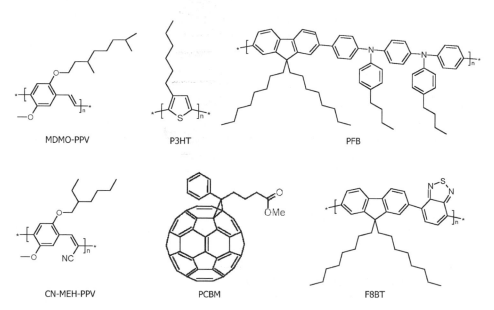

Fig. 1. Some conjugated polymers and a soluble C_{60} derivative commonly applied in polymer-based solar cells are shown.

representatives of hole-conducting donor-type polymers are MDMO-PPV (poly-[2-methoxy-5-(3,7-dimethyloctyloxy)]-1,4-phenylenevinylene), P3HT (poly(3-hexylthiophene-2,5-diyl)) and PFB (poly(9,9′-dioctylfluorene-co-bis-N,N'-(4-butylphenyl)-bis-N,N'-phenyl-1,4-phenylenediamine)). They are shown together with electron-conducting acceptor polymers like CN-MEH-PPV (poly-[2-methoxy-5-(2′-ethylhexyloxy)]-1,4-(1-cyanovinylene)-phenylene) and F8TB (poly(9,9′-dioctylfluoreneco-benzothiadiazole)), and a soluble derivative of C_{60}, namely PCBM (C61-butyric acid methyl ester). All of these materials are solution processible due to side-chain solubilization and the polymers show pronounced photo- and electroluminescence.

The potential of conjugated polymers to absorb light in the UV-visible part of the solar spectrum and to transport charge carriers is due to the sp^2-hybridization of carbon atoms. The electron in the p_Z-orbital of each sp^2-hybridized carbon atom forms π-bonds with neighboring p_Z electrons throughout the polymer. Owing to the isomeric effect these π-electrons are of a delocalized nature, resulting in high electronic polarizability. Because of the Peierls instability, the originally half-filled p_Z-"band" splits up into two: the π- and π*-bands. Upon light absorption electrons may be excited from the bonding π- into the anti-bonding π*-band. This absorption corresponds to the first optical excitation from the highest occupied molecular orbital (HOMO) to the lowest unoccupied molecular orbital (LUMO). The optical band gap of these conjugated polymers is usually around and larger 2 eV.

To display the fraction of the sunlight, which can contribute to energy conversion in polymer solar cells, absorption coefficients of films of some materials are shown in comparison with the AM 1.5 solar spectrum in Fig. 2. While the silicon absorption spectrum extends up to 1100 nm, the organic materials use only the blue side of the solar spectrum.

An important difference to inorganic solid state semiconductors lies in the generally poor (orders of magnitudes lower) charge carrier mobility in these materials [16], which has a large effect on the design and efficiency of organic semiconductor devices. However, as organic semiconductors exhibit relatively strong absorption coefficients (usually $\geq 10^5 \, cm^{-1}$), these low mobilities are partially balanced, when film thicknesses of a few hundred nanometers or even less are applied.

1.2. Working Principles of Organic Solar Cells

Upon illumination with light, photon absorption within the photoactive layer of the organic solar cell first leads to the creation of a bound electron–hole pair – the exciton. In a second step these Frenkel type

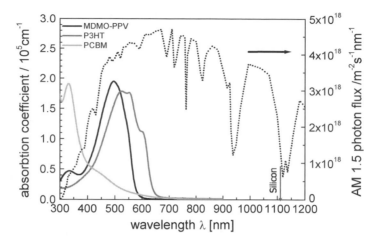

Fig. 2. Absorption coefficients of some conjugated polymers and a fullerene derivative PCBM are shown together with the AM 1.5 standard solar spectrum.

excitons may diffuse during their lifetime within the material, in which they were created. In these rather amorphous organic materials, exciton diffusion lengths are limited to about 5–20 nm. Exciton binding energies between 0.1 and 1 eV necessitate a specific exciton dissociation mechanism. In single layer organic solar cells this may be achieved by the strong electric field present within the depletion region of a Schottky contact. As will be discussed below, single layer devices do not play a role in today's solar cell development anymore. Actual solar cells rely on the donor (D)–acceptor (A) concept that is based on the photoinduced charge transfer between these materials [17]. Upon light absorption in the donor an excited state is formed from which the electron may transfer to the LUMO of the acceptor. The driving force for this charge transfer is the difference in ionization potential and electron affinity between donor and acceptor, respectively [17]. As a result of the photoinduced charge transfer the hole remains on the donor material whereas the electron is located on the acceptor. The situation together with the energetical description is illustrated in Fig. 3 for the initially investigated system of a soluble derivative of poly(paraphenylene vinylene) as a donor and C_{60} as an acceptor.

This photoinduced charge transfer takes place very rapidly within some 45 femtoseconds [18]. Since any competing process like photoluminescence (ns) and back transfer and thus recombination of the charge (μs) take place on a much larger timescale, the charge separated state is relatively stable. The

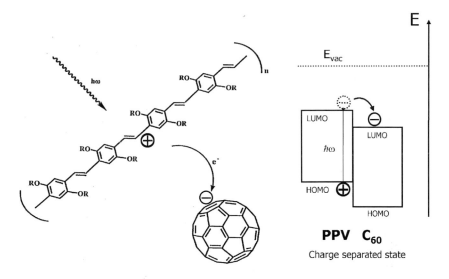

Fig. 3. The photo-induced charge transfer from a donor (PPV) to an acceptor (C$_{60}$) serves as charge separation mechanism in most organic photovoltaic devices.

possible decay pathways of the system upon excitation are displayed in Fig. 4 for comparison.

Once the charge carriers have been separated, they need to be transported to their respective electrodes for generation of an external direct current. Here the donor material serves to transport the holes whereas the electrons travel within the acceptor material. The processes of carrier generation and charge transport are highly dependent on the internal phase structure of the D–A blend. For example, percolating paths are required to ensure that the charge carriers will not experience the fate of recombination due to trapping in dead ends on an isolated material's domain.

In D–A blends, holes and electrons are separately transported within different nanophases; thus, considerably large charge carrier lifetimes are observed. As the charge carrier mobilities within organic solar cells often do not exceed 10^{-4} cm^2/Vs, these large lifetimes are indeed required for extracting all photoexcited charge carriers from the photoactive layer. The charge carrier extraction is driven by internal electric fields caused by the different work function electrodes for holes and electrons. Recently, it has been shown that all charge carriers in these devices are not indeed free and that electron–hole pairs may be slightly bound to each other across the D–A

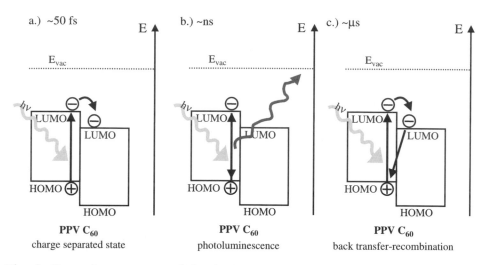

Fig. 4. Competing processes of the donor–acceptor system upon light excitation are shown. As the photo-induced charge transfer elapses on a much smaller time scale than photoluminescence and recombination, the charge separated state is relatively stable.

interface [19, 20]. In such cases the internal electric field may also serve to separate these weakly bound complexes.

Ultimately the distance d that charge carriers can travel within the device is a product of charge carrier mobility μ, charge carrier lifetime τ and the internal electric field F:

$$d = \mu \tau F$$

(1)

The internal electric field F that drives this drift current is generally originating from the difference in the electrodes work functions. For example, for gold as hole accepting electrode ($\Phi = 5.2\,eV$) and aluminum as electron accepting electrode ($\Phi = 4.3\,eV$) an internal electric field of $10^5\,V/cm$ is given for an active layer thickness of 90 nm. Assuming charge carrier mobilities of $10^{-4}\,cm^2/Vs$ and charge carrier lifetimes of 1 μs results in a distance $d = 10^{-4}\,cm = 100\,nm$ at short circuit conditions.

In general the device function of thin organic solar cells, photodiodes and even light-emitting diodes (LEDs) can be simplified by the metal–insulator–metal (MIM) picture [21]. This is valid, since the organic semiconductors in these devices are not intrinsically doped with charges and thus they represent an insulator due to the large band gap. Therefore no charge

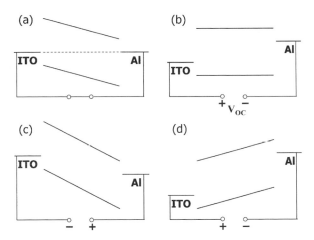

Fig. 5. Principles of device function for organic layers sandwiched between two metallic electrodes: (a) short circuit condition; (b) flat band condition; (c) reverse bias; and (d) forward bias.

depletion regions are formed within the active layer and the difference in work function of the two electrodes lead to the formation of an internal electric field.

The selectivity of charge injection through barriers into molecular HOMO or LUMO levels ensures further on the rectifying diode behavior of these organic devices [22]. The different working regimes due to externally applied voltage are shown in Fig. 5.

The different situations in Fig. 5 translate to different working regimes of the photovoltaic device, shown in Fig. 6. While (a) corresponds to the short circuit photocurrent I_{SC}, (b) usually represents the condition for the open circuit voltage V_{OC}. In (c) the internal voltage is increased, corresponding to the condition in photodetectors or blocking behavior of diodes. For the case of forward bias, efficient charge carrier injection takes place and the direction of the current inside the device is reversed. This is the condition under which organic LEDs are operating.

From Fig. 6 the calculation of the power conversion efficiency η can be derived. Only in the fourth quadrant the device delivers power. One point on the curve, denoted as maximum power point (MPP), corresponds to the largest product of current and voltage and thus power. The factor between $V_{MPP} \times I_{MPP}$ and $V_{OC} \times I_{SC}$ is called the fill factor (FF), and therefore the power output is often written in the form, $P_{Max} = V_{OC} \times I_{SC} \times FF$. Division

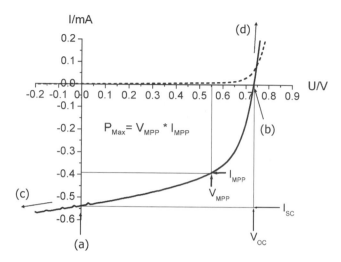

Fig. 6. Current–voltage characteristics of an organic solar cell under illumination (solid line) and in the dark (broken line). The various cases (a–d) from Fig. 5 are shown for comparison.

of the output power by the incident light power results in the power conversion efficiency η

$$\eta_{POWER} = \frac{P_{OUT}}{P_{IN}} = \frac{I_{MPP}V_{MPP}}{P_{IN}} = \frac{FFI_{SC}V_{OC}}{P_{IN}} \qquad (2)$$

As the transport of charges and thus the photocurrent is field dependent, the $\mu\tau$-product has a strong influence on the FF. Close to the open circuit voltage the internal electric field is reduced considerably and thus all generated charge carrier cannot be extracted anymore. Therefore large charge carrier mobilities and lifetimes are demanded for efficient device operation.

1.3. Device Architectures

During the last decade the device architecture of organic photovoltaic devices based on conjugated polymers has reached increasing complexity. The common design of an organic solar cell is displayed in Fig. 7. The photoactive layer is often sandwiched between an indium tin oxide (ITO)-covered glass and a reflecting aluminum electrode. As ITO is rather transparent, illumination takes place from the glass side. The two electrodes may be further modified by the introduction of PEDOT:PSS (poly[3,4-(ethylenedioxy) thiophene]:poly(styrene sulfonate)) on the ITO side and lithium fluoride (LiF) on the aluminum side, diminishing the energy barrier heights and thus improving the charge injection properties.

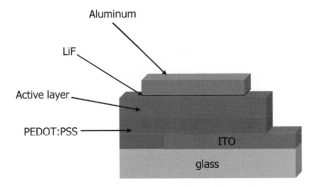

Fig. 7. General design of an organic solar cell. The photoactive layer is sandwiched between optimized electron (LiF/Al) and hole accepting (ITO) electrodes.

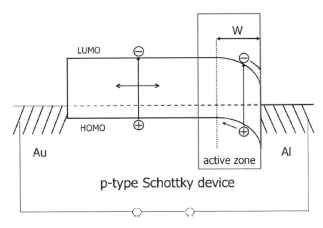

Fig. 8. In single layer devices, charge carriers can be dissociated at the Schottky junction. Only excitons generated closely to the depletion region W can contribute to the photocurrent.

The device architecture of the photoactive layer has a strong impact on charge carrier separation and transport. For example, in single layer (single material) devices, after dissociation both types of charge carriers have to travel within the same material. This enables in turn the parasitic charge carrier concentration dependent bimolecular recombination. Fig. 8 shows the situation for a single layer single material device. Owing to the limited exciton diffusion range in organic materials, only photoexcitations generated close to the depletion region W of the Schottky contact may lead to separated charge carriers. Therefore only a small region denoted as active zone contributes to the photocurrent generation.

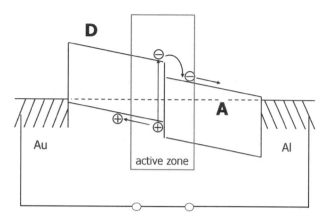

Fig. 9. In bilayer devices, charge carriers can be dissociated at the donor–acceptor (D–A) materials junction. Only excitons generated closely to the interface can contribute to the photocurrent.

A major improvement in organic solar cell design represents the introduction of the bilayer device [23]. Bilayer devices apply the D–A concept introduced above. Here the exciton is dissociated at the materials interface, leading to holes on the donor and electrons on the acceptor. The different types of charge carriers may then travel independently within separate materials and bimolecular recombination is largely depressed. Therefore, light-intensity-dependent photocurrent measurements in these systems may lead to a rather linear behavior of the photocurrent with respect to the light intensity, and monomolecular recombination processes will dominate. However, bilayer devices suffer again from an active zone limited by the exciton diffusion, as only close to the materials heterojunction photoexcitations may lead to charge carrier dissociation, as indicated in Fig. 9.

This limitation was finally overcome by the revolutionary concept of the bulk heterojunction, where the donor and acceptor materials are intimately blended throughout the bulk [9–11]. Thus excitons do not need to travel long distances to reach the D–A interface between two separate layers and for the first time, charge separation can take place throughout the whole photoactive layer. Here the active zone extends throughout the full absorption region as illustrated in Fig. 10. Hence the bulk heterojunction concept led to major improvements in the photocurrent and thus the power conversion efficiency. Today, the bulk heterojunction serves as the state-of-the-art concept for polymer-based photovoltaics, reaching power conversion efficiencies of up to 5% [24–26].

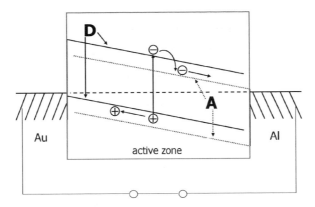

Fig. 10. In bulk heterojunction devices, charge carriers can be dissociated throughout the active layer. Thus every absorbed photon in the active layer can contribute to the photocurrent.

When the bulk heterojunction is composed of polymer–polymer or polymer–fullerene blends the donor and acceptor domains appear to be organized in a rather disordered manner. Hence the optimization of the internal nanomorphology of the blend is an important key for optimal photovoltaic behavior. The understanding and revealing of the internal nanomorphology in polymer–fullerene bulk heterojunctions will be discussed in Section 2.

Owing to molecular motion, these mixtures may further phase separate with time, which represents a morphological instability ultimately leading to device failure [27]. To overcome this and for a better control of the nanomorphology itself, several concepts have been recently introduced to construct ordered bulk heterojunctions. They span a range from using self assembled inorganic nanostructures for the infiltration of conjugated polymers to self-organizing diblock copolymers, where the two blocks carry the different functionalities of donor and acceptor, respectively. Ordered bulk heterojunctions will be further discussed in Section 3. In summary, Fig. 11 displays once again the different device architectures for comparison.

2. NANOMORPHOLOGY OF POLYMER–FULLERENE BULK HETEROJUNCTIONS

In this section, the engineering of the nanomorphology in polymer–fullerene bulk heterojunction solar cells will be discussed. The investigation of nanomorphology in polymer–polymer blends has been performed by others; the interested reader is e.g. referred to the works of Moons et al. [28] and Kim et al. [29].

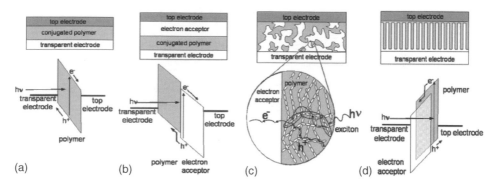

Fig. 11. Evolution of device architectures of conjugated polymer-based photovoltaic cells: (a) single layer; (b) bilayer; (c) "disordered" bulk heterojunction; and (d) ordered bulk heterojunction (reproduced from Coakley and McGehee [67] with permission, copyright 2005, American Chemical Society).

We will distinguish two separate cases: the first part focuses on the nanomorphology of polymer–fullerene blends containing amorphous polymers like PPV, while the second part reports on the use of semicrystalline polymers like P3HT. Since polymers like MEH-PPV or MDMO-PPV exhibit an asymmetric side chain substitution and thus side chain lengths, they tend to form more coiled structures than symmetrically substituted ones [30]. This feature in turn prevents a more ordered organization of the polymer backbone and thus hinders crystallization by a large extent.

2.1. Polymer–Fullerene Blends using Amorphous Polymers

Initial nanomorphology studies were performed on MEH-PPV:C_{60} bulk heterojunctions by Heeger et al. about 10 years ago [31] as method of choice transmission electron microscopy (TEM) and electron diffraction were applied. The authors reported on selectively dissolving the fullerene from blend films, both isolated and connected regions, with characteristic sizes of about 10 nm, corresponding to the C_{60} phase. Upon increasing fullerene contents they observed the formation of a bicontinuous network of C_{60} within the MEH-PPV matrix. These interpenetrating networks are demanded for a good phase percolation to yield efficient charge transport. Furthermore, it was found by electron diffraction that C_{60} is organizing in a nanocrystalline manner.

With the introduction of PCBM [32], a more soluble C_{60} derivative was developed and applied with MEH-PPV in bulk heterojunctions [23]. Since the report of 2.5% power conversion efficiency in MDMO-PPV:PCBM-based blend films [33], this system became one of the most investigated to date. Shaheen et al. [33] reported an increase of the power conversion efficiency

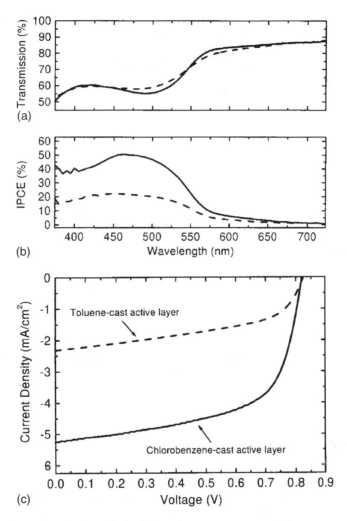

Fig. 12. Optical transmission (a), incident photon to converted electron spectra (b), and current voltage curve of either toluene- or chlorobenzene-cast (c) blend films of MDMO-PPV:PCBM 1:4 (reprinted with permission from [33], copyright 2001, American Institute of Physics).

by almost a factor of 3, when changing the spin casting solvent of which the blend films were spun from toluene to chlorobenzene. Since the spectral absorption of the films was not considerably changed, the larger photocurrents observed for the chlorobenzene-cast films were appointed to changes in the nanomorphology, which was supported by atomic force microscope (AFM) measurements. Fig. 12 shows the two current–voltage curves together with the transmission of the films and the respective spectrally resolved photocurrents (incident photon-to-collected electron, IPCE). The message that came along with this study and motivated many following investigations was a pronounced influence of the solvent on the internal nanomorphology in these polymer–fullerene bulk heterojunction films.

In another study on the MEH-PPV:C_{60} system, Liu et al. [34] correlated the solar cell device parameters with different solvents (xylene, chlorobenzene, 1,2-dichlorobenzene, chloroform and tetrahydrofuran). They claimed that nonaromatic solvents prevent an intimate contact between the MEH-PPV backbone and C_{60}, thus reducing the charge transfer efficiency and subsequently the photocurrent, but increasing the photovoltage. Using AFM measurements they found tetrahydrofuran (THF)-based devices to exhibit a larger scale of phase separation. Furthermore, the authors applied the phase image of the noncontact AFM scans to determine the ratio of C_{60} and MEH-PPV exposed to the surface of the film and correlated this to the observed open circuit voltages by a simple linear combination of the corresponding magnitude for the pristine devices.

Rispens et al. [35] have compared the surface topography of MDMO-PPV:PCBM devices by varying the solvent from xylene (XY) over chlorobenzene (CB) to 1,2-dichlorobenzene (DCB). They found a decrease in phase separation from XY over CB to DCB. Furthermore, they proposed a certain crystal packing of the PCBM molecules, with solvent molecules being introduced into the crystal lattice. These data were based on crystals grown from solution [35].

Martens et al. [36–38] have comparatively investigated the nanostructure of MDMO-PPV:PCBM bulk heterojunctions by applying TEM on films and cross-sections of films spin-cast from toluene and chlorobenzene. On increasing the PCBM concentration in the blends, the authors observed larger and larger dark clusters and attributed these to the fullerene-rich phase.

Since for the ratio 1:1 (1:2) of MDMO-PPV:PCBM for toluene (chlorobenzene)-cast films there was no phase separation visible, the authors concluded that a homogeneous blend of PCBM and MDMO-PPV exists around the PCBM clusters for the blends with a higher PCBM content. Interestingly, the chlorobenzene-based blends were able to incorporate more PCBM than the toluene-based counterparts, thus the compatibility between MDMO-PPV and PCBM in the film seems to be influenced by the choice of solvent. Systematically, the PCBM clusters in the toluene-cast films were larger in size (up to several 100 nm) as compared to chlorobenzene-cast films (less than 100 nm). In the TEM cross-sections of films spin-cast on PET the fullerene-rich clusters are visible as darker regions (cf. Fig. 13).

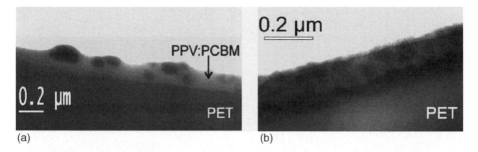

Fig. 13. TEM cross-sectional view of 1:4 MDMO-PPV:PCBM films spin-cast from toluene (a) and chlorobenzene (b) on a PET substrate. The darker regions were attributed to PCBM-rich regions (reprinted from [38], copyright 2003, with permission from Elsevier).

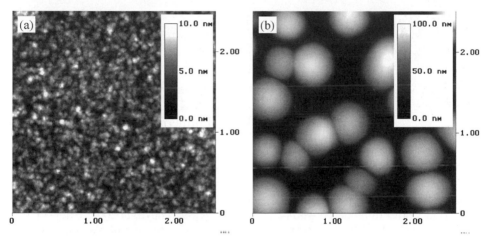

Fig. 14. Tapping mode AFM topography scans of MDMO-PPV:PCBM 1:4 (by weight) blended films, spin-cast from (a) chlorobenzene and from (b) toluene solution. Features of a few hundred nanometers in width are visible in (b), while features in (a) are only around 50 nm (reproduced with permission from [39], copyright 2004, Wiley-VCH).

Our recently published study on the nanoscale morphology of MDMO-PPV:PCBM solar cells was also triggered by the findings of Shaheen et al. [33] and aimed toward the decoding of the different phases within these MDMO-PPV:PCBM blends cast from both toluene and chlorobenzene [39]. Furthermore, the different power conversion efficiencies caused by these morphologies needed a deepened understanding. In agreement with prior studies, a large difference in the scale of phase separation could be identified as major difference between toluene- and chlorobenzene-cast blends (cf. Fig. 14).

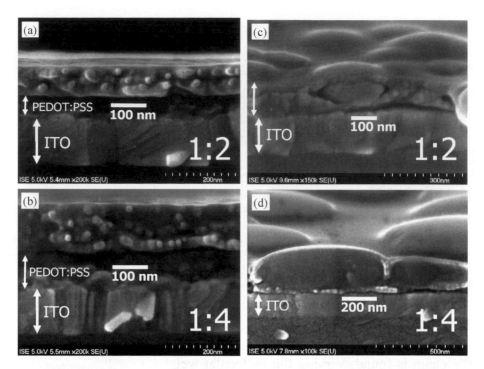

Fig. 15. SEM cross-sections of chlorobenzene-based (a, b) and toluene-based (c, d) MDMO-PPV:PCBM blends. Whereas chlorobenzene-based blends are rather homogeneous, toluene-cast blends reveal large PCBM clusters embedded in a polymer-rich matrix or skin-layer. Small features – referred to as "nanospheres" – are visible in all cases and can be attributed to the polymer in a coiled conformation. The blending ratio is depicted in the lower right corner (reproduced with permission from [39], copyright 2004, Wiley-VCH).

For the first time we used high-resolution scanning electron microscopy (HR-SEM) to image cross-sections of toluene- and chlorobenzene-cast MDMO-PPV:PCBM blends (cf. Fig. 15), whereby much smaller "nanospheres" became observable and have been assigned to the polymer MDMO-PPV in a coiled conformation. It has been suggested previously that conjugated polymers are present as little particles and thus form with solvents rather dispersions than solutions [40].

We have determined the size of the nanospheres to be of 15 nm in diameter using HR-SEM measurements. An example is shown in Fig. 16 together with an illustration of the proposed polymer conformation.

Considering a molecular mass of $m = 10^6\,\mu$ (according to the supplier), and a density of $\rho = 910\,\text{kg/m}^3$ [42] for MDMO-PPV, the volume of a single polymer chain becomes $\sim V_c = 1.8 \times 10^{-24}\,\text{m}^3$. For a sphere of the same volume

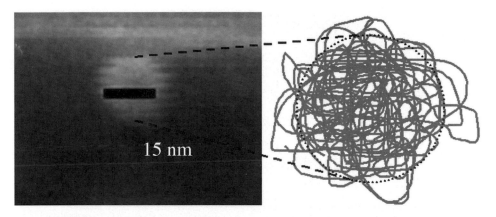

Fig. 16. Magnification of an SEM cross-sectional measurement: a typical polymer nanosphere radius of 15 nm was found in films cast from chlorobenzene-based blends. The conformation of the polymer chain is illustrated in the right side (reprinted from [41], copyright 2005, with permission from Elsevier).

the diameter calculates to ~15 nm, and an astonishing agreement between the size of one nanosphere and the calculated volume taken up by a single MDMO-PPV chain is found. Varying the molecular weight between 2.5×10^5 and 2.5×10^6 μ changes the resulting nanosphere diameter only between 10 and 20 nm. Furthermore, the spherical shape of the polymer was confirmed by tapping mode AFM measurements on phase-separated films due to thermal annealing [41].

The commonly observed larger scale of phase separation of the toluene-cast MDMO-PPV:PCBM blends has been interpreted as the main reason for the reduced photocurrents as compared to the chlorobenzene-cast blends. Especially, a lower charge carrier generation efficiency could be understood as a result of too small exciton diffusion lengths (10–20 nm) for reaching the interface between the large fullerene clusters (200–500 nm) and the polymer. Experimentally it has been identified that indeed some unquenched photo-excitations give rise for residual PCBM-photoluminescence in toluene-cast blends, whereas in chlorobenzene-cast blends the fullerene photolumines-cence could not be detected anymore (cf. Fig. 17) [39].

However, the specific photoluminescence signal of even bare PCBM films was very small (<1%) when compared with the luminescence of the pristine polymer MDMO-PPV and therefore, it is not sufficient to explain the two- to threefold difference in the photocurrents observed earlier [33]. The smaller photoluminescence efficiency of PCBM is due to the symmetrically

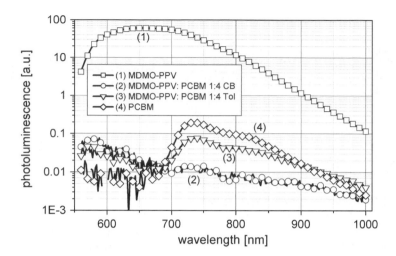

Fig. 17. PL spectra for pure MDMO-PPV, pristine PCBM and blends of the two are shown. It is clearly visible, that the PCBM peak at ~735 nm is found for the toluene-cast blend, whereas for the chlorobenzene-cast blend no such clear peak occurs (reproduced with permission from [39], copyright 2004, Wiley-VCH).

forbidden LUMO–HOMO transition in fullerenes as well as due to strong intersystem coupling to the dark triplet state. In time-dependent photo-luminescence measurements for chlorobenzene-based blends it has been shown that the lifetime of the PCBM singlet state was indeed decreased by addition of the polymer [43]. Furthermore, it should be noted that around the singlet exciton lifetime almost all of the excitons are already in the triplet state due to rapid intersystem crossing [44]. Since the total number of optically observable photoexcitations in the fullerene phase is very small, but the contribution of the fullerene to the spectral photocurrent is vital, it has to be concluded that indeed triplet excitons take part by large extent in the photocurrent generation. This statement is supported by the large triplet exciton diffusion lengths reported for C_{60}-based bilayer devices, where the photocurrent generation is indeed dominated by those triplets [44]. This, however, leads to the conclusion that this small photoluminescence signal of PCBM singlet excitons observed in toluene cast MDMO-PPV:PCBM blends cannot sufficiently explain the observed overall photocurrent loss.

Furthermore, if the large fullerene clusters or domains would be the reason for the decrease in photocurrent, the charge generation due to the absorption of the polymer should not be affected at all. However, this is in contrast to the experimental observation, that indeed the spectral photocurrent is

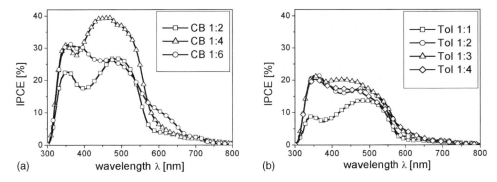

(a)   (b)

Fig. 18. IPCE or photocurrent spectra of chlorobenzene-cast blend (a) and toluene-cast blends (b). As reported earlier, the photocurrents of the toluene-cast blends are lower over the whole spectral range when compared to chlorobenzene. However, PCBM absorption is clearly present and contributes to the photocurrent (reproduced with permission from [39], copyright 2004, Wiley-VCH).

smaller over the whole wavelength range detected [33, 39]. In Fig. 18 experimentally determined spectral photocurrents – IPCE – for different MDMO-PPV:PCBM mixing ratios and for the two solvents chlorobenzene and toluene are depicted for comparison. The peak around 350 nm as well as the little kink slightly above 700 nm can be clearly assigned to the absorption of PCBM, whereas the broad absorption feature between 400 and 550 nm stems from the MDMO-PPV [45]. Therefore, it is evident that PCBM contributes by large extent to the photocurrent.

This leads to the conclusion that another loss mechanism is required to explain the lower overall photocurrents of toluene-cast polymer–fullerene bulk heterojunctions. McNeill et al. [46] resolved the local photocurrent obtained on MDMO-PPV:PCBM toluene-cast blends and revealed that the photocurrent was considerably reduced on top of the elevations caused by the PCBM clusters (cf. Fig. 19), whereas it stayed nearly constant on chlorobenzene-cast blends. The authors proposed several mechanisms to explain that, among others an electron insulating polymer shell around the fullerene clusters.

We recently resolved the proposed polymer – "skin" structure which envelopes the fullerene nanoclusters in toluene-cast films using high-resolution SEM measurements (cf. Fig. 20, right-hand side) [39, 47]. The cross-sectional view clearly shows that the fullerene clusters are embedded into a 10–30 nm thick "skin", presumably consisting of the polymer-rich phase. Using Kelvin probe force microscopy we could confirm this by the detection of a considerably increased work function on top of the embedded clusters [47]. The larger

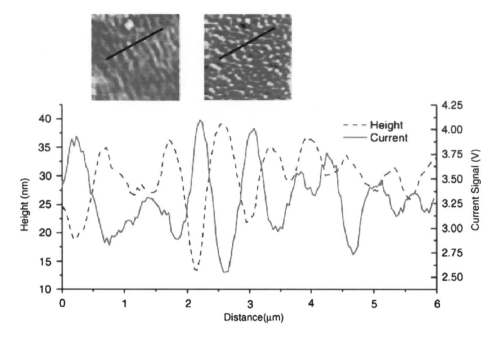

Fig. 19. Height and local photocurrent signal obtained by near-field scanning photocurrent measurements. At the top both the topographic (left) and the photocurrent (right) images are shown (reprinted from [46], copyright 2004, with permission from Elsevier).

work function on top of the clusters as compared to the polymer-rich matrix around the clusters and on chlorobenzene-based blends is a clear signature for an increased hole density at the film surface, which in turn points to the presence of the hole conducting polymer [47].

The presence of the polymer skin layer around the fullerene clusters represents now a severe loss mechanism of the photocurrent for two reasons: (a) electrons, which are accelerated to the top-aluminum electrode, have to penetrate this hole-rich layer and suffer recombination and (b) holes in the polymer skin would have to travel a too long way around the fullerene clusters to reach the hole collecting PEDOT:PSS electrode. However, the strong electric field present inside the photoactive layer accelerates the charge carriers perpendicular to the film plane. This issue is illustrated in Fig. 21 for clarity.

The exploration of the morphological differences between chlorobenzene- and toluene-cast MDMO-PPV:PCBM blend films made clear that not only the observed larger scale of phase separation but rather the difference in the material's phase percolation and thus charge transport properties influence the observed photovoltaic performance. Therefore, it becomes obvious that

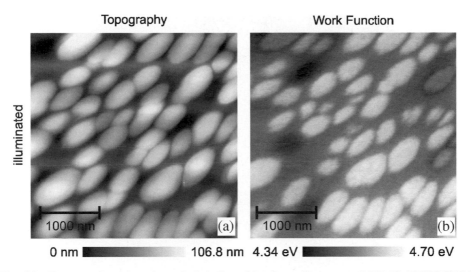

Fig. 20. Topography (a) and work function (b) of a toluene-cast MDMO-PPV:PCBM blend film measured by Kelvin probe force microscopy (KPFM). A clear correlation between the topographic hills caused by the PCBM clusters in (a) and the locally highest work functions in (b) is observed (reprinted with permission from [47], copyright 2005, American Chemical Society).

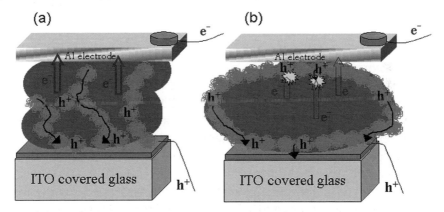

Fig. 21. The schematic displays the differences in the chlorobenzene-based (a) and toluene-based (b) MDMO-PPV:PCBM blend film morphologies when employed as photoactive layer. In (a) both the polymer nanospheres the fullerene phase offer percolating pathways for the transport of holes and electrons, respectively. In (b) electrons and holes suffer recombination, as percolation is not sufficient. Furthermore, a strong electric field across the blend film, originating from the difference in the electrode work function, forces the charge carriers to travel orthogonal to the electrodes (reprinted from [41], copyright 2005, with permission from Elsevier).

Fig. 22. Dilutions (1:50) of saturated PCBM solutions in toluene and chlorobenzene. The different PCBM content is clearly distinguishly (reproduced with permission from [48], copyright 2006, Royal Society of Chemistry).

the charge carrier mobility measured in these devices has to be a function of the nanomorphology simply for geometrical reasons.

One question remains to be answered: what is the origin of the increased phase separation in toluene-cast MDMO-PPV:PCBM blends? A possible answer is simply a different solubility of PCBM in toluene and chlorobenzene, which was experimentally identified and is shown in Fig. 22 [48]. Although chlorobenzene contained up to 4.2% w/v, toluene could only dissolve about 1% w/v.

Upon film formation the solvent is rapidly extracted from the ternary blend of polymer, fullerene, and solvent. Once the system moves below the solubility limit of the fullerene it may start falling out from solution and thus fullerene clusters could be formed supported by the set in of phase separation (cf. Fig. 23) [48].

2.2. Polymer–Fullerene Blends using Semicrystalline Polymers

Owing to their ability for crystallization, semicrystalline polymers offer a richer phase behavior in polymer–fullerene blends. For example, the degree of

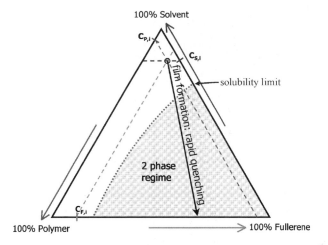

Fig. 23. Schematic ternary phase diagram of a polymer–fullerene–solvent system at constant temperature T and constant pressure p. The arrows indicate the direction of increasing concentration, $C_{S,i}$, $C_{P,i}$ and $C_{F,i}$ are the initial concentrations of solvent, polymer and fullerene in the solution. During the film formation a more or less rapid quenching of the solution toward a solid state blend takes place upon extraction of the solvent (reproduced with permission from [48], copyright 2006, Royal Society of Chemistry).

crystallinity may be adjusted by controlled annealing of the blend films. The most extensively studied system using semicrystalline polymers is based on the polythiophene P3HT in combination with PCBM. It delivered another major step of improvement of polymer–fullerene bulk heterojunction solar cells.

Padinger et al. [49] reported recently on postproduction treatments of P3HT:PCBM bulk heterojunction solar cells. After a combined heat and applied dc-voltage postproduction treatment, the power conversion efficiency could be raised to 3.5%. Applying only the thermal annealing step raised the efficiency from 0.4% to 2.5% already. However, the diode characteristics were further improved by application of the relatively strong forward dc-current at 2.7 V. The authors thereby argued that parasitic shunt currents could be burned out. In Fig. 24 the effect of postproduction treatments on the IV-characteristics are presented.

In a correlated study on the optical properties of similarly prepared P3HT:PCBM devices, the effect of increased absorption due to the postproduction treatments was investigated via optical modeling [50]. The absorption in the active layer of the different devices was determined by the reflection spectra. The strongest increase in absorption achieved by the treatments was determined to be about 40% and resulted mainly from a red-shift in the

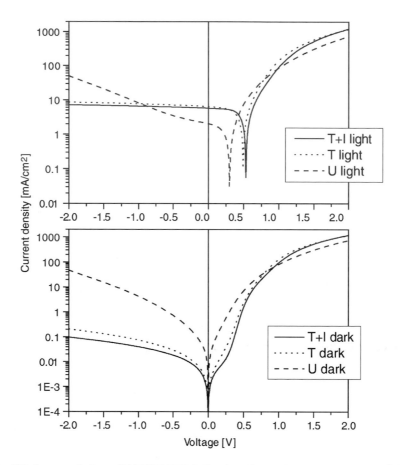

Fig. 24. IV-characteristics of P3HT:PCBM plastic solar cells under $80\,\mathrm{mW/cm^2}$ AM 1.5 solar spectrum simulation (light) and in the dark. The photocurrent and the diode characteristics improved from untreated (U) over thermal annealing (T) to thermal annealing in combination with the application of external voltage (T+I) (reproduced with permission from [49], copyright 2004, Wiley-VCH).

polythiophene absorption [50]. The calculated absorption compared rather well to the experimentally determined spectral photocurrents exhibiting a maximum of 70% [49, 50]. These values were in the same range as reported earlier for P3HT:PCBM photodetectors by Schilinsky et al. [51]. Since the efficiency increase of the treated P3HT:PCBM devices was much higher than 40%, the improved performance could not only be related to an improved absorption in the devices; the morphological changes also played a role in the charge generation and transport.

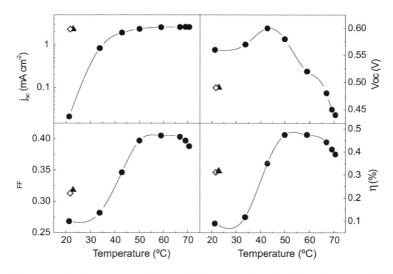

Fig. 25. Short circuit photocurrent (upper left), open circuit voltage (upper right), fill factor (lower left), and power conversion efficiency (lower right) of P3HT:fulleropyrrolidine 3:2 (by weight) blend solar cells. Full circles mark the parameters measured at elevated temperatures, full triangles denote the parameters after the first, diamonds after the second heat circle (reproduced with permission from [52], copyright 2002, Wiley-VCH).

Camaioni et al. [52] investigated the effect of a "mild thermal treatment" on the performance of poly(3-alkylthiophene):fullerene solar cells. The authors observed a three- to fourfold increase in the power conversion efficiency upon annealing at relatively low temperatures (50–60°C). This improvement has been related to an improved order in the film, especially to that of the polythiophene, which is known to crystallize upon thermal annealing [53] or chloroform vapor treatment [54]. In Fig. 25, the results are summarized in the case of a P3HT:fulleropyrrolidine 3:2 solar cell.

It is noteworthy that the induced changes in the nanomorphology remained after cooling the devices down again (cf. Fig. 25, full triangle and diamonds), i.e. the P3HT:fulleropyrrolidine blend relaxed to an energetically lower lying thermodynamic state and remained there after the first heat cycle.

Chirvase et al. [55] reported a comprehensive study on the influence of thermal annealing on the nanomorphology and performance of very similar P3HT:PCBM bulk heterojunction solar cells. The authors concluded the red shift in absorption of the annealed devices to result from molecular diffusion of PCBM out of the polythiophene matrix. Furthermore, they argued that the growth of PCBM clusters led to the formation of percolation paths and thus

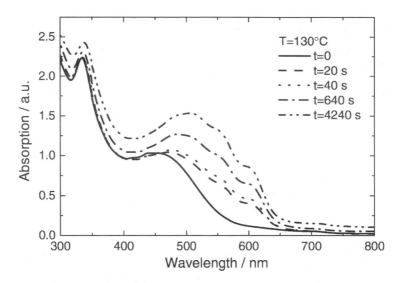

Fig. 26. Absorption spectra of a P3HT:PCBM composite film as cast (solid curve) and after four successive thermal annealing steps, as indicated in the legend. The PCBM concentration is 67% (reprinted with permission from [55], copyright 2004, Institute of Physics Publishing).

improved the photocurrent. An improved ordering of P3HT domains via interchain interaction [56] and improved interface properties due to reduced interface defects [57] have been proposed as a result of thermal annealing. Chirvase et al. [55] identified that thermal annealing of pristine PCBM or P3HT films yielded only slight changes in the absorption, whereas annealing of blends resulted in a large increase of P3HT absorption (cf. Fig. 26).

The observed growth of large micron-sized PCBM crystal domains was found to depend on the initial concentration of PCBM in the blend as well as on the duration of the annealing process. In Fig. 27, the topography of P3HT:PCBM films with and without aluminium electrode is shown for two different PCBM concentrations (50% and 75%). Clearly the dendritic structures observed for the 75% PCBM concentration are much larger in size than the aggregates found for films containing 50% PCBM.

Kim et al. [58] also reported on the effect of thermal annealing on P3HT:PCBM solar cell device efficiency. The authors suggested a vertical phase segregation between P3HT and PCBM to result from the thermal annealing, where P3HT was deposited adjacent to the PEDOT:PSS electrode. Thus the holes could be transported more efficient to the PEDOT:PSS electrode and electrons directly to the top-aluminum contact, yielding better diode properties [58].

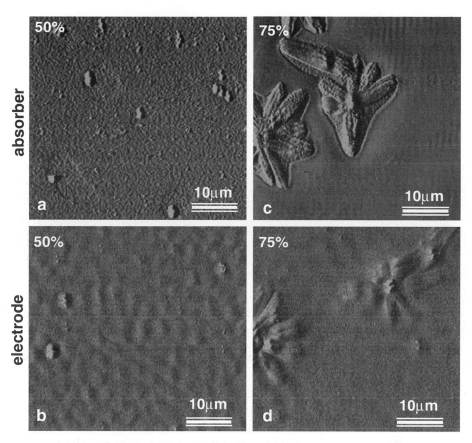

Fig. 27. Tapping mode AFM images taken on P3HT:PCBM films without (a, b) and with aluminum top electrode (c, d) at different PCBM concentrations. Large dentritic PCBM crystals are observed for the larger fullerene concentration (reprinted with permission from [55], copyright 2004, Institute of Physics Publishing).

Yang et al. [59] reported on the nanoscale morphology of P3HT:PCBM solar cells using TEM and electron diffraction. Upon annealing the blend the authors observed an increase in crystallinity not only for the P3HT phase but also for PCBM. The authors observed that due to annealing fibrillar P3HT crystals extend their length and new PCBM domains are developed. As a result the charge transport properties will improve for both types of charges due to expanding of the crystalline domains, yielding higher device efficiencies [59]. Fig. 28 shows TEM images together with electron diffraction images for the untreated and the annealed P3HT:PCBM blend. The schematic shows the proposed extension of P3HT crystallites and formation of PCBM crystallites.

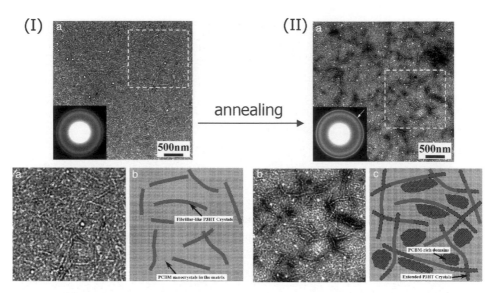

Fig. 28. TEM images in combination with electron diffraction of untreated (I) and thermally annealed (II) P3HT:PCBM blend films. The schematic indicates the mechanism for improved crystallization of both, P3HT and PCBM (reprinted with permission from [59], copyright 2005, American Chemical Society).

Remarkably the P3HT backbone was found to be oriented vertical to the P3HT-fibrils, thus the $\pi-\pi$ stacking direction itself was parallel to the long axis of the P3HT crystals [60]. As a result the charge transport followed the $\pi-\pi$ stacking direction and the improved order led to better charge carrier mobilities.

Erb et al. [61] correlated XRD measurements with P3HT:PCBM composite films. The authors showed that upon annealing P3HT crystallites of sizes of about 10 nm were grown. The polymer backbone orientation within these crystallites was found to be parallel to the substrate, whereas the side chains were oriented perpendicular to the substrate (a-axis orientation of P3HT crystallites). No other orientation of P3HT crystals could be observed in the blend films, though all orientations were verified earlier for a powder sample. The XRD signals before and after thermal annealing of P3HT:PCBM composites are shown in Fig. 29.

Very recently, Zhokhavets et al. [62] could show that the crystallinity of P3HT in blends with PCBM correlates strongly with the absorption coefficient of the films. This feature was verified for the two different solvents investigated, chlorobenzene and chloroform. This is shown in Fig. 30 for P3HT: PCBM 1:2 (by weight) blends annealed at several different temperatures.

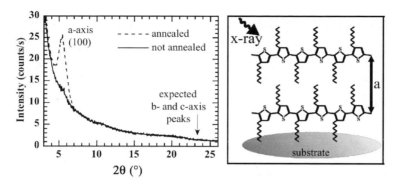

Fig. 29. Diffraction diagram (grazing incidence) of P3HT:PCBM composite films deposited on glass/ITO/PEDOT:PSS substrates (left) and schematic of the chain orientation with respect to the substrate (right) (reproduced with permission from [61], copyright 2005, Wiley-VCH).

Interestingly, the chloroform-based blend films showed a considerably higher ordering of the polythiophene.

Yang et al. [63] have shown that the crystallinity of pristine P3HT films depends indeed on the solvent from which the films were cast [63]. They found that the use of chloroform resulted in both high crystallinity as well as large charge carrier mobility in thin film field effect transistors. This can be understood, as the (100) orientation of P3HT enables efficient charge transport along the π–π stacking direction parallel to the substrate [64]. Although in several cases the highest crystallinity and adjunct charge carrier mobility was found for chloroform-cast films, the most efficient P3HT:PCBM solar cells reported up to date apply chlorobenzene [24, 25] or dichlorobenzene [26] as spin-casting solvent. Therefore, only a very good (100) orientation might be useful for charge transport in the plane – as required for field effect transistors – but in solar cells the main direction for charge transport is perpendicular to that. Thus further investigations are required to elucidate in detail the crystallinity in chlorobenzene-cast blends. As an alternative to thermal annealing, Yang et al. applied long solvent drying times for the increase of structural ordering in the P3HT:PCBM system yielding high efficiency due to high charge carrier mobility [26, 65]. They also pointed out the importance of a balanced charge transport of both electrons and holes using time of flight (TOF) measurements. They found that for blending ratios of 1:1 (by weight) the best balance was achieved showing for both types of charge carriers nondispersive transport behavior [65].

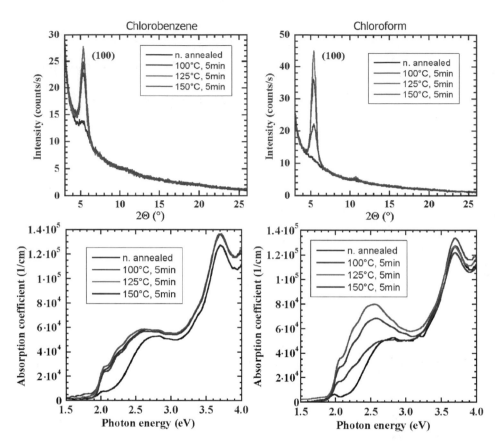

Fig. 30. Crystallinity observed by grazing incidence X-ray deflection (upper) and absorption coefficients determined by spectroscopic ellipsometry (lower) of P3HT:PCBM 1:2 (by weight) films cast from either chlorobenzene (left) or chloroform (right) are shown (reproduction with permission from [62], copyright 2006, Elsevier).

In contrast, Ma et al. [25] applied thermal annealing to improve the solar cell device efficiency. Interestingly, they observed the device efficiency to saturate but not to degrade upon annealing for up to 2 h at 150°C. They accounted this remarkable thermal stability to the use of chlorobenzene as solvent and the low-volume fraction of PCBM, leading to a suppression of PCBM aggregate overgrowth. High resolution TEM images of the blend film as cast, after 30 min and after 2 h at 150°C are depicted in Fig. 31.

Further on, the structure is considerably different to that reported by Yang et al. as shown in Fig. 28. This difference may indeed result from the use of different solvents, 1,2-dichlorobenzene and chlorobenzene. Thus also

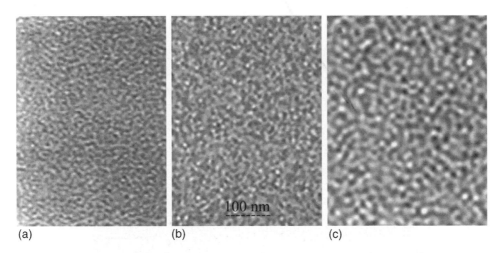

(a)   (b)   (c)

Fig. 31. High resolution TEM images of regioregular P3HT:PCBM 1:0.8 (by weight) blend films cast from chlorobenzene, (a) as cast, (b) annealed at 150°C for 30 min and (c) for 2 h. Clearly visible are the spherical nanostructures (presumably P3HT) with a size on the order of 10 nm (reproduced with permission from [25], copyright 2005, Wiley-VCH).

for the P3HT:PCBM system a dependence of nanostructure and efficiency on the spin-casting solvent may be related.

In conclusion, thermal annealing or solvent drying treatments of P3HT: PCBM-blend films provide a better ordering and increase of crystallinity for P3HT and the formation of PCBM crystallites as well. If the PCBM concentration is kept low, the overgrowth of larger PCBM dendrites can be prevented and a remarkable morphological stability achieved [25]. Yet, from Fig. 31 it is still visible that the structure is slightly coarsening, indicating limits for the morphological lifetime of the device.

3. DESIGN CONCEPTS FOR ORDERED BULK HETEROJUNCTIONS

In this section, several approaches will be discussed that improve long-time morphological stability. For example, this can be achieved by applying highly ordered inorganic nanostructures, in which the conjugated polymer is filled in with a second step. Here two general types – both grown on the substrate – may be distinguished: nanopores and nanorods. Both of these structures may serve as a kind of scaffold which then gets filled up by the conjugated polymer. The scaffold has to fulfill two functionalities: it has to accept the photoexcited electron from the polymer and it has to transport

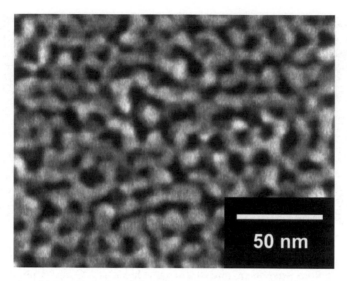

Fig. 32. Scanning electron micrograph of a mesoporous TiO$_2$ film, obtained by calcination of a precursor. The pores of these nanostructured electrodes were filled with P3HT by spin coating. This method gives high control over the pore size and thus the scale of donor–acceptor phase separation (reprinted with permission from [67], copyright 2003, American Institute of Physics).

those electrons to the negative electrode. Since photoexcitations in polymers travel only short distances between 5 and 20 nm, the dimension of the scaffold network has to be on the same nanoscale to allow for efficient use of the light absorption in the polymer.

Based on a preparation route applying triblock copolymers as structure directing agents, regular nanoporous TiO$_2$ structures could be obtained by calcination of a precursor film on top of fluorine doped tin oxide (SnO:F) covered glass (Fig. 32) [66, 67]. The nanopores were adjusted to about 10 nm in diameter, to enable harvesting of all photogenerated excitons in the polymer. The filling was done by spin coating P3HT on top of the structure followed by an annealing step to allow the polymer for settling into the pores. The device was completed by spin coating a 30 nm thick P3HT overlayer on top of the structure to prevent electrons from reaching the hole extracting gold electrode, which was finally evaporated on top of the film. Although the infiltration of P3HT into the pores worked well the photocurrent seemed to be limited by recombination of holes generated deeper inside the pores.

To increase the hole mobility within the pores, more recently Coakley et al. [68] extended their studies to the infiltration of P3HT into anodic aluminum, which forms straight channels with adjustable pore diameter. Here

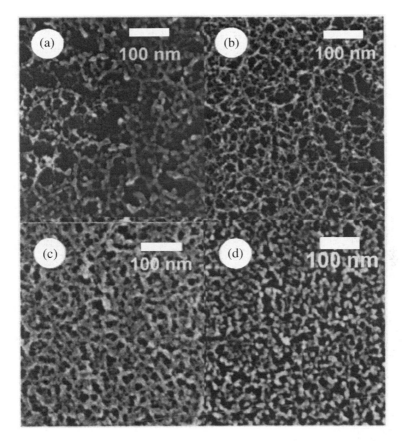

Fig. 33. SEM images of P(S-*b*-EO) and TTIP films with different TTIP concentrations annealed at 400°C for 5 h; (a) 15%, (b) 25%, (c) 35%, and (d) 45% (reprinted with permission from [69], copyright 2005, American Institute of Physics).

they could demonstrate a pore size dependent charge carrier mobility in polythiophene, which was increased compared to a neat P3HT film by a factor of 20. The authors accounted the polymer orientation in the pores as well as an increased charge carrier density at the P3HT:anodic alumina junction for this improvement.

Similar to the works of Coakley et al., Wang et al. [69] did report recently on the fabrication of a TiO2 interconnected network structure for photovoltaic applications, which was obtained using polystyrene-*block*-polyethylene oxide (P(S-*b*-EO)) diblock copolymer as the templating agent. The pore size of the TiO2 structure was controlled by the amount of Ti precursor (titanium tetraisopropoxide, TTIP) added to the structure directing copolymer (cf. Fig. 33). Thereafter MEH-PPV was infiltrated into the highly porous inorganic network

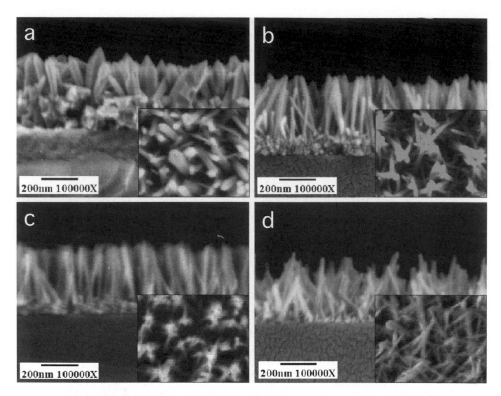

Fig. 34. Scanning electron microscopy on ZnO nanofibers grown on annealed zinc acetate nucleation layers under different conditions: nucleation layer were annealed at (a) 500°C for 60 min, (b) 300°C for 30 min, (c) 240°C for 10 min, and (d) 300°C for 5 min. These nanofibers serve as electron accepting and transporting materials in the bulk heterojunction with P3HT (reproduced with permission from [71], copyright 2006, Wiley-VCH).

and covered by a gold contact. The authors observed improved solar cell performance when using a compact TiO2 interlayer to the ITO-glass substrate.

Bartholomew and Heeger [70] reported on the infiltration of P3HT in random nanocrystalline TiO_2 networks. These inorganic networks were produced by spin coating TiO_2 nanocrystals modified by organics from dispersion. Since the porosity of the resulting film was relatively low, infiltration of the polymer appeared to be difficult. However, the amount of infiltrated P3HT could be raised considerably by using a lower molecular weight fraction of the polymer, in combination with annealing and surface modification of the TiO_2 nanocrystals using amphiphilic Ru-based dyes.

A related yet different approach to the upper ones is the intercalation of P3HT into a film of ZnO nanofibers, as demonstrated by Olson et al. [71].

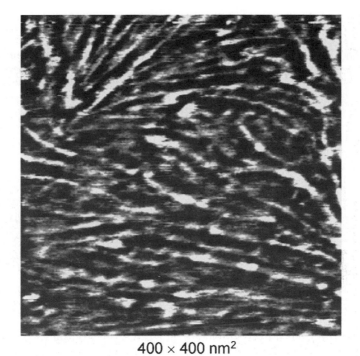

400 × 400 nm²

Fig. 35. The tapping mode AFM image of a dichlorobenzene-cast PPV-PS(stat-C$_{60}$) film shows nanostructures of about 15 nm in size. This is comparable to the double PPV-block length (reproduce with permission from [78], copyright 2001, Elsevier).

Photovoltaic devices were built on ITO-glass, which was covered first by a nucleation layer of ZnO. After annealing, ZnO nanofibers are grown from the nucleation layer using zinc nitrate solution. Thereafter, a layer of P3HT is spin-coated on top of this and covered by an evaporated silver contact for hole extraction. In comparison with bilayer ZnO–P3HT devices these nano-structured bulk heterojunctions show an increase between a factor of 3–4 in the IPCE spectrum. Fig. 34 shows several samples of ZnO nanofiber films grown under various conditions.

The application of block copolymers offers a route for the design of highly ordered self-assembled phase separated organic structures [72]. In the past few years several approaches have been pursued by incorporating often both donor and accepter functionalities onto different blocks of a single copolymer [73–76].

The concept of the Hadziioannou group is a polymer–fullerene bulk heterojunction: a donor PPV block is attached to a polystyrene block

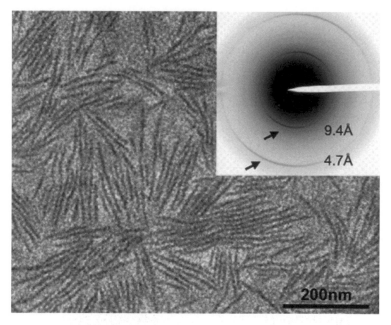

Fig. 36. Shown is a TEM micrograph of a diblock copolymer film annealed for 15 min at 90°C. The inset shows the electron diffraction signal corresponding roughly to the PPV interchain distances in the film (reproduced with permission from [80], copyright 2006, Elsevier).

with pendant C_{60} fullerenes [73, 77–79]. On comparing the photovoltaic response of films prepared from the D–A block copolymer with molecular blends of the corresponding donor and acceptor blocks, a superior response was found for the block copolymer both for photocurrent and photovoltage [78]. Furthermore, a nanometer-sized phase separated structure was obtained for the same copolymer as imaged by tapping mode AFM (cf. Fig. 35) [78]. The size of the nanophase structure of about 15 nm was found to be comparable to the double of the PPV-block length.

In a more recent study Heiser et al. [80] found out that for rod–coil block copolymers highly ordered structures were only obtained, when the block coil is characterized by a glass transition temperature significantly lower than the melting temperature of the alkyl side chains [80]. Since the PPV-based rod block showed a liquid crystalline behavior, ordered crystalline structures could be obtained for block copolymer films after annealing for 15 min at about 90°C (cf. Fig. 36). As coil block a statistical copolymer of poly(butylacrylate) (PBA) and 4-chloromethylstyrene (CMS) was used. The coil block did not

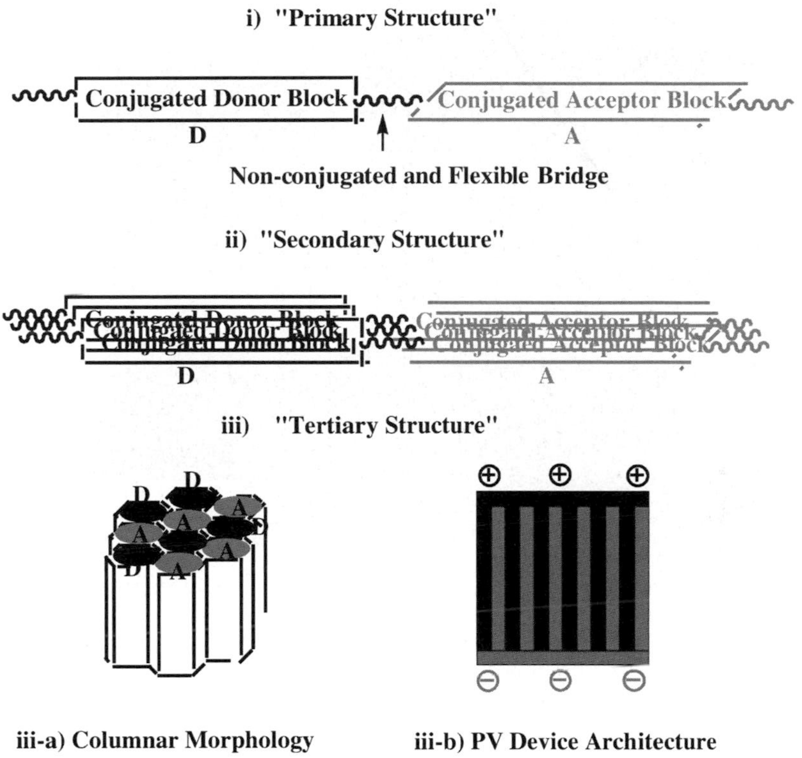

i) "Primary Structure"

Conjugated Donor Block ⫯ Conjugated Acceptor Block

D ↑ A

Non-conjugated and Flexible Bridge

ii) "Secondary Structure"

Conjugated Donor Block ⫯ Conjugated Acceptor Block

D A

iii) "Tertiary Structure"

⊕ ⊕ ⊕

⊖ ⊖ ⊖

iii-a) Columnar Morphology   **iii-b) PV Device Architecture**

Fig. 37. Proposed structural organization for a diblock copolymer donor–acceptor (D–A) solar cell (reproduced with permission from [74], copyright 2003, Elsevier).

carry the functionality of an acceptor, but the CMS allowed for attaching of C_{60} in a subsequent step.

Sun and coworkers [74, 81, 82] developed another approach by constructing a diblock copolymer with both blocks based on polyphenylenevinylene. As donor block they used an alkyloxy derivatized PPV, whereas the acceptor was constructed from a sulfone-derivatized PPV. As block lengths they used between 5 and 20 nm for both blocks and the side chain length was varied as well. A nonconjugated flexible alkyl chain bridged the donor and acceptor blocks. Since the two blocks are of the stiffer rod-type they enable for both donor and acceptor domains the development of $\pi-$stacks. As tertiary structure a hexagonal columnar structure is aimed at. On comparing the dark current voltage characteristics between a simple oligomeric D–A blend with the diblock copolymer-based film, Sun et al. [83] found an increase in the current density by two to three orders of magnitude, indicating a highly improved charge

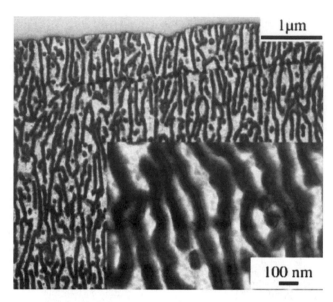

Fig. 38. TEM of cross-section from relatively thick films (5–6 μm) obtained by melting down the TPA/perylene acrylate based diblock copolymers onto a glass substrate and subsequently tempering them at 200°C for 1 h (reproduced with permission from [84], copyright 2004, American Chemical Society).

transport due to the copolymer. Furthermore, the spectral photocurrent was increased against the D–A blend by a factor of 2 [82]. Fig. 37 shows the proposed structure organization on several levels.

Lindner and Thelakkat [84] recently demonstrated nanophase separation in films of side chain functionalized diblock copolymers. The poly(vinyltriphenylamine) (PvTPA) donor block is based on the well-known hole conductor triphenalamine (TPA), whereas the acceptor block is made up of perylene bisimide acrylate, which has high electron mobility and functions also as an efficient light absorber. A low polydispersity could be maintained even for higher molecular weights, controlled by the polymerization reaction time. Owing to the tendency of the perylene unit for $\pi-\pi$ stacking, nanowires were built up within the TPA matrix in a relatively thick sample upon annealing at 200°C for 1 h. A cross-section TEM image visualizes the nanophase separation (Fig. 38) perpendicular to the film plane.

Taking the results of this section together yields an optimistic assessment and encouraging viewpoint for the development of highly ordered bulk heterojunctions at the nanoscale of exciton diffusion lengths. Key techniques have been developed and it remains a question of time, when organic photovoltaic devices of this kind will one day represent the state of the art.

4. CONCLUSION

The most successful architecture for conjugated polymer-based solar cells is relying upon the bulk heterojunction concept – introduced about 10 years ago. These intimate D–A blends require a good control over the internal nanomorphology of the phase-separated domains to enable efficient charge generation and transport. Therefore, interpenetrating and percolating donor and acceptor domains need to be formed and the intermolecular order has to be optimized as well. In disordered bulk heterojunctions the nanomorphology has to be tuned by several production parameters, for example, the choice of solvent, blending ratio, and postproduction annealing. In the case of ordered bulk heterojunctions the nanostructure will be determined to a higher degree by self-assembling processes often obtained by a block copolymer approach. These enable a precise control over domain size by parameters as polymer block length or growth conditions for inorganic nanostructures. Initial concepts introduced are encouraging milestones on the path to high efficiency organic photovoltaics.

REFERENCES

[1] "Handbook of Conducting Polymers" 2nd edition, (T.A. Skotheim, R.L. Elsenbaumer and J.R. Reynolds, eds.), Marcel Dekker, New York, 1998.

[2] Semiconducting Polymers, (G. Hadziioannou and P.F. van Hutten, eds.), Wiley-VCH, Weinheim, 2000.

[3] C.K. Chiang, J.C.R. Fincher, Y.W. Park, A.J. Heeger, H. Shirakawa, E.J. Louis, S.C. Gau and A.G. MacDiarmid, Phys. Rev. Lett., 39 (1977) 1098–1101.

[4] J.H. Burroughes, D.D.C. Bradley, A.R. Brown, R.N. Marks, K. Mackay, R.H. Friend, P.L. Burns and A.B. Holmes, Nature, 347 (1990) 539.

[5] Z.N. Bao, Y. Feng, A. Dodabalapur, V.R. Raju and A.J. Lovinger, Chem. Mater., 9 (1997) 1299.

[6] N.S. Sariciftci, D. Braun, C. Zhang, V.I. Srdanov, A.J. Heeger, G. Stucky and F. Wudl, Appl. Phys. Lett., 62 (1993) 585.

[7] S. Karg, W. Riess, V. Dyakonov and M. Schwoerer, Synth. Met., 54 (1993) 427.

[8] R.N. Marks, J.J.M. Halls, D.D.C. Bradley, R.H. Friend and A.B. Holmes, J. Phys. Condens. Mater., 6 (1994) 1379.

[9] G. Yu and A.J. Heeger, J. Appl. Phys., 78 (1995) 4510.

[10] G. Yu, J. Gao, J.C. Hummelen, F. Wudl and A.J. Heeger, Science, 270 (1995) 1789.

[11] J.J.M. Halls, C.A. Walsh, N.C. Greenham, E.A. Marseglia, R.H. Friend, S.C. Moratti and A.B. Holmes, Nature, 376 (1995) 498.

[12] C.J. Brabec, N.S. Sariciftci and J.C. Hummelen, Adv. Funct. Mater., 11 (2001) 15.

[13] H. Hoppe and N.S. Sariciftci, J. Mater. Res., 19 (2004) 1924.

[14] H. Spanggaard and F.C. Krebs, Sol. Energy Mater. Sol. Cells, 83 (2004) 125.

[15] C.J. Brabec, J.A. Hauch, P. Schilinsky and C. Waldauf, MRS Bull., 30 (2005) 50.

[16] C.D. Dimitrakopoulos and D.J. Mascaro, IBM J. Res. Dev., 45 (2001) 11.

[17] N.S. Sariciftci, L. Smilowitz, A.J. Heeger and F. Wudl, Science, 258 (1992) 1474.

[18] C.J. Brabec, G. Zerza, G. Cerullo, S.D. Silvestri, S. Luzzati, J.C. Hummelen and S. Sariciftci, Chem. Phys. Lett., 340 (2001) 232.

[19] V.D. Mihailetchi, L.J.A. Koster, J.C. Hummelen and P.W.M. Blom, Phys. Rev. Lett., 93 (2004) 216601.

[20] H.H.P. Gommans, M. Kemerink, J.M. Kramer and R.A.J. Janssen, Appl. Phys. Lett., 87 (2005) 122104.

[21] S.M. Sze, Physics of Semiconductor Devices, Wiley, New York, 1981.

[22] I.D. Parker, J. Appl. Phys., 75 (1994) 1656.

[23] C.W. Tang, Appl. Phys. Lett., 48 (1986) 183.

[24] M. Reyes-Reyes, K. Kim and D.L. Carrolla, Appl. Phys. Lett., 87 (2005) 083506; M. Reyes-Reyes, K. Kim, J. Dewald, R. Lopez-Sandoval, A. Avadhanula, S. Curran and D.L. Carroll, Org. Lett., 7 (2005) 5749.

[25] W. Ma, C. Yang, X. Gong, K. Lee and A.J. Heeger, Adv. Funct. Mater., 15 (2005) 1617.

[26] G. Li, V. Shrotriya, J. Huang, Y. Yao, T. Moriarty, K. Emery and Y. Yang, Nature, Mater., 4 (2005) 864.

[27] X. Yang, J.K.J. van Duren, R.A.J. Janssen, M.A.J. Michels and J. Loos, Macromolecules, 37 (2004) 2151.

[28] E. Moons, J. Phys. Condens. Mater., 14 (2002) 12235–12260.

[29] J.-S. Kim, P.K.H. Ho, C.E. Murphy and R.H. Friend, Macromolecules, 37 (2004) 2861.

[30] M. Kemerink, J.K.J. van Duren, P. Jonkheijm, W.F. Pasveer, P.M. Koenraad, R.A.J. Janssen, H.W.M. Salemink and J.H. Wolter, Nano Lett., 3 (2003) 1191.

[31] C.Y. Yang and A.J. Heeger, Synth. Met., 83 (1996) 85.

[32] J.C. Hummelen, B.W. Knight, F. LePeq, F. Wudl, J. Yao and C.L. Wilkins, J. Org. Chem., 60 (1995) 532.

[33] S.E. Shaheen, C.J. Brabec, N.S. Sariciftci, F. Padinger, T. Fromherz and J.C. Hummelen, Appl. Phys. Lett., 78 (2001) 841.

[34] J. Liu, Y. Shi and Y. Yang, Adv. Funct. Mater., 11 (2001) 420.

[35] M.T. Rispens, A. Meetsma, R. Rittberger, C.J. Brabec, N.S. Sariciftci and J.C. Hummelen, Chem. Commun., 17 (2003) 2116.

[36] T. Martens, J. D'Haen, T. Munters, L. Goris, Z. Beelen, J. Manca, M. D'Olieslaeger, D. Vanderzande, L.D. Schepper and R. Andriessen, "The influence of the microstructure upon the photovoltaic performance of MDMOPPV:PCBM bulk heterojunction organic solar cells," presented at the MRS Spring Meeting, p. P7.11.1, Materials Research Society, San Francisco, 2002.

[37] T. Martens, Z. Beelen, J. D'Haen, T. Munters, L. Goris, J. Manca, M. D'Olieslaeger, D. Vanderzande, L.D. Schepper and R. Andriessen, Morphology of MDMO-PPV:PCBM bulk heterojunction organic solar cells studied by AFM, KFM and TEM, in Proceedings of the SPIE (Z.H. Kafafi and D. Fichou, eds.), Vol. 4801, p. 40, Organic Photovoltaics III, SPIE-Int. Soc. Opt. Eng., 2003.

[38] T. Martens, J. D'Haen, T. Munters, Z. Beelen, L. Goris, J. Manca, M. D'Olieslaeger, D. Vanderzande, L.D. Schepper and R. Andriessen, Synth. Met., 138 (2003) 243.

[39] H. Hoppe, M. Niggemann, C. Winder, J. Kraut, R. Hiesgen, A. Hinsch, D. Meissner and N.S. Sariciftci, Adv. Funct. Mater., 14 (2004) 1005.

[40] B. Wessling, Chem. Innov., 31 (2001) 34.

[41] H. Hoppe, T. Glatzel, M. Niggemann, W. Schwinger, F. Schaeffler, A. Hinsch, M.C. Lux-Steiner and N.S. Sariciftci, Thin Solid Films, 511–512 (2006) 587–592.

[42] C.W.T. Bulle-Lieuwma, W.J.H. van Gennip, J.K.J. van Duren, P. Jonkheijm, R.A.J. Janssen and J.W. Niemantsverdriet, Appl. Surf. Sci., 547 (2003) 203–204.

[43] J.K.J. van Duren, X. Yang, J. Loos, C.W.T. Bulle-Lieuwma, A.B. Sieval, J.C. Hummelen and R.A.J. Janssen, Adv. Funct. Mater., 14 (2004) 425.

[44] P. Peumans, A. Yakimov and S.R. Forrest, J. Appl. Phys., 93 (2003) 3693.

[45] H. Hoppe, N. Arnold, D. Meissner and N.S. Sariciftci, Sol. Energy Mater. Sol. Cells, 80 (2003) 105.

[46] C.R. McNeill, H. Frohne, J.L. Holdsworth and P.C. Dastoor, Synth. Met., 147 (2004) 101.

[47] H. Hoppe, T. Glatzel, M. Niggemann, A. Hinsch, M.C. Lux-Steiner and N.S. Sariciftci, Nano Lett., 5 (2005) 269.

[48] H. Hoppe and N.S. Sariciftci, J. Mater. Chem., 16 (2006) 45.

[49] F. Padinger, R.S. Rittberger and N.S. Sariciftci, Adv. Funct. Mater., 13 (2003) 1.

[50] H. Hoppe, N. Arnold, D. Meissner and N.S. Sariciftci, Thin Solid Films, 451–452 (2004) 589–592.

[51] P. Schilinsky, C. Waldauf and C.J. Brabec, Appl. Phys. Lett., 81 (2002) 3885.

[52] N. Camaioni, G. Ridolfi, G. Casalbore-Miceli, G. Possamai and M. Maggini, Adv. Mater., 14 (2002) 1735.

[53] Y. Zhao, G.X. Yuan, P. Roche and M. Leclerc, Polymer, 36 (1995) 2211.

[54] M. Berggren, G. Gustafsson, O. Inganas, M.R. Andersson, O. Wennerstrom and T. Hjertberg, Appl. Phys. Lett., 65 (1994) 1489.

[55] D. Chirvase, J. Parisi, J.C. Hummelen and V. Dyakonov, Nanotechnology, 15 (2004) 1317–1323.

[56] P.J. Brown, D.S. Thomas, A. Köhler, J. Wilson, J.S. Kim, C. Ramsdale, H. Sirringhaus and R.H. Friend, Phys. Rev. B, 67 (2003) 064203.

[57] T. Ahn and H.L. Sein-HoHa, Appl. Phys. Lett., 80 (2002) 392.

[58] Y. Kim, S.A. Choulis, J. Nelson, D.D.C. Bradley, S. Cook and J.R. Durrant, Appl. Phys. Lett., 86 (2005) 063502.

[59] X. Yang, J. Loos, S.C. Veenstra, W.J.H. Verhees, M.M. Wienk, J.M. Kroon, M.A.J. Michels and R.A.J. Janssen, Nano Lett., 5 (2005) 579.

[60] K.J. Ihn, J. Moulton and P. Smith, J. Polym. Sci. Polym. Phys., 31 (1993) 735.

[61] T. Erb, U. Zhokhavets, G. Gobsch, S. Raleva, B. Stühn, P. Schilinsky, C. Waldauf and C.J. Babec, Adv. Funct. Mater., 15 (2005) 1193.

[62] U. Zhokhavets, T. Erb, G. Gobsch, M. Al-Ibrahim and O. Ambacher, Chem. Phys. Lett., 418 (2006) 343.

[63] H. Yang, T.J. Shin, L. Yang, K. Cho, C.Y. Ryu and Z. Bao, Adv. Funct. Mater., 15 (2005) 671.

[64] H. Sirringhaus, P.J. Brown, R.H. Friend, M.M. Nielsen, K. Bechgaard, B.M.W. Langeveld-Voss, A.J.H. Spiering, R.A.J. Janssen, E.W. Meijer, P. Herwig and D.M. de Leeuw, Nature, 401 (1999) 685.

[65] J. Huang, G. Li and Y. Yang, Appl. Phys. Lett., 87 (2005) 112105.

[66] K.M. Coakley, Y. Liu, M.D. McGehee, K. Frindell and G.D. Stucky, Adv. Funct. Mater., 13 (2003) 301.

[67] K.M. Coakley and M.D. McGehee, Appl. Phys. Lett., 83 (2003) 3380.

[68] K.M. Coakley, B.S. Srinivasan, J.M. Ziebarth, C. Goh, Y. Liu and M.D. McGehee, Adv. Funct. Mater., 15 (2005) 1927.

[69] H. Wang, C.C. Oey, A.B. Djurisic, M.H. Xie, Y.H. Leung, K.K.Y. Man, W.K. Chan, A. Pandey, J.-M. Nunzi and P.C. Chui, Appl. Phys. Lett., 87 (2005) 023507.

[70] G.P. Bartholomew and A.J. Heeger, Adv. Funct. Mater., 15 (2005) 677.

[71] D.C. Olson, S.E. Shaheen, R.T. Collins and D.S. Ginley, Adv. Funct. Mater., 16 (2006) in print.

[72] C. Park, J. Yoon and E.L. Thomas, Polymer, 44 (2003) 6725.

[73] G. Hadziioannou, MRS Bull., 27 (2002) 456.

[74] S.-S. Sun, Sol. Energy Mater. Sol. Cells, 79 (2003) 257.

[75] J.A. Gratt and R.E. Cohen, J. Appl. Polym. Sci., 91 (2004) 3362–3368.

[76] F.C. Krebs, O. Hagemann and M. Jorgensen, Sol. Energy Mater. Sol. Cells, 83 (2004) 211.

[77] U. Stalmach, B. de Boer, C. Videlot, P.F. van Hutten and G. Hadziioannou, J. Am. Chem. Soc., 122 (2000) 5464.

[78] B. de Boer, U. Stalmach, P.F. van Hutten, C. Melzer, V.V. Krasnikov and G. Hadziioannou, Polymer, 42 (2001) 9097.

[79] M.H. van de Veen, B. de Boer, U. Stalmach, K.I. van de Wetering and G. Hadziioannou, Macromolecules, 37 (2004) 3673.

[80] T. Heiser, G. Adamopoulos, M. Brinkmann, U. Giovanella, S. Ould-Saad, C. Brochon, K. van de Wetering and G. Hadziioannou, Thin Solid Films, (2006) in press.

[81] S.S. Sun, Z. Fan, Y. Wang, C. Taft, J. Haliburton and S. Maaref, Synthesis and characterization of a novel -D-B-A-B- block copolymer system for potential licht harvesting applications, in Proceedings of the SPIE (Z.H. Kafafi and D. Fichou, eds.), Vol. 4801, p. 114, Organic Photovoltaics III, SPIE-Int. Soc. Opt. Eng., 2003.

[82] S. Sun, Z. Fan, Y. Wang and J. Haliburton, J. Mater. Sci., 40 (2005) 1429.

[83] S.-S. Sun, Z. Fan, Y. Wang, K. Winston and C.E. Bonner, Mater. Sci. Engin. B, 116 (2005) 279.

[84] S.M. Lindner and M. Thelakkat, Macromolecules, 37 (2004) 8832.

Nanostructured Materials for Solar Energy Conversion
T. Soga (editor)

Chapter 11

Nanostructured Organic Bulk Heterojunction Solar Cells

Yoshinori Nishikitani, Soichi Uchida and Takaya Kubo

8 Chidoricho, Naka-ku, Yokohama 231-0815, Japan

1. INTRODUCTION

More than three decades of research on organic solar cells based on π-conjugated materials has led to steadily increasing efficiency of solar cells. A timeline of organic solar cell conversion efficiency is shown in Fig. 1 [1–3]. The solar cell structure proposed first was a Schottky junction,

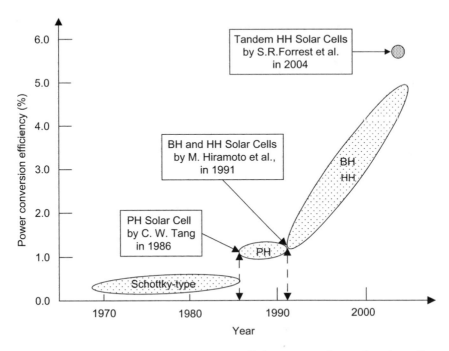

Fig. 1. Improvement of the power conversion efficiency, η_p, of organic solar cells.

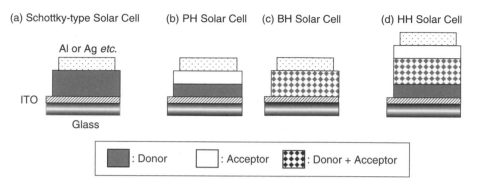

Fig. 2. Structures of typical organic solar cells. (a) Schottky-type solar cell; (b) Planar heterojunction (PH) solar cell; (c) Bulk heterojunction (BH) solar cell, and (d) Hybrid-type heterojunction (HH) solar cell.

shown in Fig. 2(a). The conversion efficiency of solar cells stayed below 0.1% for many years. In 1986, Tang [4, 5] introduced a planar donor (D), acceptor (A), heterojunction (PH) as shown in Fig. 2(b), and successfully increased the conversion efficiency up to 1% in a CuPc (copper phthalocyanine)/perylene derivative heterojunction solar cell. In 1991, Hiramoto et al. [6] expanded on Tang's idea and developed a concept of bulk heterojunction (BH) as depicted in Fig. 2(c).

Recently, Forrest et al. [7] reported conversion efficiency over 5% in organic solar cells based on hybrid bulk heterojunction (HH). This relatively high conversion efficiency and a potentiality of low production cost make organic solar cells more attractive and a promising candidate not only for an alternative to conventional silicon-based solar cells but also for many applications in emerging niche markets.

This article, focusing on small molecular organic materials and their solar cells, comprises the following three sections.

(1) Photophysical properties of organic π-conjugated materials
(2) Characteristics of simple bulk heterojunction solar cells
(3) Characteristics of hybrid-type heterojunction solar cells

2. PHOTOPHYSICAL PROPERTIES OF ORGANIC π-CONJUGATED MATERIALS

The photoconversion mechanisms in organic solar cells and in conventional inorganic solar cells are considered to be different in their respective fundamental photoexcitation processes. Light absorption in organic solar cells leads to the formation of excitons and their dissociation to generate free carriers. In contrast, free electron–hole pairs are directly photogenerated in inorganic solar cells. Thus, understanding of the nature of excitons in organic π-conjugated molecules is important to establish a guiding principle to improve the conversion efficiency of organic solar cells.

An exciton is the photoexcited state of an electron–hole pair bound together by the Coulomb attraction. Excitons are broadly classified in three classes depending on the strength of the Coulomb attraction [8]. They are: the Mott–Wannier exciton, the Frenkel exciton, and the charge-transfer exciton. The Mott–Wannier exciton is weakly bound and is found in inorganic semiconductors with a small band gap and a high dielectric constant, because the Coulomb attraction is weak in the material. On the other hand, the Frenkel (localized excitation on a single molecule) and charge-transfer excitons are produced by light absorption, since the noncovalent electronic interaction, such as van der Waals interaction, between organic molecules is weak compared to the strong covalent electronic interaction between each atom of an inorganic semiconductor. The Frenkel exciton plays a dominant role in organic π-conjugated molecules, whereas the charge-transfer exciton is important in the case of molecular solids such as anthracene.

The exciton binding energy is one of the key parameters that govern the photoconversion mechanism in organic solar cells. This section focuses on the exciton binding energy, particularly on the exciton binding energy of Frenkel excitons in organic π-conjugated molecules. The argument on the exciton binding energy is based on the paper published by Knupfer [9]. The exciton binding energy is expressed as $E_{\mathrm{B}} = E_{\mathrm{B}}^{\mathrm{intra}} + U - W$. E_{B} is the exciton binding energy in the solids that is the energy difference between a bound electron–hole pair on one molecular unit and a free electron and hole on different units. $E_{\mathrm{B}}^{\mathrm{intra}}$ is the intra-molecular exciton binding energy, and schematically explained in Fig. 3(a). W is the kinetic energy of a free electron and hole, corresponding to the inter-molecular bandwidth as shown in Fig. 3(b). U is the Coulomb repulsion energy, i.e., the charging energy, corresponding to the step by which one of the charges of the exciton is taken from a molecule (N) and put it on another molecular unit (either $N+1$ or $N-1$) far apart. Here, the Coulomb repulsion U is defined as $U = E(N+1) + E(N-1) - 2E(N)$, and obtained by $I - A - \Delta_{\mathrm{HL}}$, where I is the ionization potential, A the electron affinity, and Δ_{HL} the HOMO-LUMO gap in the ground state.

A quantum-chemical calculation of energy levels in various π-conjugated molecules has revealed that $E_{\mathrm{B}}^{\mathrm{intra}}$ is relatively small and estimated to be less

(a) Molecule

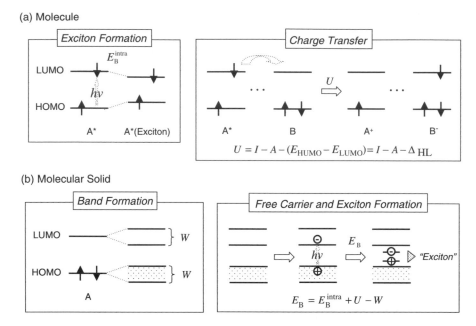

Fig. 3. Energy level diagram of different photophysical steps (a) in a molecule and (b) in a molecular solid.

than 0.3 eV, since the correlation energy in the final exciton state is small. This corresponds to the fact that the HOMO-LUMO gap in many π-conjugated molecules is attributed to the lowest-lying optical excitation. The kinetic energy W is also reported to be small, ranging from 0.2 to 0.3 eV. This value is reasonable since the interaction of any molecules in solids is rather weak and the kinetic energy of each carrier is small. Consequently, the exciton binding energy E_B in organic molecular solids is given by the Coulomb repulsion energy plus a small constant: $E_B \approx U + 0.1$ eV. The Coulomb repulsion energy U can be calculated on the basis of a simple classical electrostatic calculation. If a molecule is assumed to be a charge reservoir and has a capacitance C, the charging energy is given by $e^2/2C$. C depends on molecular shape. Its explicit forms for sphere, disk, and ellipsoid are as follows:

sphere: $4\pi\varepsilon_0\varepsilon_r d$
disk: $4\pi\varepsilon_0\varepsilon_r d^2$
ellipsoid: $2\pi\varepsilon_0\varepsilon_r z/\ln(2z/d)$,

where d is the diameter of a sphere or disk, z the length of an ellipsoid, ε_0 the vacuum permittivity, and ε_r the relative dielectric constant. The Coulomb repulsion U is the energy difference of an $N-1$ site and $N+1$ site, and given by $U = e^2/C$. Then its molecular shape dependence is obtained as shown below.

sphere: $e^2/4\pi\varepsilon_0\varepsilon_r d$
disk: $e^2/4\pi\varepsilon_0\varepsilon_r d^2$
ellipsoid: $e^2/2\pi\varepsilon_0\varepsilon_r z/\ln(2z/d)$.

In the case of sphere- and disk-shaped molecules, the smaller the molecule diameter becomes, the larger the U. For instance, the U of C_{60} and C_{70} are 1.4–1.6 and 1.0 eV, respectively. In the case of ellipsoid-shaped molecules, the U becomes larger as the molecule length gets shorter.

To produce a photovoltaic effect, an electrically neutral exciton must dissociate into free carriers, i.e., free electrons and holes. Therefore, a small E_B corresponding to large π-conjugated molecules would be desirable from the standpoint of effective charge generation. In addition, the introduction of highly polarizable side chains to the molecules screening the charging energy U has been proposed. In this case, however, the side chains should not interfere with free carrier transport by putting molecular solids into disorder. One of the most effective means to dissociate an exciton is to use a D–A heterointerface. The band offset between HOMOs of D and A molecules and/or that between their LUMOs provides excess energy for the interfacial dissociation of an exciton. The optimization of the organic solar cell structure to utilize a D–A heterointerface effectively is crucial to improving its conversion efficiency.

3. CHARACTERISTICS OF SIMPLE BULK HETEROJUNCTION SOLAR CELLS

This section explains the working mechanism of organic solar cells and key parameters relating to their conversion efficiency, and then discusses C_{60}/CuPc bulk heterojunction solar cells.

The working mechanism of organic thin-film solar cells is broken down as follows:

Step 1: Generation of an exciton in D and/or A molecules by light absorption.
Step 2: Diffusion of the exciton to a D–A interface.
Step 3: Exciton dissociation by the charge transfer at the D–A interface into free carriers, i.e., free electrons and holes.

Step 4: Migration of each free carrier to the corresponding electrode and
subsequent collection by an external circuit.

Eq. (1) gives the external quantum efficiency, η_{EQE}, of a solar cell.

$$\eta_{EQE} = \eta_A \times \eta_{IQE} = \eta_A \times \eta_{ED} \times \eta_{CT} \times \eta_{CC} \tag{1}$$

$$\eta_{IQE} = \eta_{ED} \times \eta_{CT} \times \eta_{CC} \tag{2}$$

Here, η_A is the absorption efficiency, η_{ED} the exciton diffusion efficiency, η_{CT} the charge transfer efficiency, η_{CC} the carrier collection efficiency, and η_{IQE} the internal quantum efficiency. Each value is calculated as follows.

η_A = [the number of photogenerated excitons]/[the number of incident photons].

η_{ED} = [the number of diffusing excitons to a D–A interface]/[the number of photogenerated excitons].

η_{CT} = [the number of dissociated excitons at a D–A interface]/[the number of diffusing excitons to a D–A interface].

η_{CC} = [the number of free charge carriers at the electrodes]/[the number of dissociated excitons at a D–A interface].

Fig. 4 shows the electro and optical processes taking place for three different types of organic solar cells. In the case of PH solar cells, η_{CT} and η_{CC} are considered to be almost unity, because most of the photogenerated excitons dissociate into free carriers at a D–A interface once the excitons reach the interface, and because the free carriers reach the electrodes without recombination. This indicates that η_{EQE} is determined by $\eta_A \times \eta_{ED}$. Therefore, the conversion efficiency of PH solar cells is limited by the exciton diffusion to the D–A interface and/or photon absorption processes. The exciton diffusion length ($L_D < 100\,\text{Å}$) is about one tenth of the optical absorption length ($L_A > 1000\,\text{Å}$). If the thicknesses of D and/or A films are made thicker than L_D in order to increase η_A, η_{ED} in turn decreases, since some of the photogenerated excitons annihilate before reaching the D–A interface. Thus, there is a trade-off between η_A and η_{ED}. Consequently, the highest value of η_{EQE} is obtained when each thickness of D and A layers is balanced such that $\eta_A \times \eta_{ED}$ gives a maximum value.

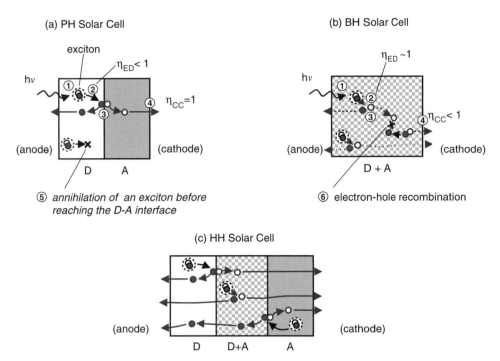

Fig. 4. Working mechanism for three different types of organic solar cells. Process ①: Photogeneration of excitons; process ②: Diffusion of excitons; process ③: Dissociation of excitons; process ④: Collection of charged carriers.

On the other hand, the photoactive layer of a BH solar cell has an interpenetrating network of D and A molecules on the nanometer scale. Therefore, the distance between every exciton generation site and the D–A interface lies within a distance L_D, and η_{ED} becomes almost unity. Thus, unlike PH solar cells, η_A is increased by making the bulk heterojunction layer thicker since η_{ED} is not affected by its thickness, d_m. There is, however, a possibility that η_{CC} may decrease because the recombination of free electrons and holes is presumed to occur in the BH layer, as shown in Fig. 4. The bulk heterojunction layer thickness d_m and η_{CC} should be well balanced to achieve high efficiency in BH solar cells.

Carrier collection efficiency, η_{CC}, is calculated as a function of d_m based on a parallel model used for *pin* amorphous-silicon solar cells by Crandall [10].

This model assumes that an electric field E is applied only to an insulator region, which corresponds to the bulk heterojunction layer of the solar cells. On this assumption, the carrier drift length (ℓ) is given by $\ell = \mu \tau E$. Here E is expressed as

$$E = \frac{V_{bi} - V}{d_m} \tag{3}$$

where V_{bi} is the built-in potential and V the applied potential. The carrier drift length ℓ decreases with increasing d_m and/or approaching V to V_{bi}. Based on these considerations, η_{CC} is obtained as follows:

$$\eta_{CC}(V) = \frac{L_c(V)}{d_m}\left[1 - \exp\left\{-\frac{d_m}{L_c(V)}\right\}\right] \tag{4}$$

$$L_C(V) = (\tau_p \mu_p + \tau_n \mu_n)(V_{bi} - V)/d_m = L_{C_0}(V_{bi} - V)/V_{bi} \tag{5}$$

$$L_{C_0} = (\tau_p \mu_p + \tau_n \mu_n)V_{bi}/d_m = L_C(V = 0) \tag{6}$$

where L_C is the collection length corresponding to the sum of the hole and electron drift lengths, τ_p and τ_n are the hole and electron lifetimes, and μ_p and μ_n are the respective mobilities. This equation implies that the bulk heterojunction layer needs to be thinner than the corresponding collection length to increase η_{CC}.

The current density vs. voltage (JV) characteristics of a BH solar cell is obtained as follows:

$$J(V) = J_{dark}(V) - J_{photo}(V) = J_s\left\{\exp\left\{\frac{q(V - JR_s)}{nkT}\right\} - 1\right\} - J^0_{photo}\eta_{CC}(V) \tag{7}$$

where $J_{dark}(V)$ is the JV characteristics in the dark of a conventional *pn* diode with a series resistance, J_s the reverse-bias saturation current, n the ideality factor, R_s the series resistance, q the electron charge, k the Boltzmann's constant, T the temperature, and J^0_{photo} the photocurrent obtained when η_{CC} is unity.

The performance of the CuPc/C$_{60}$ BH solar cell given in Fig. 5 is examined. Fig. 6 shows the JV characteristics of a CuPc/C$_{60}$ BH solar cell.

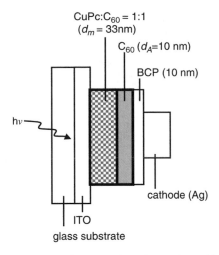

Fig. 5. Structure of the CuPc/C$_{60}$ bulk heterojunction solar cell.

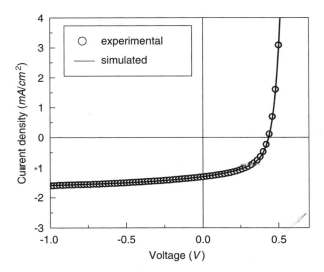

Fig. 6. The *JV* characteristics of CuPc/C$_{60}$ bulk heterojunction solar cell. The experimental *JV* characteristics of the cell are best fitted with Eq. (7) when $L_{C_0} = 48$ nm.

Numerical values for η_{CC} and L_{C_0} are evaluated by curve fitting the *JV* characteristics with Eq. (7). Eq. (7) can reproduce the experimental *JV* characteristics, when $L_{C_0} = 48$ nm. Eq. (4) together with the known values of $d_m (=33 \, \text{nm})$ and L_{C_0} yields η_{CC} to be 0.73.

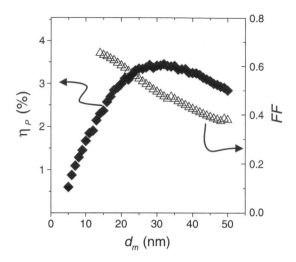

Fig. 7. d_m dependence of η_p and *FF*. Maximum η_p of 3.5% can be obtained when d_m is ca. 30 nm, while *FF* monotonically decreases with d_m.

The conversion efficiency (η_p) dependence on d_m can also be estimated using this model, provided that J_{photo}^0 is proportional to d_m. Fig. 7 shows how η_p changes with d_m. The maximum η_p of 3.5% can be obtained when d_m is ca. 30 nm. In the region of d_m thicker than 30 nm, both η_p and *FF* decrease with increasing d_m. This is because the free carrier recombination becomes dominant over the free carrier generation. The maximum of η_p agrees well with the conversion efficiency obtained experimentally. It turns out that increasing d_m alone gives no further improvement of η_p of the solar cell.

4. CHARACTERISTICS OF HYBRID-TYPE HETEROJUNCTION SOLAR CELLS

One way to improve the conversion efficiency of BH solar cells is to use a hybrid planar-mixed molecular heterojunction (HH) structure. As shown in Fig. 2(d), a HH solar cell is composed of a bulk heterojunction layer sandwiched between D and A thin films. Exciton dissociation occurs in a mixed layer as well as at the interfaces between the mixed layer and the homogeneous D and A layers. In the mixed layer, η_{ED} becomes close to unity, because the distance between every exciton generation site and D–A interface lies well within a distance L_D as mentioned before. Thus, a HH solar cell is expected to give an even higher η_p than PH or BH solar cells,

Table 1
Layer thicknesses of solar cells and their performance (short-circuit current density J_{SC}, open-circuit voltage V_{OC}, fill factor FF, and power conversion efficiency η_p) under AM1.5G solar illumination. The charge recombination zone in the tandem-HH cell consists of a 5-nm PTCBI layer, a 0.5-nm Ag nanocluster layer, and a 5-nm m-MTDATA doped with 5 mol% F_4-TCNQ

Cell	d_D (nm)	d_m (nm)	d_A (nm)	P_0 (suns)	J_{SC}/P_0 (A/W)	FF	V_{OC} (V)	η_p (%)
PH	20	–	40	1.3	11.8	0.61	0.51	3.7
BH	–	33	10	0.9	15.4	0.46	0.50	3.5
HH(A)	10	20	30	1.2	17.1	0.50	0.51	4.4
HH(B)	15	10	35	1.2	15.0	0.61	0.54	5.0
Tandem-HH	7.5 6	12.5 13	8 16	1.2	8.6	0.61	1.07	5.7

assuming selection of a proper mixed-layer thickness that balances η_{ED} and η_{CC}.

Hybrid-type heterojunction solar cells fabricated with C_{60} and CuPc can be optimized on the basis of these considerations. Table 1 summarizes the performance of four types of solar cells depicted in Fig. 8. All the cells except for the BH solar cell have the same photoactive layer thickness of 60 nm. The HH(B) solar cell achieved $\eta_p = 5.0\%$ under 1–4 suns, which is approximately 40% higher than for the PH or BH solar cells. The thicknesses of the D and A layers (d_D and d_A) of the HH(B) were 15 and 35 nm, respectively. These values are quite close to the exciton diffusion length of CuPc and C_{60}: L_{DCuPc} and $L_{DC_{60}}$ are 10 and 40 nm, respectively. Moreover, η_{CC} of a mixed layer 10 nm thick is estimated at 0.90 by using Eq. (4). This value is fairly high compared to η_{ED}'s of the D and A layers.

The conversion efficiency of a single hybrid-type heterojunction solar cell can be increased even further by stacking two cells in series with an ultra-thin metal layer between them (referred to as tandem-HH hereafter). The metal layer works as a charge recombination zone. The photovoltage of a tandem-HH solar cell is the sum of the photovoltage produced by the two subcells. The photocurrent is equal to the smallest of the currents produced by the subcells. To achieve maximum efficiency in a tandem-HH solar cell, the photocurrents generated by the subcells must be equal. Otherwise, an accumulation of photogenerated charge in the tandem-HH solar cell deteriorates the JV characteristics, thus leading to decreasing η_p. In trying to optimize the solar cell structure, optical field intensity distribution across the

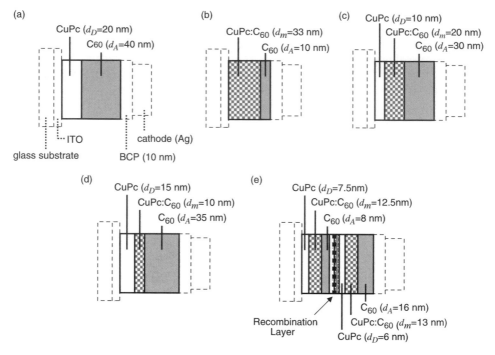

Fig. 8. Four different types of solar cells: (a) PH solar cell; (b) BH solar cell; (c) and (d) HH solar cells; (e) Tandem-HH solar cell.

thickness of the solar cell is taken into account. This is because the optical field intensity is proportional to the exciton concentration, thus corresponding to the numbers of photogenerated charges. The distribution is caused by optical interference between the incident light and that reflected from the metal cathode. Optical field intensity resulting from the interference has a maximum intensity at a distance of about $\lambda/4n$ from the surface of the cathode illuminated by the incident light, where λ is the incident light wavelength and n the average refractive index of organic layers at that wavelength.

The tandem-HH solar cell given in Fig. 8(e) is composed of CuPc, C_{60}, and the photogenerated charge recombination layer of Ag nanoclusters. This cell has an "asymmetric" tandem structure. The front cell has absorption tuned to the longer-wavelength spectral region, and the back cell efficiently absorbs the photon energy of shorter wavelengths, since the maximum optical intensity at about $\lambda/4n$ is shifted from Ag cathode to indium-tin-oxide (ITO) anode as the incident light wavelength becomes longer. The front cell has a thicker homogeneous CuPc layer and thinner C_{60} layer than the back cell, because the absorption regions of CuPc and C_{60} are 500–750 nm and

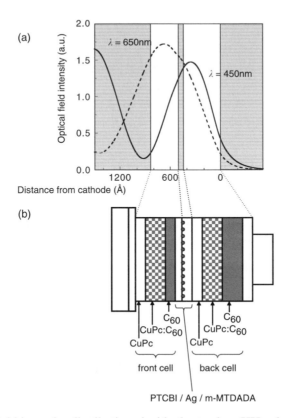

Fig. 9. Optical field intensity distributions inside the tandem-HH solar cell.

$\lambda = 350$–550 nm, respectively. Fig. 9 shows optical intensities at $\lambda = 450$ and 650 nm calculated as functions of the distance from the cathode in the tandem cell. This result implies that a tandem cell having a front cell rich in longer-wavelength absorbing material, CuPc, and a back cell rich in shorter-wavelength absorbing material, C_{60}, gives rise to high conversion efficiency. Fig. 10 gives η_{EQE} for the front and back cells. Actually, the cell was confirmed to achieve a maximum conversion efficiency of $\eta_p = (5.7 \pm 0.3)\%$ under 1 sun (AM1.5G) solar illumination [11].

5. SUMMARY

The HH solar cell was shown to have a high conversion efficiency compared to the PH and BH solar cells. The HH solar cell fabricated with CuPc/C_{60} was confirmed to achieve a conversion efficiency η_p of 5%. The tandem

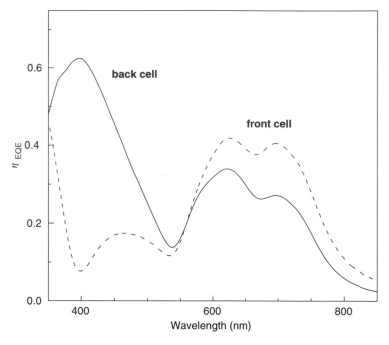

Fig. 10. External quantum efficiency, η_{EQE}, for the front and back cells.

solar cell with a HH solar cell was verified to give a η_p of 5.7%. Moreover, there is a possibility that applying anti-reflecting coatings to the tandem cell could increase efficiency to over 6% [11].

The design of the BH solar cell structure has been discussed so far by focusing on the exciton diffusion and charge carrier collection efficiencies only.

The product of the four key factors, $\eta_A \times \eta_{ED} \times \eta_{CT} \times \eta_{CC}$, determines the conversion efficiency of an organic solar cell. Therefore, all four quantum efficiencies must be optimized to achieve conversion efficiency in excess of 10–15%. In doing so, the molecular engineering is very important. New donor and acceptor materials must be tailored to have optimum energy levels of HOMO and LUMO so that the materials can absorb a broader solar-spectral region and have lower exciton-binding energy.

This continuous and fundamental research will mark a new epoch in organic solar cells, which are expected to offer low-cost solar energy conversion and be environmental friendly.

REFERENCES

[1]   H. Spanggaard and F.C. Krebs, Sol. Energy Mater. Sol. Cells, 83 (2004) 125.

[2]   P. Peumans, A. Yakimov and S.R. Forrest, J. Appl. Phys., 93 (2003) 3693.

[3]   C.J. Brabec, N.S. Sariciftci and J.C. Hummelen, Adv. Funct. Mater., 11 (2001) 15.

[4]   C.W. Tang, Appl. Phys. Lett., 48 (1986) 183.

[5]   C.W. Tang, US patent, 4,281,053.

[6]   M. Hiramoto, H. Fujiwara and M. Yokoyama, Appl. Phys. Lett., 58 (1991) 1062.

[7]   J. Xue, B.P. Rand, S. Uchida and S.R. Forrest, Adv. Mater., 17 (2005) 66.

[8]   M. Pope and C.E. Swenberg, Electronic Processes in Organic Crystals and Polymers, 2nd ed., Oxford University Press, Oxford, 1999.

[9]   M. Knupfer, Appl. Phys., A 77 (2003) 623.

[10] R.S. Crandall, J. Appl. Phys., 53 (1982) 3350.

[11] J. Xue, S. Uchida, B.P. Rand and S.R. Forrest, Appl. Phys. Lett., 85 (2004) 5757.

Nanostructured Materials for Solar Energy Conversion
T. Soga (editor)
© 2006 Elsevier B.V. All rights reserved.

Chapter 12

The Application of Photosynthetic Materials and Architectures to Solar Cells

J.K. Mapel and M.A. Baldo

Department of Electrical Engineering and Computer Science, Massachusetts Institute of Technology, Cambridge, MA 02139, USA

1. WHY PHOTOSYNTHESIS?

Widespread adoption of solar cells remains limited by their high cost per watt of generated power [1]. This is due in part to the expensive equipment and energy hungry processes required in the manufacture of conventional semiconductor-based photovoltaic (PV) cells. On the other hand, PV cells made from organic semiconductors such as films of molecules or polymers hold the promise of low-cost production. For example, one class of suitable molecular PV materials, the phthalocyanine pigments [2], are currently produced in quantities exceeding 80000 tons annually [3]. In addition, this inexpensive feedstock is compatible with high throughput web processing. The printing, paint, and packaging industries routinely spray coat, stamp, and evaporate molecular and polymeric materials onto flexible plastics and foils [1]. If similar web-based processing is realized for organic PV cells, organic devices need only reach performance levels commensurate to inorganic PV technologies to decrease the cost per watt of PV power.

Organic PV power efficiencies have steadily improved, reaching approximately 5% in recent results [4, 5], still substantially below that of more mature conventional technologies [6]. But conventional semiconductor solar cells are not necessarily the most appropriate model for the development of organic PV. The physics of organic PV cells is much closer to that other, much older and more sophisticated, example of organic electronics: photosynthesis.

Photosynthesis efficiently converts solar to electrical energy, which then drives a series of chemical reactions. This ubiquitous, time-tested energy transduction method is the source of all current biomass and, over geologic timescales, all the fossil fuels relied upon today [7]. Photosynthetic plants and bacteria utilize organic molecules similar to those used in organic PV to fix more than 100 Gtons of carbon annually, equivalent to 100 TW, a feat accomplished without high temperature processing or huge initial energetic expenditures. From a manufacturing standpoint, the utilization of photosynthetic organisms represents the ultimate in low-cost processing. A field of soybeans, for example, can be grown at very low cost but produces the raw material equivalent of several times its area in PV cells *annually* [8].

In Section 2, we will compare organic PV to photosynthesis. The principal challenge in organic PV is to absorb sufficient light in the vicinity of charge generation interfaces. We discuss the different architectures employed in photosynthesis and organic PV to address this problem. The direct integration of photosynthetic protein structures into photoelectric devices constitutes one route toward achieving efficient and low-cost organic PV. In Section 3, we summarize work in hybrid solid-state photosynthetic devices. In Section 4, we discuss the implementation of photosynthetic architectures with separate light absorption and charge generation structures in synthetic organic PV cells. Finally, we discuss the prospects for photosynthetic materials and architectures in organic PV.

2. ORGANIC PHOTOVOLTAIC AND PHOTOSYNTHESIS COMPARED

2.1. Organic Photovoltaics

We begin by briefly reviewing the processes and structures commonly used in organic semiconductor heterostructure PV. For an in-depth review of these devices, see Peumans et al. [9]. Similar to their inorganic counterparts, organic PV devices are comprised of donor and acceptor semiconducting regions sandwiched between conducting electrodes. Usually, these materials are different semiconductors, as reliable doping to control majority carrier type is difficult to achieve.

The sequence of processes yielding light to electrical energy transduction in organic PV can be divided into four phases, as summarized in Fig. 1. In the first, upon optical excitation in one or both organic materials, localized Frenkel or charge transfer excitons are generated [10, 11]. These tightly bound, charge-neutral species diffuse until they recombine or dissociate. Excitons that reach an interface between the donor and acceptor layers will dissociate if the energetic offsets favor the process. For large offsets, dissociation occurs

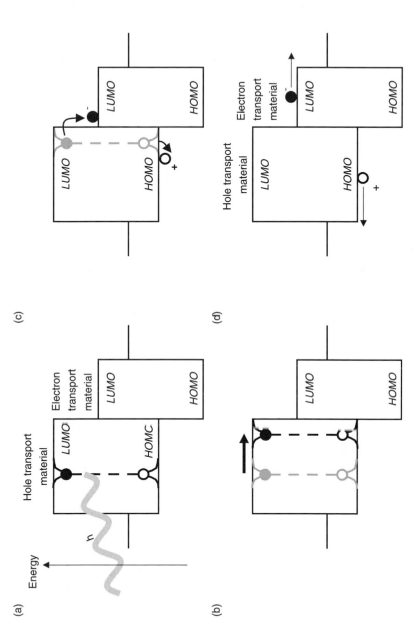

Fig. 1. Summary of processes in organic PV resulting in photocurrent generation. (a) Optical absorption in one or more active semiconducting layers creates an exciton, an electron-hole pair localized on a single molecule. (b) Excitons diffuse in the thickness of the film. (c) Those that reach the interface between the donor and acceptor layers can dissociate. In this example, an excited molecule in the donor hole transport material reduces a nearby acceptor molecule in the adjacent electron transport material. (d) The separated free electrons and holes diffuse out towards the metal electrodes, completing the energy transduction process. After Peumans et al [9].

over timescales of a few hundred femtoseconds [12] and results in free electrons in the lowest unoccupied molecular orbital (LUMO) of the electron transport material and free holes in the highest occupied molecular orbital (HOMO) of the hole transport material. These free carriers diffuse out toward the contact and are available to perform electrical work.

The useful thickness of an organic PV cell is restricted to the distance that excitons can travel before recombining, typically on the order of 10 nm [9]. Within this region the internal quantum efficiency (the ratio of charge extracted to absorbed photons) can be 100%. But the quantum efficiency drops dramatically in thicker devices due to exciton recombination losses [13]. Thus, despite optical absorption coefficients exceeding 10^5 cm^{-1} averaged over the visible spectrum, organic PV is limited by an inability to absorb enough light. Several classes of solar cells have emerged whose device architectures address this concern, including dye-sensitized nanostructured oxide cells [14], bulk organic heterojunction cells [5, 15], and organic–inorganic hybrid composites [16–18]. These approaches share the characteristic of increased surface area of the exciton dissociation interface, increasing the useful thickness of the cell.

Photosynthesis also maximizes its active surface area by embedding charge generation components into a flexible membrane. But in contrast to organic PV, the architecture of photosynthesis employs separate components for light absorption and charge generation, allowing these two functions to be optimized independently. Overall, photosynthesis can be divided into at least three distinct phases: (1) light absorption and energy transport by antenna systems, (2) energy collection and charge separation in reaction centers (RCs), and (3) stabilization by secondary reactions for use in the synthesis of sugars. The first two components are the biological equivalent of a PV cell, albeit with a very different architecture; see Fig. 2 [19].

2.2. Photosynthetic Antenna Complexes

All photosynthetic organisms contain light-gathering antenna systems; as such, they are remarkably diverse. Antenna types can be divided into several categories: (1) light-harvesting complexes of purple bacteria, (2) light-harvesting complexes of plants and algae, (3) phycobilisomes of cyanobacteria and red algae, (4) peridinin-chlorophyll proteins of dinoflagellate algae, and (5) chlorosomes of green bacteria. We refrain from an extensive discussion of all antenna types, as excellent reviews can be found elsewhere [20].

Antennas contain high concentrations of pigment molecules, including chlorophylls, bilins, carotenoids, and their derivatives. Photons captured by these pigments generate excitons, as in organic PV. But unlike the

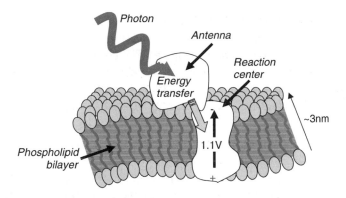

Fig. 2. Photosynthetic membrane, showing the spatial distribution of the light-harvesting antenna and reaction center, the sites of photon absorption and exciton dissociation, respectively. After Purves et al. [18].

semiconducting films in organic PV which rely on diffusion, antenna complexes are designed to guide excitons to RCs. For example, phycobilisomes possess pigments at the periphery of the complex that absorb at higher energies than those at the core. Excitons at the periphery travel via Förster energy transfer toward the core where they are coupled to the RC.

Most antenna systems are comprised of pigment–protein complexes where the photoactive pigment cofactors are positioned by a protein matrix, altering their optical properties and controlling energy transfer. Chlorosomes are an exception: as perhaps the only example of solid-state semiconductors in nature, they are of particular interest to organic PV. They are unique in that they are largely composed of pigments (> 50% by dry weight [20]) and constitute the most efficient light-harvesting complexes found in nature [21]. The green photosynthetic bacteria that possess chlorosomes are frequently found in volcanic hot springs where the ambient temperature reaches 47°C [22]. Such extreme conditions may have contributed to the unique structure of chlorosomes, but they also appear especially well adapted to conditions of extremely low-light flux. Compared to other photosynthetic antennas, they have very high-absorption cross sections.

While several models have been proposed for the pigment organization in chlorosomes, a common characteristic is the existence of aggregates of bacteriochlorophyll c, either rods of 5–10 nm in diameter and 100–200 nm in length [23–27] or semicrystalline lamellar sheets [21]. The regular structure of Bchl c van der Waals bonded aggregates leads to strong exciton coupling and a redshift in absorption. Crystallinity over 100 nm length scales and exciton delocalization make the Bchl *c* aggregates highly desirable for use as organic semiconductors in organic PV; materials with these characteristics are currently under development for use in organic electronic devices [28–31].

2.3. Photosynthetic Reaction Centers

In photosynthesis, the role of the donor–acceptor interface is performed by the photosynthetic reaction centre (RC). The dissociation of excitonic energy states and formation of separated charges occurs at the RC via a series of electron transfer reactions. The RC is a membrane-bound, multi-subunit, pigment–protein complex which incorporates chlorophyll derivatives and other electron transfer cofactors such as quinones. The pigments and cofactors are held together by van der Waals interactions with the protein matrix; their positioning and orientations are important in facilitating electron transfer.

The ultimate collection point for excitons from neighboring antenna complexes is a chlorophyll dimer in the RC known as the special pair. This is the lowest energy site in the photosynthetic optical circuit. It is also the primary electron donor for the subsequent electron transfer cascade that carries the electron across the membrane while the hole remains at the special pair, thereby separating the exciton into isolated charges; see Fig. 3 [32]. Recombination,

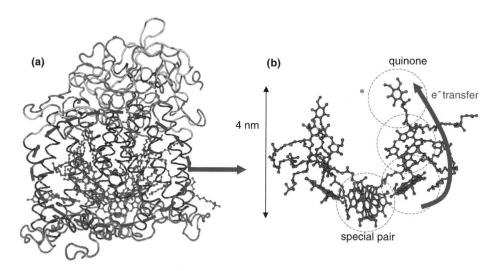

Fig. 3. Structure of the reaction center complex of *Rhodobacter sphaeroides*. (a) Entire complex, including the L, M, and H cofactors. (b) Cofactors only. The special pair is the primary electron donor of the electron transfer cascade, illustrated by the arrow. Figure produced from the Protein Data Bank file *1AIJ* using VMD [31].

or the back transfer of the electron to the special pair, is prevented by the electron transfer cascade which occurs in a series of very fast (1–100 ps) electron transfer reactions, rapidly separating the charges to ~3 nm and strongly reducing the rate of recombination. Exciton dissociation in RCs thus proceeds with high efficiency; the quantum yield of photons to charge pairs is nearly unity [33]. The potential of the separated charges varies from approximately 0.5 V in primitive purple bacteria, to approximately 1.1 V in more advanced systems [34]. The secondary reactions that follow stabilize the oxidized and reduced species, yielding a chemical potential across the photosynthetic membrane that can then be used to drive cellular metabolism. The rapid, multi-step spatial separation achieved in RCs may reduce their recombination losses relative to less sophisticated donor–acceptor interfaces in organic PV.

Unlike antenna complexes, RC complexes are remarkably well preserved across plants and photosynthetic bacteria [33]. All RCs follow the above described general structure of electron transfer cofactors embedded in a protein matrix. In plants and cyanobacteria, two special RCs called photosystems I (PSI) and II (PSII) operate in tandem to split water and create molecular oxygen, a highly energetic reaction since water is an extremely poor electron donor. Oxygen produced by photosynthesis is the source of oxygen in the atmosphere and fundamentally affected the development of life on Earth.

3. INTEGRATION OF PHOTOSYNTHETIC COMPLEXES IN ORGANIC PHOTOVOLTAIC

Much of the initial work on organic PV utilized thin films of molecules structurally similar to the chlorophyll pigments of photosynthesis [35]. Such devices, however, have not approached the efficiency of photosynthesis. Equating the potential developed across a photosynthetic RC to the open-circuit voltage of a solid-state solar cell yields a photosynthetic power conversion efficiency of approximately 20% and potentially competitive with the best silicon devices. This high performance is a consequence of the unique molecular-scale engineering of photosynthetic complexes. Thus, the prospect of using photosynthetic complexes directly is tempting, as agriculture constitutes a far less expensive manufacturing route compared to semiconductor foundries.

3.1. Self-assembly of Photosynthetic Complexes

The first step in the construction of a device containing photosynthetic complexes is the assembly of the biological structures on a substrate. The self-assembled structure may then be employed as an electrode in an electrochemical cell or integrated in a solid-state PV cell. The goals for this self-assembly are: (1) to uniformly orient complexes to minimize recombination losses, and (2) form an optically dense film to increase absorption.

Self-assembly technology for photosynthetic complexes was pioneered in wet electrochemical cells. This section is not intended as a comprehensive description of the history of photosynthetic materials in electrochemical cells, but the work of Katz on the self-assembly of bacterial reaction centers is especially notable [36]. Using cysteine binding to RC complexes, Katz demonstrated wet electrochemical cells with internal quantum efficiencies as high as 60%. Following the work of Katz, Lebedev et al. [37] investigated the self-assembly of oriented films of photosynthetic complexes on transparent and conductive indium tin oxide (ITO) surfaces using Ni^{2+}–NTA binding to His_6 tags on genetically engineered RCs from the *R. sphaeroides* strain SMpHis, [38] shown schematically in Fig. 4. Lebedev et al. [39] found that binding to His_6 tags increased the photocurrent despite a theoretical increase in the length of the linker molecule connecting the RC to the substrate. A typical tapping mode atomic force microscopy (TM–AFM) image of a His_6–RC self-assembled monolayer on atomically flat Au-on-mica substrates is shown in Fig. 4(c). Although there is significant disorder in the film, it is relatively closely packed.

The self-assembly technology of PSI is less developed. Greenbaum et al. [40] have demonstrated preferential orientation of PSI by engineering the surface chemistry of gold. By controlling the surface charge and hydrophobicity,

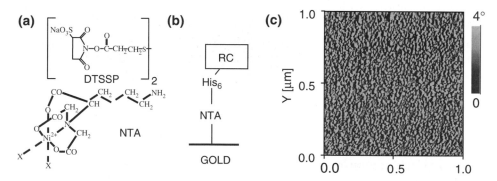

Fig. 4. Method for oriented assembly of reaction centers onto Au. (a) The substrate is treated with 3,3′-Dithiobis[sulfosuccinimidylpropionate] (DTSSP) and nickel 2^+ nitrilotriacetic acid (Ni-NTA) (b) A polyhistidine (6) tag on the reaction center expressed from *R. sphaeroides* chelates the charged Ni ion of the Ni-NTA, immobilizing and orienting the complex on the substrate. (c) Atomic force microscopy phase image of assembled reaction centers on gold.

they demonstrated several possible orientations of PSI on modified gold. An alternative technique allows a single His$_6$ tag to be introduced to native PSI complexes in a three-step process; see Fig. 5(a). Minai et al. [41] have demonstrated that the native psaD subunit of PSI may be exchanged and replaced by a genetically engineered psaD with His$_6$ tagged onto the C-terminus.

To investigate the orientation of PSI bound by psaD exchange, we performed TM–AFM phase imaging [42] in the intermittent contact mode and varied the potential between the AFM tip and the ITO/Au substrate [43]. The phase angle of the driven vibration of the cantilever in TM–AFM is related

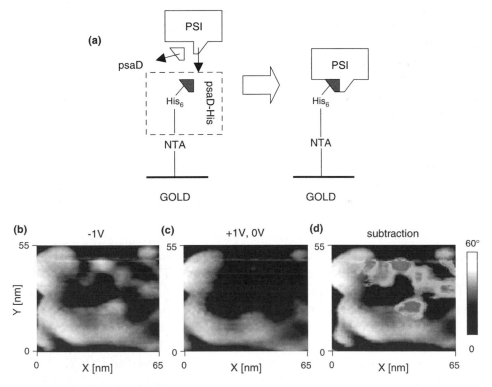

Fig. 5. Method for oriented assembly of PSI onto Au. (a) As with reaction centers, the substrate is treated with DTSSP and Ni-NTA. A polyhistidine tag is introduced to PSI by assembly of an engineered psaD subunit, then exposing the surface to native PSI such that the subunit is substituted [40]. TM–AFM phase images of assembled PSI on Au at (b) −I V, (c) 0, and +I V. The image difference (d) shows localized regions of increased phase, signifying a shange in dissipative energy corresponding to tip interactions with positively charged PSI complexes.

to the energy dissipated in the tip–sample interaction [42]. Thus, phase images of biological materials provide a map of the dissipative part of their mechanical response. When a potential is applied to the AFM tip, we can alter its mechanical interactions with polar or charged samples by, for example, aligning polar molecules in the electric field [44]. Voltage-dependent phase scans of a likely PSI particle is shown in Figs 5(b–d). Phase scans taken at $+1$ V and 0 V show little difference, but phase scans taken at -1 V exhibit the appearance of localized regions of increased phase. The increase in phase in the -1 V scan corresponds to an increase in the attractive forces between the tip and the sample [42] and indicates the presence of a positive charge trapped on the surface of PSI, mostly likely at P700. Thus, the voltage dependence of TM–AFM phase imaging is consistent with the expected rectifying characteristics [40] of PSI in the orientation prescribed by the self-assembly technique in Fig. 5(a). The packing density of PSI is, however, far less than optimal, most likely due to incomplete exchange of psaD.

3.2. The Stability of Photosynthetic Complexes in Solid State

The rinsing and drying steps required during fabrication of solid-state photosynthetic devices are particularly prone to damage the photosynthetic complexes. The integrity of these large complexes can be increased with the use of surfactant stabilizers [45]. To quantify the effect of stabilization, the low temperature fluorescent spectrum is measured. After excitation by a pump laser at $\lambda = 408$ nm with intensity 0.5 mW/cm^2, protein degradation is recognized by wavelength shifts in fluorescence. The chlorophyll molecules associated with the PSI complex provide an intrinsic steady-state fluorescence spectrum at $T = 20$ K between $650 < \lambda < 800$ nm that reflects the organization of the pigment–protein interactions. Thick, vacuum-dried films of PSI prepared directly on glass substrates prior to functionalization exhibit a large blueshift of the fluorescence maxima from $\lambda \sim 735$ to $\lambda \sim 685$ nm, indicating a disruption in light-harvesting subunit organization. Polyelectrolytes such as polyethylene glycol that have been used to preserve dried biological materials [46], were not found to improve the stability of PSI. In contrast, incubating PSI with the peptide surfactants A$_6$K/V$_6$D [47–50] was found to almost entirely preserve its low-temperature fluorescent spectrum [45]; see Fig. 6. The $\lambda = 735$ nm fluorescent peak of peptide-stabilized films stored in an ambient environment exhibited a gradual blueshift over several weeks, indicative of gradual structural changes in the light-harvesting antennae of PSI [51]. The low-temperature fluorescent data demonstrate that PSI can be successfully integrated in a solid-state environment.

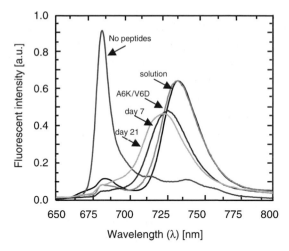

Fig. 6. Fluorescence measurements ($T = 10$ K) of assembled PSI on Au. Comparison between the fluorescence spectrum of PSI solution as extracted from spinach, with washed and dried films of PSI, demonstrates that PSI may be protected against degradation after washing and drying steps by stabilizing the complex with surfactant peptides A_6K and V_6D [44]. The excitation source was a pump laser at $\lambda = 408$ nm with intensity 0.5 mW/cm^2. The 50 nm blueshift of assembled PSI without peptide surfactants shows the disruption of PSI light-harvesting unit organization. The stabilizing action of A_6K/V_6D is preserved for several weeks for dried films left in ambient conditions.

3.3. Solid-State Integration of Bacterial Reaction Centers

To date, wet electrochemical implementations of photosynthetic PV cells have not succeeded in realizing efficient devices. In many PV applications, wet cells require additional packaging and are hampered by stability concerns [52]. Furthermore, diffusion-limited charge transport in the electrolyte increases the series resistance, lowering the fill factor. Consequently, it is desirable to demonstrate technology for integrating biological protein–molecular complexes with solid-state electronics.

The simplest model of a solid-state photosynthetic device consists of uniformly oriented photosynthetic protein–molecular complexes deposited between two metallic contacts. After absorption of a photon and rapid charge separation within a complex, a potential of up to 1.1 V can be developed across the metal contacts [34, 53]. However, this model of a solid-state photosynthetic device must overcome several practical obstacles. First, the optical cross section of a single layer of photosynthetic complex is fairly low; second, deposition of the top metallic contact may cause damage to biological materials; and finally, defects in the layer of photosynthetic complexes may permit electrical shorts between the metallic contacts. The two latter problems are circumvented by depositing a thin (<100 nm) layer of an amorphous organic semiconductor between the photosynthetic complexes and the top metal contact. The semiconductor transports the photogenerated electrons to the cathode of the cell.

The energy-level structure of an RC-based PV cell is shown in Fig. 7(a). RCs are oriented using a His_6 tag with the electron-accepting special pair facing the substrate. Fabrication of solid-state cells begins with self-assembly of the A_6K/V_6D stabilized photosynthetic complexes as in electrochemical cells. But after the complexes are self-assembled on a functionalized electrode, they must be washed with deionized water to remove unbound material and excess salt and detergent from the buffer. Since solid-state devices are much thinner than electrochemical cells, they are less tolerant of debris on the substrate. The RC-based PV cell employs a 60 nm-thick protective layer of the fullerene C_{60}. C_{60} was chosen because of its relatively deep LUMO energy of 4.7 eV [54] that should enhance electron transfer from the electron acceptor in the RC. It is observed that C_{60} transports electrons in its LUMO far more readily than holes in its HOMO. Consequently, C_{60} is employed as an electron transport layer (ETL). After C_{60}, a 12 nm-thick

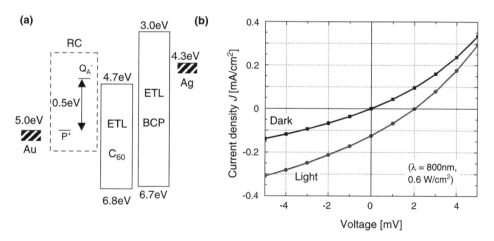

Fig. 7. RC solar cell. (a) Device energy levels for the reaction center solar cell. After photoexcitation, electrons localized on the QA anion reduce nearby C_{60} molecules and electrons conduct out the cathode. The energy levels of C_{60} and BCP are from Refs. 53 and 54. (b) The current–voltage characteristics show that under 0.6 W/cm^2 illumination, the device has an open circuit voltage of 2 mV and short-circuit current density of 1.2 mA/cm^2. The fill factor is approximately 25%.

layer of a second ETL 2,9-dimethyl-4,7-diphenyl-1,10-phenanthroline (batho-cuproine or BCP) [55] is deposited and finally, an 80 nm-thick layer of Ag is deposited through a 1-mm-diameter shadow mask. Thermally evaporated films of C_{60}, BCP, and Ag were deposited at a rate of ~0.3 nm/s in a vacuum of $<10^{-6}$ Torr. The Ag deposition likely damages the thin BCP layer, facilitating electron extraction [56]. But even in a damaged layer, the deep HOMO of BCP effectively prevents the injection of holes into the device, markedly improving the device's reverse bias characteristics [43].

The current–voltage characteristics of the RC-based PV cell are shown in Fig. 7(b). Under illumination at $\lambda = 808$ nm, where C_{60} and BCP are transparent [9] the device exhibits photocurrent in reverse bias, i.e. the ITO is negative relative to the top Ag contact. Most notably, the device exhibits PV behavior, albeit weak, with an open-circuit voltage that varies slightly between devices but is typically ~0.10 V and a short-circuit current density of 0.12 mA/cm^2 under an excitation intensity of 0.6 W/cm^2 at $\lambda = 808$ nm. Assuming a perfectly formed RC monolayer of density 8×10^{-12} mol/cm^2 and given an extinction coefficient of 2.9×10^5 M^{-1}/cm, [57] we calculate the optimum photocurrent as 2 mA/cm^2, where we have ignored possible interference effects due to reflections from the ITO/Au electrode, and assumed 100% reflection of the optical pump by the Ag cathode. Thus, at a bias of -1 V, a conservative estimate of the internal quantum efficiency of the device is 6%. The solid-state quantum efficiency of 0.03% at an excitation intensity of 0.6 W/cm^2 at $\lambda = 808$ nm is similar to a photoelectrochemical cell with an external quantum efficiency of 0.016% under an excitation intensity of 6 mW/cm^2 at $\lambda = 800$ nm [37].

In Fig. 8, verification of the activity of RCs is confirmed by spectrally resolving the short-circuit current using a Ti-Sapphire CW laser tunable between $\lambda = 790$ and 890 nm. The photocurrent spectrum is compared with both the solution absorption spectrum of the RC complexes, and a photocurrent spectrum of identical RC complexes in a photoelectrochemical cell reproduced from Ref. [37]. With the exception of a region near $\lambda = 860$ nm the spectra overlap closely.

4. SYNTHETIC IMPLEMENTATIONS OF PHOTOSYNTHETIC ARCHITECTURES

As noted in Section 2, the organizational architecture of the initial phases of photosynthesis is different from that of organic PV in at least one major respect. In photosynthesis, light absorption and exciton dissociation occur in

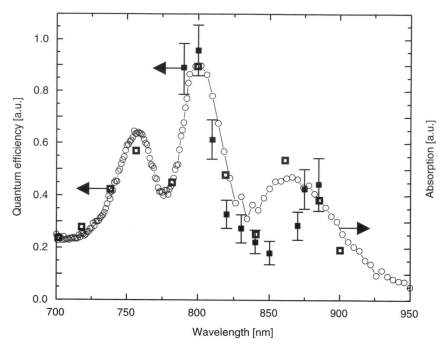

Fig. 8. Optical performance of the RC solar cell. The external quantum efficiency (■) calculated from the short-circuit photocurrent qualitatively matches both the absorption spectrum of RCs in solution (○) and electrochemical RC photoelectrochemical cell (□) from Ref. 36.

the spatially separated components of the antenna and RC complexes. In contrast, absorption, exciton dissociation, and charge extraction all occur in the organic semiconductors that comprise the active donor and acceptor layers in organic PV. This characteristic frustrates materials selection for organic PV, as the organic semiconductors must simultaneously satisfy several constraints: (1) strong broadband optical absorption with an extinction coefficient of at least $10^5 \, cm^{-1}$ across the visible spectrum, (2) efficient long-range exciton transport, (3) optimal energy-level alignment for rapid exciton dissociation efficiency, and (4) high electron and hole mobilities and continuous charge pathways to the two electrodes to minimize recombination losses.

Akin to photosynthesis, organic PV may benefit from separating the functions of light absorption and exciton dissociation into two spatially distinct structures, allowing individual optimization of each. We demonstrate separation of optical and electrical functions by utilizing guided wave-mediated energy transfer across thin metal films. In such a device, energy

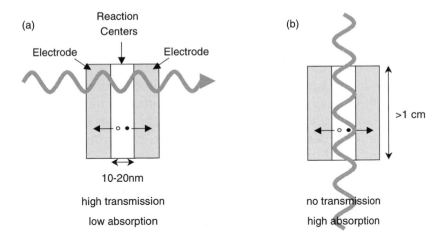

Fig. 9. Device excitation routes. Excitation of solar cells under normal (perpendicular) (a) and parallel surface excitation (b). The interaction distance of the electromagnetic fields and the absorbing artificial RC ratio between the two differ by several orders of magnitude. For very thin PV, high absorption and no transmission is preferred.

transduction proceeds by photon absorption in an 'artificial antenna'. Excited molecular dipoles in the antenna either radiate into waveguide modes or non-radiatively couple to surface plasmon polariton (SPP) modes in the multilayer structure.

A major advantage of coupling into guided modes is that these modes are absorbed even in very thin organic PV cells, optimized for maximum internal quantum efficiency. Guided modes propagate in the plane of the device, parallel to the charge generation interface. The dimensions of the cell in this plane are on the order of 10^{-2} m, rather than $\sim 10^{-7}$ m perpendicular to the interface. The maximum distance of interaction between a RC and a guided mode is thus the distance that these modes travel at visible frequencies. For both SPPs and waveguide modes, this can be several orders of magnitude greater than the thickness of the RC, increasing the likelihood that they will be absorbed; see Fig. 9.

Energy which propagates in these guided modes is absorbed in the 'artificial reaction center' of the PV, after which the processes of exciton diffusion, dissociation, and charge collection occur as in Conventional Organic PV.

4.1. Energy Transfer via Guided Modes

The oscillating electric field of the radiative dipole at an excited molecule in the antenna layer can be dampened by several mechanisms, resulting in energy transfer. These mechanisms are: (1) non-radiative decay into phonons, (2) radiation of photons into free space modes not guided within the PV, (3) radiation into dielectric waveguide modes in the antenna/PV stack, and (4) non-radiative energy transfer into SPP modes at nearby metal interfaces. Photons in waveguide modes interact with the absorbing active layers in the artificial RC identically to normal light illumination.

Non-radiative decay is minimized in efficient antenna dye molecules. Thus, radiation into free space modes is the dominant process for an *isolated* oscillating dipole on an efficient dye molecule. But within a multilayer stack composed of metals and dielectrics, this process can be minimized. The rate of photon emission is described by Fermi's golden rule and depends on the photonic mode density. For example, near a metal film, the photonic mode density drops dramatically as visible light is strongly absorbed by the free charges of the metal.

Within a multilayer stack, energy transfer to guided electromagnetic modes is preferred. The most important guided modes are surface plasmons polaritons and waveguide modes. The stack acts as a waveguide since its refractive index, $n \sim 2$, higher than air or the glass substrate. Plasmons are quasiparticles comprised of the collective oscillation of the conduction electrons in metals. SPPs are a unique class of electromagnetic excitations associated with interfaces between metals and dielectrics. SPPs propagate along the interface with electromagnetic fields, energy, and charges highly localized within the interface area. Their properties depend strongly on characteristics of both the metal (complex dielectric function, corrugations, roughness) and the dielectric (refractive index). The surface character of SPPs leads to a dramatic electric field enhancement at the interface that supports the SPP, making absorption of its energy in nearby dissipative materials strong. In the absence of the adjacent artificial RC, SPPs are internally dampened by joule heating in the metal film.

To summarize, there are several advantages to the biomimetic approach of separating light absorption and exciton dissociation in organic PV:

1. By decoupling the optical and electrical components of the solar cell, the artificial RC can be made thinner than the exciton diffusion length, ensuring that all excitons are generated close to the location of exciton dissociation. The efficiency of this process should approach unity, resulting in internal quantum efficiencies approaching unity as well, as the efficiency of charge transfer and charge collection is known to be highly efficient [15, 58].

2. Molecular excitonic states exhibit highly structured absorption spectra. Thus, to increase the photocurrent in organic PV, one must choose a combination of active materials that absorb evenly across the visible spectrum. In contrast, separating the optical and electrical functions allows the RC to be optimized at a single-peak wavelength corresponding to the emission of the antenna.

3. Since the light-absorbing antenna layer no longer needs to transport charge, new classes of solar cell materials can be used. The ideal antenna layer should be highly absorptive and have a high efficiency for photoluminescence such that reemission is strong. Candidate materials include those which absorb strongly like J-aggregates, nanometallic particles, quantum dots, and photosynthetic complexes that possess high-quantum photoluminescent efficiency such as phycobilisomes from cyanobacteria and red algae. While quantum dots and nanometallic particles have been embedded as active layer of solar cells previously [59, 60], their poor charge transport characteristics have decreased overall device performance.

4. The energetic funneling that biological antennas like cholorosomes employ can be utilized in mixed antenna layers. In mixed layers, light can be absorbed in a host material and energy is funneled to a less absorptive, highly luminescent material for reemission into the bound modes.

4.2. Simulation of Energy Transfer from Antenna Excitons to Surface Modes

Energetic transfer from excited molecules to SPP modes can occur with high efficiency to metallic slabs [61, 62] and thin films [63]. The theoretical basis for dipole coupling to modes in a multilayer stack is well understood [64] and agrees well with experiments [65]. To examine dipole coupling to thin silver films comprising the cathode of an organic PV, we use the method of Chance et al. [64] to simulate classical damping of an oscillating charge distribution near a multilayer stack to investigate energy transfer to our artificial RC. Energy transfer is calculated directly from the Poynting vector [66].

In Fig. 10, we show the dispersion relation for guided SPP modes, propagating parallel to the electrode plane in a typical PV cell with external antenna. Three guided modes are identified in this structure and the mode intensity profile of each is shown in the insets. Each of the guided modes has significant overlap with the charge generation layers sandwiched between

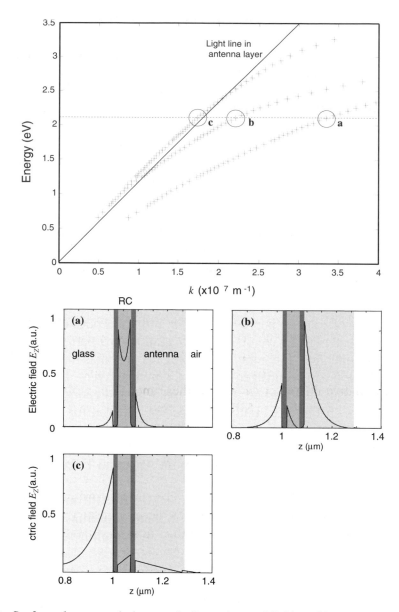

Fig. 10. Surface plasmon polariton mode dispersion and field profile. (a) Calculated dispersion relation of electromagnetic modes in the multilayer stack. The curves correspond to SPP modes shifted to the right of the photon dispersion light lines in the dielectrics that partially support the SPP. The electric filed profile in the thickness direction for $E = 2.1\,eV$ shows the field localization at the interfaces that support the SPP modes. Mode (a) is strongest in the RC semiconductor. Mode (b) is strongest in the antenna, and mode (c) is strongest in the glass substrate. Because it has the highest intensity in the antenna, mode (b) dominates energy transfer from the antenna to the RC. The structure of the simulated device is: glass/Ag(20 nm)/CuPc(45)/PTCBI(25)/BCP(13)/Ag(30)/air.

the metal electrodes. Mode (a) is based in the RC, (b) in the antenna, and (c) is based in the glass substrate.

We calculate the dipole energy dissipation to the multilayer stack in a technologically relevant device geometry as a function of normalized wavevector and distance to the antenna–silver layer interface in Fig. 11. The normalization factor for the wavevector is k_0, the wavevector in the antenna layer; normalized wavevectors with $u < 1$, correspond to radiative modes; $u > 1$ corresponds to non-radiative energy transfer. Since the energy coupling is dependent on the transition dipole orientation with respect to the plane of the interface, we consider the cases of perpendicular (Fig. 11(a)) and parallel (Fig. 11(b)) orientation separately. At a given dipole distance, integration of the energy dissipation across wavevectors u yields unity. If the molecules are randomly oriented, the transition dipoles will be 1/3 perpendicular and 2/3 parallel.

Energy transfer to the stack confirms that of the four decay mechanisms listed above, guided electromagnetic modes are dominant. For perpendicularly oriented dipoles (Fig. 11(a)), prevailing energy transfer is to the non-radiative SPP mode with normalized propagation constant $u = 1.2$. This corresponds to mode (b) in Fig. 10. Mode (a) is also visible at $u = 2.3$. The non-radiative character ($u > 1$) of these modes describes their interfacial localization. Coupling to SPPs is especially strong near the interface, as expected. For dipoles parallel to the interface, both waveguide and SPPs modes are significant. The waveguided photon modes exist in the antenna layer, where the nodes are set by the reflection conditions at the adjacent silver layer and the neighboring air interface for the condition of total internal reflection. For the structure modeled here, only the primary mode exists, however, the number of modes increase as the luminescence wavelength of the dipole decreases and/or the antenna thickness increases.

The efficiency of energy transfer from the antenna to the active layers within the RC is shown in Fig. 12 for various dipole orientations. The efficiency was calculated directly from the Poynting vector. The structure is glass/Ag (25 nm)/RC (50, modeled by copper phthalocyanine, CuPc)/Ag (25)/antenna (200, $n = 1.7$)/air. We also assume an antenna with a free space photoluminescent efficiency of 70% and emission at $\lambda = 620$ nm. For antennas comprised of molecules with isotropic transition dipole moments, the efficiency of energy transfer to the RC is typically greater than 50%.[1]

---

[1] Note that molecules with transition dipoles oriented perpendicularly absorb the least incident radiation. The ideal antenna should transfer energy from parallel dipoles, which have the highest absorption, to perpendicular dipoles.

(a)

Logarithm of perpendicular dipole energy dissipation fraction

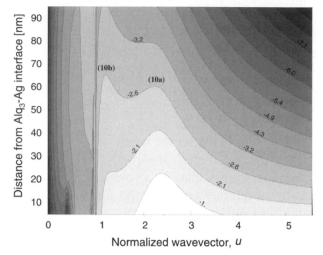

(b) Logarithm of parallel dipole energy dissipation fraction

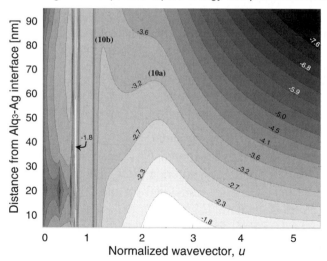

Fig. 11. Logarithmic contour plot of dipole energy dissipation for (a) perpendicular and (b) parallel orientation. The peaks labeled (10a and 10b) correspond to the guided modes in Fig. 10. Dipole energy dissipation is greatest for perpendicularly oriented dipoles into modes with $u > 1$, corresponding to SPPs. Also, note that coupling to waveguide modes is strongest for dipoles oriented parallel to the Ag–antenna interface. The structure modeled here is air/Alq3(115 nm)/Ag(1)/BCP(15)/PTCBI(20)/CuPc(30)/Ag(45)/glass. The photoluminescent wavelength, λ, and free space quantum efficiency, q, of the dipole are 650 nm and 70%, respectively. Energy dissipation is plotted as a logarithm to facilitate visual interpretation.

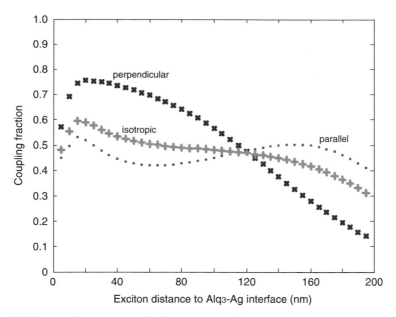

Fig. 12. Efficiency of energy transfer from excitons in the antenna to the RC as a function of the exciton position and orientation in the antenna. Stronger energy coupling across the metal film occurs for perpendicularly oriented dipoles due to their stronger emission into SPP modes (see Fig. 11(a)). The structure modeled here is glass/Ag(25 nm)/RC(50, modeled by CuPc)/Ag(25)/antenna(200, $n = 1.7$)/air. The photoluminescent wavelength, λ, and free space quantum efficiency, q, of the dipole are 620 nm and 70%, respectively. Over the first 100 nm, the mean exciton coupling fraction to the RC is 60%.

4.3. Experimental Verification of Energy Transfer from Artificial Antennas to Reaction Centers

Finally, we experimentally demonstrate energy transfer from the antenna in Fig. 13. In this proof of concept, two antennas were fabricated. The first, with a photoluminescent efficiency of approximately 30% employed a 200 nm-thick film of tris (8-hydroxyquinoline) aluminum (Alq_3). In the second antenna, the Alq_3 was doped with 1% of the laser dye DCM2, increasing the photoluminescent efficiency of the antenna to approximately 70%. Both antennas absorb light in the blue and near UV, and the excitons are randomly oriented. The structure of the solar cell here is Ag (20 nm)/CuPc(40)/PTCBI(20)/BCP(10)/Ag(30)/Antenna(200). For wavelengths above $\lambda = 450$ nm, the quantum efficiency spectra are nearly identical, showing that the antenna does not perturb the diode performance at frequencies where the antenna is inactive. However, the efficiency exhibits a modest increase

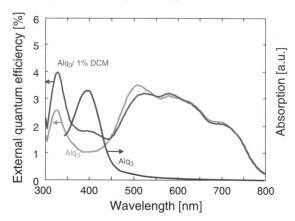

Fig. 13. Quantum efficiency spectra. The photocurrent spectra of PV cells with external Alq_3-based antennas show enhanced photocurrent at wavelengths where Alq_3 absorbs. An increase in quantum efficiency for the Alq_3:DCM2 antenna coincides with the absorption maximum of Alq_3, demonstrating energy coupling from the artificial antenna to artificial reaction center. The device structure is Ag(20 nm)/CuPc(40)/PTCBI(20)/BCP(10)/Ag(30)/Antenna(200).

around $\lambda = 390$ nm, corresponding to the Alq_3 absorption peak. The correlation between an increased photoluminescence quantum efficiency and an increased external quantum efficiency, localized to the narrow absorption peak of the antenna, demonstrates that energy coupling from the antenna layer has occurred. The quantum efficiency is low overall since the antenna-less solar cell suffers from a low internal quantum efficiency. Higher efficiencies are possible by improving the efficiency of the RC.

5. CONCLUSION

Photosynthetic materials and structures deserve investigation because of the promise of harnessing the high efficiency of photosynthesis. Device performance of initial demonstrations of solid-state solar cells with integrated photosynthetic proteins is limited by an inability to absorb enough incident light. This problem is endemic to thin films of even strongly absorbing materials and constitutes a limitation that must be addressed with alternative techniques.

The local environment of the solid state is drastically different to the aqueous solution where proteins usually preside. Since the structural stability and hence functionality of proteins hinge on local environment, it is necessary to inquire whether the harsh environment of solid matter is too destructive for proteins to withstand. We find that solid-state integration necessitates stabilization by surfactants and have demonstrated retained functionality over the timescale of weeks. To be useful in practical devices, significant work in increasing stability is needed. Indeed, the stability of some photosynthetic RCs is poor even in their native thylakoid membranes; the half-life of PSII can be as short as 30 min [67]. Plants survive this damage through an energetically costly and complicated repair process of degradation, resynthesis, and replacement.

Separation of the functions of light absorption and exciton dissociation constitutes a significant photosynthetic redesign, unaccompanied by the limitations of traditional organic PV. Initial device performances are promising. The separation of optical and electrical functionalities discussed here represents a completely synthetic implementation where the active materials of the artificial antenna and RC are amorphous films of pigment semiconductors. However, it is possible to construct devices where one or both components are biological in origin. The excellent absorption characteristics of chlorosomes and charge-separation characteristics of RCs are tempting, the trade off between performance and stability may dictate which type of devices yield high performance and reliability.

REFERENCES

[1] J.R. Sheats, J. Mater. Res., 19 (2004) 1974–1989.
[2] N.B. McKeown, Phthalocyanine Materials: Synthesis, Structure, and Function, Cambridge University Press, Cambridge, UK; New York, 1998.
[3] D. Wohrle, Macromol. Rapid Commn., 22 (2001) 68–97.
[4] J.G. Xue, S. Uchida, B.P. Rand and S.R. Forrest, Appl. Phys. Lett., 85 (2004) 5757–5759.
[5] G. Li et al., Nature Mater, 4 (2005) 864–868.
[6] M.A. Green, K. Emery, D.L. King, Y. Hisikawa and W. Warta, Prog. Photovoltaics: Res. Appl., 14 (2006) 46–51.
[7] US Department of Energy, Basic Research Needs for Solar Energy Utilization, 2005. Available at http://www.sc.doe.gov/bes/reports/files/SEU_rpt.pdf.
[8] J.K. Mapel, The Application of Photosynthetic Materials and Architectures to Solar Cells, Thesis, Available at http://dspace.mit.edu, Massachusetts Institute of Technology, Cambridge, 2006.
[9] P. Peumans, A. Yakimov and S.R. Forrest, J. Appl. Phys. 93 (2003) 3693–3723.
[10] M. Pope and C. Swenberg, Electronic Processes in Organic Crystals, Oxford University Press, Oxford, 1982.

[11] E.A. Silinsh and V. Capek, Organic Molecular Crystals: Interaction, Localization, and Transport Phenomena, AIP Press, New York, 1994.

[12] G. Zerza, C.J. Brabec, G. Cerullo, S. De Silvestri and N.S. Sariciftci, Synth. Met., 119 (2001) 637–638.

[13] S.R. Forrest, MRS Bul., 30 (2005) 28–32.

[14] M. Gratzel, Inorg. Chem., 44 (2005) 6841–6851.

[15] P. Peumans, S. Uchida and S.R. Forrest, Nature, 425 (2003) 158–162.

[16] W.U. Huynh, J.J. Dittmer and A.P. Alivisatos, Science, 295 (2002) 2425–2427.

[17] K.M. Coakley and M.D. McGehee, Appl. Phys. Lett., 83 (2003) 3380–3382.

[18] M. Law, L.E. Greene, J.C. Johnson, R. Saykally and P.D. Yang, Nature Mater, 4 (2005) 455–459.

[19] W.K. Purves, Life, the Science of Biology, Sinauer Associates; W.H. Freeman and Co., Gordonsville, 2004.

[20] B.R. Green and W.W. Parson, Light-Harvesting Antennas in Photosynthesis, Kluwer Academic, Dordrecht, 2003.

[21] J. Psencik et al., Biophys. J., 87 (2004) 1165–1172.

[22] B. Mukhopadhyay, E.F. Johnson and M. Ascano, Appl. Environ. Microbiol., 65, (1999) 301–306.

[23] K. Matsuura, M. Hirota, K. Shimada and M. Mimuro, Photochem. Photobiol., 57, (1993) 92–97.

[24] A.R. Holzwarth and K. Schaffner, Photosynthesis Res., 41 (1994) 225–233.

[25] T. Nozawa et al., Photosynth. Res. 41 (1994) 211–223.

[26] V.I. Novoderezhkin and Z.G. Fetisova, Mol. Biol., 31 (1997) 435–440.

[27] B.J. van Rossum et al., Biochemistry, 40 (2001) 1587–1595.

[28] C.D. Dimitrakopoulus, A.R. Brown and A. Pomp, J. Appl. Phys., 80 (1996) 2501–2508.

[29] J. Fraxedas, Adv. Mater., 14 (2002) 1603–1614.

[30] M.S. Shtein, J.K. Mapel, J.B. Benziger and S.R. Forrest, Appl. Phys. Lett., 81, (2002) 268–270.

[31] G. Witte and C. Woll, J. Mater. Res., 19 (2004) 1889–1916.

[32] W. Humphrey, A. Dalke and K. Schulten, J. Mol. Graph., 14 (1996) 33–38.

[33] R.E. Blankenship, Molecular Mechanisms of Photosynthesis, Blackwell Science, Oxford, 2002.

[34] A.J. Hoff and J. Deisenhofer, Phys. Rep., 287 (1997) 1–247.

[35] C.W. Tang and A.C. Albrecht, J. Chem. Phys., 62 (1975) 2139–2149.

[36] E. Katz, J. Electroanal. Chem., 365 (1994) 157–164.

[37] S.A. Trammell, L. Wang, J.M. Zullo, R. Shashidhar and N. Lebedev, Biosensors Bioelectron., 19 (2004) 1649–1655.

[38] M.C. Smith, T.C. Furman, T.D. Ingolia and C. Pidgeon, J. Biol. Chem., 263 (1988) 7211–7215.

[39] M.H. Moore, S.A. Trammell, S.K. Pollack, N. Lebedev and J.G. Kushmerick, Abstr. Papers Am. Chem. Soc., 229 (2005) U664–U664.

[40] I. Lee, J.W. Lee and E. Greenbaum, Phys. Rev. Lett., 79 (1997) 3294–3297.

[41] L. Minai, A. Fish, M. Darash-Yahana, L. Verchovsky & R. Nechushtai, Biochemistry, 40 (2001) 12754–12760.

[42] J.P. Cleveland, B. Anczykowski, A.E. Schmid and V.B. Elings, Appl. Phys. Lett., 72 (1998) 2613–2615.

[43] R. Das et al., Nano Letters 4 (2004) 1079–1083.

[44] T. Renger, V. May and O. Kuhn, Phys. Rep., 343 (2001) 137–254.

[45] P. Kiley et al., PLoS Biol., 3 (2005) e230.

[46] Y.L. Mi, G. Wood, L. Thoma and S. Rashed, PDA J. Pharm. Sci. Technol., 56 (2002) 115–123.

[47] S. Vauthey, S. Santoso, H. Gong, N. Watson and S. Zhang, Proc. Nat. Acad. Sci. USA, 99 (2002) 5355–5360.

[48] S. Santoso, W. Hwang, H. Hartman and S. Zhang, Nano Letters, 2 (2002) 687–691.

[49] G. von Maltzahn, S. Vauthey, S. Santoso and S. Zhang, Langmuir, 19 (2003) 4332–4337.

[50] S. Zhang, Nat. Biotechnol., 21 (2003) 1171–1178.

[51] T. Morosinotto, J. Breton, R. Bassi and R. Croce, J. Biol. Chem., 278 (2003) 49223–49229.

[52] P. Wang et al., Nat. Mater, 2 (2003) 402–407.

[53] J. Barber and B. Andersson, Nature, 370 (1994) 31–34.

[54] G. Dutton and X.-Y. Zhou, J. Phys. Chem. B 106 (2002) 5975–5981.

[55] I.G. Hill and A. Kahn, J. Appl. Phys., 86 (1999) 4515–4519.

[56] P. Peumans and S. Forrest, Appl. Phys. Lett., 79 (2001) 126–128.

[57] S.C. Straley, W.W. Parson, D.C. Mauzerall and R.K. Clayton, Biochim. Biophys. Acta, 305 (1973) 597–609.

[58] C.J. Brabec et al., Chem. Phys. Lett., 340 (2001) 232–236.

[59] B.P. Rand, P. Peumans and S.R. Forrest, J. Appl. Phys., 96 (2004) 7519–7526.

[60] E.H. Sargent, Adv. Mater., 17 (2005) 515–522.

[61] W.H. Weber and C.F. Eagen, Opt. Lett., 4 (1979) 236–238.

[62] W.L. Barnes, J. Mod. Opt., 45 (1998) 661–699.

[63] P. Andrew and W.L. Barnes, Science, 306 (2004) 1002–1005.

[64] R.R. Chance, A. Prock and R. Silbey, Adv. Chem. Phys., 37 (1978) 1.

[65] D.H. Drexhage, Progress in Optics XII (E. Wolf, ed.), North-Holland, Amsterdam, 1974.

[66] J.K. Mapel, T.D. Heidel, K. Celebi, M. Singh, M.A. Baldo, Nature Materials (forthcoming) 2006.

[67] F. Mamedov and S. Styring, Physiol. Plant., 119 (2003) 328–336.

Nanostructured Materials for Solar Energy Conversion
T. Soga (editor)

Chapter 13

Fullerene Thin Films as Photovoltaic Material

E.A. Katz

Department of Solar Energy and Environmental Physics, Jacob Blaustein Institutes for Desert Research, Ben-Gurion University of the Negev, Sede Boqer, 84990 Israel

1. INTRODUCTION

Carbon is a unique element existing in a wide variety of stable forms ranging from insulator/semiconducting diamond to metallic/semimetallic graphite. The discovery [1] of *buckminsterfullerene*, C_{60}, a new variety of carbon, together with an easy method [2] to produce macroscopic quantities of the material, have generated enormous interest in many areas of physics, chemistry, and material science.

Crystals and thin films of pristine C_{60} are found to exhibit n-type semiconductor-like behavior in their optical and electronic properties while, at the same time, retaining their molecular character. In particular, relatively high photoconductivity (in comparison with the dark conductivity values) was observed in C_{60} thin films [3]. Early studies of solid C_{60} as a material for inorganic solar cells [4–5] were instigated by the theoretical prediction [6] that a C_{60} crystal has a direct band gap of 1.5 eV. On the one hand, this value is close to the experimentally detected fundamental edge in the optical absorption (at $\sim$1.6 eV [7]) and photoconductivity (at $\sim$1.7 eV [8]) spectra of C_{60} thin films. On the other hand, this is the optimal value for high efficiency photovoltaic devices of the single junction type [9]. Recent experimental studies [10–11] demonstrated that the electronic structure of a C_{60} crystal is more complicated. Specifically, the band gap value is suggested to be about 2.3 eV (the mobility gap) but the optical absorption extends from the gap energy to the lower energy side (the optical gap is of $\sim$1.6 eV). However, even in this case, the electronic structure and optical properties of C_{60} thin films are suitable for the use of this material in efficient heterojunction solar cells.

There is probably no more environmentally benign semiconductor than C_{60} in that it can be synthesized from graphite using nothing more than a beam of concentrated solar energy, and subsequently purified and crystallized using the same energy source.

Furthermore, it turns out that C_{60} is only the most abundant member of an entire class of fully conjugated, all-carbon molecules – the fullerenes (C_{20}, C_{24}, C_{26}, ..., C_{60}, ..., C_{70}, C_{72}, C_{74}, ..., carbon nanotubes). It is already clear from the vast and increasing literature on fullerene-based materials that much complex and fascinating physics lies behind the geometrical simplicity of these structures. In addition to the interesting properties of pristine fullerenes, by doping these materials in an appropriate manner, not only can their electronic properties be "tuned" to coincide with those of semiconductors and conductors but even superconductivity may also result [12].

Various examples of structural transformations of solid fullerenes to other forms of carbon nanomaterials have been demonstrated experimentally. In particular, an ion impanation of C_{60} thin film was shown to cause a transformation of some of C_{60} molecules to amorphous carbon (α–C) while the band gap of the resultant C_{60}/α–C composite can be widely vary by a control of the implantation dose [13–16]. These results together with the control of the depth-implantation profile by the ion energy variation open a possibility to produce a C_{60}/α–C composite with a variable depth-profile optimized for maximum sunlight absorption and cell efficiency. Combining this strategy with a feasibility of band gap engineering for an α–C [17–18] and carbon nanotubes [19–20] can result in future in production of efficient *all-carbon* nanostructured multijunction solar cells.

Organic solar cells are becoming now a serious alternative to conventional inorganic photovoltaic devices due to a number of potential advantages, such as their light-weight, flexibility and low-cost fabrication of large areas. For organic donor-acceptor solar cells, a C_{60} molecule is an ideal candidate due to strong acceptor properties of the C_{60} molecule, which can accept as many as six electrons [12]. Fullerene thin films and nanostructures in contact with organic semiconductors, most of which are of donor-like p-type, form a variety of effective donor-acceptor molecular junction cells.

All of these combined reasons make fullerene thin films very attractive for multidisciplinary research towards future cost-effective photovoltaics. A summary of such research to date is provided in this chapter.

2. DISCOVERY OF C$_{60}$ AND MOLECULAR STRUCTURE OF FULLERENES

In 1970 Osawa suggested that a C$_{60}$ molecule might be chemically stable [21]. In 1973 Bochvar and Gal'pern, using Hückel calculations, studied electronic structure of the C$_{60}$ molecule and showed that it should have a large electronic gap between the highest occupied molecular orbital (HOMO) and the lowest unoccupied molecular orbital (LUMO) [22]. These early theoretical suggestions were not widely appreciated, and were rediscovered only after the experimental work of Kroto et al. [1], who in 1985 established the stability of C$_{60}$ molecules in the gas phase.

Kroto was at the time active in microwave spectroscopy and radioastronomy research of gas in space and particularly of carbon-rich giant stars. Meanwhile, Smalley had developed a special laser-supersonic cluster beam apparatus and studied the design and distribution of the clusters of different atoms and carbon clusters in particular. Kroto's idea was to compare spectroscopic readings from space with those obtained from well-characterized materials in the laboratory. This motivation led to a joint Kroto–Smalley experiment [1]. Studying laser beam ablation of graphite in helium gas they recorded by mass spectrometry the predominance of a single peak with a mass of 720 amu, i.e., the formation of a stable molecule with exactly 60 carbon atoms.

The fundamental question for the team was how to construct a molecular structure that satisfies normal bonding (four bonds per carbon atom) and consists of exactly 60 carbon atoms. The collaborators proposed a molecular structure of a truncated icosahedron, a polyhedron, with 20 hexagonal surfaces and 12 pentagonal surfaces (Fig. 1). This polyhedron has 60 vertices (carbon atoms) and 90 edges (C–C bonds). Each carbon atom site is equivalent but C–C bonds are of two different kinds. One is the fusion between 2 hexagons (double bond), and the other is the fusion between a pentagon and a hexagon (single bond). The newly discovered structure was named *buckminsterfullerene* after the American architect Buckminster Fuller who designed the geodesic dome in the shape of such polyhedron.

It should be noted that the Kroto–Smalley experiment did not prove directly the structure of buckminsterfullerene. It became possible only five years later when the discovery of a simple technique of C$_{60}$ synthesis by arc discharge vaporization of graphite [2] (see Section 3) provided the material in macroscopic quantities. The original Kroto–Smalley hypothesis was completely confirmed by nuclear magnetic resonance (NMR) [23–24] and infrared

Fig. 1. A molecular structure of buckminsterfullerene, C_{60}.

Table 1
Structural parameters of C_{60} molecule

Parameter	Value (Å)
C–C bond length on a pentagon	1.46
C–C bond length on a hexagon	1.40
Average C–C distance	1.44
C_{60} mean ball diameter	7.10

(IR) [25] spectroscopy, neutron scattering [26] and other spectroscopic and microscopic methods. Table 1 summarizes structural parameters of the C_{60} molecule provided by these experimental data.

Now, term *fullerenes* is used for various all-carbon closed-cage molecules in the form of convex polyhedra containing only hexagonal and pentagonal faces. Structural description of the entire class of fullerenes could be done on the basis of Euler's theorem for convex polyhedra [27]

$$F - E + V = 2 \tag{1}$$

where F, E, and V are, respectively, the numbers of faces, edges, and vertices in the polyhedra.

If one considers polyhedra formed by h hexagonal and p pentagonal faces, then

$$F = p + h \tag{2}$$

$$2E = 5p + 6h \tag{3}$$

$$3V = 5p + 6h \tag{4}$$

A simple accounting then yields

$$6(F - E + V) = p = 12 \tag{5}$$

$$V = 20 + 2h = 2(10 + h) \tag{6}$$

Now we can conclude that

(a)   Any fullerene molecule must have 12 pentagonal faces and the number of hexagonal faces is arbitrary;

(b)   Any fullerene molecule must have even number of carbon atoms V;

(c)   The smallest possible fullerene is C_{20}, which would form a regular dodecahedron with 12 pentagonal and no hexagonal faces; and

(d)   The entire class of fullerene molecules is: C_{20}, C_{24}, C_{26}, C_{28} ...C_{60}, ..., C_{70}, ... etc.

Topological analysis shows that there can be no convex polyhedron with 12 pentagons and 1 hexagon, i.e., fullerene C_{22}, but all other h numbers are permissible [28].

C_{60} is the most abundant and stable fullerene because of the following reasons:

(1) C_{60} belongs to the group of molecules described by the icosahedral point group I_h with the highest degree of symmetry of any known molecule. The symmetry operations of C_{60} consist of the identity operation, 6 fivefold rotation axes through the centers of the 12 pentagonal faces, 10 threefold axes through the centers of the 20 hexagonal faces, and 15 twofold axes through the centers of the 30 edges joining 2 hexagons (Fig. 2). Each of the 60 rotation symmetry operations can be compounded with the inversion operation resulting in the 120 symmetry operations of the I_h group.

(2)   It is energetically and chemically unfavourable for two pentagons in a fullerene structure to be adjacent to each other. C_{60} is the smallest fullerene to satisfy the "isolated pentagon rule" [29, 30]. Indeed, every pentagon of C_{60} is surrounded by 5 hexagons. The next largest fullerene to do so is C_{70}.

For every even number $B \geq 70$ there is at least one possible fullerene structure satisfying to the isolated pentagon rule [31] while structures of all small

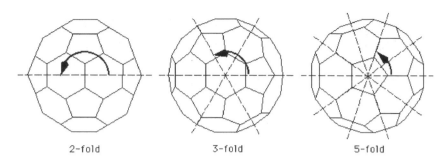

2-fold   3-fold   5-fold

Fig. 2. Rotation symmetry operations of C_{60} molecule.

fullerenes ($B < 60$) contradict to the rule. Therefore the smallest fullerene synthesized by graphite vaporization in mass-quantities was C_{36} [32]. C_{20}, the smallest and most unstable fullerene consisting only of 12 pentagons, was recently produced by a multistep organic synthesis via $C_{20}H_{20}$ and $C_{20}Br_{20}$ stabilized structures [33].

There are a number of ways to design a molecular structure of giant fullerenes. For example, it is possible to increase successively the number of atoms and the diameter of fullerene molecules with icosahedral symmetry [12]. Another way is illustrated by Fig. 3. Cutting a C_{60} molecule in half and inserting 10 more carbon atoms in the breach, one can get a molecule of C_{70} [34] (Fig. 3b). Then, adding another belt of 10 carbon atoms will result in one of several possible C_{80} structures (Fig. 3c). By repeating this process indefinitely, one can create a carbon nanotube of unlimited length and diameter of C_{60}.

The same nanotube structure can be produced by rolling up a two-dimensional graphene sheet into a cylinder and adding caps of a half of fullerene molecules at each end of the cylinder. Three types of single-walled carbon nanotubes are possible depending on how the two-dimensional graphene sheet is rolled up with respect to its hexagonal lattice [35]: (1) armchair tubes (if C_{60} molecule is bisected normal to a fivefold axis, Fig. 3d), (2) zigzag nanotubes (if C_{60} is bisected normal to a threefold axis, Fig. 3e), and (3) a variety of chiral nanotubes with a screw axis along the axis of the tubule (Fig. 3f).

The nanotube structure can be specified mathematically in terms of the tube chiral angle θ and chiral vector C_h, which are shown in Fig. 4a, where

$$C_h = na_1 + ma_2 \tag{7}$$

Here a_1 and a_2 are unit vectors in the two-dimensional hexagonal graphene lattice, and n and m are integers. θ is the angle between C_h and a_1. It is equal to 0 and 30° for the zigzag and armchair tubes, respectively.

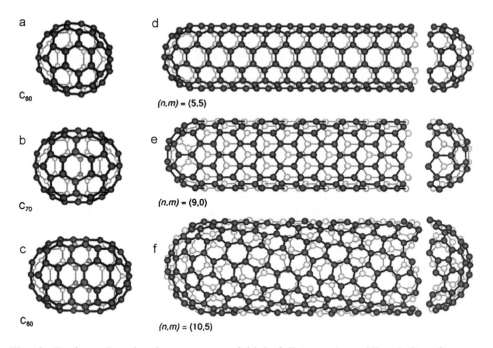

Fig. 3. Design of molecular structure of high fullerencs by adding belts of atoms. (a) C_{60}; (b) C_{70}; (c) C_{80}; (d) armchair nanotube; (e) zigzag nanotube; (f) chiral nanotube. (D-f reprinted from Ref. 35, with permission from Elsevier Science Ltd © 1995).

An ensemble of possible chiral vectors is shown in Fig. 4b. The properties of nanotubes are determined by their diameter d_t and θ [35] both of which depend on n and m as

$$d_t = C_h/\pi = \sqrt{3}\, a_{\text{C–C}} (n^2 + m^2 + nm)^{1/2}/\pi \qquad (8)$$

$$\theta = \tan^{-1}[\sqrt{3}\, m/(2n+m)], \qquad (9)$$

where $a_{\text{C–C}}$ is the nearest-neighbour C–C distance (1.421 Å in graphite).

The nanotube diameters and the chiral angles measured with scanning tunnelling microscopy (STM) and high resolution transmission electron microscopy (HRTEM) [12, 36] confirmed this theoretical picture. Many of the experimentally observed carbon nanotubes are multiwall consisting of capped or open concentric cylinders separated by ~3.5 Å [37].

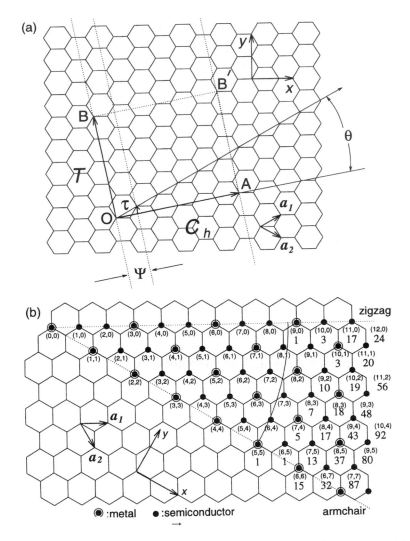

Fig. 4. A carbon nanotube is based on a two-dimensional graphene sheet. (a) The chiral vector C_h is defined as $C_h = na_1 + ma$, where a_1 and a_2 are unit vectors, and n and m are integers. The chiral angle θ, is measured relative to the zigzag axis (a_1). This particular diagram has been constructed for $(n, m) = (4, 2)$, and the unit cell of this nanotube is bounded by OAB'B. To form the nanotube, imagine that this cell is rolled up so that O meets A and B meets B', and the two ends are capped with half of a fullerene molecule. (b) Zigzag nanotubes correspond to $(n, 0)$ or $(0, m)$ and have a chiral angle of 0°, armchair nanotubes have (n, n) and a chiral angle of 30°, while chiral nanotubes have general (n, m) values and a chiral angle of between 0° and 30°. The encircled dots denote metallic nanotubes while the small dots are for semiconducting nanotubes. (Reprinted from Ref. 35, with permission from Elsevier Science Ltd © 1995).

Furthermore, another set of the remarkable theoretical prediction [35, 38, 39] on the electronic structure and properties of single-wall carbon nanotubes was then completely confirmed by the experimental observations [19, 20]. The theory explains electronic properties of carbon nanotubes by the quantum confinement of electrons normal to the nanotube axis. In the radial direction, electrons are confined by the monolayer thickness of the graphene sheet. Around the circumference of the nanotube, periodic boundary conditions come into play. Because of this quantum confinement, electrons can only propagate along the nanotube axis, and so their wavevectors point in this direction. The resulting number of one-dimensional conduction and valence bands effectively depends on the standing waves that are set up around the circumference of the nanotube. These simple ideas were used to calculate the dispersion relations of the one-dimensional bands, which link wavevector to energy, from the well-known dispersion relation in a graphene sheet. The results show that the choice of the tube n and m parameters determines whether the nanotube is metallic or semiconducting, while the chemical bonding between the carbon atoms is exactly the same in both cases. One-third of small-diameter nanotubes are metallic, while the rest are semiconducting, depending on their d_t and θ. In general, an (n, m) carbon nanotube will be metallic when $n - m = 3q$, where q is an integer. All armchair nanotubes are metallic, as are one-third of all possible zigzag nanotubes (Fig. 4b). The band gap in semiconducting nanotubes is inversely proportional to the tube diameter. Based on these unique properties of carbon nanotubes, all-carbon metal-semiconductor hetero-junction devices, made without doping, were predicted [40] and then practically realized [41].

3. SYNTHESIS OF FULLERENES

In 1990 Krätschmer et al. [2] discovered a simple technique to produce fullerenes. The method uses arc discharge between two graphite electrodes in a helium atmosphere to vaporize graphite and produce fullerene-containing soot, and separation of C_{60} from the soot. The soot contains a variety of different fullerenes and hydrocarbon species. The first step of the fullerene purification involves the extraction of soluble fullerenes from the insoluble species of the soot by an organic solvent. The fullerene extract consists mainly of C_{60} ($\sim 75\%$) but also C_{70} and small amount of higher fullerenes ($<1\%$). The next step is chromatographic separation of isolated fullerenes, pure C_{60}, for example. At present, the most of fullerene production techniques involve different modifications of thermal evaporation of graphite in an inert atmosphere (e.g., AC or DC arc discharge between graphite electrodes, laser ablation of graphite) [12].

The electric arc process requires use of high quality, electrically conductive graphite rods. On the other hand, the yield of fullerene production from the collected soot is low (5–10%) due to photodestruction of fullerene molecules by UV radiation of the electric arc [42]. Furthermore, because the yield is low, excessive amounts of toluene are required to separate the fullerenes from the soot. These factors present serious limitations on the minimum cost that can be obtained by this process.

Even though the yield from the laser ablation process can be high, it is generally accepted that this process will not be cost-effective for scaling up to large production levels [43]. The search for a more efficient and environmentally benign method led to the use of concentrated solar energy to evaporate the carbon and efficiently produce fullerenes [42–58] and carbon nanotubes [51, 59–67]. In all reported solar techniques, concentrated sunlight ($\geq 1100\,W/cm^2$) is focused on a graphite target resulting in its heating up to temperatures in excess of 3500 K. Controlling vaporization and condensation conditions leads to the formation of fullerene-rich soot and/or carbon nanotubes (if the target consists of a mixture of graphite and metal catalysts). The UV component in sunlight is very small (compared with irradiation of an electric arc). Therefore, the photodestruction process is very weak and the yield of fullerenes is as high as 20% [51]. Increased yield leads to less solvent use, which reduces cost. Furthermore, the relaxation of the requirement to use conductive graphite rods opens the possibility to use less expensive forms of graphite, including mineral graphite powders. A preliminary cost analysis [44] suggested that solar production could be less expensive than the arc process by at least a factor of four. It is also noteworthy that fullerenes were demonstrated to be purified and crystallized via differential sublimation [12]. These technological steps are certainly possible to perform in a solar furnace.

Fullerenes can be also extracted from the soot made by either combustion [68–69] or pyrolysis of aromatic hydrocarbons [70, 71]. Recently, Frontier Carbon Corporation/Mitsubishi (FCC) [72] organized the first large-scale enterprise for fullerenes synthesis (40,000 kg/year). The FCC technology [73] is based on the combustion method for the soot generation with high yield of fullerenes using a continuous and easily scalable process similar to that employed for commercial carbon black production.

Carbon nanotubes were reported to produce by vaporization of graphite [37, 51, 59–67] and an entire battery of catalytic methods including a catalytic disproportionation of carbon monoxide, pyrolysis of aromatic hydrocarbons, etc. A comprehensive review of such methods is suggested by Harris [36].

4. CRYSTALLINE STRUCTURE OF PRISTINE AND DOPED C_{60} SOLIDS

4.1. C_{60} Fullerite

Crystalline C_{60} is the most prominent member of a family of solid fullerenes, or *fullerites*. Solid C_{60} is a molecular crystal in which C_{60} molecules occupy the lattice sites of a face-centered cubic (*fcc*) structure (Fig. 5) [74, 75]. Packed in the *fcc* crystal C_{60} molecules leave large holes (interstitials) around 4 octahedral and 8 tetrahedral sites of the unit cell. The radii of the voids, $r_{octa} = 3.06\,\text{Å}$, $r_{tetra} = 1.12\,\text{Å}$, are such that this fullerite can accommodate

Fig. 5. The unit cell of C_{60} *fcc* crystal. (Reprinted from Ref. 75, with permission from Institute of Physics Publishing © 2001).

almost all elements from the periodic table and even small molecules, conserving the regular crystal structure (see Section 4.2).

While the carbon atoms within each C_{60} molecule are held together by strong covalent bonding, van der Waals interactions are the dominant intermolecular forces in C_{60} crystals [76].

Fullerites are unusual solids in many aspects. Usually, molecular (including organic) crystals consist of molecules that are rigidly fixed in the crystal array. The fullerites, and in particular solid C_{60}, consist of (rotationally) mobile entities and in the crystal form exhibit variations of the charge transport characteristics as response to changes in the rotational state of the solid and in the mutual orientations of neighboring molecules. Near the critical temperature $T_c = 250-260$ K, the C_{60} crystal is known to undergo a first-order phase transition associated with changes in molecular rotations. Above T_c, C_{60} molecules rotate almost freely and therefore are equivalent, thus resulting in an *fcc* structure. Below T_c, the molecular rotations are partially locked with the fivefold symmetry axes having specific orientations. As a result, the molecules stay in their positions but acquire four different orientations, and the *fcc* structure transforms into a Pa3 structure with four interpenetrating simple cubic (*sc*) sublattices [77]. Now the average anisotropic forces are nonzero and negative, which corresponds to an extra constricting force and which thus results in a negative (upon cooling) jump in the lattice parameter (from 14.154 Å in the *fcc* to 14.111 Å in the *sc* phase) [74, 77].

4.2. C_{60} Fullerides

Structures and properties of the doped fullerene crystals, or *fullerides*, are described in a number of comprehensive reviews [12, 75, 78–79]. The unique structure of the C_{60} molecule and crystal allow the possibility of doping in, at least, four different ways.

(1) *Intercalation*: dopants are located between the C_{60} molecules in the interstitial positions of the host crystal structure. Intercalated fullerides may be produced by the simultaneous vacuum evaporation of C_{60} molecules and dopant atoms [80, 81] or diffusion of dopant atoms into pristine C_{60} crystal. Impurity diffusion may occur as a spontaneous process or it can be induced by an external stimulus, like vapor pressure or an electric field applied to the sample [78]. There has been a considerable research effort to study M_3C_{60} compounds (where M is an alkali or alkali-earth metal) since the discovery of superconductivity in these compounds [82]. Several stable crystalline phases for C_{60} intercalated with alkaline or alkaline-earth metals have been reported [12, 75]. The resultant solid may retain the *fcc* structure of the pristine

crystal or transform into *bct, bcc* or *sc* structure. However, charge transfer from the dopant to the C_{60} host is observed only in vacuum conditions because of the fast oxidation of the system. Intercalation of C_{60} fullerite by metals, other than alkaline or alkaline-earth [83–86], nonmetal elements (S [80], Te [81], halogens [79]), inorganic (H_2, N_2, H_2O, CO, $SbCl_5$, ASF_5, $InCl_3$) [79] and even organic (cationic dye "pyronin B (PyB)" [87]) molecules have been also reported. Spontaneous intercalation of C_{60} thin films by oxygen and its effect on the electrical and photoelectrical properties of the material will be discussed in Section 6.2.

(2) *Endohedral doping*: the dopant goes inside a fullerene molecule (Fig. 6a). Since initial discovery in 1985 that La atoms might be trapped inside the molecule to form endohedral $La_2@C_{60}$ and $La@C_{60}$ [34], the synthesis of many endofullerenes has been reported (e.g., $M@C_{60}$, where M = Ca, Y, Ba, Ce, Pr, Nd, Gd, Er, Eu, Dy; $R@C_{60}$, where R = N, P; $X@C_{60}$, where X = He, Ne, Ar, Kr, Xe) [12, 75, 79]. In the most of techniques endofullerenes are produced at the technological step of the molecule formation. Another approach which is developed by Cambell et al. suggests endofullerene synthesis by low energy ion implantation of C_{60} [88]. Growth of thin films of $Li@C_{60}$ produced by this technique have been already demonstrated [89]. In Sections 5 and 6.3, we will briefly discuss growth, crystalline structure and semiconductor properties of thin films of some endofullerenes which could be used in photovoltaic solar cells, for example, $Dy@C_{82}$ (see Section 10.2). Here we would like to note only that

a)   b)

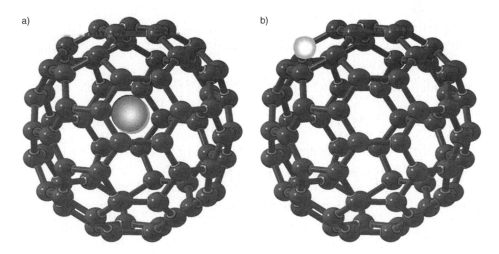

Fig. 6. Endohedral (a) and substitutional, (b) doping of C_{60} molecule.

properties of endoffulelrenes could be strongly affected by the dopant atom position with respect to a fullerene cage. For example, at least three $Dy@C_{82}$ isomers with different properties are known [90]. In the remainder of the chapter (Sections 5, 6.3, and 10.2) we will discuss only the major isomer of $Dy@C_{82}$ (isomer I) in which the Dy atom lies along the C_2 axis on the six-membered ring of the C_{82} cage with C_{2v} symmetry [90].

(3) *Substitutional doping*: noncarbon atom replaces one (Fig. 6b) or more of the carbon atom on the surface of the molecule. This kind of doping is common for Group IV semiconductors, such as Si or Ge. Already in 1991 Smalley's group reported on the preparation and mass spectroscopy identification of *heterofullerene* $C_{59}N$ and $C_{59}B$ molecules [91]. Shortly thereafter, the electronic structure for these compounds was calculated [92]. An electronic behavior similar to that of deep donor and acceptor levels in doped semiconductors was predicted. The publication on a method for the production of $C_{59}N$ in bulk quantities [93] provided a possibility to grow thin films and investigate the electronic structure and properties of the material. Unfortunately, such a study [94] concluded that a dimer formation which prevents donation of an extra electron is observed in the solid state. However, recently such novel heterofullerenes as $C_{48}N_{12}$ [95] and $C_{50}Cl_{10}$ [96] were synthesized in large quantities. It was demonstrated [97–98] that the HOMO–LUMO band gap for $C_{48}N_{12}$ is of ~1eV and *fcc* solid of $C_{48}N_{12}$ is an n-type semiconductor material. Since $C_{48}N_{12}$ is a good electron donor, it was suggested that a molecular rectifier can be formed in a contact with an acceptor $C_{48}B_{12}$ [99]. Photovoltaic cell with a heterojunction $C_{48}N_{12}/C_{48}B_{12}$ was also theoretically suggested [98].

(4) *Adductive*, or *exohedral*, bonding to the outside of the C_{60} molecule is demonstrated to produce the exohedral complexes of C_{60} with metals [79] and an extremely wide variety of organic molecules and can be considered as the fourth type of doping. Fig. 7 illustrates molecular structures of a number of such compounds which are used in fullerene-based solar cells and will be discussed in Sections 8 and 9: (a) [6,6]-Phenyl C_{61}-butyric acid methyl ester (PCBM) [100], (b) *trans-3* C_{60} [101], (c) azafulleroid [102], and (d) ketolactam [102].

5. GROWTH OF FULLERENE THIN FILMS

Most results on deposition of C_{60} thin films have been published for vapor growth methods in vacuum [103–141] or in a N_2 or Ar atmosphere [142–144]. The simplest technique of C_{60} film vacuum deposition [103–110, 112, 115–117, 119–121, 123–126] uses a sublimation of microcrystalline C_{60} powder

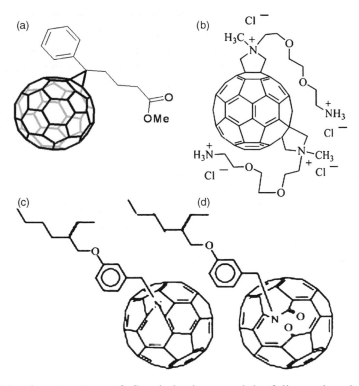

Fig. 7. Molecular structures of C_{60} derivatives used in fullerene-based solar cells (a) [6,6]-Phenyl C_{61}-butyric acid methyl ester (PCBM), (b) *trans-3* C_{60}, (c) azafulleroid, (d) ketolactam. ((a), (c) and (d) reprinted from Ref. 102, with permission from Elsevier Science Ltd © 2002. (b) Reprinted from Ref. 101, with permission from the Royal Soc. of Chemistry © 2002).

that is heated up to 600°C. Then C_{60} is vapor transported and condenses on a substrate the temperature of which is controlled independently.

Well-ordered *fcc* polycrystalline C_{60} films can be grown on substrates with a weak surface bonding (GaAs [103], GaN [122], GeS [104], mica [105, 115–116], MoS_2 [111], Au [106–110], Ag [109–110], NaCl [111–114], KCl [112–114], KBr [124], LiF [112], VSe_2 [117], highly oriented pyrolitic graphite [123], Bi [118] and Sb [119–121] surfactant sublayers. Another necessary condition for such growth is that the substrate temperature should be high enough to provide high mobility of C_{60} molecules during nucleation of the first monolayer and the subsequent film growth. The degree of crystallinity for C_{60} films increases with decreasing deposition rate and increasing substrate temperature [125–126]. Since a temperature of desorption of C_{60} multilayers is known to be in the range of 230–300°C [104, 107], the well-ordered films are, in general, obtained at the substrate temperature of 180–200°C.

In contrast, Katz et al. suggest that deposition of well-ordered C_{60} thin films onto a substrate with weak surface bonding also requires a combination of high values of C_{60} deposition rate and substrate temperature (near to the temperature of equilibrium "adsorption (deposition) $\leftrightarrow$ desorption" for C_{60} molecules) [109–110, 116]. Using this approach, C_{60} films with homogeneous thickness have been grown with an extremely high deposition rate (up to 10 Å/sec) on glass substrates partially predeposited with an Ag layer and held at temperatures up to 300°C [109–110]. It should be stressed that the structure of the C_{60} film grown onto the Ag part of the substrate was found to differ substantially from that deposited onto the clear glass part of the same substrate. While the film grown onto the glass part of the substrate was almost amorphous, the C_{60} film grown on the Ag part of the same substrate had an *fcc* polycrystalline and strong (111)-textured structure with a grain size of ~200 nm. Crystalline structure of C_{60} thin films grown under the same conditions on a mica substrate was even better [116]. The sizes of crystalline domains of (111)-textured C_{60} films were of 500–1500 nm.

Hebard et al. demonstrated another example of the substrate effect [125]. They deposited C_{60} films on a silicon substrate. The high density of dangling bonds on a Si surface results in strong C_{60}–Si bonding. As a result the C_{60} film was almost amorphous. However, after hydrogen passivation of dangling bonds on a silicon surface one can grow highly crystalline films with (111) texture. Similar strategy was used by Katz et al. [4, 145–147] for production of a C_{60}/Si heterojunction (Section 8.1.2).

Highly oriented polycrystalline C_{60} films with large grain sizes or, in some cases, even monocrystalline films are reported to be grown using more sophisticated techniques of vacuum deposition such as molecular beam [111, 113–114, 122] and hot wall [127–133] epitaxy, discrete evaporation in a quasi-closed volume [118, 148–149], ionized cluster beam [134–136] and pulsed supersonic molecular beam [137–141] deposition.

Kano et al. demonstrated a growth of fullerene film on a Ni substrate using vaporization of graphite powder by a continuous-wave CO_2 laser in a flow of Ar or He gas at 30 or 200 Torr [150]. The film produced in the Ar flow at 200 Torr showed the Raman bands of C_{60} and C_{70}, and the C_{60}/C_{70} ratio was estimated to be 5/1 by mass spectrometry. It should be stressed that in this process the fullerene films are produced directly by graphite vaporization without any intermediate stage of C_{60} powder synthesis and purification. These results suggest that it also may be done with a concentrated sunlight [42–58].

In the context of photovoltaic applications it is important to note that polycrystalline thin films of C_{60} [151–152], such C_{60} derivative as PCBM [153], such high fullerenes as C_{76}, C_{78}, C_{84} [154] can be obtained by solution growth or, for example, the Langmuir–Blodgett technique [155–156]. Thin films of Dy@C_{82} were reported to have been grown by both vacuum deposition [157–158] and the Langmuir–Blodgett [160] techniques. It was shown that films of Dy@C_{82} (isomer I) have an *sc* polycrystalline structure with a lattice constant of 15.78 Å [158].

6. ELECTRICAL AND PHOTOELECTRICAL PROPERTIES OF FULLERENE THIN FILMS

6.1. Electronic Structure of C_{60} Solid, Optical and Photoelectrical Properties of C_{60} Thin Films

C_{60} solid is a semiconductor with a minimum of the energy gap at the X point of the Brillouin zone [79]. The value of the band gap is still under debate. As was already mentioned (Section 4.1), fullerenes form molecular solids. Thus, the electronic structure of C_{60} crystal is expected to be closely related to the electronic levels of the isolated fullerene molecule and to have narrow bands. Saito and Oshiyama [6] using a one-electron model calculated HOMO–LUMO gap of 1.92 eV for the free C_{60} molecule and a direct band gap of 1.5 eV for the crystal. The latter value is close to the experimental value of $\sim$1.6 eV for the fundamental edge in the optical absorption spectra [7] and photoconductivity spectra [8] of C_{60} thin films. However, the analysis of photoemission and inverse photoemission spectra of C_{60} films [10] suggested that the band gap value is about 2.3 eV but the optical absorption extends from the gap energy to the lower energy side. These experimental works stimulated new theoretical calculations of the band gap value of 2.15 eV using a "many-body" approach [161]. The results of the combined experiment of photoconductivity, photoluminescence, photoinduced absorption excitation spectroscopy, resonant Raman spectroscopy and electroabsorption [162] and the comparative study of optical absorption and photoconductivity spectra [163] confirmed that the main threshold in optical absorption and photoinduced carrier generation of C_{60} films is at 2.3 eV. However, Wei et al. [162] and Konenkamp et al. [163] attributed this signal to the lowest charge-transfer excitation state in solid C_{60} (formed by a Coulomb-bound electron and hole located on different molecules) while the single-particle band gap of solid C_{60} is expected to be at 2.6 eV. The origin of near-gap tail in the optical absorption of C_{60} films [7] is still under discussion. A series of weak absorption structures have been

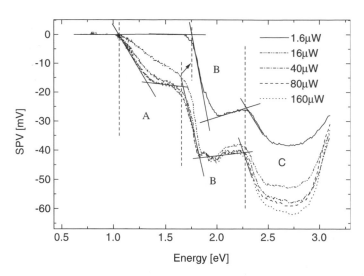

Fig. 8. SPV spectra for an as-grown C_{60} film at different illumination intensities (illumination intensities are indicated). (Reprinted from Ref. 11, with permission from Elsevier Science Ltd © 1997).

observed in the tail of the optical and photoacoustical spectra [164, 165]. These structures have been ascribed to either exciton levels or the splitting due to vibronic modes of the molecules. Specifically, the absorption at 1.65 eV has been related to $h_u \rightarrow t_{1u}$ intramolecular Frenkel excitation [166–168].

The detailed electronic structure of C_{60} films was studied by surface photovoltage (SPV) spectroscopy [11, 169, 170]. SPV formation requires both photogeneration and transport of charge carriers while the band gap and gap-state energy positions can be determined from the positions of slope changes in the SPV spectrum caused by photon assisted population or depopulation of the states. Fig. 8 represents characteristic SPV spectra for an as-grown C_{60} film. The negative sign of the SPV indicates the n-type of photoconductivity of C_{60} films. There are four sharp changes in the spectrum slope: at about 0.9–1.0, 1.3 (regions A_1 and A_2), 1.6 (region B) and 2.3 eV (region C). To describe the tail in optical absorption and, in particular, region B of the SPV spectra the authors adopted an approach used in amorphous semiconductor physics and considered the progressive shift of the region B to higher photon energies recorded in the experiment with a reduction of light intensity as evidence for band tails of localized electronic states extending towards the forbidden gap (to 1.6 eV). This approach is consistent with those used for analysis of optical absorption and the light intensity dependence of photoconductivity

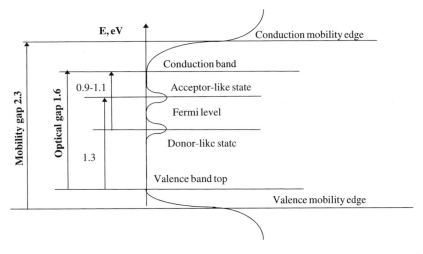

Fig. 9. Electronic structure of C_{60} thin films. (Reprinted from Ref. 11, with permission from Elsevier Science Ltd © 1997).

[171, 172] in C_{60} solids. The corresponding model of the electronic structure of C_{60} film including a mobility gap of about 2.3 eV and a photoconduction gap of 1.6–1.7 eV is shown in Fig. 9. The features at 0.9–1.1 and 1.3 eV indicate, respectively, the contributions of depopulation and population of donor and acceptor gap states. Each of the mentioned SPV signals (A_1, A_2, B and C) coincides with the corresponding well-defined features in the photoconductivity spectra of C_{60} films [3, 8].

6.2. Conductivity and Charge Carrier Mobility in C_{60} Thin Films

Pristine C_{60} thin films exhibit a semiconductor-like activated dark conductivity of n-type [12]. However, there is extremely wide dispersion (up to 10 orders of magnitude) in the reported values for the room temperature conductivity of C_{60} thin films [173–174]. The range of reported activation energies is also very wide (from 0.3 to 1.1 eV) [173]. Authors reporting lower conductivity also tend to report higher activation energy. We believe that two effects are responsible for this wide dispersion. First is the difference in crystalline structure of the films. Another factor is oxygen effect on the electronic transport in C_{60} films which has been shown by a variety of experiments. Some studies [175–176] have revealed a drastic but reversible reduction of dark and photo-conductivity in C_{60} films upon their exposure to air. Given the large amount of interstitial volume in the C_{60} crystal molecular oxygen from the air readily diffuses into this solid [177]. It quenches the conductivity but does not react chemically with the C_{60} molecules. On the other hand, illumination of C_{60} films in air causes larger and irreversible changes in conductivity [174, 175]. It has been suggested [175] that air/light exposure promotes C–O binding that damages the C_{60} molecules, producing dangling bonds or other defects with deep levels in the gap. Redistribution between fast and slow components of photoconductivity under the air/light exposure of C_{60} films [178] also points to a generation of recombination centers with deep levels. In a joint EPR/SPV experiment Katz et al. demonstrated [169] that the air/light exposure of C_{60} films leads to the generation of the paramagnetic centers and deep acceptor states at $E_v + 1.3\,eV$. These acceptors act as recombination and/or scattering centers. The paramagnetic and recombination/scattering centers were suggested to have the same origin. Their density is controlled by the oxygen diffusion while the rate of the diffusion decreases with a rise in the grain sizes of the polycrystalline C_{60} films [179].

Curves 1 and 2 in Fig. 10 display Arrhenius plots of temperature dependences of the dark and photoconductivity for an oxygen-free (111)-textured polycrystalline C_{60} film [180]. One can observe a qualitatively similar behavior

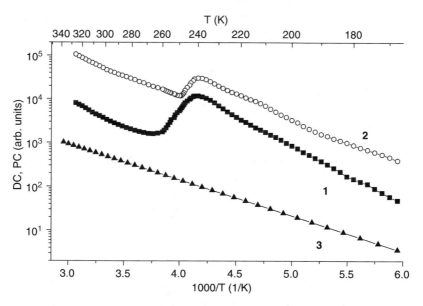

Fig. 10. Temperature dependence of the dark conductivity (Curves 1, 3) and photoconductivity (Curve 2) for a textured large-grained polycrystalline C_{60} film. Curves 1 and 2 are for the oxygen-free sample. Curve 3 is for the same sample after exposure to an oxygen atmosphere at a pressure of ~100 Bar for 15 min. (Reprinted from Ref. 180, with permission from the American Institute of Physics © 2003).

for the dark and photoconductivity with two activated parts separated by the crossover region near the critical temperature $T_c = 250$–260 K of the first-order phase transition associated with changes in rotations of C_{60} molecules (see Section 4.1). Both the dark and photoconductivity increase rapidly as the sample temperature is lowered through the crossover region. *In situ* exposure of this sample to molecular oxygen gas under a pressure of about 100 Bar for 15 min was found to reduce the dark conductivity by 1–2 orders of magnitude and to suppress any critical behavior near T_c (Curve 3 in Fig. 10). The temperature dependence of the photoconductivity after such an exposure (not shown in Fig. 10) was similar to the behavior observed for the dark conductivity.

Photoelectrical measurements [163] revealed the charge carrier mobility of oxygen-free C_{60} films and their degradation as the samples were exposed to air. The initial values for the room-temperature electron and hole drift mobility were 1 and 10^{-4} $sm^2 V^{-1} s^{-1}$, respectively. These mobility values are consistent with those obtained by Dember effect measurements [181]. The highest values of electron mobility measured in oxygen-free C_{60}-based n–channel field effect transistor were reported to be in the range of 0.4–1 $sm^2 V^{-1} s^{-1}$ [133, 182–184]. The effects of oxygen [163, 184] and the 260 K phase transition [185] on the charge carrier mobility in C_{60} films were experimentally demonstrated.

6.3. Semiconductor Properties of Doped Fullerene Films

Reproducible increase, by orders of magnitude, in the n-type semiconductor-like conductivity and decrease in its activation energy have been reported for C_{60} films intercalated by Li, Na, and Mg [186, 187], Ag [188], Au [86, 189], S [80], Te [81] and such organic molecules as pyronin B (PyB) [87]. It should be stressed that C_{60} intercalation with Mg and pyronin B was used to improve photovoltaic performance of fullerene-based solar cells (see Sections 8.2.2, 9.1.4, 9.2). Increase in the n-type room-temperature conductivity by four orders of magnitude has been demonstrated for films of endofullerene Li@C_{60} [89].

Electronic transport in thin film of high metallofullerenes Dy@C_{82} and La@C_{82} also shows an n-type semiconducting behavior. The energy gap E_g values of 0.2 and 0.3 eV were estimated from the experimentally observed temperature dependence of conductivity for Dy@C_{82} and La@C_{82}, respectively [158]. Accordingly, normally-on type characteristics were reported for the n–channel field effect transistors based on Dy@C_{82} [159, 190] and La@C_{82} [191] thin films. However, the electron mobility values in these transistors were found to be considerably lower than those in the C_{60}-based devices. This result was attributed [190] to the low crystallinity of the endofullerene thin films. Promising results were reported for C_{60} field effect transistors with interfacial electrode/fullerene regions modified by La@C_{82} [192]. The interfacial surface modifications on the electrodes using the endofullerene with the narrow band gap and, in turn, the large number of carriers, are effective in reducing the trapping levels at the interface between C_{60} thin film and the gold electrodes. The transistor operation was observed without any annealing processes even once the fabricated devices were exposed to air, which has not ever seen in the conventional C_{60} devices. Possible benefits of Dy@C_{82} incorporation in donor-acceptor photovoltaic devices will be discussed in Section 10.2.

7. GENERAL FEATURES OF VARIOUS FULLERENE-BASED PHOTOVOLTAIC SOLAR CELLS

Photovoltaic solar cells convert solar radiation directly into electricity due to the photovoltaic effect which was first observed by Becquerel in 1839 [193]. A photon can interact with matter by losing its energy and using this energy to promote an electron to a higher energy level. If a conventional *inorganic* semiconductor solar cell is illuminated with light of photon energy higher than a semiconductor band gap energy of E_g, pairs of *free* electrons and holes are generated. To separate these photoinduced charge carriers and generate photovoltage, a solar cell should have a junction between two materials across which there is an electrochemical potential difference in equilibrium.

There are at least three main types of such junctions: homo- and hetero-p–n junctions, formed between p- and n-types of the same or different semiconductors, respectively, and the Schottky barrier, i.e., a rectifying interface between a semiconductor and a metal. Other types of solar cells may include a p–i–n junction, formed from two doped and one intrinsic semiconductor layer, or metal-insulator-semiconductor (MIS) configuration.

Most inorganic photovoltaic devices to date have been formed using a p–n junction [194–195]. The different Fermi levels, E_F, in p- and n-type regions result in the formation of a potential barrier $q\varphi_{eff}$ and depletion region at the junction interface where this difference is accommodated by a built-in electric field (Fig. 11). Photoinduced free carriers, generated within approximately one diffusion length from the depletion region, may diffuse to this region, where they will be separated by the built-in electric field (e.g., the diffusion length of free electrons in the p-base of industrial silicon solar

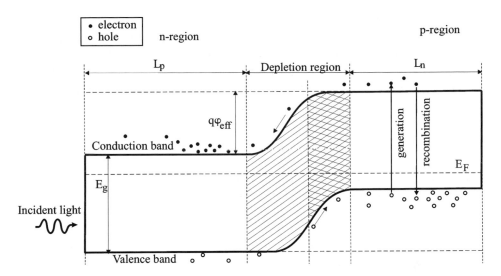

Fig. 11. Energy band diagram for a p–n junction under illumination.

cells are of ~100 μm or even higher [196]). Electrons and holes move to the n- and p-regions, respectively. A potential difference thus appears between the external metal electrodes attached to the p- and n-regions, and a current may flow through a connected load resistor.

A current-voltage, *I–V*, characteristics of an illuminated solar cell (shown by the lower curve in Fig. 12a) is described by a standard one-exponential model as [195]:

$$I - I_o \left[\exp\left(\frac{q(V + IR_s)}{nkT} \right) - 1 \right] + \frac{V + IR_s}{R_{sh}} - I_p \qquad (10)$$

where I_p is the photocurrent, I_o and n respectively denote the reverse saturation current and diode quality factor of the p-n junction, R_s and R_{sh} are lumped series and shunt resistances of the cell, respectively. The first two terms in Eq. (10) define the cell *I–V* characteristics in the dark (the upper curve in Fig. 12a) while I_p is independent on voltage and proportional to the power of incident light P_{in}. Fig. 12b represents the part of the irradiated solar cell *I–V* curve from the fourth quadrant just by its inversion around the voltage axis. For the remainder of the present chapter we will represent the experimental *I–V* curves in this way as it is accepted by photovoltaic community. The quantities I_{max} and V_{max}, corresponding to the current and voltage respectively, for the maximum power output $P_{max} = I_{max} \times V_{max}$ are also defined in Fig. 12b.

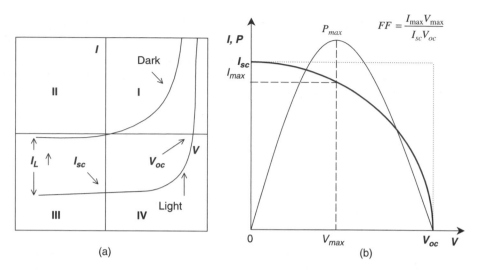

Fig. 12. The dark and light I–V curves (a) and the general representation of the light I–V curve in the first quadrant showing maximum power, P_{max} (b).

Now we can define such principle cell parameters as the short-circuit current, I_{sc}, the open-circuit voltage V_{oc}, fill factor FF, and the power conversion efficiency, η.

For $V = 0$, $I = -I_L = I_{sc}$, and for $I = 0$, $V = V_{oc}$. The FF is defined by the relation

$$FF = \frac{I_{max} \times V_{max}}{V_{oc} \times I_{sc}} \tag{11}$$

Finally, the power conversion efficiency is given by

$$\eta = \frac{I_{max} \times V_{max}}{P_{in}} = FF \frac{V_{oc} \times I_{sc}}{P_{in}} \tag{12}$$

The main difference in operation mechanism of fullerene-based solar cells and inorganic photovoltaic devices lies in the basic properties of photogenerated excitations in such molecular crystals as fullerites and organic solids (polymers, dyes, etc). Indeed, intermolecular van der Waals forces in molecular solids are weak compared to bonds in inorganic crystals and much weaker than the intramolecular bonds. As a consequence, photogenerated excitations ('excitons') are strongly bound and do not spontaneously disso-ciate into separate charges. (Dissociation requires an input of energy of ~100 meV compared to a few meV for an inorganic semiconductor [197]). This means that generation of free carriers does not necessarily result from the absorption of light. Contrary to conventional solar cells with an inorganic semiconductor p–n junction, as described above, in fullerene-based mole-cular solar cells neutral excitons are first created by light. Finally, charges should be separated in the narrow interface region due to donor-acceptor electron transfer interactions, built-in electric field of the heterojunction, etc. The thickness of the photoactive region is limited by low values of the exciton diffusion length in fullerene films and/or organic semiconductors (typically <10–40 nm [198–201]).

These fundamental aspects of the photovoltaic operation of fullerene- and organic-based devices referred as 'excitonic solar cells' are presently under strong debate [202–209]. It was demonstrated that contrary to inorganic solar cells (1) I_p shows a considerable voltage dependence and at short-circuit conditions only a part of electron-hole pairs are separated; (2) V_{oc} can be con-trolled mostly by chemical potential energy gradient, which is created by the interfacial exciton dissociation process, and in some cases can exceed the $q\varphi_{eff}$ values (electrical potential energy difference); and (3) the effect of inter-facial surface area on the photovoltaic performance is controlled by a trade-off between enhancing exciton dissociation and interfacial recombination.

This kind of behaviour requires a sophisticated nanostructural engineer-ing of fullerene-based photovoltaic materials and devices compromising the requirements of thin photoactive layers with an effective light absorption and photogeneration of free carriers. In the remainder of the chapter we will describe the particular examples of such engineering. In order to classify a variety of the fullerene-based devices we divide them into two large groups: inorganic and organic fullerene-based solar cells. This classification reflects nothing more than that the cells are based on junction(s) of fullerene layer(s) with either inorganic or organic semiconductors. Table 2 summarizes the photovoltaic parameters of various types of devices.

8. INORGANIC C_{60}-BASED SOLAR CELLS

8.1. Solar Cells Based on Pristine C_{60}

8.1.1. Metal-Insulator-Semiconductor and Schottky Barrier Devices

In 1992 Hebard et al. patented an idea for the utilization of the pho-toconductivity and photovoltaic properties of fullerites, and in particular,

Table 2

Photovoltaic parameters of fullerene-based solar cells. For comparison, the parameters of amorphous carbon cells are listed too

Solar cell design strategy	Cell structure	V_{oc} (mV)	J_{sc} (mA/cm^2)	FF	η (%)	Test conditions	Ref.
Pure C_{60}/inorganic semiconductor heterojunction	C_{60}/p-Si	306	1.1×10^{-4}	0.55	2.2×10^{-5}	AM 1.5, 100 W/cm^2, in air	[220]
C_{60} derivative/nanocrystalline inorganic semiconductor heterojunction	ITO/CIS/PCBM/LiF/Al	792	0.26	0.44	0.09	White light by a Steuernagel solar simulator, 80 mW/cm^2	[230]
C_{60}/inorganic semiconductor heterojunction deposited onto a nanoporous TiO_2	FTO/TiO$_2$/C$_{60}$/CuSCN/C/Au/FTO	350	0.225	0.4	0.12	White light, 26 mW/cm^2	[235]
Schottky barrier with doped C_{60} layer	ITO/C$_{60}$:Mg/Mg/Al	400	1.04	0.4	0.15	AM 1.5, 100 W/cm^2, in air	[216]
Ion-implanted C_{60}/inorganic semiconductor heterojunction	B ion-implanted C$_{60}$/n-Si	170	0.33	0.41	0.023	AM 1.5, 100 mW/cm^2, in air	[13]
	α-C/C$_{60}$/Si p-i-n cell	90	4	0.29	0.1	AM 1.5, 100 mW/cm^2, in air	[242]
Conjugated polymer/C$_{60}$ heterojunction	ITO/ZnO/C$_{60}$/P3HT/Au	420	5.8	0.41	1	AM 1.5, 100 mW/cm^2	[291]
Conjugated polymer/doped C$_{60}$ heterojunction	Mg/C$_{60}$: Mg/MEH-PPV/ITO	680	2.57	0.31	0.54	White light, 100 mW/cm^2	[248]
Bulk heterojunction between C$_{60}$ derivative and conjugated polymer	ITO/PEDOT/MDMO-PPV:PCBM/LiF/Al	820	5.25	0.61	2.5	AM 1.5, 100 mW/cm^2, T = 50°C	[264]
	ITO/PEDOT/MDMO-PPV:PCBM/LiF/Al	870	4.9	0.60	2.55	STC (AM 1.5G, 100 mW/cm^2, T = 25°C)	[273]

Bulk heterojunction between C_{60} derivative and conjugated polymer after the post-production annealing	ITO/PEDOT/P3HT:PCBM/LiF/Al	550	8.5	0.6	3.5	AM 1.5, 80 mW/cm²	[282]
	ITO/PEDOT/P3HT:PCBM/LiF/Al	640	11.1	0.55	4.9	AM 1.5, 80 mW/cm²	[277]
CNT/polymer bulk heterojunction	ITO/P3OT: single-walled CNT/Al	750	0.2	0.4	0.06	AM 1.5, 100 mW/cm²	[336]
Fullerene/polymer bulk gradient heterojunction	ITO/PEDOT/P3OT: C_{60}/Al	360	0.27	0.57	1.5	Monochromatic irradiation, 3.8 mW/cm² λ = 470 nm	[290]
"Double cable" donor-acceptor solar cells	ITO/PEDOT:PSS/methanofullerene covalently attached to hybrid of PPV and poly (p-phenylene ethynylene)/Al	830	0.42	0.29	0.1	AM 1.5, 100mW/cm²	[286]
Bulk heterojunction between C_{70} derivative and conjugated polymer	ITO/PEDOT-PSS/[70] PCBM: MDMO-PPV//LiF/Al	770	7.6	0.51	3	STC (AM 1.5G, 100 mW/cm², T = 25°C)	[316]
Small molecule/ C_{60} bi-layer heterojunction with an exciton-blocking layer	ITO/PEDOT:PSS/CuPc/C_{60}/BCP/Al	580	18.8	0.52	3.6	AM 1.5, 150 mW/cm²	[297]
	ITO/CuPc/C_{60}/BCP/Al ITC/	609	49.72	0.61	4.2	AM 1.5, 440 mW/cm²	[302]
	PEDOT:PSS/tetracene/C_{60}/BCP/Al	580	7	0.57	2.3	AM 1.5, 100 mW/cm²	[313]

Table 2 (*continued*)

Solar cell design strategy	Cell structure	V_{oc} (mV)	J_{sc} (mA/cm^2)	FF	η (%)	Test conditions	Ref.
Small molecule/ C_{60} bi-layer heterojunction with an exciton-blocking layer	ITO/pentacene/ C_{60}/BCP/Al	363	15	0.5	2.7	Broadband light, 100 mW/cm^2	[314]
Phthalocyanine/ C_{60} *p–i–n* structure	ITO/PEDOT:PSS/ ZnPc/ZnPc:C_{60}/ C_{60}/BCP/Mg:Ag	570	5.36	0.5	1.5	AM 1.5, 100 mW/cm^2, in air	[305]
	ITO/ *p*-MeOTPD/ ZnPc:C_{60} /C_{60}/ *n*-doped C_{60}/Al	440	8.44	0.45	1.67	AM 1.5, 100 mW/cm^2	[306]
	ITO/CuPc/CuPc: C_{60}(3:1)/CuPc:C_{60} (1:1)/ CuPc:C_{60} (1:3)/C_{60}/BCP/Al	510	4.94	0.54	1.36	AM 1.5, 100 mW/cm^2	[308]
	ITO/CuPc/CuPc: C_{60}(1:1)/C_{60}/BCP/Ag	540	18.2	0.61	5	AM 1.5, 120 mW/cm^2	[310]
Tandem cells based on phthalocyanines and C_{60}	ITO/ *p*-MeOTPD/ ZnPc:C_{60} /*n*-doped C_{60}/ Au/*p*-MeOTPD/ZnPc: C_{60}/*n*-doped C_{60} /Al	850	6.6	0.53	2.4	AM 1.5, 125 mW/cm^2, 50°C	[306]
	ITO/ *p*-MeOTPD/ ZnPc:C_{60}/*n*-doped C_{60}/ Au/*p*-MeOTPD/ZnPc: C_{60}/*n*-doped C_{60} /Al	990	10.8	0.47	3.8	AM 1.5, 135 mW/cm^2, 40°C	[311]
	ITO/CuPc/CuPc:C_{60}/ C_{60}/PTCBI/Ag/m-MTDATA/CuPc/CuPc: C_{60}/C_{60}/BCP/Ag	1040	9.13	0.6	5.7	AM 1.5, 100 mW/cm^2	[309]

fabrication of fullerene-based solar cells and "the generation of a current by illumination of appropriate fullerene interfaces" [210]. Although a photovoltaic response requires fabrication of a device structure with rectifying properties, the authors described a symmetrical (non-rectifying) $Ag/C_{60}/Ag$ device structure. Such device can exhibit photoconductivity but no photo-voltaic effect.

First fullerene-based solid state device, displaying a remarkable rectify-ing effect in the dark and photoresponse under illumination, was demonstrated by Yohehara and Pac [211]. Polycrystalline C_{60} film with a thickness of 100 nm was deposited by the vacuum evaporation technique on a substrate of Al coated glass. This Al layer was used as the front electrode. Sandwich cells were com-pleted by vacuum evaporation of an Al or Au back electrode (0.5 cm^2 in area). The devices without any exposure of the front electrode to oxygen exhibited no rectification and photoresponse. The devices with a front Al electrode exposed to oxygen (before C_{60} film deposition) exhibited a rectification ratio of 66 (at ± 2 V) and greatly enhanced photocurrent (the quantum yield of photocurrent for monochromatic irradiation of $\lambda = 400$ nm and 0.1 V forward bias was about 53%). The excitation profile of photocurrent was observed to follow the optical absorption of a C_{60} film. The authors explained the results by the formation of an oxide layer at the front Al electrode interface and the MIS configuration rather than a Schottky barrier device. MIS devices with a fullerene layer as an active semiconductor contact were also studied by Pichler et al. [212]. Solar cells with a Schottky barrier at the C_{60}/Mg-In [213], C_{70}/Ca [213], C_{60}/Ag [4, 145], C_{60}/Au [214], C_{60}/Pt [214, 215], $C_{60}/$indium tin oxide (ITO) [216] interfaces were demonstrated later.

8.1.2. Solar Cells with a $C_{60}/$Inorganic Semiconductor Heterojunction

A heterojunction between a C_{60} thin film and another semiconductor with a high rectifying ratio in the dark ($>10^4$ at ± 2 V) [217] and photovolt-age generation [218] were demonstrated, first, for a C_{60}/p-Si interface. The properties of such a heterojunction were then extensively studied in the dark and under irradiation of various light sources [4, 145, 146, 219–223]. Although different groups studied various device configurations ($Nb/C_{60}/Si$, $Ti/C_{60}/Si$ [217, 219], $Au/C_{60}/Si$, $ITO/C_{60}/Si$ [218, 221], $Al/C_{60}/Si$ [4, 145, 146, 220]) all authors seem to agree that potential barrier formation at the C_{60}/p-Si rather than at metal/C_{60} interface is responsible for the strong rectifying properties of the heterostructures. X-ray photoelectron spectroscopy (XPS), UV photo-electron spectroscopy (UPS) [224, 225], low-energy electron diffraction (LEED) and Auger electron spectroscopy (AES) [225] have revealed an abrupt character of the C_{60}/Si(111) interface and no chemical reaction or diffusion between the C_{60} overlayer and Si(111) substrate.

Later, a rectifying heterojunction was reported for the C_{60}/n-Si [221], C_{60}/n-GaAS [226] and C_{60}/n-GaN [227] interfaces as well.

Dark current–voltage (*I–V*) curves of a Ti/C_{60}/p-Si/Al device were studied in Ref. 219. Polycrystalline *fcc* C_{60} film with a thickness of 200 nm was deposited by a vacuum deposition technique onto a chemically terminated (to remove any surface oxide) substrate of (111) p-type Si (2–4 Ω cm) with a back ohmic contact of Al. Finally, Ti electrode dots of area 5.03×10^{-3} cm^2 were deposited on the C_{60} surface. The device was found to be forward-conducting when the Si substrate was positively biased relative to the C_{60} film, the rectification ratio being greater than 10^4 at ± 2 V. Similar results were reported for Nb/C_{60}/p-Si/Al [217] and Al/C_{60}/p-Si/Al [146, 220, 221] device configurations.

It was demonstrated by measurements of capacitance–voltage (C–V) characteristics of heterojunctions between a Si-substrate and C_{60} films of various thickness (from 200 to 900 nm), that the entire C_{60} films and near-interface region of the Si wafer behave as a depletion region. This means that an internal electric field necessary for a photovoltaic effect exists over the entire thickness of the C_{60} films [228]. This fact is very important for fullerene-based solar cells because of excitonic mechanism of the carrier photogeneration in solid C_{60} and the limited exciton diffusion (see Section 7).

Photovoltage generation of C_{60}/p-Si heterojunction was demonstrated by Wen et al. [218]. C_{60} film with a thickness of 1500 nm was vapor-deposited in vacuum on a chemically cleaned p-Si substrate (20 Ω cm) with a back ohmic contact of Au. Finally, a front ITO electrode was sputtered on the C_{60} surface. An Au front electrode (transparent to light) was also used for the measurements to confirm that there is no photoelectric effect of the C_{60}/p-Si interface. A difference between the results with ITO and Au electrodes was not observed. Although the rectification ratio in the dark of the ITO/C_{60}/p-Si/Au was not reported, the direction of the rectification was the same as that reported in [4, 217, 219–221]. The V_{oc} value was found to be about 0.2 V under irradiation by a 150 W Xe lamp (longer than 200 nm in wavelength). Unfortunately, other solar cell parameters as well as the active area of the device were not reported.

Comprehensive studies of the photovoltaic properties were performed for C_{60} thin film/p-Si heterojunction solar cells [4, 145–147]. Their dark *I–V* characteristics exhibited strongly rectifying properties: the rectification ratio being approximately 10^4 at ± 2 V. Under the AM 1.5 solar irradiation, the short-circuit current density, J_{sc}, was found to be 42 μA/cm^2, open-circuit

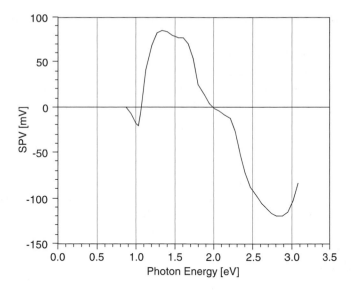

Fig. 13. SPV spectrum of an Al/C$_{60}$/p-Si/Al heterojunction solar cell. (Reprinted from Ref. 145, with permission from Materials Research Society © 1997).

voltage V_{oc} = 322 mV, *FF* = 0.3. The low values of J_{sc} and *FF* were attributed to the high resistivity of pristine C$_{60}$ films. SPV spectroscopy characterization of the devices (Fig. 13) suggested that the C$_{60}$ film acts as an active layer in the device, in particular for the conversion of short-wavelength components of sunlight. Indeed, it is evident that the positive SPV-signal at 1.1 eV originates from the valence band-to-conduction band transition in p-Si while the signals at 1.6 eV and 2.25 eV are attributable to the SPV transitions in the n–C$_{60}$ layer (see Section 6.1 and Fig. 8). This conclusion is consistent with the experimental spectral response of similar Al/C$_{60}$/p-Si/Al solar cells [220].

The maximum V_{oc} of a conventional inorganic heterojunction corresponds to the effective barrier height of the heterojunction $q\varphi_{eff}$ that depends on the difference in the Fermi-levels of both semiconductors forming the junction, in this case, of C$_{60}$ thin film and Si. The Fermi-level of Si can be controlled by changing the type and concentration of dopant. In Ref. 221, photovoltage generation by a C$_{60}$/Si heterojunction was studied with four different types of Si substrates: heavily doped p-Si (8×10^{18} cm^{-3} of B), lightly doped p-Si (1.5×10^{15} cm^{-3} of B), lightly doped n-Si (5×10^{14} cm^{-3} of P), and heavily doped n-Si (5×10^{18} cm^{-3} of P). The lightly doped n-Si/C$_{60}$ junction was measured to have the highest V_{oc}, 0.4 V, while the lightly doped p-Si/C$_{60}$ junction was measured to have the lowest, 0.12 V. This result implies that the Fermi-level of C_{60} should be more than 0.4 eV below the Fermi-level of lightly doped n-Si and more than 0.12 eV above that of lightly doped p-Si. The Fermi-level of C_{60} was estimated to be about 4.7 eV below the vacuum level. In our opinion, further research is needed to understand the true electronic structure and photogeneration mechanisms in such heterojunctions.

The restriction of the photoactive region thickness by low values of the exciton diffusion length in fullerene films in comparison with the diffusion length of free carriers in inorganic semiconductors (as discussed in Section 7) and the very low intrinsic conductivity of C_{60} crystals constitute the main limiting factors for the solar cell efficiency. Therefore, nanostructuring of photoactive junctions and doping of C_{60} layers should be developed for production of high efficiency fullerene-based solar cells.

8.1.3. Nanostructured Fullerene/inorganic Semiconductor Heterojunctions

Key features responsible for intensive investigations of semiconductor nanoparticles (NP) include (1) the high surface to volume ratio and (2) tunability of their optical and electrical properties by utilizing size quantization effects [229]. Accordingly, high solar cell efficiency is expected if the NP can be prepared as rough nanostructured layers.

Arici et al. reported a preparation and characterization of solar cells with a flat interface heterojunction between the C_{60} derivative and nanocrystalline film of $CuInS_2$ (CIS) [230]. The latter is a classical photovoltaic material with a high absorbance coefficient ($\alpha \sim 10^5\,cm^{-1}$) and photoconductivity. The CIS NPs with size of about 8 nm were synthesized by a colloidal route. Thin (~200 nm) film of the nanocrystalline CIS was deposited by spin coating onto a glass substrate precoated subsequently with an ITO electrode and a layer of poly (3,4-ethylene-dioxythiophene)/poly (styrenesulfonate) (PEDOT/PSS) [231] of about 100 nm thickness. PEDOT/PSS is widely used in organic electronics in order to smoothen the ITO surface and to improve the hole injection at the ITO anode. The absorbance spectra of the nanocrystalline CIS films exhibited a high shoulder at 700 nm indicating a significant blue-shift, relative to that of the bulk material (at 825 nm), due to the size quantization. The heterojunction was formed between the CIS and such soluble C_{60} derivative, as [6,6]-Phenyl C_{61}-butyric acid methylester (PCBM) (Fig. 7a) [100]. PCBM is widely used in the film and device preparation by solution methods due to its enhanced solubility compared to pure C_{60} (see Section 9.1.2). PCBM was spin-coated onto the CIS layer and then a very thin (0.6 nm) film of LiF and an Al counter-electrode (300 nm) were evaporated to complete the solar cell structure. LiF/Al bi-layer contact was chosen instead of pristine Al in order to guarantee a good ohmic contact between the metal and PCBM. Recently, it was shown that insertion of a thin LiF interlayer significantly enhances electron injection from an Al electrode to PCBM [232, 233]. Measuring short-circuit photocurrent spectra (incident photon-to-current conversion efficiency, IPCE) of ITO/CIS/PCBM/LiF/Al cells in comparison to those for the devices with a single photoactive layer (ITO/CIS/LiF/Al) the authors clearly demonstated the effect of the fullerene/CIS heterojunction. The integrated IPCE value increases by a factor of 25. Current–voltage curves of the fullerene/CIS heterojunction cells, measured under $80\,mW/cm^2$ white light by a Steuernagel solar simulator, are characterized by $V_{oc} = 792\,mV$, $J_{sc} = 0.26\,mA/cm^2$, $FF = 0.44$ and $\eta = 0.09\%$.

Guldi et al. used NPs of another AIIBVI semiconductor, CdTe (for review, see the chapter by V.P. Singh et al. in the present volume), in order to build a more sophisticated fullerene-based photovoltaic structure with layer-by-layer assembled heterojunctions [101]. Nanometer-sized hybrid assemblies were demonstrated to be formed by van der Waals interactions between a water-soluble C_{60} derivative, *trans-3* C_{60} (Fig. 7b), and three different size-quantized CdTe NPs – green (2.4 nm), yellow (3.4 nm), and red (5.0 nm). Based on this result, the authors suggested a novel layer-by-layer (LBL) technique for preparation of sandwich-like NP/C_{60} multilayer nanostructures (Fig. 14). Main LBL deposition cycles include subsequent dipping of a chemically pretreated substrate (ITO, in this case) in solutions of the NPs and C_{60} derivative. Despite some limitations imposed by the interpenetration of the

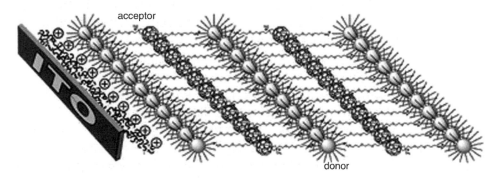

Fig. 14. Schematic illustration of $\{NP/C_{60}\}_n$ multijunction cell grown on a chemically pretreated ITO substrate. (Reprinted from Ref. 101, with permission from the Royal Soc. of Chemistry © 2002).

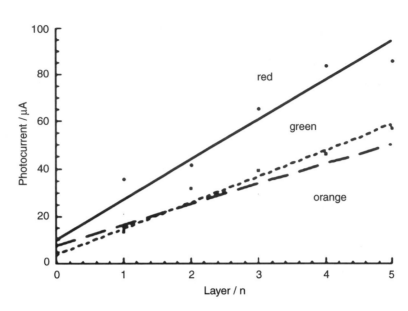

Fig. 15. Layer dependence of photocurrent for $\{NP/C_{60}\}_n$ multijunction cell grown on a chemically pretreated ITO substrate, with n ranging from 1 to 5, under deoxygenated photoelectrochemical conditions (i.e., an aqueous solution containing 0.1 M NaH_2PO_4 as supporting electrolyte and 50 mM ascorbic acid – no bias voltage applied). (Reprinted from Ref. 101, with permission from the Royal Soc. of Chemistry © 2002).

adjacent layers, this method constitutes a potential alternative to traditional thin film techniques in fabricating photoactive molecular devices due to its simplicity and universality combined with high quality of films. A solid state photovoltaic device was not fabricated but photocurrent generated by the ITO/$\{NP/C_{60}\}_n$ multilayer structure was characterized in a photoelectrochemical cell under deoxygenated conditions with 1 mM ascorbate/0.1 M NaH_2PO_4 solutions. The most interesting result of this investigation is a linear increase of the photocurrent of the $\{NP/C_{60}\}_n$ device with number of heterojunctions n ranging from 1 to 5 (Fig. 15). The highest IPCE of the NP/C_{60} monolayer device was of 1.7%. Strong amplification of the IPCE values was revealed for $\{NP/C60\}_n$ – with n ranging from 1 to 5 – yielding the highest IPCE value of 5.4% for red NPs with $n = 5$. For orange and green NPs the highest ICPE values were as large as 3.0% and 3.5%, respectively. Further optimization of structure and morphology of the multijunctions is in progress. In the author's opinion, this strategy makes possible the surpassing the performance records reported for molecular photovoltaic systems.

An increase in the photoactive heterojunction area using deposition of on/into porous substrates is another approach for the efficiency improvement. In order to use this idea, Feng and Miller deposited a C_{60} monolayer on a nanostructured substrate of p-type porous silicon [234]. The authors reported an increase in J_{sc} (by 100 times) and V_{oc} in comparison with the corresponding values for a planar configuration of C_{60}/p-Si heterojunction. Unfortunately, the absolute values of these parameters were not reported.

Senadeera and Perera produced a solar cell by electro-depositing of C_{60} layer ($\sim$1000 nm) onto a nanoporous TiO_2 film followed by depositing of a p-type CuSCN layer from solution [235]. The charge transferring mechanism is described as the formation of a C_{60} anion from the excited C_{60} molecule by donating a hole to CuCNS and then, injecting the electron from the C_{60} anions into the conduction band of TiO_2 (CuSCN is known to be used as a hole collector in dye-sensitized solid state photovoltaic cells [236]). Under a white light of 26 mW/cm^2, the device delivered $J_{sc} = 225$ μA/cm^2 and $V_{oc} = 350$ mV and was characterized by $FF = 0.4$ and $\eta = 0.12$ %. The highest IPTCE of $\sim$17% was observed at about 480 nm.

8.2. Photovoltaic Devices Based on the Doped C_{60}-Films
8.2.1. Ion Implantation into C_{60} Thin Films

Rectifying device structures based on doped C_{60} films were reported, for the first time, in Ref. 222 and 237 where an ion implantation of phosphorus (P) into C_{60} films was demonstrated to enhance their n-type conductivity. The same group observed p-type conductivity in C_{60} films doped with an Al by its simultaneous sputtering during the film deposition [222]. Photovoltaic parameters of C_{60}/Si solar cells were found to be improved by doping. In our opinion, the authors' conclusion about such p-doping of C_{60} by Al is quite speculative because of the known fact of the very high electron affinity of C_{60} molecule.

The idea of ion implantation of P and B in C_{60} films was then developed and applied to solar cell production by Yamaguchi and coauthors [13–16]. Implantation of B$^+$ (with energy of 50–80 keV and dose of 10^{14} cm^{-2}) into C_{60} films, grown on an n-Si substrate, was shown to lead to structural changes from the crystalline C_{60} to amorphous carbon, and a dramatic increase in film conductivity. Hall effect measurements of the implanted films indicate p-type conduction. Boron implantation improved the parameters of the resultant p–C/n-Si solar cells [13]. Especially, the cell series resistance R_s was improved by the implantation from 35 to 370 Ω. J_{sc} for the implanted cells was found to be 0.33 mA/cm^2, $V_{oc} = 0.17$ V, $FF = 0.415$ and efficiency $\eta = 0.023\%$.

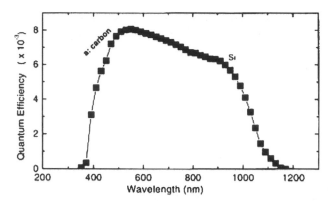

Fig. 16. Spectral response of a boron ion-implanted C_{60} (α–C)/n-Si heterojunction solar cell. (Reprinted from Ref. 13, with permission from the American Institute of Physics © 1999).

The spectral response of these cells (Fig. 16) includes two broad features: at 370–600 nm due to α–C layer and 700–1150 nm due to Si [13]. This result is in agreement with the experimental finding that the optical gap of the C_{60} films decreased after B^+ implantation [14]. Furthermore, it is possible to control the band gap by variation of the implantation dose. The optical gap was found to decrease gradually with the ion dose. For example, the gap was reported to be reduced continuously from 1.6 eV for nonimplanted C_{60} films to 0.8 eV after implantation with the dose of 8×10^{14} cm^{-2} [14]. The authors explained this effect by the structural transformation from C_{60} to α–C. The intermediate optical gap values can be attributed to a composite with various concentrations of C_{60} and α–C. In our opinion, this result is even more important than the improvement of the solar cell parameters (which are still lower than those for other α–C-based solar cells [238–240]). Indeed, using this effect, together with the control of the depth-implantation profile by the ion energy variation [15], it might be possible to produce a C_{60}/α–C composite with a variable depth-profile optimized for maximum sunlight absorption and cell efficiency. Combining this strategy with a feasibility of band gap engineering over a wide range, by control of sp^2/sp^3 ratios [17–18], and n- and or p–type doping [241] of an α–C can result in future in production of efficient all-carbon multijunction solar cells. The possibility of producing α–C/C_{60}/Si (p–i–n) solar cells has been already demonstrated [16]. The efficiency of these devices was found to be 0.1% under AM 1.5 irradiation which is five times higher than the best cells fabricated using the boron ion implanted fullerenes without the insulating layer. A similar α–C/C_{60}/Si (p–i–n) photovoltaic device with α–C layers,

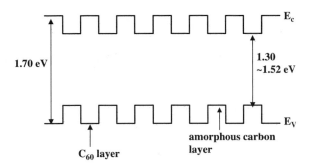

Fig. 17. Band diagram of C_{60}/ amorphous carbon supperlattice structure. (Reprinted from Ref. 243, with permission from the IEEE Inc. © 2000).

grown by rf plasma enhanced chemical vapour deposition and magnetron co-sputtering, has been demonstrated by Soga [242]. The author discussed the possibility of p–α–C/C_{60}/n–α–C all-carbon heterojunction solar cells for the future photovoltaic applications. Furthermore, a novel approach for all-carbon quantum well solar cells with a C_{60}/α–C supperlattice structure (Fig. 17), produced by an intermittent supply of nitrogen ions during C_{60} deposition, was recently suggested [243]. The periodic structure of the resulted films was confirmed by X-ray diffraction [244] while the UV-visible reflectance/transmittance spectroscopy revealed the energy shift of fundamental absorption edge with a variation of thickness of the α–C and C_{60} layers in the supperlattice structures [245]. Use of hydrogen plasma in C_{60} amorphization process, instead of nitrogen, reduces the defect density in the α–C layers by two orders of magnitude and improves dark and photoconductivity of the superlattices [246].

8.2.2. Solar Cells Based on C_{60} Films Intercalated with Metals

As we already mentioned (see Section 6.3) C_{60} films intercalated with Mg atoms exhibit semiconductor properties [186, 187] with enhanced n-type conductivity [247, 248]. Taima et al. used this effect in order to improve photovoltaic performance of Schottky-type cells with an ITO/C_{60}/Al structure [216]. C_{60} film of $\sim$100 nm thickness was grown, by vacuum evaporation, onto a glass substrate predeposited with an ITO Schottky-type electrode. An ohmic Al electrode was deposited onto the surface of C_{60} layer. The resulting ITO/C_{60}/Al device exhibited asymmetric *I–V* characteristics in the dark but did not show definite photovoltaic effects. A thin Mg interlayer (10 nm) was inserted at the C_{60}/Al interface to examine enhancement of the photovoltaic parameters. The ITO/C_{60}/Mg/Al cell showed, under illumination of AM1.5 ($100\,mW/cm^2$), $J_{sc} = 1.04\,mA/cm^2$, $V_{oc} = 0.40\,V$, $FF = 0.4$ and efficiency $\eta = 0.15$ %. The latter is the highest efficiency observed for inorganic C_{60}-based solar cells (see Table 2). These photovoltaic parameters were measured in air without any device encapsulation. If the cells were protected against oxygen they would show higher η. Based on the experimental results the authors concluded that (1) the improved photovoltaic performance of the device was originated from the doping of the C_{60} layer by diffusion of Mg during the Mg layer deposition; (2) the charge separation region was near to the ITO/C_{60} interface; and (3) the Mg doping did not cause considerable changes in the absorption spectra of the C_{60} layer. Recently the same group, using secondary-ion mass spectroscopy with depth profiling and conductivity measurements, directly demonstrated that Mg atoms strongly diffuse into C_{60} film during Mg deposition [247] increasing the dark conductivity by two orders of magnitude [248].

Intercalation of C_{60} film by Mg was also shown to improve markedly (by 360 times) the efficiency of organic solar cells based on a heterojunction between C_{60} film layer and conjugated polymer MEH-PPV (see Section 9.1.4) [248]. All these results imply that the efficiencies of a variety of fullerene-based solar cells could be improved by using such doping technique.

Some improvement of photovoltaic properties of Schottky-type ITO/C_{60}/Al solar cells was demonstrated by doping of C_{60} layer with Ni [249] and In [216]. However, the cell parameters were considerably lower than those for the Mg intercalated devices.

8.2.3. Solar Cells with an Oxygenated C_{60} Film

All of the research attempts described in the previous two sections were aimed at the increasing the low intrinsic conductivity of C_{60} films by doping. If one uses a fullerene film as an insulator layer in the MIS solar cells, the opposite problem should be solved. Indeed in this case decrease of the fullerene film conductivity by oxygenation (see Section 6.2) will increase the solar cell efficiency. Yang and Mieno [250] have suggested using this effect in MIS solar cells with Au/C_{60}/$C_{60}O_x$/Au structure in which C_{60} and $C_{60}O_x$ films act as a semiconductor and insulator layer, respectively. Highly oxygenated fullerene films ($C_{60}O_x$, with $x = 8, 16, 24, 38, 50$) were obtained using RF O_2 plasma with a cooling system. The experimental results revealed that the highest photocurrent was obtained when the total thickness of the oxygenated fullerene film and the pure fullerene film was about 200 nm. Under this condition, the photocurrent was observed to become 30 times higher than that without the $C_{60}O_x$ layer. Unfortunately, the absolute values of the photovoltaic parameters of these MIS solar cell have not been reported.

9. ORGANIC C_{60}-BASED PHOTOVOLTAIC DEVICES

9.1. Fullerene-Polymer Solar Cells

9.1.1. Photoinduced Electron Transfer from Conjugated Polymers onto Fullerenes

The discovery of semiconducting, conjugated polymers and the ability to dope these polymers over the full range from insulator to metal has resulted in the creation of a new class of materials that combines the electronic and optical properties of semiconductors and metals with the attractive mechanical properties and processing advantages of polymers [251]. Molecular structures of some semiconducting conjugated polymers are summarized in Fig. 18.

Conjugated polymers in their undoped states are electron donors upon photoexcitation (electrons being promoted to the antibonding π^* band). If one uses this property in conjunction with a molecular electron acceptor, long-lived charge separation may be achieved. Once the photoexcited electron is transferred to an acceptor unit, the resulting cation radical (positive polaron) species are known to be highly delocalized and stable. On the other hand, as already mentioned above, the C_{60} molecule is an excellent electron acceptor capable of taking on as many as six electrons [12]. In the beginning of the 1990s, independently, Heager's group [252] and Yoshino's group [253] reported the discovery of a photoinduced electron transfer from semiconducting conjugated polymers onto C_{60} molecules in solid films. The observed phenomenon is schematically illustrated by Fig. 19. The kinetics of this process has recently been time-resolved to occur within 40 fs [254]. Since the time scale of this process is more than 10^3 times faster than the radiative or nonradiative decay of photoexcitations, the quantum efficiency for charge transfer and charge separation is estimated to be close to unity [255]. All of these findings provided a new approach to low-cost and high-efficiency photovoltaics.

9.1.2. Bi-Layered and Bulk Heterojunction Fullerene-Polymer Solar Cells

Immediately after the discovery of photoinduced electron transfer from a conjugated polymer to a C_{60} molecule, this molecular effect was used for the preparation of a heterojunction between poly[2-methoxy, 5-(2'-ethyl-hexyloxy)-1,4-phenylene-vinylene] (MEH-PPV) (Fig. 18c) and a C_{60} thin film [256]. The dark *I–V* characteristics of such a device consisting of successive layers of ITO/MEH-PPV/C_{60}/Au revealed the rectification ratio at ± 2 V

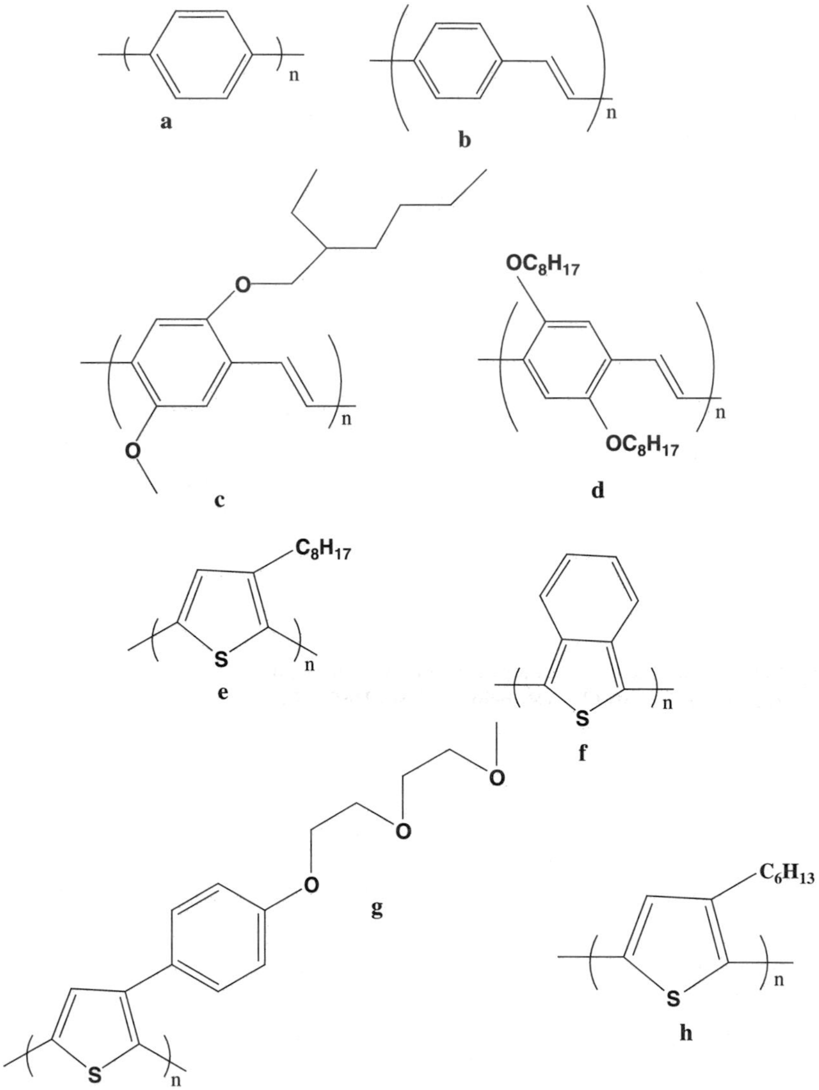

Fig. 18. Molecular structures of some semiconducting conjugated polymers used in fullerene-based solar cells. a – poly(paraphenylene) (PPP), b – poly(phenylene-vinylene) (PPV), c – poly[2-methoxy,5-(2′-ethyl-hexyloxy)-1,4-phenylene-vinylene] (MEH-PPV), d – poly(2,5-dioctyloxy-p-phenylene-vinylene) (OOPPV), e – poly (3-octylthiophene) (P3OT), f – poly(isothianaphthene) (PITN), g – poly[3-(4′-(1″,4″, 7″-trioxaoctyl)-penyl-thiophene] (PEOPT), h – regioregular poly(3-hexylthiophene) (P3HT).

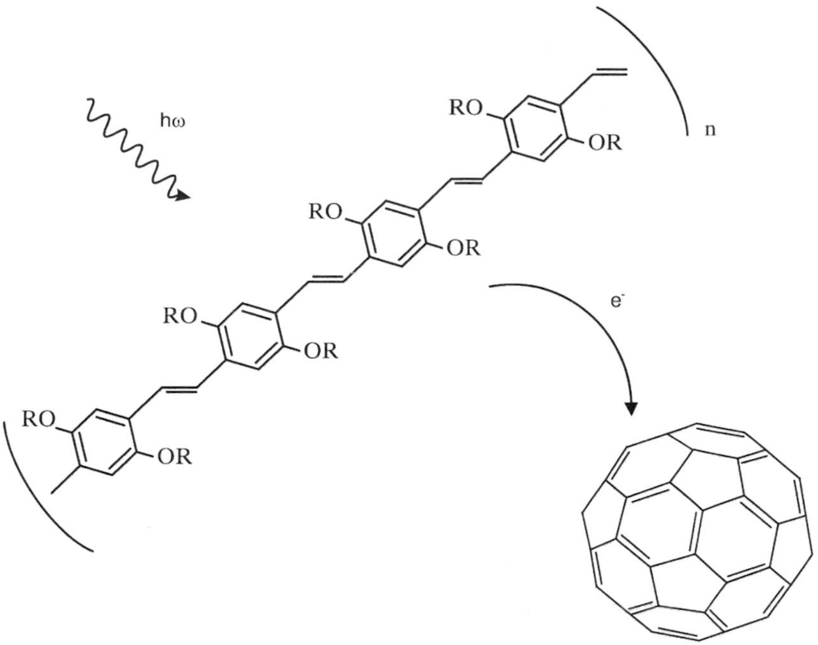

Fig. 19. Schematic illustration of photoinduced electron transfer from semiconducting conjugated polymers onto C_{60}. (Reprinted from Ref. 255, with permission from Elsevier Science Ltd © 1995).

of $\sim 10^4$. Analysis of this result suggests formation of a heterojunction at the MEH-PPV/C_{60} interface, with rectifying properties analogous to those of a semiconductor p–n junction. The device demonstrated photovoltaic behavior. The values of J_{sc} were reasonably linear over five decades of light intensity up to approximately 1 W/cm^2, i.e., one order of magnitude higher than the terrestrial solar intensity (AM 1.5). The following photovoltaic parameters (measured for a cell area of 0.1 cm^2 under illumination of an argon ion laser with wavelength $\lambda = 514.5$ nm and a light power of 1 mW/cm^2) were reported: $J_{sc} = 2\,\mu$A/cm^2, $V_{oc} = 0.5$ V, $FF = 0.48$ and $\eta = 0.04\%$.

Later, solar cells with a heterojunction between C_{60} thin films and donor layers of such other polymers, as (poly(paraphenylene) (PPP) (Fig. 18a) [257], poly(phenylene-vinylene) (PPV) (Fig. 18b) [198–199], poly(2,5-dioctyloxy-p-phenylene-vinylene) (OOPPV) (Fig. 18d) [258], poly(3-octylthiophene) (P3OT) (Fig. 18e) [259], poly(isothianaphthene) (PITN) (Fig. 18f) [260], poly(3-(4'-(1'',4'',7''-trioxaoctyl)penyl)-thiophene) (PEOPT) (Fig. 18g) [261]), were demonstrated as well.

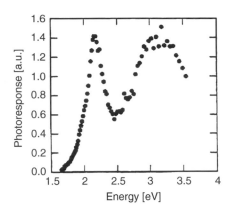

Fig. 20. Spectral response of the photocurrent in an ITO/MEH-PPV/C$_{60}$/Au solar cell at reverse bias of −1 V. (Reprinted from Ref. 256, with permission from the American Institute of Physics © 1993).

The main factor limiting photovoltaic parameters of these bilayered donor-acceptor solar cells originates from the excitonic mechanism of the photogeneration in these devices and the particular fact that the thickness of the photoactive layer is limited by the distance of exciton diffusion length from the donor-acceptor interface (see Section 7). The above statement is well illustrated by the spectral response of an ITO/MEH-PPV/C$_{60}$/Au solar cell with illumination through the ITO/MEH-PPV side of the device [256] (Fig. 20). The onset of photocurrent at a photon energy of 1.7 eV follows the absorption of MEH-PPV, which initiates the photoinduced electron transfer. However, the minimum in the photocurrent at a photon energy of 2.5 eV corresponds to the region of maximum absorption in the polymer layer. The MEH-PPV layer, therefore, acts as a filter which reduces the number of photons reaching the MEH-PPV/C$_{60}$ interface while the only very thin layer at the heterojunction interface contributes to the cell photocurrent. This results in a relatively low collection efficiency of such cells.

Significant improvement of the collection efficiency has been achieved using composite material with a network of internal heterojunctions between fullerene and conjugated polymer, forming *bulk heterojunctions*. Through control of the morphology of the phase separation into an interpenetration network, one can achieve a high interfacial area within the bulk material. If any point in the bi-continuous network is within an exciton diffusion length from a donor-acceptor interface, wherever an exciton is photogenerated in either material, it is likely to diffuse to the interface and break-up. Then the

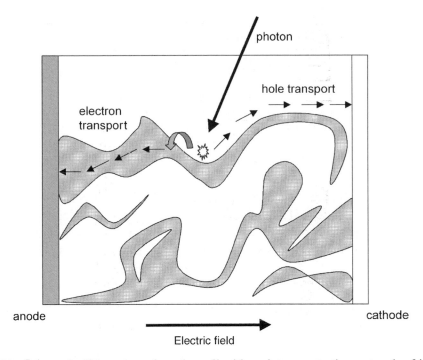

Fig. 21. Schematic illustration of a solar cell with an interpenetration network of internal heterojunctions (bulk heterojunction) (Reprinted from Ref. 197, with permission from Elsevier Science Ltd © 2002).

separated charge carriers may travel to the contacts and deliver current to an external circuit (Fig. 21).

Using this principle of the material construction, a solar cell based on MEH-PPV/fullerene derivative composite was demonstrated [262]. The authors believed that a built-in electric field was created by asymmetric metallic contacts (with different work functions), one of which was transparent ITO. The idea of such internal electric field generation in organic devices was first described by Parker [263]. In Ref. 262, uniform films of such soluble fullerene derivative, as PCBM (Fig. 7a) and MEH-PPV were cast from 1,2-dichlorobenzene solutions containing 1:4 weight ratio MEH-PPV: PCBM. The enhanced solubility of PCBM compared to pure C_{60} allows such high fullerene/conjugated polymer ratio and strongly supports the formation of bulk donor-acceptor heterojunctions. Using this film as a photoactive layer as well as Ca transparent ITO films as asymmetric electrodes, bulk heterojunction solar cells, with $J_{sc} = 0.5\,\text{mA/cm}^2$ under 20mW/cm^2 monochromatic illumination, were demonstrated. This J_{sc} value corresponds to a collection efficiency of 7.4% electrons per incident photon which is about 2 orders of magnitude higher than that of the bilayered MEH-PPV/C_{60} solar cells described in the previous section.

Starting with Ref. 262, substantial progress has been made toward efficient solar cells based on fullerene/conjugated polymer blends [231, 232, 264–274]. In particular, a significant breakthrough has been achieved by realizing that the morphology of the photoactive composite layer plays an important role for charge carrier mobility [264, 275] and solar cell performance (Table 2) [232, 264, 268, 269]. In particular, Shaheen et al. [264] demonstrated the efficiency of 2.5% under AM 1.5 irradiation for the following device configuration. Poly [2-methoxy,5-(3′,7′-dimethyl-octyloxy)]-p-phenylene-vinelyne (MDMO-PPV) [233] was used as the electron donor in these cells while the electron acceptor was PCBM. The thickness of the spin-cast MDMO-PPV:PCBM active layer with the improved morphology was about 100 nm. As electrodes, a transparent ITO/ PEDOT on one side and a LiF/Al bi-layer contacts on the other side were used. The device structure of the ITO/PEDOT/MDMO-PPV:PCBM/LiF/Al solar cell is shown in Fig. 22. The reported photovoltaic parameters of such solar cells under AM 1.5 irradiation, at 50°C, were as $J_{sc} = 5.25$ mA/cm^2, $V_{oc} = 0.82$ V, $FF = 0.61$, $\eta = 2.5\%$ (Table 2). These high parameters have been approved by accurate in-door [273] and out-door [274] photovoltaic characterization of such solar cells, the results of which had been appropriately adjusted to Standard Test Conditions (STC). These correspond to a radiant intensity of 100 mW/cm^2 with a spectral distribution defined as "AM 1.5G" (IEC 904-3) and a cell temperature of 25°C. The I–V characteristics of a ITO/PEDOT/MDMO-PPV: PCBM/LiF/Al solar cell measured under appropriately corrected simulated STC conditions are shown in Fig. 23 [273]. The correction [273–274] is based on the determination of the simulator spectral mismatch factor using spectral responses of the cell under test and of the reference cell (Fig. 24) as well as the simulator spectrum and the AM1.5G standard solar spectrum. In spite of the existence of such a standard, all kinds of efficiencies have been reported for fullerene solar cells, based on measurements performed under a wide variety of test conditions (Table 2).

Since V_{oc} of fullerene/polymer bulk heterojunction solar cells is related to the energy difference between the LUMO level of the acceptor and the HOMO level of the donor components of the active layer, Brabec et al. [102, 276] compared two new fullerene derivatives, ketolactam and azafulleroid (Fig. 7c, d), to C_{60} and PCBM, as acceptors of different strengths in bulk heterojunction solar cells of similar device architecture and morphology of

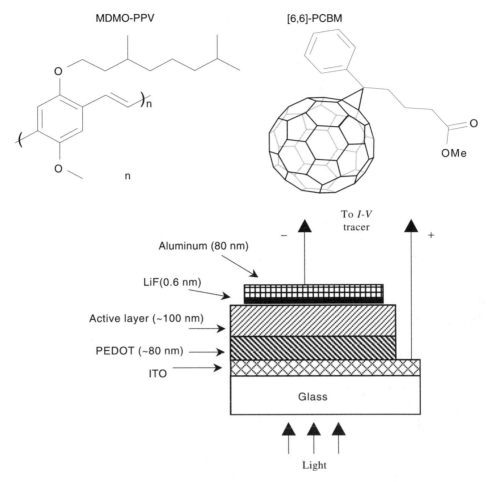

Fig. 22. Device structure of an ITO/PEDOT/MDMO-PPV:PCBM/LiF/Al solar cell, together with chemical structure of compounds used for the cell active layer. (Reprinted from Ref. 274, with permission from the American Institute of Physics © 2001).

the photoactive layer. It was found that PCBM, which has the lowest electron affinity, i.e., acceptor strength, resulted in the highest V_{oc} values. For further increase of the V_{oc}, new acceptors with lower electron affinity may help.

The only location where substantial recombination can occur in a bulk heterojunction solar cell is at the exciton-dissociating donor-acceptor interface, since this is the only location where substantial numbers of electrons and holes coexist [202]. Bulk recombination, the major recombination process in conventional solar cells, can usually be neglected in such photovoltaic devices

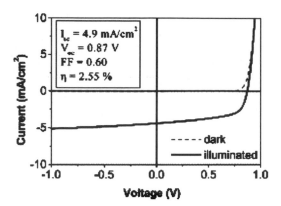

Fig. 23. The *I-V* characteristics of a 0.1 cm² ITO/PEDOT/MDMO-PPV:PCBM/LiF/Al solar cell measured under appropriately corrected simulated STC conditions. (Reprinted from Ref. 273, with permission from Elsevier Science Ltd © 2002).

because the bulk density of minority carriers is insignificant. Therefore, all else being equal, the greater the interface area, the faster the interfacial recombination rate. This trade-off between enhancing exciton dissociation by increasing interfacial surface area and enhancing recombination (see Section 7) is a fundamental controversy of bulk heterojunction solar cells that limits J_{sc} and efficiency of the devices. The following research strategies have been suggested in order to overcome this limit.

(1) *Additional improvement of the morphology of the blended active layer through the controlled postdeposition annealing of the cells.* Recently Reyes–Reyes et al. [277] clearly demonstrated strong positive effect of such annealing on the photovoltaic parameters of bulk heterojunction solar cells with the blended active layer of PCBM and poly(3-hexylthiophene) (P3HT). It should be noted that P3HT (Fig. 18h) is one of the best candidates for the polymer donor moiety in such devices because of its high electron mobility, negligible recombination losses in the blends with PCBM and relatively low band gap [278]. Since J_{sc}, *FF* and even V_{oc} are known to depend considerably on fullerene content in the blend [279–280], Reyes–Reyes et al. [277] determined optimal concentrations of PCBM and P3HT, for unannealed devices at a specific photoactive film thickness, and then examined the annealing effect in the "optimally filled" devices. In the unannealed devices, an optimal PCBM:P3HT ratio of 0.8:1 was found in rough agreement with other sources [280]. For the unannealed ITO/PEDOT/P3HT:PCBM/LiF/Al cells with the optimal P3HT:PCBM ratio the photovoltaic parameters,

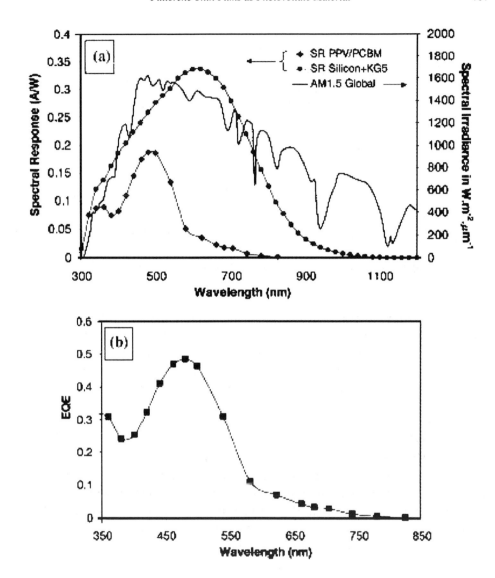

Fig. 24. (a) Spectral response (*SR*) of an ITO/PEDOT/MDMO-PPV:PCBM/LiF/Al solar cell and of a filtered (KG5) silicon reference cell as well as the AM1.5G solar spectrum; (b) the External Quantum Efficiency (*EQE*) of an ITO/PEDOT/MDMO-PPV:PCBM/ LiF/Al solar cell as calculated from the *SR*. (Reprinted from Ref. 273, with permission from Elsevier Science Ltd © 2002).

measured under AM1.5G irradiation of $80\,mW/cm^2$, were: $V_{oc} = 0.65\,V$, J_{sc} = $3.86\,mA/cm^2$, $FF = 0.34$ and $\eta = 1.11\%$. Then the cells were annealed at 80°C, 105°C, and 155°C for different times. The best parameters were observed for the annealing at 155°C for 5°min. They were: $V_{oc} = 0.64\,V$, $J_{sc} = 11.1\,mA/cm^2$, $FF = 0.55$ and $\eta = 4.9\%$. This is, to the best of our knowledge, the highest efficiency reported for the fullerene/polymer bulk heterojunction solar cells. Similar efficiency has recently been announced by Brabec et al. [281]. Reyes–Reyes et al. [277] attributed the improved photovoltaic performance of the annealed cells to the changes in the photoactive film crystallinity and aggregation within the PCBM nanophase. Positive effect of postgrowth treatment on the morphology of PCBM:P3HT blends and the cell performance (but with lower efficiency, see Table 2) was also reported in Ref. 282.

(2) *Bulk gradient heterojunction* (see below Section 9.1.3).

(3) *Preparation of "double cable" polymers, i.e., supramolecules, in which the polymeric electron donors and the fullerene electron acceptors are covalently linked* [283–286]. In this supramolecular photovoltaic approach, the desired nanoscopic bi-continuous network have been achieved by synthesizing novel π-conjugated polymers with pendant fullerenes that prevents any problem arising from interface defects, as observed for fullerene/polymer blends. Photovoltaic measurements of the solar cells based on the "double cable" polymer under white-light illumination ($100mW/cm^2$) revealed J_{sc} value of $0.42\,mA/cm^2$, V_{oc} of $830\,mV$ and FF of 0.29 [286]. Further research is required in order to optimize the efficiency of the devices.

Architecture and photovoltaic performance of bulk heterojunction solar cells are summarized in a number of detailed reviews [255, 284, 287, 288] including the chapters by Hoppe and Sariciftci and Nishikitani et al. in the present volume. In the remainder of the present chapter we will focus only on multilayered cells containing a fullerene film as one of the active layers (in some cases, in combination with bulk heterojunctions) or those prepared from multilayered structures containing a fullerene film (see the next section).

9.1.3. Solar Cells with Gradients of Fullerene and Polymer Concentrations

Drees et al. [289, 290] suggested another approach for improvement of the morphology of the solar cell active layer and the following enhancement of the charge transfer and improvement of the charge transport. In this study, a bilayer system, consisting of spin-cast MEH-PPV or P3OT and sublimed C_{60} layers, was heated at elevated temperatures in order to induce the thermally controlled interdiffusion of the fullerene and polymer layers, resulting in a concentration gradient structure. With this process, a controlled, *bulk gradient heterojunction* was created. Because the fullerene acceptor was thus distributed throughout the film, exciton recombination was dramatically reduced. This was observed as a decrease in the photoluminescence by an order of magnitude and an increase in the photocurrent by an order of magnitude throughout much of the visible spectrum [289]. The optimal conditions of the annealing include heating of C_{60} (80 nm)/P3OT (160 nm) heterostructure at 130°C for 5 min [290]. This treatment resulted in the intended concentration gradient of donor and acceptor throughout most of the active layer and monochromatic (470 nm) efficiency of the resultant cell of 1.5% (see Table 2). Unfortunately, the photovoltaic parameters of such solar cells measured under the proper conditions were not reported. Recently, Hayashi et al. reported similar results [290]. Heating of C_{60}/MEH-PPV bi-layer at 80°C led to the donor-acceptor interdiffusion and yielded $\eta = 0.2\%$ for the TiO$_2$/PEDOT/ C_{60}:MEH-PPV/Al cells irradiated by AM 1.5 illumination (100 mW/cm^2).

9.1.4. Advanced Structures with Bi- and Multilayered Fullerene-Polymer Heterojunctions

Effect of doping of C_{60} film by Mg on the photovoltaic performance of simple bilayer C_{60}/MEH-PPV solar cells was investigated in Ref. 248. Mg diffusion to C_{60} film was found to occur during the deposition of the Mg top electrode onto the C_{60} layer. The solar cells were markedly improved by the doping. η of the Mg-doped device under 100 mW/cm^2 white light illumination was 0.54% (see Table 2 for other cell parameters), which was approximately 400 times larger than that of the nondoped device (mostly due to the J_{sc} and *FF* enhancement). We already mentioned the stable n-type semiconductor doping of fullerite by Mg (Section 6) and its strong positive effect on the efficiency of inorganic fullerene-based solar cells (Section 8). Below (see Section 9.2) we will discuss the effect of doping of C_{60} film on the efficiency of multilayered solar cells with small organic molecules.

Yoshino's group has demonstrated a high monochromic external quantum efficiency, of over 70% at the peak wavelength, and $\eta = 1.0\%$ under AM 1.5 illumination (other cell parameters are listed in Table 2) for a cell based on a C_{60}/P3HT heterojunction [291–292]. The cell with structure ITO/ ZnO/C_{60}/P3HT/Au was fabricated by spin-coating a chloroform solution of P3HT onto the C_{60} thin film (50–150 nm) thermally evaporated in vacuum on ZnO-coated ITO. The improved photogeneration in the device was attributed to the fact that C_{60}/P3HT heterojunction fabricated using this method differs from a flat interface produced by depositing C_{60} layer onto a polymer film. During spin-coating of the chloroform solution of P3HT onto the C_{60} thin film, the latter is dissolved slightly, resulting in the increase of the active interface area. The ZnO layer also plays an important role increasing, to the authors's opinion, a built-in potential in the devices.

Stübinger and Brütting developed another approach for improvement of the efficiency of bilayer heterojunction cells [293]. They investigated the influence of the layer thickness of the electron donor materials on the photocurrent spectra and efficiency of the ITO/PPV/C_{60}/Al cells. By systematically varying the layer thickness of the C_{60} layer the authors proved the effect of an optical interference effect due to a superposition of the incident light with that reflected from the Al electrode. Thickness of the donor and acceptor layers was shown to influence significantly not only the photocurrent spectra but also the efficiencies of these heterolayer devices. With optimized donor and acceptor layer thickness, $\eta > 0.5\%$ measured under AM 1.5 illumination was achieved even for this, extremely simple, cell structure.

Yoshino et al. [287] have suggested the insertion of an additional *middle excitonic layer* between the polymeric donor and fullerene acceptor layers, in a three-layered molecular solar cell which is reminiscent of a p–i–n structure in inorganic solar cells. To test this idea, the authors fabricated a three-layered cell utilizing OOPPV, octaethylporphine (OEP) and C_{60} films as donor, middle excitonic and acceptor layers, respectively. Photocurrent yield spectra, measured under irradiation from different sides of the cell, were interpreted by light absorption mainly in the OEP layer, exciton migration and charge generation at both heterojunctions. Both interfacial regions of OOPPV/OEP and OEP/C_{60} were found to contribute to the charge generation by excitonic dissociation. Photocurrent enhancement and broadening of spectral sensitivity were observed for the OOPPV/OEP/C_{60} solar cells, as compared with the OEP single layer and OOPPV/C_{60} bilayered solar cells. The same research team suggested an idea of fullerene/conjugated polymer multilayered solar cells with quantum well structure (Fig. 25) [294].

In our opinion, serious experimental research is still needed for further clarification of the advantages of either the double-heterojunction or the multilayered solar cells, in comparison with the most efficient bulk heterojunction devices.

9.2. Solar Cells Based on C_{60}/Phthalocyanine Heterojunctions

The first solar cell with a heterojunction between C_{60} thin films and donor layers of metal phthalocyanines was demonstrated in 1996 [295]. Starting

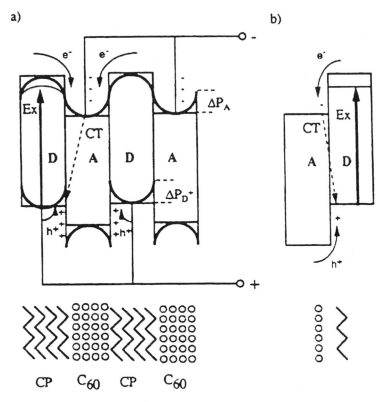

Fig. 25. Schematic diagram of fullerene/conjugated polymer multilayered solar cells with quantum well structure. (Reprinted from Ref. 294, with permission from Elsevier Science Ltd © 1995.)

with this work, a wide variety of metal phthalocyanines were used to build fullerene-based heterojunction solar cells [296–311] (Fig. 26).

The Princeton group has experimentally demonstrated C_{60}/copper phthalocyanine (CuPc) bilayered solar cells, incorporating an *exciton-blocking layer*. These devices showed efficiency of 3.6% under AM 1.5 spectral illumination of $150\,mW/cm^2$ (1.5 suns) [297]. The cells were fabricated on glass substrates predeposited with an ITO anode contact. The ITO film was then spin-coated with an $\sim 320\,$Å-thick film of PEDOT:PSS. Then the photoactive layers were grown at room temperature in high vacuum ($\sim 10^{-6}$ Torr) in the following sequence: a 50–$400\,$Å-thick film of the donor-like CuPc, followed by a 100–$400\,$Å-thick film of the acceptor-like C_{60}. Next, a 50–$400\,$Å-thick exciton-blocking layer of bathocuproine (BCP) was deposited. Finally, an Al cathode was deposited by thermal evaporation.

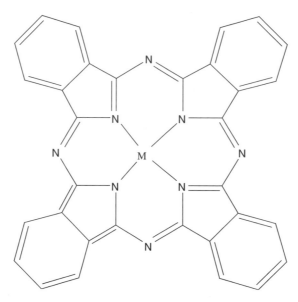

Fig. 26. Molecular structures of phthalocyanines (MPc). Fullerene-based solar cells have been demonstrated with MPc, where M = Cu [297, 302, 308–310], TiO [295], Zn [296, 299, 303–307, 311].

(Previously [312], a large band gap ($E_g > 3$eV) and, therefore, a transparent BCP layer had been found to transport electrons to the cathode from the adjoining acceptor layer while effectively blocking excitons in the lower-energy-gap acceptor layer from recombining at the cathode.) Recently, Heutz et al. argued that the positive effect of BCP is mainly due to its mechanical role in protection of the active layer form Al diffusion [308]. Highest photovoltaic parameters were observed for the ITO/PEDOT: PSS/200 Å CuPc/400 Å C_{60}/120 Å BCP/Al solar cells. At an illumination intensity of 150 mW/cm², the values of $\eta = 3.6\%$, $V_{oc} = 0.58$ V, of $I_{sc} = 18.8$ mA/cm² and $FF = 0.52$ were demonstrated. The specific series resistance of these cells was estimated as 6 Ω/cm². Then the same group succeeded in further improvement of the performance of similar cells by reducing their specific series resistance down to 0.1 Ω/cm² [302]. A high FF of 0.61 was achieved, which was only slightly reduced at very intense illumination. As a result, η was found to increase with the incident light intensity, reaching a maximum of ~4.2% under 4–12 suns simulated AM 1.5G illumination (Table 2). The cell structure was ITO/200 Å CuPc/400 Å C_{60}/100 Å BCP/Al, i.e., contrary

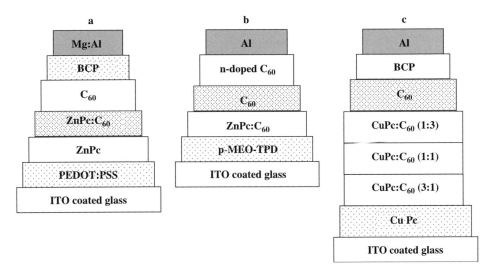

Fig. 27. Configuration of *p–i–n* solar cells based on C$_{60}$ and phthalocyanines. (a) Reprinted from Ref. 305, with permission from the American Institute of Physics © 2004; (b) reprinted from Ref. 306, with permission from Springer-Verlag (c) reprinted from Ref. 308, with permission from Elsevier © 2004.

to the cells described in Ref. 297, these devices do not include the PEDOT: PSS layer. On the other hand, the ITO anode was treated in UV-ozone for 5 min immediately before the CuPc vacuum deposition in order to improve hole injection from this electrode.

In a number of studies [303–308, 310] the C$_{60}$/MePc p–n heterojunction cells have been modified to p–i–n photovoltaic devices where films of pristine or doped C$_{60}$ and MePc or another donor molecular material constitute, respectively, n- and p-layers while a composite film of C$_{60}$ and MePc form a bulk-heterojunction-like i-layer (Fig. 27). The introduction of the i-layer can increase the practical width of a thin photoactive layer in excitonic solar cells however its thickness should not be higher than the diffusion length of photogenerated carriers. Taima et al. investigated effects of the i-layer thickness on the photovoltaic parameters of C$_{60}$/C$_{60}$:ZnPc/ZnPc p–i–n cells with a BCP exciton-blocking layer [305]. The device was produced by a successive vacuum deposition of the semiconductor layers on an ITO substrate with a spin-coated PEDOT:PSS layer (Fig. 27a). The i-layer was formed by a codeposition of ZnPc and C$_{60}$ with the volume ration of 1:1. The total p–i–n thickness was 50 nm. The thickness of the codeposition layer was changed from $x = 0$ nm (*p–n* heterojunction) to $x = 50$ nm (only i-layer). Dependence

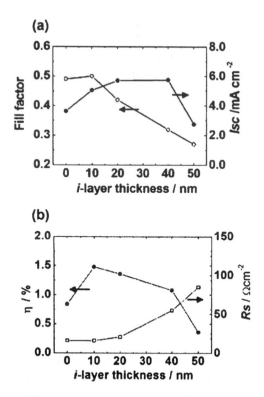

Fig. 28. Dependence of *FF* and J_{sc} (a) and R_s and η (b) for the ITO//PEDOT:PSS/ ZnPc/ ZnPc:C_{60}/C_{60}/BCP/Mg:Ag solar cells on *i*-layer thickness. (Reprinted from Ref. 305, with permission from the American Institute of Physics © 2004).

of the cell parameters measured under solar simulated light (AM 1.5, 100 mW/cm²) on the i-layer thickness is shown in Fig. 28. One could see that, for low *x* values, J_{sc} tends to increase with increasing the thickness while *FF* shows the opposite tendency. The latter could be attributed to the corresponding increase in R_s (Fig. 28b). V_{oc} was found to be reasonably independent on *x*. As a result, η exhibits a maximum (of 1.5%) at *x* = 10 nm. The optimized *x* seems to reflect the diffusion length of photogenerated carriers in the i-layer. It should be noted that such cell parameters (Table 2) were obtained in air without any cell protection against oxygen. The cell encapsulation could further improve its performance.

Maennig et al. further developed p–i–n approach using thermally evaporated doped C_{60} and *wide-gap* N,N,N′,N′-tetrakis(4-methoxyphenyl)-benzidine (MeO-TPD) as the best candidates for n- and p-layers, respectively

[306]. The fullerene film doping was realized by controlled co-evaporation of C_{60} and a cationic dye 'pyronin B (PyB)' [87] (see also Sections 4.2 and 6.3). Such doping resulted in the conductivity of C_{60} film of 10^{-4} S/cm, which is sufficient for low ohmic losses in device operation under one-sun illumination conditions. The thermally evaporated p-layer was doped by tetrafluoro-tetracyano-quinodimethane (F4-TCNQ). The photoactive i-layer was a 1:1 mixture of phthalocyanine zinc (ZnPc) and C_{60}. The i-layer was mainly amorphous while n and p transport layers had a polycrystalline structure. With the help of optical multilayer modeling and consideration of interference of the incident light and that reflected from the metal electrode, the optical properties of the cells were optimized by placing the active i-region at the maximum of the optical field distribution. The modeling results were confirmed by the experimental findings. The sequence and thickness of the layers for the optimized p–i–n cell was ITO/p-MeOTPD (55 nm)/ZnPc:C_{60} (32 nm)/C_{60} (10 nm)/n-doped C_{60}(30 nm)/Al (Fig. 27b). For an optically optimized device, an internal quantum efficiency of 82% and $J_{sc} = 8.44$ mA/cm^2, $V_{oc} = 0.44$ V, $FF = 0.45$ and $\eta = 1.67\%$, measured under simulated AM 1.5 sunlight of 100 mW/cm^2, were reported. The authors used this cell as an optimized building block for development of highly efficient tandem solar cells (see below).

In addition to the optimization of the layer thickness in a p–i–n cell Heutz et al. varied a composition of the C_{60}:CuPc i-layer [308]. The exact composition control was afforded by organic molecular beam deposition technique. A series of ITO/CuPc/CuPc:C_{60}/C_{60}/BCP/Al devices were fabricated with the composition ratio of the mixed layer varied between 25% and 75% of CuPc. The thickness of this layer was 50 nm as determined from the initial experiment on the layer thickness optimization. $I–V$ curves, measured under simulated AM 1.5 sunlight of 100 mW/cm^2, are shown in Fig. 29a. The efficiencies for the entire series are plotted in the inset and Fig. 29b shows the variation of I_{sc} and V_{oc} with CuPc content. For low CuPc content the absorption was low, leading to only a small number of photogenerated charges and hence a low I_{sc}. As the CuPc content was increased, the creation and dissociation of excitons increased leading to a higher I_{sc}, peaking at 60% CuPc content, but remaining high from 50% to 75%. At higher CuPc content, though the light absorption increased, I_{sc} rapidly decreased, probably due to the smaller interfacial area, more rapid charge recombination and reduced electron transport through the small fraction of C_{60}. V_{oc} displayed a similar trend: at 25% CuPc content, V_{oc} was low, probably due to the poor morphology. As the CuPc content increased, V_{oc} also increased, peaking at 75%. The FF remained similar across the center of the series and varied between 0.44 and 0.47 for 25–75%

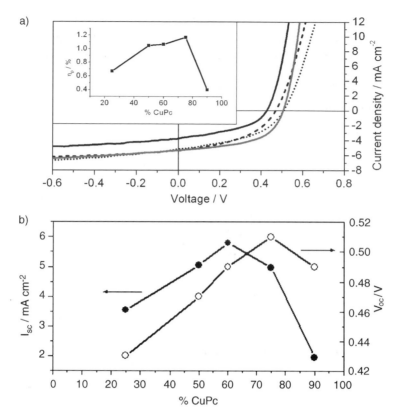

Fig. 29. *I–V* characteristics for a series of *p–i–n* ITO/CuPc/CuPc:C_{60}/C_{60}/BCP/Al solar cells, with varying *i*-layer composition: (a) *I–V* curves for the *i*-layer compositions of 25% CuPc (solid line), 50% CuPc (dashed line) and 75% CuPc (dotted line). The cell with a gradient of donor-acceptor compositions in the *i*-layer is shown in grey. Inset: the effect of composition on power conversion efficiency. (b) The effect of composition on V_{oc} and I_{sc}. (Reprinted from Ref. 308, with permission from Elsevier © 2004).

CuPc content. This was consistent with the morphological and charge transport properties of the layers which were independent of mixed-layer composition. At 90% CuPc content the *FF* decreases to 0.40, due to the increased resistivity of CuPc compared to C_{60}. These values result in the efficiency peak of $\eta = 1.17\%$ at 75% CuPc composition. Further improvements in the device performance were seen in a multiple mixed-layer device, incorporating three mixed layers with increasing acceptor composition between pure CuPc and C_{60} layers (Fig. 27c). The *I–V* curve of this cell (Fig. 29a, solid grey line) indicates an increase in *FF* to 0.54. It was attributed to the increased charge collection

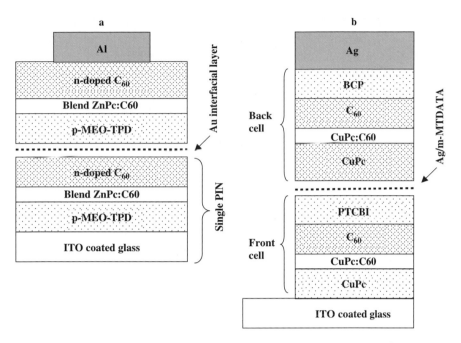

Fig. 30. Device structure of tandem C_{60}/MePc based solar cells: (a) ITO/*p*-MeOTPD/ ZnPc:C_{60}/*n*-doped C_{60}/Au/*p*-MeOTPD/ZnPc:C_{60}/*n*-doped C_{60}/Al; (b) ITO/CuPc/CuPc: C_{60}/C_{60}/PTCBI/Ag/m-MTDATA/CuPc/CuPc:C_{60}/C_{60}/BCP/Ag. ((a) Reprinted from Ref. 311 with permission from the American Institute of Physics © 2005; (b) reprinted from Ref. 309 with permission from the American Institute of Physics © 2004).

efficiency due to the gradient of donor acceptor compositions driving charges to the appropriate electrodes. The efficiency of the device was $\eta = 1.36\%$. The authors mentioned that these devices were not optimized in terms of electrode treatment and supposed that such treatment can improve the efficiency.

The Princeton group reported the highest efficiency for p–i–n cells [310] combining its experience in optical and electrical optimization of CuPc–C_{60} heterojunctions [297, 302] and introduction of a bulk-heterojunction-like CuPc:C_{60} interlayer. The device with a structure of ITO/CuPc/CuPc:C_{60} (1:1)/ C_{60}/BCP/Ag exhibited efficiency of $\eta = 5\%$ under 1 to 4 suns of simulated AM 1.5 sunlight (see also Table 2).

Next step towards efficiency improvement uses a strategy of a *tandem* solar cell. Maennig et al. stacked in serious two ITO/p-doped MeOTPD/ ZnPc:C_{60}/n-doped C_{60}/Al p–i–n cells, similar to those described above, in order to increase optical absorption (Fig. 30a) [306, 311]. An ultra-thin gold layer was introduced between the cells. As expected, the tandem cell showed an $V_{oc} = 0.85$ V, which is almost doubled value of a single cell, and a significantly higher power efficiency of 2.4% (under simulated AM 1.5 sunlight of 125 W/cm^2). However, $I_{sc} = 6.6$ mA/cm^2 was only about half the value of the single p–i–n device. The authors assumed that such limiting is rather due to the first cell which was – in contrast to the second cell – not placed in an interference maximum of the optical field. Very recently they optimized the layer configuration and further improved the efficiency [311]. For the optimized tandem cell $J_{sc} = 10.8$ mA/cm^2, $V_{oc} = 10.8$ V, $FF = 0.47$ and $\eta = 3.8\%$ were reported for simulated AM 1.5 sunlight of 130 mW/cm^2.

However, the main disadvantage of the above mentioned tandem cells is the fact that both subcells have identical structure and absorption spectra. This drawback has been overcome by the Princeton group in an *asymmetric* tandem cell [309]. The cell was produced by stacking in series two p–i–n subcells similar to those described in Ref. 310. The photoactive region in each subcell consisted of a mixed $CuPc:C_{60}$ layer sandwiched between homogeneous thermally evaporated CuPc and C_{60} thin films. Thin films of PTCBI and BCP were used as the exciton-blocking layers in the front and back subcells, respectively. The two subcells were connected in series by a charge recombination zone for electrons generated in the front cell and holes generated in the back cell. The recombination centers were Ag nanoclusters ($\sim$5 Å average thickness) buried in a 50 Å thick 4, 48,49-tris(3-methyl-phenyl-phenyl-amino)triphenylamine (m-MTDATA) p-doped with 5 mol% tetrafluoro-tetracyanoquinodimethane. Thus the entire device structure was ITO/CuPc/$CuPc:C_{60}/C_{60}$/PTCBI/Ag/m-MTDATA/CuPc/$CuPc:C_{60}/C_{60}$/BCP/Ag (Fig. 30b). Considering that CuPc absorbs between $\lambda = 550$ nm and 750 nm, and C_{60} between $\lambda = 350$ nm and 550 nm, an optimum asymmetric tandem cell structure was obtained when the front subcell has a thicker homogeneous CuPc layer and a thinner C_{60} layer than the back subcell. The photocurrents in the two subcells were balanced by a trade-off between the homogeneous and mixed layer thicknesses. This optical engineering resulted in the increased efficiency of $\eta = 5.7\%$, measured under 1 sun simulated AM 1.5G solar illumination. To the best of our knowledge this is the highest value ever reported for fullerene-based solar cells. The open-circuit voltage was 1.04 *V* approaching 1.2 V under 10 suns, approximately twice of that of a single p–i–n $CuPc/C_{60}$ cell [310]. Analytical model reported by the authors suggested that power conversion efficiencies exceeding 6.5% can be obtained by this architecture. By applying antireflection coatings to the glass substrates, an additional 10% improvement to efficiencies are possible, suggesting that the tandem cell structure can attain efficiencies in excess of 7%. Furthermore, the main advantage of the asymmetric multijunction solar cell lies in its general ability to incorporate different donor-acceptor material combinations in the subcells in order to cover the full solar spectrum.

9.3. Solar Cells with Heterojunctions between C_{60} Film and Pentacene or Tetracene Layers

Solids of such small molecules as tetracene (Fig. 31a) and pentacene (Fig. 31b) are known to be molecular semiconductors of p-type with relatively high exciton diffusion length and carrier mobility [313, 314]. Very promising results were obtained in experiments with a simple device configuration containing a heterojunction between C_{60} thin films and donor layers of tetracene [313] and pentacene [314]. In both cases a BCP film was used as an exciton-blocking layer. In particular, an ITO/PEDOT:PSS/tetracene (80 nm)/C_{60}(30 nm)/BCP/Al cell exhibited $V_{oc} = 0.58$ V $J_{sc} = 7$ mA/cm^2, $FF = 0.57$ and $\eta = 2.3\%$ under a simulated AM 1.5 sunlight of 100 mW/cm^2 [313]. Meanwhile, an ITO/pentacene (45 nm)/C_{60} (50 nm)/BCP/Al cell was demonstrated to have the following photovoltaic parameters under the illumination of the broadband light: $V_{oc} = 0.363$ V $J_{sc} = 15$ mA/cm^2, $FF = 0.5$ and $\eta = 2.7\%$ [313]. Using the measured spectral response of the cell the authors estimated efficiency under the 1 sun AM 1.5 light as high as 1.5%. These preliminary results suggested that crucial improvements in the efficiency might be provided by optimizing the heterojunction morphology, growth rate of thin film and film thickness as well as incorporation of C_{60}/tetracene (pentacene) junctions in multijunction solar cells similar to those described above.

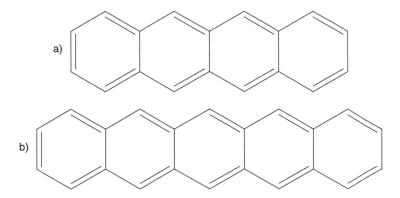

Fig. 31. Molecular structures of tetracene (a) and pentacene (b).

10. PHOTOVOLTAIC DEVICES WITH HIGH FULLERENES

10.1. C_{70}-Based Solar Cells

Low absorption coefficients of C_{60} and its derivatives (e.g., PCBM) in the visible region of the spectrum, that limit the cell photocurrent, originate from the high degree of symmetry of the C_{60} molecule (see Section 2). Such symmetry makes many of the low-energy transitions forbidden and hence of low intensity. When the C_{60} moiety is replaced by a less symmetrical fullerene, such as C_{70} (Fig. 3b) these transitions become allowed and a dramatic increase in light absorption is expected [315]. Wienk et al. [316] have developed synthesis of [70] PCBM (which is similar to PCBM (Fig. 7a) but incorporates C_{70} instead of C_{60} (Fig. 32) and produced [70] PCBM/MDMO-PPV bulk heterojunction solar cells. Fig. 33a compares spectra of the external quantum efficiency (EQE) of ITO/PEDOT-PSS/[70] PCBM:MDMO-PPV (4.6:1)/LiF/ Al cells with an active layer spin coated from chlorobenzene and ODCB (the better morphology) and the EQE spectrum of the best PCBM:MDMO-PPV devices. As a result of the increased absorption by [70] PCBM, the increase in the maximum EQE from ~50% (PCBM) to 66% ([70] PCBM) as well as the spectral response broading are in evidence. Accordingly the cell photocurrent was found to increase by 50% and the efficiency reached 3.0% (Fig. 33b and Table 2), comparing with $\eta = 2.5\%$ for the best similar PCBM devices [264]. We believe that this idea could be used for the efficiency improvement in all kinds of fullerene-based solar cells (including the most efficient multijunction devices). For a long time, utilization of C_{70} was limited by the low production yield and the high prices of this fullerene; however, the

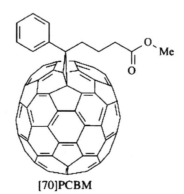

[70]PCBM

Fig. 32. Molecular structures of [70] PCBM. (Reprinted from Ref. 316 with permission from Wiley–VCH Verlag GmbH & Co © 2003).

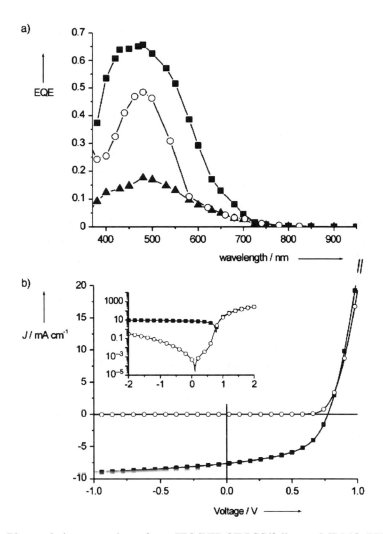

Fig. 33. Photovoltaic properties of an ITO/PEDOT-PSS/fullerene:MDMO-PPV/LiF/Al device with an active area of 0.1 cm². (a) External quantum efficiency (EQE) of [70]PCBM:MDMO-PPV cells, spin coated from chlorobenzene (▲) and ODCB (■), and of [60]PCBM:MDMO-PPV devices spin coated from chlorobenzene (○); (b) current–voltage characteristics of [70] PCBM:MDMO-PPV devices, spin coated from ODCB, in the dark (○) and under illumination (AM1.5, 100 mW/cm²; ■). The inset shows the dark and illuminated *I–V* curves in a semilogarithmic plot. (Reprinted from Ref. 316 with permission from Wiley–VCH Verlag GmbH & Co © 2003).

recent Frontier Carbon Corporation/Mitsubishi breakthrough [72–73] in the synthesis of fullerene-containing soot in tones and the corresponding price reduction (see Section 3) creates serious perspective for a cost-effective utilization of high fullerenes.

10.2. Solar Cells with an Endohedral Metallofullerene

As mentioned in Sections 4.2 and 6.3, the encapsulation of the metal atoms into the fullerene cage gives rise to a number of unique properties of these new compounds – *endofullerenes* or *endohedral metallofullerenes*. For example, the mono-metallofullerenes $M@C_{82}$ ($M =$ lanthanide elements) possess not only a greater electron accepting ability but also a greater electron donating ability than C_{60}. In particular, the first reduction potential of $Dy@C_{82}$, which in solid state is an n-type semiconductor of narrow gap [158] (see Section 6.3), are even more positive than that of C_{60}, and consequently, this metallofullerene is a stronger electron acceptor than C_{60} [317]. To our knowledge, no investigation has been reported so far of solid-state photovoltaic devices with endohedral metallofullerenes. However, a number of studies on photoelectrochemical solar cells with $Dy@C_{82}$ have been recently published [318–321]. Although fullerene-containing photoelectrochemical solar cells are not within the scope of the present review, we will briefly mention structures and properties of the $Dy@C_{82}$ devices due to a potencial importance of such structures for photovoltaics.

In particular, stable Langmuir-Blodgett films of $Dy@C_{82}$ and its mixtures with molecules of long-chain arachidic acid (AA) were demonstated in Ref. 318. The metallofullerene films exhibited a higher anodic photocurrent quantum yield than the corresponding C_{60} films on ITO electrodes in a photoelectrochemical cell. Furthermore, oxygen was shown to quench the excited state and photoelectrical properties of $Dy@C_{82}$ in much less extent than in C_{60}. Mixing the metallofullerene $Dy@C_{82}$ with CuPc and ZnPc, the monolayers and thin films of new, donor-acceptor type complexes with the enhanced photocurrent generation were successfully constructed [319]. Moreover, a dramatic enhancement (by factor of 20) of the cathodic photocurrent of a photoelectrochemical cell based on the Langmuir–Blodgett film of P3HT was achieved by doping the films with $Dy@C_{82}$ [320–321]. The photocurrent was found to be composition-dependent. It is important to note that the photocurrent and the cell efficiency was the highest when the molar ratio of P3HT:$Dy@C_{82}$ was 20:1. It means that very small amounts of the metallofullerene are needed for the efficient device operation. All of these results allude to serious benefits of $Dy@C_{82}$ incorporation in donor-acceptor photovoltaic devices and especially in "red" layers of multijunction cells (due to the low band gap of $Dy@C_{82}$, which was estimated of $\sim0.2\,eV$ from resistivity measurements [158] or $0.8\,eV$ as obtained from the absorption spectra [322]).

10.3. Photovoltaic Solar Cells Containing Carbon Nanotubes

Longest fullerene molecules, carbon nanotubes, may provide unique advantages for photovoltaics. They offer a wide range of band gaps [35, 38, 39] (see also Section 2) to match the solar spectrum, enhanced optical absorption [19, 20] and reduced carrier scattering for hot carrier transport [323, 324]. The latter may even result in a near-ballistic transport in the nanotubes with submicron-meter lengths [325].

Accordingly, one of the initial motivations to use carbon nanotubes in photovoltaic devices was to improve electronic transport of the fullerene channel in bulk heterojunction solar cells. Being blended with conjugated polymers, carbon nanotubes may not only act as electron acceptors but also allow the transferred electrons to be efficiently transported along their length, thus providing percolation paths. Indeed, the extremely high surface area, $\sim1600\,m^2/g$, reported for purified single-walled nanotube (SWNT) [326] offers a tremendous opportunity for exciton dissociation. Since SWNTs have diameters of $\sim1\,nm$ and lengths of $\sim1\text{--}10\,\mu m$, these materials exhibit very large aspect ratios ($>10^3$). Thus, percolation pathways could be established at low doping levels, providing the means for high electron mobility. Electrical conductivity data has validated that SWNT-doped polymer composites demonstrate the extremely low percolation threshold. For SWNT-epoxy composites, for example, the electrical conductivity has been claimed to rise by nearly 10^5 for SWNT concentration of only 0.1–0.2% [327].

Electrical and photoelectrical properties of CNT/conjugated polymer composites and interfaces have been investigated since 1996 [328–333]. Recently, bulk heterojunction solar cells based on conjugated polymers blended with multiwalled carbon nanotubes (MWNT) [334] and SWNTs [335–337] have been reported. For 1% SWNT/P3OT bulk heterojunction solar cells (Fig. 34) high values of $V_{oc} = 0.75V$ were achieved and reasonably explained in terms of HOMO–LUMO electronic structures of P3OT and SWNTs [336]. However, the efficiency of the best SWNT/P3OT devices is well below 1% (Table 2) due to low photocurrent mostly limited by incomplete phase separation and the lack of light absorption. Improvement of the light absorption was achieved by a dye (naphthalocyanine, NaPc) coating of CNTs that were blended with P3OT in a bulk heterojunction configuration

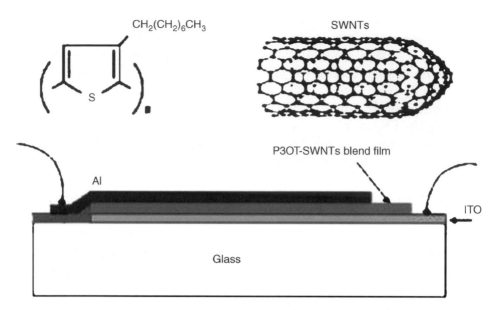

Fig. 34. Schematic representation of a bulk heterojunction solar cell based on conjugated polymer blended with single-walled CNTs, as well as the chemical structures of the compounds in its active layer. (Reprinted from Ref. 335, with permission from the American Institute of Physics © 2002).

[338]. For the same purpose, Jin and Dai suggested to build a cell that, instead of randomly mixing CNTs with polymers, will contain a network of vertically aligned CNTs separated by vertical polymer layers [339]. The authors reviewed various techniques for growth of vertically aligned CNT networks that could be used in such solar cells.

Significant interest has recently been shown in establishing synthetic strategies and characterization of composite films of SWNTs decorated by nanocrystals (quantum dots, QDs) of such inorganic semiconductors as CdSe [337, 340], CdS [341, 342], ZnS [343] (Fig. 35). SWNTs decorated by CdS nanocrystals have been shown to undergo charge transfer interactions under excitations with visible light and generate the photocurrent in a photoelectrochemical cell [341, 342].

Landi et al. suggested to utilize bundles of SWNTs decorated by CdSe QDs in a *cascade* bulk heterojunction solar cell [337, 340]. The covalent attachment of CdSe QDs to SWNTs through an organic coupling reaction has been confirmed by FTIR spectroscopy, atomic force microscopy (AFM), and transmission electron microscopy (TEM) [340]. The ability to capture a

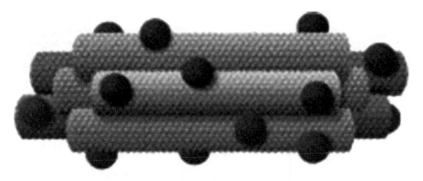

Fig. 35. Schematic representation of bundles of SWNTs decorated by semiconductor quantum dots (Reprinted from Ref. 340, with permission from Elsevier © 2005).

larger energy range of the solar spectrum through incorporation of CdSe QDs and SWNTs in P3OT was demonstrated. However, the photovoltaic characterization of the cascade solar cells did not show the efficiency improvement in comparison with the best SWNT/P3OT devices [336, 337]. The authors attributed this efficiency limiting to strong recombination and surface effects and formulated the following strategy for the further research in order to build an optimized cascade cell. The ideal cascade of energy transitions for an optimal cell would include photon absorption by the components over the entire solar spectrum. Fig. 36a illustrates the energy levels in relation to the vacuum level for the valence, conduction, and exciton binding levels of the P3OT polymer, CdSe QDs, and semiconducting SWNTs (S-SWNT). Also, the work-function of metallic SWNTs (M-SWNT) indicates an appropriate potential energy level for exciton dissociation when used in conjunction with the P3OT polymer, CdSe QDs, and S-SWNTs. Although the P3OT's exciton binding energy is relatively high, 0.5 eV [344], each of the nanomaterials discussed would have sufficient electron affinity to dissociate the electron-hole pair. Since the QDs rely on a hopping conduction to transport the electrons to the negative electrode, an alternative would be a ballistic conductor like SWNTs. In comparison to the polymer, dissociation of the excitons generated in CdSe QDs shows a reduced energy barrier (0.1 eV [345]) and the electron transport through the SWNTs would presumably be most efficient with appropriate coupling of the QDs to the SWNTs. The cascade of energy transitions, exciton dissociation, and carrier transport for a QD–SWNT-polymer cell is depicted in Fig. 36b. Although the schematic represents a series of planar junctions, the reality in a nano-composite is actually a complex three-dimensional network of junctions. The photo-induced excitons in the polymer are expected to be

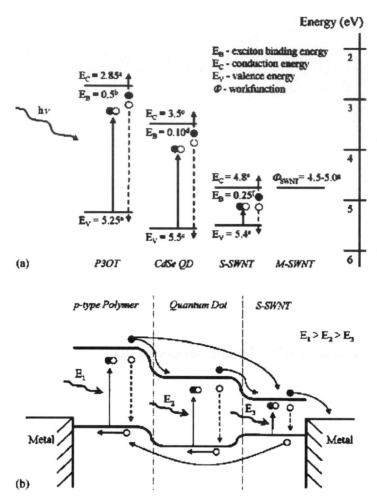

Fig. 36. Potential energy level diagram for a cascade QD–SWNT-Polymer solar cell. (a) The diagram adjusted in relation to the vacuum level for the P3OT polymer, CdSe QDs, semi-conducting SWNTs (S-SWNT), and metallic SWNTs (M-SWNT). (b) Schematic which illustrates a cell equilibrated at the Fermi energy. The corresponding electronic transitions from optical absorption, exciton dissociation, and carrier transport are shown for each of the components. The references in (a) are the following: a [336], b [344], c [351], d [345], e [352], f [353] and g [336, 354]. (Reprinted from Ref. 340, with permission from Elsevier © 2005).

dissociated by the nearest high electron affinity material, either the QD or SWNT. Ultimately, the holes are transported by the polymer to the positive electrode and the dominant electron path is through the percolating SWNTs to the negative electrode. In addition to the absorption by the polymer and QDs, it has been shown that semiconducting SWNTs can also absorb light and contribute to the photoconductivity in these types of devices [323]. Since V_{oc} in a bulk heterojunction cell is, as already mentioned, controlled by the energetic difference in the HOMO level of the polymer and the LUMO level of the acceptor material, it is important to note that the energetic structure of the QD and SWNT components can be tailored by their diameter distribution control. Due to the fact that the polymer, QDs, and SWNTs may each absorb in a different spectral region, it is very reasonable that these nanomaterials could be combined in such a way as to provide the most efficient combination for light absorption, exciton dissociation and carrier transport.

Another exciting research direction could be instigated by the recent report on the discovery of photovoltaic effect in an SWNT-based p–n junction diode [346]. Such diode was prepared by electrostatic doping of SWNTs using a device structure shown in the inset of Fig. 37a. Different bias polarities on the split gate electrostatically couple to form separate regions of electron and hole doping along an individual SWNT. This is possible because the metal contacts make Schottky barriers, allowing either electrons or holes to tunnel into the nanotube. The fabrication process followed standard lithography and etch techniques, and SWNTs were grown on top of the S and D metal electrodes. A large number of devices were fabricated and many were found with only a single semiconducting SWNT between S and D contacts. Fig. 37a shows a typical dark I–V curve of the diode that was analyzed using a classical equation (that coincides with the first term in Eq. (10) in Section 7) [194]

$$I - I_o\left[\exp\left(\frac{q(V+IR_s)}{nkT}\right)-1\right] \tag{13}$$

In general, the diode quality factor n in Eqs. (10) and (13) is precisely equal to 1 for an ideal p–n junction (without any recombination in its depletion region) but approaches 2 or even higher values for materials with defects. n is intimately related to the solar cell efficiency: the larger the n, the lower the η [195]. The most important result from Lee [346] is that the experimental dark I–V curve is well fitting with $n = 1$. This indicates an ideal character of the diode and highlights the potential of SWNTs for efficient photovoltaic devices.

To check the photovoltaic effect in this ideal diode its I–V characteristics was measured under illumination of a 1.5 μm (0.8 eV) laser diode (Fig. 37b). The inset shows the expected linear irradiance dependence of the I_{sc}. The author analyzed the V_{oc} vs. Log I_{sc} on the base of Eq. (4) and confirmed the

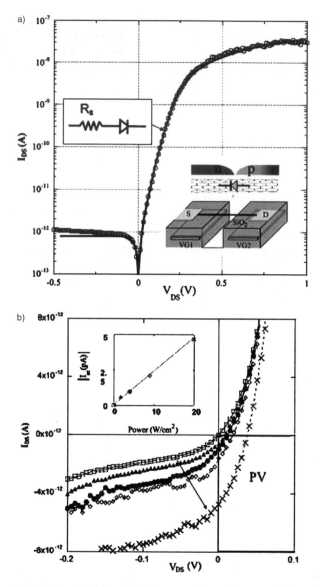

Fig. 37. Typical dark (a) and light (b) current–voltage characteristics of an ideal diode based on an electrostatic doped SWNT, at $T = 300$ K. The dark I–V curve was fitted by Eq. (13) using $I_o = 8.0 \times 10^{-13}$ A, $n = 1.0$, $R_s = 18 \times 10^6 \, \Omega$. The inset in (a) shows the split gate device where VG1 and VG2 are biased with opposite polarities (VG1 $= -$VG2 $= +10$ V) to form a p–n junction along a SWNT. The arrow in (b) indicates the increased light intensity. The inset in (b) shows the expected linear increase in the I_{sc} with illuminated power. Reprinted from Ref. 346, with permission from the American Institute of Physics © 2005.

excellent agreement of the experimental data with $n = 1$. Using an expression for the reverse saturation current of an ideal diode $I_o \sim e^{-qEg/kT}$ [194], the band gap values of $E_g = 0.6$–0.8 eV were estimated for the SWNTs under the study. The estimated FF and η were, correspondently, 0.52 and 0.2%. The latter is significant considering only a small fraction of the incident light was absorbed. The author estimated $\eta > 5\%$ for reasonably absorbed solar power. A significant improvement in η can be expected by considering a 0.8 nm diameter SWNT with $E_g = 1.4$ eV [347], which is close to the optimal theoretical E_g value for a single junction solar cell [9]. For such SWNTs, based on the expected reduction in the I_o by at least 10^{10} compared to SWNTs with $E_g = 0.8$ eV, the author approximately calculated $V_{oc} > 0.8$ V and FF > 0.8. These values are comparable or even higher than those for the best inorganic cells with a single junction [348]. In our opinion, it is too early to tell if (1) a diode based on chemically doped SWNTs preserves an ideal character; (2) *cloth-like* films are possible to produce on the base of network of such tubular diodes. If so, further improvements in η could be expected due to a strong electromagnetic coupling in the SWNT networks and scattering between the tubes [349–350].

11. CONCLUSION

In this chapter we review structure, properties, and prospects of fullerene thin films for photovoltaics and present a comprehensive picture of the state-of-the-art and future directions of fullerene-based solar cell research. Device structures and photovoltaic parameters of inorganic and organic fullerene-based solar cells are presented and summarized in Table 2. A special attention is paid to the excitonic mechanism of photogeneration in these devices that requires a sophisticated nanostructural engineering of fullerene-based photovoltaic materials and devices compromising the requirements of thin photoactive layers with an effective light absorption, photogeneration and transport of free carriers. Examples of device architecture that may satisfy to these challenging requirements include fullerene/amorphous carbon multi-heterojunctions and quantum wells, fullerene/conjugated polymer bulk hetero-junctions with controlled morphology, "double-cable" donor-acceptor systems, fullerene/phthalocyanine p–i–n single and multijunctions, carbon nanotube and quantum dots based solar cells. Efficiencies of ~5% have now been reported [277, 310]. Based on current trends, efficiency up to 10% appears to be within reach in the near future.

REFERENCES

[1]   H.W. Kroto, J.R. Heath, S.C. O'Brien, R.F. Curl and R.E. Smalley, Nature, 318 (1985) 162–163.

[2]   W. Kratschmer, L.D. Lamb, K. Fostipoulos and D.R. Huffman, Nature, 347 (1990) 354–358.

[3]   N. Minami and S. Kazaoui, in "Optical and electronic properties of fullerenes and fullerne-based materials" (J. Shinar, Z.V. Vardeny and Z.H. Kafafi, eds.), pp. 261–292, Marcel Dekker, New York, 2000.

[4]   D. Faiman, S. Goren, E. Katz, M. Koltun, E. Kunoff, A. Shames, S. Shtutina and B. Uzan, in "Proceedings of the 13th European Photovoltaic Solar Energy Conference" (W. Freiesleben, W. Palz, H.A. Ossenbrink and P. Helm, eds.), Vol. 2, pp. 1284–1286, H.S. Stephens & Associates, Bedford, UK, 1995.

[5]   M. Koltun, D. Faiman, S. Goren, E.A. Katz, E. Kunoff, A. Shames, S. Shtutina and B. Uzan, Sol. Energy Mater. Sol. Cells, 44 (1996) 485–491.

[6]   S. Saito and A. Oshiyama, Phys. Rev. Lett., 66 (1991) 2637–2640.

[7]   A. Skumanich, Chem. Phys. Lett., 182 (1991) 486–490.

[8]   M. Hosoya, K. Ichimura, Z.H. Wang, G. Dresselhaus and P.C. Eklund, Phys. Rev. B, 49 (1994) 4981–4986.

[9]   J.J. Loferski, J. Appl. Phys., 27 (1956) 777.

[10] R.W. Lof, M.A. van Veenendaal, B. Koopmans, H.T. Jonkman and G.A. Sawatzky, Phys. Rev. Lett., 68 (1992) 3924–3927.

[11] B. Mishori, E.A. Katz, D. Faiman and Y. Shapira, Solid State Commun., 102 (1997) 489–492.

[12] M.S. Dresselahaus, G. Dresselahaus and P.C. Eklund, Science of Fullerenes and Carbon Nanotubes, Academic Press, NY, 1996.

[13] K.L. Narayanan and M. Yamaguchi, Appl. Phys. Lett., 75 (1999) 2106–2108.

[14] N. Dharmasu, T. Sakai, N. Kojima, M. Yamaguchi, Y. Ohshita and K.L. Narayanan, J. Appl. Phys., 89 (2001) 318–322.

[15] K.L. Narayanan, O. Goetzberger, A. Khan, N. Kojima and M. Yamaguchi, Sol. Energy Mater. Sol. Cells, 65 (2001) 29–35.

[16] K.L. Narayanan and M. Yamaguchi, Sol. Energ. Mater. Sol. Cells, 75 (2003) 345–350.

[17] T. Soga, T. Jimbo, K.M. Krishna and M. Umeno, Int. J. Mod. Phys. B, 14 (2000) 206–217.

[18] F. Xiong, Y.Y. Wang, V. Leppert and R.P.H. Chang, J. Mater. Res., 8 (1993) 2265–2272.

[19] J.W.G. Wildöer, L.C. Venema, A.G. Rinzler and R.E. Smally, Nature, 391 (1998) 59–62.

[20] T.W. Odom, J.-L. Huang, P. Kim and C.M. Lieber, Nature, 391 (1998) 62–64.

[21] E. Osawa, Kagaku (Kyoto), 25 (1970) 854.

[22] D.A. Bochvar and E.G. Gal'pern. Proc. Acad. Sci. USSR, 209 (1973) 239.

[23] R. Taylor, J.P. Hare, A.K. Abdul-Sada and H.W. Kroto, J. Soc. Chem. Commun., 20 (1990) 1423–1424.

[24] R.D. Jonson, D.S. Bethune and C.S. Yannoni, Accounts of Chem. Res., 25 (1992) 169–175.

[25] W. Kratschmer, K. Fostiropoulos and D.R. Huffman, Chem. Phys. Lett., 170 (1990) 167–170.

[26] W.I.F. David, R.M. Ibberson, J.C. Matthewman, K. Prassides, T.I.S. Dennis, J.P. Hare, H.W. Kroto, R. Taylor and D.R. Walton, Nature, 353 (1991) 147–149.

[27] L. Euler, Novi corumentarii academie Petropolitanae 4 (1752/1753) 109.

[28] B. Grünbaum and T.S. Motzkin, Can. J. Mathematics, 15 (1963) 744–751.

[29] H.W. Kroto, Nature, 329 (1987) 529–531.

[30] T.Z. Schmalz, W.A. Seitz, D.J. Klein, G.E. Hite, Chem. Phys. Lett., 130 (1986) 203–207.

[31] Y.L. Voytekhovsky and D.G. Stepenshchikov, Acta Crystallogr., A60 (2004) 278–280.

[32] C. Piskoti, J. Yarger and A. Zettl, Nature, 393 (1998) 771–773.

[33] H. Prinzbach, A. Weiler, P. Landenberger, F. Wahl, J. Wörth, L. Scott, M. Gelmont, D. Olevano and B. von Issendorff, Nature, 407 (2000) 60–63.

[34] J.R. Heath, S.C. O'Brien, Q. Zhang, Y. Liu, R.F. Curl, H.W. Kroto, F.K. Tittel and R.E. Smalley, J. Am. Chem. Soc., 107 (1985) 7779–7780.

[35] M.S. Dresselhaus, G. Dresselhaus and R. Saito, Carbon, 33 (1995) 883–891.

[36] P.J.F. Harris, Carbon Nanotubes and Related Structures, pp 279, Cambridge University Press, Cambridge, UK, 1999.

[37] S. Iijima, Nature, 354 (1991) 56–58.

[38] N. Hamada, S. Sawada and A. Oshiyama, Phys. Rev. Lett., 68 (1992) 1579–1581.

[39] R. Saito, M. Fujita, M.S. Dresselhaus and G. Dresselhaus, Phys. Rev. B, 46 (1992) 1804–1811.

[40] L. Chico, V.H. Crespi, L.X. Benedict, S.G. Louie and M.L. Cohen, Phys. Rev. Lett., 76 (1996) 971–974.

[41] P.G. Collins, A. Zettl, H. Bando, A. Thess and R.E. Smally, Science, 278 (1997) 100–103.

[42] L.F.P. Chibante, A. Tess, J. Alford, M.D. Dierner and R.E. Smalley, J. Phys. Chem., 97 (1993) 8696–8700.

[43] J.R. Pitts, D. Mischler, C.L. Fields, A. Lewandowski and C. Bingham, Solar Chem. News, 6 (1998) 1–2.

[44] C.L. Fields, D.H. Parker, J.R. Pitts, M.J. Halle, C. Bingham, A. Lewandowski and D.E. King, J. Phys. Chem., 97 (1993) 8701–8702.

[45] P. Bernier, D. Laplaze, J. Auriol, L. Barbedette, G. Flamant, M. Lebrun, A. Brunelle and S. Dellanegra, Synth. Met., 70 (1995) 1455–1456.

[46] D. Laplaze, P. Bernier, G. Flamant, M. Lebrun, A. Brunelle and S. DellaNegra, Synth. Met., 77 (1996) 67–71.

[47] D. Laplaze, P. Bernier, G. Flamant, M. Lebrun, A. Brunelle and S. Della-Negra, J. Phys. B, 29 (1996) 4943–4954.

[48] D. Laplaze, P. Bernier, C. Journet, V. Vie, G. Flamant and M. Lebrun, Synth. Met., 86 (1997) 2295–2296.

[49] D. Laplaze, P. Bernier, C. Journet, J.L. Sauvajol, D. Bormann, G. Flamant and M. Lebrun, J. de Physique III, 7 (1997) 463–472.

[50] E. Anglaret, N. Bendiab, T. Guillard, C. Journet, G. Flamant, D. Laplaze, P. Bernier and J.L. Sauvajol, Carbon 36 (1998) 1815–1820.

[51] G. Flamant, A. Ferriere, D. Laplaze and C. Monty, Sol. Energ., 66 (1999) 117–132.

[52] C.L. Fields, J.R. Pitts and A. Lewandowski, Abstr. Pap. Am. Chem. Soc., 212 (1996) 26.

[53] T. Guillard, G. Flamant and D. Laplaze, J. Sol. Energ. Eng., 123 (2001) 153–159.

[54] J.M. Gordon, D. Feuermann, M. Huleihil and E.A. Katz, Proc. of SPIE, 5185 (2003) 99–108.

[55] T. Guillard, L. Alvarez, E. Anglaret, J.L. Sauvajol, P. Bernier, G. Flamant, D. Laplaze, J. de Physique IV, 9 (1999) 399–404.

[56] L. Alvarez, T. Guillard, G. Olalde, B. Rivoire, J.F. Robert, P. Bernier, F. Flamant and D. Laplaze, Synth. Met., 103 (1999) 2476–2477.

[57] L. Alvarez, T. Guillard, J.L. Sauvajol, G. Flamant and D. Laplaze, Chem. Phys. Lett., 342 (2001) 7–14.

[58] G. Flamant, J.F. Robert, S. Marty, J.M. Gineste, J. Giral, B. Rivoire, D. Laplaze, Energy, 29 (2004) 801–809.

[59] T. Guillard, S. Cetout, L. Alvarez, J.L. Sauvajol, E. Anglaret, P. Bernier, G. Flamant and D. Laplaze, European Phys. J. Appl. Phys., 5 (1999) 251–256.

[60] M.J. Heben, T.A. Bekkedahl, D.L. Schulz, K.M. Jones, A.C. Dillon, C.J. Curtis, C. Bingham, J.R. Pitts, A. Lewandowski and C. Fields, in "Recent Advances in the Chemistry and Physics of Fullerenes and Related Materials III" (K.M. Kadish and R.S. Ruoff, eds.), PV 96–10, pp. 803–811, Electrochemical Society, Pennington, NJ, 1996.

[61] D. Laplaze, P. Bernier, W.K. Maser, G. Flamant, T. Guillard and A. Loiseau, Carbon, 36 (1998) 685–688.

[62] T. Guillard, G. Flamant, Jean-F. Robert, B. Rivoire, J. Giral and D. Laplaze, J. Sol. Energ. Eng., 124 (2002) 22–27.

[63] L. Alvarez, T. Guillard, J.L. Sauvajol, G. Flamant and D. Laplaze, Appl. Phys. A, 70 (2000) 169–173.

[64] D. Laplaze, L. Alvarez, T. Guillard, J.M. Badie and G. Flamant, Carbon, 40 (2002) 1621–1634.

[65] W.K. Maser, A.M. Benito and M.T. Martinez, Carbon, 40 (2002) 1685–1695.

[66] D. Luxembourg, G. Flamant, A. Guillot and D. Laplaze, Mat. Sci. Eng. B, 108 (2004) 114–119.

[67] D. Luxembourg, G. Flamant and D. Laplaze, Carbon, 43 (2005) 2302–2310.

[68] J.B. Howard, J.T. McKinnon, Y. Makarovsky, A.L. Lafleur and M.E. Jonson, Nature, 352 (1991) 139–141.

[69] M. Bachmann, J. Griesheimer and K.-H. Homann, Chem. Phys. Lett., 223 (1994) 506–510.

[70] R.F. Curl, Carbon, 30 (1992) 1149–1155.

[71] R. Taylor, G.J. Langley, H.W. Kroto, D.R.M. Walton, Nature, 366 (1993) 728–731.

[72] http://www.f-carbon.com/eng/.

[73] H. Murayama, S. Tomonoh, J.M. Alford and M.E. Karpuk, Fullerene, Nanotubes and Carbon Nanostructures, 12 (2004) 1–9.

[74] R. Tycko, G. Dabbagh, R.M. Fleming, R.C. Haddon, A.V. Makhia and S.M. Zahurak, Phys. Rev. Lett., 67 (1991) 1886–1889.

[75] L. Forró and L. Mihály, Rep. Prog. Phys., 64 (2001) 649–699.

[76] J.P. Lu, X.-P. Li and R.M. Martin, Phys. Rev. Lett., 68 (1992) 1551–1554.

[77] P.A. Heiney, J.E. Fisher, A.R. McGhhie, W.J. Romanow, A.M. Denestejn, J.P. McCauley Jr., A.B. Smith III and D.E. Cox, Phys. Rev. Lett., 66 (1991) 2911–2914.

[78] L. Firley, Condens. Matter News, 8 (2000) 22–52.

[79] R. Tycko, G. Dabbagh, R.M. Fleming, R.C. Haddon, A.V. Makhia and S.M. Zahurak. Phys. Rev. Lett., 67 (1991) 1886–1889.

[80] J. Yao, Y. Zou, D. He and G. Chen. Mater. Lett., 33 (1997) 27–30.

[81] T.L. Makarova, V.G. Melekhin, I.T. Serenkov, V.I. Sakharov, I.B. Zakharova and V.E. Gasumyants. Phys. Solid State, 43 (2001) 1393–1399.

[82] A.F. Hebard, M.J. Rosseinsky, R.C. Haddon, D.W. Murphy, S.H. Glarum, T.T.M. Palstra, A.P. Ramirez and A.R. Kortran, Nature, 350 (1991) 600–601.

[83] D. Sarkar and N.J. Halas, Appl. Phys. Lett., 63 (1993) 2438–2450.

[84] L. Firlej, N. Kirova and A. Zahab, Phys. Rev. B, 59 (1999) 16028–16032.

[85] E.A. Katz, D. Faiman, S. Shtutina, N. Froumin, M. Polak, A.P. Isakina, K.A. Yagotintsev, M.A. Strzhemechny, Y.M. Strzhemechny, V.V. Zaitsev and S.A. Schwarz, Phys. B, 304 (2001) 348–356.

[86] E.A. Katz, D. Faiman, S.M. Tuladhar, S. Shtutina, N. Froumin, M. Polak, Y. Strzhemechny, Sol. Energ. Mater. Sol. Cells, 75 (2003) 421–426.

[87] A.G. Werner, F. Li, K. Harada, M. Pfeiffer, T. Fritz and K. Leo, Appl. Phys. Lett., 82 (2003) 4495–4497.

[88] E.E.B. Campbell, R. Tellegmann, N. Krawez and I.V. Hertel, J. Phys. Chem. Sol., 58 (1997) 1763–1769.

[89] V.N. Popok, I.I. Azarko, A.V. Gromov, M. Jönsson, A. Lassesson and E.E.B. Campbell, Solid State Commun., 133 (2005) 499–503.

[90] S. Iida, Y. Kubozono, Y. Slovokhotov, Y. Takabayashi, T. Kanbara, T. Fukunaga, S. Fujiki, S. Emura and S. Kashino, Chem. Phys. Lett., 338 (2001) 21–28.

[91] T. Guo, C. Jin and R.E. Smalley, J. Phys. Chem., 95 (1991) 4948–4950.

[92] W. Andreoni, F. Gygi and M. Parrinello, Chem. Phys. Lett., 190 (1992) 159–162.

[93] J.C. Hummelen, B. Kringht, J. Pavlovich, R. Gonzalez and F. Wudl. Science, 269 (1995) 1554–1556.

[94] T. Pichler, M. Knupfer, M.S. Golden, S. Haffner R. Freidlen J. Fink, W. Andreoni, A. Curioni, M. Keshavarz-K, C. Bellavia-Lund, A. Sastre, J.C. Hummelen and F. Wudl, Phys. Rev. Lett., 78 (1997) 4249–4252.

[95] L. Hultman, S. Stafstrom, Z. Czigany, J. Neidhardt, N. Hellgren, I.F. Brunell, K. Suenaga and C. Cooliex, Phys. Rev. Lett., 87 (2001) 225503.

[96] S.Y. Xie, F. Gao, X. Lu, R.B. Huang, C.R. Wang, X. Zhang, M.L. Liu, S.L. Deng and L.S. Zheng, Science, 304 (2004) 699.

[97] R.H. Xie, G.W. Bryant, L. Jensen, J. Zhao and V.H. Smith, Jr. J. Chem. Phys., 118 (2003) 8621–8635.

[98] R.H. Xie, G.W. Bryant, G. Sun, M.C. Nicklaus, D. Heringer, Th. Frauenheim, M.R. Manaa, V.H. Smith, Jr., Y. Araki and O. Ito, J. Chem. Phys., 120 (2004) 5133–5147.

[99] R.H. Xie, G.W. Bryant, J. Zhao, V.H. Smith, Jr., A. Di Carlo and A. Pecchia, Phys. Rev. Lett., 90 (2003) 206602.

[100] J.C. Hummelen, B.W. Knight, F. Lepec and F. Wudl, J. Org. Chem., 60 (1995) 532–538.

[101] D.M. Guldi, I. Zilbermann, G. Anderson, N.A. Kotov, N. Tagmatarchis and M. Prato, J. Mater. Chem., 15 (2005) 114–118.

[102] C.J. Brabec, A. Cravino, D. Meissner, N.S. Sariciftci, M.T. Rispens, L. Sanchez, J.C. Hummelen and T. Fromherz, Thin Solid Films, 403–404 (2002) 368–372.

[103] P.J. Benning, F. Stepniak and J.H. Weaver, Phys. Rev. B, 48 (1993) 9086–9096.

[104] G. Gensterblum, K. Hevesi, B.-Y. Han, L.-M. Yu, J.-J. Pireaux, P.A. Thiry, R. Caudano, A.-A. Lucas, D. Bernaerds, S. Amelinckx, G. Van Tendeloo, G. Bendele, T. Buslaps, R.L. Jonson, M. Foss, R. Feidenhans'l and G. Le Lay, Phys. Rev. B, 50 (1994) 11981–11995.

[105] D. Schmicker, S. Schmidt, J.G. Skofronick, J.P. Toennies and R. Vollmer, Phys. Rev. B, 44 (1991) 10995–10997.

[106] J.K. Gimzewski, S. Modesti and R.R. Schlittler, Phys. Rev. Lett., 72 (1994) 1036–1039.

[107] E.I. Altman and R.J. Colton, Surf. Sci., 279 (1992) 49–67.

[108] E.I. Altman and R.J. Colton, Phys. Rev. B, 48 (1993) 18244–18249.

[109] E.A. Katz, D. Faiman, S. Shtutina and A. Isakina, Thin Solid Films, 368 (2000) 49–54.

[110] E.A. Katz, US Patent No: 5,876,790, 1999.

[111] K. Tanigaki, S. Kuroshima and T.W. Ebbesen, Thin Solid Films, 257 (1995) 154–165.

[112] Z. Dai, H. Naramoto, K. Narumi, S.Yamamoto, J. Phys.-Cond. Mat., 11 (1999) 6347–6358.

[113] Z. Dai, H. Naramoto, K. Narumi, S. Yamamoto and A. Miyashita, Appl. Phys. Lett., 74 (1999) 1686–1687.

[114] Z. Dai, H. Naramoto, K. Narumi, S. Yamamoto and A. Miyashita, Thin Solid Films, 360 (2000) 28–33.

[115] J.K.N. Lindner, S. Henke, B. Rauschenbach and B. Stritzker, Thin Solid Films, 279 (1996) 106–109.

[116] E.A. Katz, D. Faiman, S. Shtutina, A. Isakina, K. Yagotintsev and K. Iakoubovskii, Solid State Phenomena, 80–81 (2001) 15–20.

[117] R. Schwedhelm, J.-P. Schlomka, S. Woedtke, R. Adelung, L. Kipp, M. Tolan, W. Press and M. Skibowski, Phys. Rev. B, 59 (1999) 13394–13400.

[118] V.E. Pukha, V.V. Varganov, I.F. Mikhailov, A.N. Drozdov. Phys. Sol. State. 46 (2004) 1574–1576.

[119] W.T. Xu, J.G. Hou, Z.Q. Wu, Appl. Phys. Lett., 73 (1998) 1367–1369.

[120] W.T. Xu and J.G. Hou, J. Appl. Phys., 86 (1999) 4660–4667.

[121] W.T. Xu and J.G. Hou, J. Cryst. Growth, 208 (2000) 365–369.

[122] H. Takashima, M. Nakaya, A. Yamamoto and A. Hashimoto, J. Cryst. Growth, 227–228 (2001) 829–833.

[123] D.J. Kenny and R.E. Palmer, Surf. Sci., 447 (2000) 126–132.

[124] K. Narumi and H. Naramoto, Diamond and Relat. Mater., 10 (2001) 980–983.

[125] A.F. Hebard, O. Zhou, Q. Zhong, R.M. Fleming and R.C. Haddon, Thin Solid Films, 257 (1995) 147–153.

[126] D. Faiman, S. Goren, E.A. Katz, M. Koltun, N. Melnik, A. Shames and S. Shtutina, Thin Solid Films, 295 (1997) 283–287.

[127] D. Stifter and H. Sitter, Appl. Phys. Lett., 66 (1995) 679–681.

[128] J.L. Zeng, Y. Wang, Y.Q. Li and J.G. Hou, Physics C, 282 (1997) 739–740.

[129] D. Stifter, H. Sitter and T.N. Manh, J. Cryst. Growth, 174 (1997) 828–836.

[130] D. Stifter, H. Sitter and T.N. Manh, Thin Solid Films, 306 (1997) 313–319.

[131] J.G. Hou, J. Zeng, Y. Li and Z.Q. Wu, Thin Solid Films, 320 (1998) 179–183.

[132] G. Chen and G. Ma. Thin Solid Films, 323 (1998) 309–316.

[133] Th.B. Singh, N. Marjanović, G.J. Matt, S. Günes, N.S. Sariciftci, A. Montaigne Ramil, A. Andreev, H. Sitter, R. Schwödiauer and S. Bauer, Org. Electron., 6 (2005) 105–110.

[134] H. Gao, Z. Xue and Sh. Pang, J. Phys. D, 29 (1996) 1868–1872.

[135] Z.M. Ren, X.X. Xiong, Y.C. Du, Z.F. Ying, F.M. Li and L.Y. Chen, J. Appl. Phys., 77 (1995) 4142–4144.

[136] X. Zou, S. Zhu, J. Xie and J. Feng, J. Cryst. Growth, 200 (1999) 441–445.

[137] M.A. Khodorkovskii, A.L. Shakhmin, S.V. Murashov, A.F. Alekseev, Y.A. Golod and A.N. Fedorov, Tech. Phys. Lett., 24 (1998) 379–380.

[138] A.L. Shakhmin, M.A. Khodorkovskii, S.V. Murashov, T.O. Artamonova and Y.A. Golod, Technol. Phys. Lett., 27 (2001) 87–89.

[139] M.A. Khodorkovskii, S.V. Murashov, T.O. Artamonova, Y.A. Golod, A.L. Shakhmin, V.L. Varentsov and L.P. Rakcheeva, Technol. Phys., 48 (2003) 523–526.

[140] M.A. Khodorkovskii, S.V. Murashov, T.O. Artamonova, A.L. Shakhmin and A.A. Belyaeva, Tech. Phys. Lett., 30 (2004) 129–130.

[141] M.A. Khodorkovskii, S.V. Murashov, T.O. Artamonova, A.L. Shakhmin, A.A. Belyaeva and V.Y. Davydov, Tech. Phys., 49 (2004) 258–262.

[142] X. Li, H. Wang, W.N. Wang, Y.J. Tang, H.W. Zhao, W.S. Zhan and J.G. Hou, J. Phys.-Cond. Mat., 13 (2001) 3987–4000.

[143] H.Y. Zhang, C.Y. Wu, L.Z. Liang, Y.Y. He, Y.J. Zhu, Y.M. Chen, N. Ke, J.B. Xu, S.P. Wong, A.X. Wei and S.Q. Peng, J. Vac. Sci. Techn. A, 19 (2001) 1018–1021.

[144] H. Zhang, Y. Ding, S. Zhong, C. Wu, X. Ning, Y. He, Y. Liang and J. Hong, Thin Solid Films, 492 (2005) 41–44.

[145] E.A. Katz, D. Faiman, S. Shtutina, A. Shames, S. Goren, B. Mishori and Yoram Shapira, in "Thin Films Structures for Photovoltaics", (E.D. Jones, J. Kalejs, R. Noufi and B. Sopori, eds.), Materials Research Society Series, Warrendale, PA, 1997.

[146] E.A. Katz, D. Faiman, S. Goren, S. Shtutina, B. Mishori and Y. Shapira, Full. Sci. Tech. 6 (1998) 103–111.

[147] E.A. Katz, D. Faiman, S. Goren, S. Shtutina, B. Mishori and Y. Shapira. in "Proceedings of the 14th European Photovoltaic Solar Energy Conference", (H.A. Ossenbrink, P. Helm and H. Ehmann, eds.), Vol. 2, pp. 1777–1779, H.S. Stephens & Associates, Felmersham, Bedford, UK, 1997.

[148] T.L. Makarova, N.V. Seleznev, I.B. Zakharova and T.I. Zubkova, Mol. Cryst. Liq. Cryst. Sci. Tech. C, 10 (1998) 105–110.

[149] T.L. Makarova, A.Ya. Vul', I.B. Zakharova and T.I. Zubkova, Phys. Sol. State, 41 (1999) 319–323.

[150] S. Kano, M. Kohno, K. Sakiyama, S. Sasaki, N. Aya and H. Shimura, Chem. Phys. Lett., 378 (2003) 474–480.

[151] C.A. Mirkin and W.B. Caldwell, Tetrahedron, 52 (1996) 5113–5130.

[152] O.F. Pozdnyakov, B.P. Redkov, B.M. Ginzburg and A.O. Pozdnyakov, Tech. Phys. Lett., 24 (1998) 916–918.

[153] X.N. Yang, J.K.J. van Duren, M.T. Rispens, J.C. Hummelen, R.A.J. Janssen, M.A.J. Michels and J. Loos, Adv. Mater., 16 (2004) 802–806.

[154] W. Kutner, K. Noworyta and F. D'Souza, AIP Conf. Proc., 591 (2001) 53–56.

[155] N.C. Maliszewskyj, P.A. Heiney, D.R. Jones, R.M. Strongin, M.A. Cichy and A.B. Smith III, Langmuir, 9 (1993) 1439–1441.

[156] E.D. Mishina, T.V. Misuryaev, A.A. Nikulin, V.R. Novak, Th. Rasing and O.A. Aktsipetrov, J. Opt. Soc. Am. B: Optical Phys., 16 (1999) 1692–1696.

[157] X.-H. Jiang, X.-T. Zhang, Y.-C. Li, Ya-B. Huang, P.Y. Zhang, D.-J. Wang and Z.-L. Du, Acta Phys.-Chim. Sin., 21 (2005) 209–213.

[158] Y. Kubozono, Y. Takabayashi, K. Shibata, T. Kanbara, S. Fujiki, S. Kashino, A. Fujiwara and S. Emura, Phys. Rev. B, 67 (2003) 115410.

[159] T. Nishikawa, S.-I. Kobayashi, T. Nakanowatari, T. Mitani, T. Shimoda, Y. Kubozono, G. Yamamoto, H. Ishii, M. Niwano and Y. Iwasa, J. Appl. Phys., 97 (2005) 104509.

[160] H. Huang and S. Yang, J. Organomet. Chem., 599 (2000) 42–48.

[161] E.L. Shirley and S.G. Louie, Phys. Rev. Lett., 71 (1993) 133–136.

[162] X. Wei, D. Dick, S.A. Jeglinski and Z.V. Vardeny, Synth. Met., 86 (1997) 2317–2320.

[163] R. Konenkamp, G. Priebe and B. Pietzak, Phys. Rev. B, 60 (1999) 11804–11808.

[164] C. Reber, Y. Lee, J. McKiernan, J.I. Zink, R.S. Williams, W.M. Tong, D.A.A. Ohlberg, R.L. Whetten and F. Diederich, J. Phys. Chem., 95 (1991) 2127–2129.

[165] S. Matsuura, T. Tsuzuki, T. Ishiguro, H. Endo, K. Kikuchi, Y. Achiba and I. Ikemoto, J. Phys. Chem. Solids, 55 (1994) 835–841.

[166] R.R. Hung and J.J. Grabowski, J. Phys. Chem., 95 (1991) 6073–6075.

[167] M. Terazima, N. Hirota, H. Shinohara and Y. Saito, J. Phys. Chem., 95 (1991) 9080–9085.

[168] C. Wen, T. Aida, I. Homna, H. Komiyama and K. Yamada, J. Phys.: Condens. Matter, 6 (1994) 1603–1610.

[169] E.A. Katz, D. Faiman, B. Mishori, Y. Shapira, A.I. Shames, S. Shtutina and S. Goren, J. Appl. Phys., 84 (1998) 3333–3337.

[170] E.A. Katz, D. Faiman, B. Mishori, Y. Shapira, A. Isakina and M.A. Strzhemechny, J. Appl. Phys. 94 (2003) 7173–7177.

[171] A. Hamed, R. Esculante and P.H. Hor, Phys. Rev. B, 50 (1994) 8050–8053.

[172] A. Hamed, R. Esculante and P.H. Hor, Solid State Commun., 94 (1995) 141–145.

[173] B. Pevzner, A.F. Hebard and M.S. Dresselhaus, Phys. Rev. B, 55 (1997) 16439–16449.

[174] T. Asakawa, M. Sasaki, T. Shiraishi and H. Koinuma, Jpn. J. Appl. Phys., 34 (1995) 1958–1962.

[175] A. Hamed, Y.Y. Sun, Y.K. Tao, R.L. Meng and P.H. Hor, Phys. Rev. B, 47 (1993) 10873–10880.

[176] S. Kazaoui, R. Ross and N. Minami. Solid State Commun., 90 (1994) 623–628.

[177] R.A. Assink, J. Schirber, D. Loy, B. Morosin and G.A. Carlson, J. Mat. Res., 7 (1992) 2136–2143.

[178] E.A. Katz, V.M. Lyubin, D. Faiman, S. Shtutina, A. Shames and S. Goren, Solid State Commun., 100 (1996) 781–784.

[179] E.A. Katz, A.I. Shames, D. Faiman, S. Shtutina, Y. Cohen, S. Goren, W. Kempinski and L. Piekara-Sady. Physica B, 273 & 274 (1999) 932–935.

[180] E.A. Katz, D. Faiman, K. Iakoubovskii, A. Isakina, K.A. Yagotintsev, M.A. Strzhemechny and I. Balberg, J. Appl. Phys., 93 (2003) 3401–3406.

[181] D. Sarkar and N.J. Halas, Solid Sate Commun., 90 (1994) 261–265.

[182] S. Kobayashi, T. Takenobu, S. Mori, A. Fujiwara and Y. Iwasa, Appl. Phys. Lett., 82 (2003) 4581–4583.

[183] N.J. Haddock, B. Domercq and B. Kippelen, Electronics Lett., 41 (2005) 444–446.

[184] A. Tapponnier, I. Biaggio and P. Günter, Appl. Phys. Lett., 86 (2005) 112114.

[185] E. Frankevich, Y. Maruyama and H. Ogata, Chem. Phys. Lett., 214 (1993) 39–44.

[186] Y. Chen, F. Stepniak, J.H. Weaver, L.P.F. Chibante and R.E. Smalley, Phys. Rev. B, 45 (1992) 8845–8848.

[187] R.P. Gupta and M. Gupta, Physica C, 219 (1994) 21–25.

[188] D. Sarkar and N.J. Halas, Appl. Phys. Lett., 63 (1993) 2438–2440.

[189] L. Firlej, N. Kirova and A. Zahab, Phys. Rev. B, 59 (1999) 16028–16032.

[190] T. Kanbara, K. Shibata, S. Fujiki, Y. Kubozono, S. Kashino, T. Urisu, M. Sakai, A. Fujiwara, R. Kumashiro and K. Tanigaki, Chem. Phys. Lett., 379 (2003) 223–229.

[191] S. Kobayashi, S. Mori, S. Iida, T. Takenobu, Y. Taguchi, A. Fujiwara, A. Taninaka, H. Shinohara and Y. Iwasa, J. Am. Chem. Soc., 125 (2003) 8116–8117.

[192] N. Hiroshiba, K. Tanigaki, R. Kumashiro, H. Ohashi, T. Wakahara and T. Akasaka, Chem. Phys. Lett., 400 (2004) 235–238.

[193] A.E. Becquerel, C. R. Acad. Sci., 9 (1839) 561.

[194] M. Sze, Physics of Semiconductor Devices, Wiley, New York, 1981.

[195] M. Green. Solar Cells, University of South Wales, Kensington, 1986.

[196] I. Tobías, C. del Canizo and J. Alonso, in "Handbook of Photovoltaic Science and Engineering" (A. Luque and S. Hegedus eds.), pp. 255–306, Wiley, New York, 2003.

[197] J. Nelson, Mater. Today, 5 (5) (2002) 20–27.

[198] J.J.M. Halls, K. Pichler, R.H. Friend, S.C. Moratti and A.B. Holmes, Appl. Phys. Lett., 68 (1996) 3120–3122.

[199] J.J.M. Halls, K. Pichler, R.H. Friend, S.C. Moratti and A.B. Holmes, Synth. Met., 77 (1996) 277–280.

[200] D.E. Markov, E. Amsterdam, P.W.M. Blom, A.B. Seival, J.C. Hummelen, J. Phys. Chem. A, 109 (2005) 5266–5274.

[201] D.E. Markov, J.C. Hummelen, P.W.M. Blom, A.B. Seival, Phys. Rev. B, 72 (2005) 045216.

[202] B.A. Gregg, J. Phys. Chem. B, 107 (2003) 4688–4698.

[203] B.A. Gregg, M.C. Hanna, J. Appl. Phys., 93 (2003) 3605–3614.

[204] J. Nelson, J. Kirkpatrick, P. Ravirajan, Phys. Rev. B, 69 (2004) 035337.

[205] V.D. Mihailetchi, L.J.A. Koster, J.C. Hummelen and P.W.M. Blom, Phys. Rev. Lett., 93 (2004) 216601.

[206] L.J.A. Koster, C.P. Smits, V.D. Mihailetchi and P.W.M. Blom, Phys. Rev. B, 72 (2005) 085205.

[207] L.J.A. Koster, V.D. Mihailetchi, R. Ramaker, P.W.M. Blom, Appl. Phys. Lett., 86 (2005) 123509.

[208] H.H.P. Gommans, M. Kemerink, J.M. Kramer, R.A.J. Janssen, Appl. Phys. Lett., 87 (2005) 122104.

[209] N.K. Persson and O. Inganäs, in "Organic Photovoltaics: Mechanisms, Materials and Devices" (S.-S. Sun, N.S. Sariciftci eds.), pp. 107–138, Taylor & Francis, Boca Raton, London, (2005).

[210] A.F. Hebard, B. Miller, J.M. Rosamilia and W.L. Wilson, US Patent No, 5, 17, 1373, 1992.

[211] H. Yohehara and Ch. Pac, Appl. Phys. Lett., 61 (1992) 575–577.

[212] K. Pichler, M.G. Harisson, R.H. Friend and S. Pekker, Synth. Met., 56 (1993) 3229–3234.

[213] S. Curran, J. Callaghan, D. Weldon, E. Bourdin, K. Cazini, W.J. Blau, E. Waldron, D. McGoveran, M. Delamesiere, Y. Sarazin and C. Hogrel, in: "Electronic Properties of Fullerenes" (H. Kuzmany, J. Fink, M. Mehring and S. Roth, eds.), Vol. 117, pp. 427–433, Springer Series in Solid-State Sciences, 1993.

[214] I. Hiromitsu, M. Kitano, R. Shinto and T. Ito, Solid State Commun., 113 (2000) 165.

[215] I. Hiromitsu, M. Kitano, R. Shinto and T. Ito, Synth. Met., 121 (2001) 1539–1540.

[216] T. Taima, M. Chikamatsu, R.N. Bera, Y. Yoshida, K. Saito and K. Yase, J. Phys. Chem. B, 108 (2004) 1–3.

[217] K.M. Chen, Y.Q. Jia, S.X. Jin, K. Wu, X.D. Zhang, W.B. Zhao, C.Y. Li and Z.N. Gu, J. Phys.: Condens. Matter., 6 (1994) L367–L372.

[218] C. Wen, T. Aida, I. Honma, H. Komiyama and K.Yamada, Denki Kagaku, 62 (1994) 264–267.

[219] K.M. Chen, Y.Q. Jia, S.X. Jin, K. Wu, W.B. Zhao, C.Y. Li, Z.N. Gu and X.H. Zhou, J. Phys.: Condens. Matter., 7 (1995) L201–L207.

[220] N. Kojima, M. Yamaguchi and N. Ishikawa, Jap. J. Appl. Phys. Part 1, 39 (2000) 1176–1179.

[221] K. Kita, C. Wen, M. Ihara and K.Yamada, J. Appl. Phys., 79 (1996) 2798–2800.

[222] D.J. Fu, Y.Y. Lei, J.C. Li, M.S. Ye, H.X. Guo, Y.G. Peng and X.J. Fan, Appl. Phys. A, 67 (1998) 441–445.

[223] Y. Shi, C.M. Xiong, X.S. Wang, C.H. Lei, H.X. Guo and X.J. Fan, Appl. Phys. A, 63 (1996) 353–357.

[224] J.S. Zhu, X.M. Liu, S.H. Xu, J.X. Wu and X.F. Sun, Solid State Commun., 98 (1996) 417.

[225] O. Janzen and W. Mönch, J. Phys.: Condens. Matter., 11 (1999) L111–L118.

[226] K.M. Chen, Y.X. Zhang, G.G. Qin, S.X. Jin, K. Wu, C.Y. Li, Z.N. Gu and X.H. Zhou, Appl. Phys. Lett., 69 (1996) 3557–3559.

[227] K.M. Chen, W.H. Sun, K. Wu, C.Y. Li, G.G. Qin, Y.X. Zhang, X.H. Zhou and Z.N. Gu, J. Appl. Phys., 85 (1999) 6935–6937.

[228] K. Kita, M. Ihara, K. Sakaki and K. Yamada, J. Appl. Phys., 81 (1997) 6246–6251.

[229] A.P. Alivisatos, Science, 217 (1996) 933–937.

[230] E. Arici, N.S. Sariciftci and D. Meissner, Mol. Cryst. Liq. Cryst., 385 (2002) 249–256.

[231] T. Fromherz, F. Padinger, D. Gebeyehu, C. Brabec, J.C. Hummelen and N.S. Sariciftci, Sol. Energ. Mat. Sol. Cells, 63 (2000) 61–68.

[232] C.J. Brabec, S.E. Shaheen, C. Winder, N.S. Sariciftci, and P. Denk, Appl. Phys. Lett., 80 (2002) 1288–1290.

[233] W.J.H. van Gennip, J.K.J van Duren, P.C. Thüne, R.A.J. Janssen and J.W. Niemantsverdriet, J. Chem. Phys., 117 (2002) 5031–5035.

[234] W. Feng and B. Miller, Electrochem. Solid-State Lett., 1 (1998) 172–174.

[235] G.K.R. Senadeera and V.P.S. Perera, Chinese J. Phys., 43 (2005) 384–390.

[236] G.R.R. A Kumara, A. Konno, G.K.R. Senadeera, P.V.V. Jayaweera, D.B.R.A. de Silva and K. Tennakone, Sol. Energ. Mat. Sol. Cells, 69 (2001) 195–199.

[237] Y. Shi, C.M. Xiong, Y.X. He, H.X. Guo, Y.G. Peng and X.J. Fan, Full. Sci. Tech., 4 (1996) 963–975.

[238] M. Rusop, X.M. Tian, S.M. Mominuzzaman, T. Soga, T. Jimbo and M. Umeno, Sol. Energ., 78 (2005) 406–415.

[239] X. Tian, T. Soga, T. Jimbo and M. Umeno, J. Non-Crystalline Solids, 336 (2004) 32–36.

[240] K.M. Krishna, M. Umeno, Y. Nukaya, T. Soga and T. Jimbo, Appl. Phys. Lett., 77 (2000) 1472–1474.

[241] K. Mukhopadhay, K.M. Krishna and M. Sharon, Carbon, 34 (1996) 251–264.

[242] T. Soga, in "Technical Digest of the 14th International Photovoltaic Science and Engineering Conference" (K. Kirtikara Ed.) pp. 687–690, Bangkok, Thailand, 2004.

[243] N. Kojima, O. Goetzberger, Y. Ohshita and M. Yamaguchi, in "Proceedings of the 28th IEEE Photovoltaic Specialists Conference" (A. Rohatgi, ed.), pp. 873–876, IEEE Inc., 2000,

[244] N. Kojima, Y. Ohshita and M. Yamaguchi, AIP Conf. Proc., 590 (2001) 349–352.

[245] N. Kojima, Y. Sugiura, T. Terayama and M. Yamaguchi, in "Technical Digest of the 14th International Photovoltaic Science and Engineering Conference" (K. Kirtikara Ed.), pp. 697–698, Bangkok, Thailand, 2004.

[246] N. Kojima, T. Terayama, H. Suzuki,T. Imaizumi and M. Yamaguchi, in Proceedings of the 31st IEEE Photovoltaic Specialists Conference, pp. 118–120, 3–7 January, Lake Buena Vista, FL, USA, 2005,

[247] M. Chikamatsu, S. Nagamatsu, T. Taima, Y. Yoshida, N. Sakai, H. Yokokawa, K. Saito and K. Yase, Appl. Phys. Lett., 85 (2004) 2396–2398.

[248] M. Chikamatsu, T. Taima, Y. Yoshida, K. Saito and K. Yase, Appl. Phys. Lett., 84 (2004) 127–129.

[249] X. Liu, Y. Jia, L. Guo and G. Wang, Sol. Energ. Mater. Sol. Cells, 87 (2005) 5–10.

[250] S.-C. Yang and T. Mieno, Jpn. J. Appl. Phys., 40 (2001) 1067–1069.

[251] T.A. Skotheim (ed.), Handbook of Conductive Polymers, Vols. 1 & 2, Marcel Dekker, New York, 1986.

[252] N.S. Sariciftci, L. Smilowitz, A.J. Heeger and F. Wudl, Science, 258 (1992) 1474–1476.

[253] S. Morita, A.A. Zakhidov and K. Yoshino, Sol. State Commun., 82 (1992) 249–252.

[254] C.J. Brabec, G. Zerza, N.S. Sariciftci, G. Cerullo, S. DeSilvestri, S. Luzatti and J.C. Hummelen, Chem. Phys. Lett., 340 (2001) 232–236.

[255] N.S. Sariciftci, Prog. Quant. Electr., 19 (1995) 131–159.

[256] N.S. Sariciftci, D. Braun, C. Zhang, V.I. Srdanov, A.J. Heeger, G. Stucky and F. Wudl, Appl. Phys. Lett., 62 (1993) 585–587.

[257] S.B. Lee, P.K. Khabillaev, A.A. Zakhidov, S. Morita and K. Yoshino, Synth. Met., 71 (1995) 2247–2248.

[258] K. Yoshino, K. Yoshimoto, K. Tada, H. Araki, T. Kawai, M. Ozaki and A.A. Zakhidov, in "Fullerenes and Photonics II" (Z.H. Kafafi, eds.), Vol. 2530, pp. 60–75, Proc. SPIE, Bellingham, WA, US, 1995.

[259] S. Morita, A.A. Zakhidov and K. Yoshino, Jpn. J. Appl. Phys., 32 (1993) L873–874.

[260] K. Tada, S. Morita, T. Kawai, M. Onoda, K. Yoshino and A.A. Zakhidov, Synth. Met., 70 (1995) 1347–1348.

[261] L.A.A. Pettersson, L.S. Roman and O. Inganäs, J. Appl. Phys., 86 (1999) 487–496.

[262] G. Yu, J. Gao, J.C. Hummelen, F. Wudl and A.J. Heeger, Science, 270 (1995), 1789–1791.

[263] I.D. Parker, J. Appl. Phys., 75 (1994) 1656–1666.

[264] S.E. Shaheen, C.J. Brabec, N.S. Sariciftci, F. Padinger, T. Fromherz and J.C. Hummelen, Appl. Phys. Lett., 78 (2001) 841–843.

[265] L.S. Roman, M.R. Andersson and O. Inganäs, Adv. Mater., 9 (1997) 1164–1168.

[266] J. Gao, F. Hide and H. Wang, Synth. Met., 84 (1997) 979–980.

[267] C.J. Brabec, F. Padinger, N.S. Sariciftci and J.C. Hummelen, J. Appl. Phys., 85 (1999) 6866–6872.

[268] J.J.M. Halls, A.C. Arias, J.D. MacKenzei, W. Wu, M. Inbasekaran, E.P. Woo and R.H. Friend, Adv. Mater. Res., 12 (2000) 498–502.

[269] D. Gebeyehu, C.J. Brabec, F. Padinger, T. Fromherz, J.C. Hummelen, D. Badt, H. Schindler and N.S. Sariciftci, Synth. Met., 118 (2001) 1–9.

[270] N. Camaioni, L. Garlaschelli, A. Geri, M. Maggini, G. Possami and G. Ridolphi, J. Mater. Chem., 12 (2002) 2065–2070.

[271] T. Aernouts, W. Geens, J. Poortmans, P. Heremans, S. Borghs and R. Mertens, Thin Solid Films, 403–404 (2002) 297–301.

[272] V. Dyakonov, Physica E, 14 (2002) 53–60.

[273] J.M. Kroon, M.M. Wienk, W.J.H. Verhees and J.C. Hummelen, Thin Solid Films, 403–404 (2002) 223–228.

[274] E.A. Katz, D. Faiman, S.M. Tuladhar, J.M. Kroon, M.M. Wienk, T. Fromherz, F. Padinger, C.J. Brabec and N.S. Sariciftci, J. Appl. Phys., 90 (2001) 5343–5350.

[275] W. Geens, S.E. Shaheen, B. Wessling, C.J. Brabec, J. Poortmans and N.S. Sariciftci, Org. Electron., 3 (2002) 105–110.

[276] C.J. Brabec, A. Cravino, D. Meissner, N.S. Sariciftci, T. Fromherz, M.T. Rispens, L. Sanchez and J.C. Hummelen, Adv. Funct. Mat., 11 (2001) 374–380.

[277] M. Reyes-Reyes, K. Kim and D.L. Carroll, Appl. Phys. Lett., 87 (2005) 083506.

[278] P. Schilinsky, C. Waldauf, C.J. Brabec, Appl. Phys. Lett., 81 (2002) 3885–3887.

[279] D. Chirvase, J. Parisi, J.C. Hummelen and V. Dyakonov, Nanotechnology, 15 (2004) 1317–1323.

[280] J.K.J. van Duren, X. Yang, J. Loos, C.W.T. Bulle-Lieuwma, A.B. Sieval, J.C. Hummelen and R.A.J. Janssen, Adv. Funct. Mater., 14 (2004) 425–434.

[281] C.J. Brabec, J.A. Hauch, P. Schilinsky, C. Waldauf and M.R.S. Bulletin, 30 (2005) 50–52.

[282] F. Padinger, R.S. Rittberger and N.S. Sariciftci, Adv. Funct. Mater., 13, (2003) 85–88.

[283] A. Cravino and N.S. Sariciftci, J. Mat. Chem., 12 (2002) 1931–1943.

[284] H. Hoppe, N.S. Sariciftci, in "Organic Photovoltaics: Mechanisms, Materials and Devices" (S.-S. Sun and N.S. Sariciftci, eds.), pp. 217–238. Taylor & Francis, Boca Raton, London, 2005.

[285] F. Zhang, M. Svensson, M.R. Andersson, M. Maggini, S. Bucella, E. Menna and O. Inganäs, Adv. Mater., 13 (2001) 1871–1874.

[286] A.M. Ramos, M.T. Rispens, J.K.J. van Duren, J.C. Hummelen and R.A.J. Janssen, J. Am. Chem. Soc., 123 (2001) 6714–6715.

[287] K. Yoshino, K. Tada, A. Fujii, E.M. Conwell and A.A. Zakhidov, IEEE Trans. Electron Devices, 44 (1997) 1315–1324.

[288] N.S. Sariciftci, in "Optical and Electronic Properties of Fullerenes and Fullerene-Based Materials" (J. Shinar, Z.V. Vardeny and Z.H. Kafafi, eds.), pp. 333–366, Marcel Dekker, New York, 2000.

[289] M. Drees, K. Premaratne, W. Graupner, J.R. Heflin, R.M. Davis, D. Marciu and M. Miller, Appl. Phys. Lett., 81 (2002) 4607–4609.

[290] M. Drees, R.M. Davis, R. Heflin, J. Appl. Phys., 97 (2005) 036103.

[291] T. Shirakawa, T. Umeda, Y. Hashimoto, A. Fujii and K. Yoshino, J. Phys. D: Appl. Phys., 37 (2004) 847–850.

[292] T. Umeda, Y. Hashimoto, H. Mizukami, T. Shirakawa, A. Fujii, K. Yoshino, Synth. Met., 152 (2005) 93–96.

[293] T. Stübinger and W. Brütting, J. Appl. Phys., 90 (2001) 3632–3641.

[294] K. Yoshino and A.A. Zahidov, Synth. Met., 71 (1995) 1875–1876.

[295] H. Yohehara and Ch. Pac, Thin Solid Films, 278 (1996) 108–113.

[296] Ch. Pannemann, V. Dyakonov, J. Parisi, O.Hild and D. Wöhrle, Synth. Met., 121 (2001) 1585–1586.

[297] P. Peumans and S.R. Forrest, Appl. Phys. Lett., 79 (2001) 126–128.

[298] J. Xue, S. Uchida, B.P. Rand and S.R. Forrest, Appl. Phys. Lett., 84 (2004) 3013–3015.

[299] M. Murgia, F. Biscarini, M. Cavalini, C. Taliani and G. Ruani, Synth. Met., 121 (2001) 1533–1534.

[300] D. Godovsky, L. Chen, L. Pettersson, O. Inganäs and J.C. Hummelen, Adv. Mater. Opt. Electron., 10 (2000) 47–54.

[301] Y.J. Ahn, G.W. Kang and C.H. Lee, Mol. Cryst. Liq. Cryst., 377 (2002) 301–304.

[302] J. Xue, S. Uchida, B.P. Rand, S.R. Forrest, Appl. Phys. Lett., 84 (2004) 3013–3015.

[303] M. Murgia, F. Biscarini, M. Cavalini, C. Taliani, G. Ruani, Synth. Met., 121 (2001) 1533–1534.

[304] J. Rostalski and D. Meissner, Sol. Energ. Mater. Sol. Cells, 61 (2000) 87–95.

[305] T. Taima, M. Chikamatsu, Y. Yoshida, K. Saito and K. Yase, Appl. Phys. Lett., 85 (2004) 6412–6414.

[306] B. Maennig, J. Drechsel, D. Gebeyehu, P. Simon, F. Kozlowski, A. Werner, F. Li, S. Grundmann, S. Sonntag, M. Koch, K. Leo, M. Pfeiffer, H. Hoppe, D. Meissner, N. S. Sariciftci, I. Riedel, V. Dyakonov and J. Parisi, Appl. Phys. A: Mater. Sci. Process., 79 (2004) 1–14.

[307] J. Drechsel, B. Männig, D. Gebeyehu, M. Pfeiffer, K. Leo and H. Hopp, Org. Electron., 5 (2004) 175–186.

[308] S. Heutz, P. Sullivan, B.M. Sanderson, S.M. Schultes and T.S. Jones, Sol. Energ, Mat. Sol. Cells, 83 (2004) 229–245.

[309] J. Xue, S. Uchida, B.P. Rand, S.R. Forrest, Appl. Phys. Lett., 85 (2004) 5757–5759.

[310] J. Xue, B.P. Rand, S. Uchida and S.R. Forrest, Adv. Mater., 17 (2005) 66–71.

[311] J. Drechsel, B. Männig, F. Kozlowski, M. Pfeiffer, K. Leo and H. Hoppe, Appl. Phys. Lett., 86 (2005) 244102.

[312] P. Peumans, V. Bulović and S. R. Forrest, Appl. Phys. Lett., 76, 2650 (2000).

[313] C.-W. Chu, Y. Shao, V. Shrotriya and Y. Yang, Appl. Phys. Lett., 86 (2005) 243506.

[314] S. Yoo, B. Domercq and B. Kippelen, Appl. Phys. Lett., 85 (2004) 5427–5429.

[315] J.W. Arbogast and C.S. Foote, J. Am. Chem. Soc., 113 (1991) 8886–8889.

[316] M.M. Wienk, J.M. Kroon, W.J.H. Verhees, J. Knol, J.C. Hummelen, P.A. van Hal and R.A.J. Janssen, Angew. Chem. Int. Ed., 42 (2003) 3371–3375.

[317] L. Fan, S.F. Yang and S.H. Yang, Chem. Eur. J., 9 (2003) 5610–5617.

[318] S.F. Yang and S.H. Yang, J. Phys. Chem. B, 105 (2001) 9406–9412.

[319] S.F. Yang and S.H. Yang, J. Phys. Chem. B, 107 (2003) 8403–8411.

[320] S.F. Yang and S.H. Yang, J. Phys. Chem. B, 108 (2004) 4394–4404.

[321] S.F. Yang and S.H. Yang, Chem. Phys. Lett., 388 (2004) 253–258.

[322] J.Q. Ding and S.H. Yang, J. Phys. Chem. Solids, 58 (1997) 1661–1667.

[323] M. Freitag, Y. Martin, J.A. Misewich, R. Martel and P. Avouris, Nano Lett., 3 (2003) 1067–1071.

[324] J. Guo, C. Yang, Z.M. Li, M.Bai, H.J. Liu, G.D. Li, E.G. Wang, C.T. Chan, Z.K. Tang, W.K. Ge and X. Xiao, Phys. Rev. Lett., 93 (2004) 017402.

[325] A. Javey, J. Guo, Q. Wang and H. Dai, Nature, 424 (2003) 654–657.

[326] M. Cinke, J. Li, B. Chen, A. Cassell, L. Delzeit, J. Han and M. Meyyappan, Chem. Phys. Lett., 365 (2002) 69–74.

[327] M.J. Biercuk, M.C. Llaguno, M. Radosavljevic, J.K. Hyun, A.T. Johnson and J.E. Fischer, Appl. Phys. Lett., 80 (2002) 2767–2769.

[328] D.B. Romero, M. Carrard, W. de Heer and L. Zuppiroli, Adv. Mater., 8 (1996) 899–903.

[329] K. Yoshino, H. Kajii, H. Araki, T. Sonoda, H. Take and S. Lee, Full. Sci. Tech., 7 (1999) 695–711.

[330] H. Ago, M.S.P. Shaffer, D.S. Ginger, A.H. Windle and R.H. Friend, Phys. Rev. B, 61 (2000) 2286–2290.

[331] S.B. Lee, T. Katayama, H. Kajii, H. Araki and K. Yoshino, Synth. Met., 121 (2001) 1591–1592.

[332] H.S. Woo, R. Czerw, S. Webster, D.L. Carroll, J.W. Park and J.H. Lee, Synth. Met., 116 (2001) 369–370.

[333] L.M. Dai and A.W.H. Mau, Adv. Mater., 13 (2001) 899–913.

[334] H. Ago, K. Pettrish, M.S.P. Shaffer, A.H. Windle and R.H. Friend, Adv. Mater., 11 (1999) 1281–1285.

[335] E. Kumakis and G.A.J. Amaratunga, Appl. Phys. Lett., 80, 112–114 (2002).

[336] E. Kumakis, I. Alexandrou and G.A.J. Amaratunga, J. Appl. Phys., 93 (2003) 1764–1768.

[337] B.J. Landi, R.P. Raffaelle, S.L. Castro and S.G. Bailey, Prog. Photovolt: Res. Appl., 13 (2005) 165–172.

[338] E. Kumakis and G.A.J. Amaratunga, Sol. Energ. Mater. Sol. Cells, 80 (2003) 465–472.

[339] M.H.-C. Jin and L. Dai, in "Organic Photovoltaics: Mechanisms, Materials and Devices" (S.-S. Sun and N.S. Sariciftci, eds.), pp. 579–598, Taylor & Francis, Boca Raton, London, 2005.

[340] B.J. Landi, S.L. Castro, H.F. Ruf, C.M. Evans, S.G. Bailey and R.P. Raffaelle, Sol. Energ. Mater. Sol. Cells, 87 (2005) 733–746.

[341] L. Sheeney-Haj-Ichia, B. Basnar and I. Willner, Angew. Chem. Int. Ed., 44 (2005) 78–83.

[342] I. Robel, B.A. Bunker and P.V. Kamat, Adv. Mater., 17 (2005) 2458–2463.

[343] J. Du, L. Fu, Z. Liu, B. Han, Z. Li, Y. Liu, Z. Sun and D. Zhu, J. Phys. Chem. B, 109 (2005) 12772–12776.

[344] H. Bassler, V.I. Arkhipov, E.V. Emelianova, A. Gerhard, A. Hayer, C. Im and J. Rissler, Synth. Met., 135–136 (2003) 377–382.

[345] U.E.H. Laheld and G.T. Einevoll, Phys. Rev. B, 55 (1997) 5184–5204.

[346] J.U. Lee, Appl. Phys. Lett., 87 (2005) 073101.

[347] S.M. Bachilo, M. Strano, C. Kittrell, R.H. Hauge, R.E. Smalley and R.B. Weisman, Science, 298 (2002) 2361–2366.

[348] M.A. Green, K. Emery, D.L. King, S. Igari and W. Warta, Progr. Photovolt: Res. Appl., 13 (2005) 49–54.

[349] M.Y. Sfeir, F. Wang, L. Huang, C.C. Chuang, J. Hone, S.P. O'Brien, T.F. Heinze and L.E. Brus, Science, 306 (2004) 1540–1543.

[350] J. Garica-Vidal, J.M. Pitarke and J.B. Pendry, Phys. Rev. Lett., 78 (1997) 4289–4292.

[351] E. Kucur, J. Riegler, G.A. Urban and T. Nann, J. Chem. Phys., 119 (2003) 2333–2337.

[352] S. Kazaoui, N. Minami, N. Matsuda, H. Kataura and Y. Achiba, Appl. Phys. Lett., 78 (2001) 3433–3435.

[353] T.G. Pedersen, Carbon, 42 (2004) 1007–1010.

OTHER NANOSTRUCTURES

Nanostructured Materials for Solar Energy Conversion
T. Soga (editor)

Chapter 14

Nanostructured ETA-Solar Cells

Claude Lévy-Clément

LCMTR, CNRS, 2-8 rue Henri Dunant, 94320 Thiais, France

1. INTRODUCTION

Over the last 15 years, there has been a steady exploration of new ideas to make solar cells based on nanostructured materials; large interfacial areas in nanostructured materials present significant advantages for light absorption and charge separation, the two critical steps for solar-to-electric energy conversion. Specifically, dye-sensitized solar cells (DSSC) that make use of nanostructured wide bandgap semiconductor films photosensitized with a metal-organic dye have emerged as a promising and potentially low-cost alternatives to the traditional photovoltaic devices based on the p–n junction [1]. A 11% solar energy conversion efficiency has been obtained at the laboratory level for photoelectrochemical cells using TiO_2-based anatase sensitized with a ruthenium (N_3) dye, opening a large new domain of research [2]. But for commercial application, problems related to reliability, long-term stability and engineering of integrated cell components have to be solved. In particular, photochemical degradation of sealants, solvents and dyes is proving to be an important bottleneck in view of extended (>5 years) stability for DSSC [3]. A possible alternative to solve such problems is to replace the organic liquid electrolyte with solid-state or quasi solid-state hole conductor such as p-type semiconductor, ionic liquid electrolyte or polymer electrolyte leading to the generation of solid-state DSSC [4–7]. However, thus far the majority of them suffer similar instability problems [8]. In order to avoid further instability problems, a new variant of the solid-state solar cell has been proposed in which the dye is replaced by a visible-light absorber made from inorganic semiconductor, and the electrolyte is replaced by a solid-state hole acceptor such as a transparent wide bandgap p-type inorganic semiconductor. As a result, the concept of the extremely thin absorber (ETA) solar cell has been recently proposed, based on an inorganic semiconducting,

ETA sandwiched between two transparent ($E_g > 3\,\text{eV}$), highly interpene-trated and nanostructured semiconductors [9]. Conversion efficiency on the order of 2–5% has been achieved, demonstrating the validity of the ETA-solar cell concept [10, 11]. In this chapter, the recent progress on ETA-solar cells is discussed with the factors that decrease their efficiency. Additionally, several potential paths towards higher efficiency are suggested.

2. ETA-SOLAR CELL PRINCIPLE

The ETA-solar cell consists of a nano- or micro-structured layer deposited on a conductive glass substrate, which also serves as an n-type window layer to the cell ($E_g > 3\,\text{eV}$), an absorber ($1.1 < E_g < 1.8\,\text{eV}$) conformally deposited on this layer, and a void filling p-type material with a metallic back contact (Fig. 1a). The physical bases of the ETA-solar cell are a merg-ing of the sensitization (light absorbing) and the p–n junction (charge sepa-ration) concepts as can be seen in Fig. 1(b).

The design principle of an ETA-solar cell (the confinement of a thin layer between wide bandgap n- and p-type layers) should lead to reduced bulk recombination in the absorber because in this configuration the carriers are always produced in the vicinity (or immediate proximity) of the n and p-type layers, respectively, into which they can be injected quasi instanta-neously after photogeneration and before any possible decay. In this case, carrier diffusion length requirements are relaxed in contrast to what is the case for conventional p–n junction cells. Therefore, the quality requirements for the absorber material are significantly lowered. The surface enlargement, by a factor between 10 and 100 relative to flat films allows a reduction in

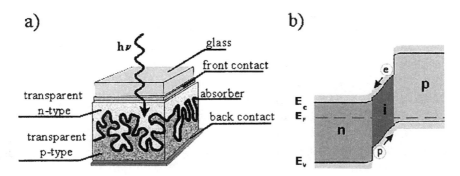

Fig. 1. (a) Schemes of the ETA-solar cell and (b) band diagram of a cell containing n- and p-type semiconductors embedding an ideal adsorber (from Ernst et al. [13]).

the absorber thickness of the same order of magnitude (a few tens of nanometers at the maximum). This reduction in the absorber thickness significantly improves the collection probability, due to the shorter transport path length for the excited carriers. At the same time, scattering at the internal interfaces of the structure will increase the optical path through the sample and thereby enhance the photon absorption. This phenomenon is known as the light-trapping effect [12]. In principle, the sensitized heterojunction solar cells offer great flexibility because the light absorber and the charge transport materials can be selected independently to obtain optimized solar energy harvesting and high photovoltaic output.

In addition to the geometry, the choice of a suitable combination of semiconductors is also important. As can be inferred from the band diagram (Fig. 1b), at the n-type/absorber interface the junction field and the band alignment should be such that electron transfer can occur only from the absorber to the n-contact, while hole transfer is blocked due to the large valence band offset. The conduction band of the p-type semiconductor should be well above that of the n-type semiconductor. Hole transport can then occur only from the absorber to the p-type layer. Therefore, the device enforces the separation of excess charge, driven by the internal electric field and the band alignment.

In the case of an ETA-solar cell, where the absorber is also a good p-type conductor ($CuInS_2$ for example), a simpler architecture can be formed in which the transparent p-type hole conductor is omitted and the porous nanostructure is filled with a p-type absorber instead. In this structure, the average distance from within the absorber to the interface remains on the nanometer scale, and thus one of the key features of the ETA concept is preserved. This kind of nanostructured solar cell is termed 3D-heterojunction or a two-component ETA-solar cell (Fig. 2) by analogy with the three-component ETA-solar cell (Fig. 1a).

ETA-solar cells present formidable challenges in terms of finding deposition techniques for thin (absorber) layers inside porous films, developing new materials (e.g. n- and p-type transparent semiconductors, visible light absorbers considering band alignment, structural, defect chemistry, structural matches and defect chemistry), and solving problems of charge-carrier recombination at the interface.

3. DEPOSITION TECHNIQUES

The basic concept of ETA-solar cells allows the use of low-quality, inexpensive inorganic semiconducting materials, which generally have not been

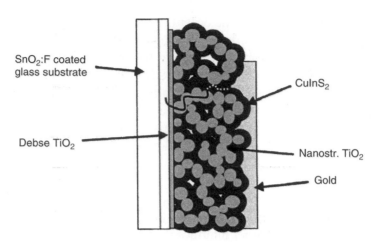

SnO$_2$:F coated
glass substrate

CuInS$_2$

Debse TiO$_2$

Nanostr. TiO$_2$

Gold

Fig. 2. Schema of the two-component ETA-solar cell (from Lenzmann et al. [14]). The electron pathway is indicated as a black line, the hole pathway as a white dotted line.

extensively studied (except for TiO$_2$) by the photovoltaic community. The special properties of nanostructured surfaces require skilful synthesis methods. In all ETA-solar cells one must have the nanostructured interface, currently the initial step is deposition of a highly developed surface area n-type but it could be the p-type. One important factor in the cell assembly is that the absorbing semiconductor and the hole conductor must be deposited throughout the mesoporous or highly nanostructured n-type semiconductor layer, which is usually several microns thick. This condition requires a method that allows infiltration of the reactants into the pores of the n-type semiconductor. Since deposition of a conformal film having intimate electrical contact between the different layers is desired, growth from the liquid phase appears advantageous. Simple methods have been developed which allow controlled infiltration of mesoporous materials or deposition following the contours of nonplanar surfaces. Some of these methods related to ETA-solar cell assembly are summarized below. Generally, a postdeposition annealing at high temperature is necessary to crystallize the film.

The most frequently used materials as window and n-type materials for the ETA-solar cell are porous TiO$_2$ [15, 16] and ZnO nanowires [11, 17–20]. As inorganic absorber materials, CdS [21], CdTe [19, 22], CdHgTe [23, 24], CdSe [11, 18], CuInS$_2$ [10, 25, 26], *a*-Si:H [17] and Se [27] have been used. As a void filling and p-type material, CuSCN [15, 25, 27] and ZnTe [13] have been used.

3.1. Sol-Gel Method

Charge-stabilized nanoparticulate suspensions (sols) are formed by acid- or base-catalyzed hydrolysis and condensation of a metal precursor (most often a titanium isopropoxide), typically in aqueous or aqueous/alcoholic solutions. In aqueous systems, the nanoparticles tend to be stable in suspension against aggregation due to electrostatic repulsion. As water/solvent is removed from the sol, the particles are forced to aggregate into a water-rich gel (hydrogel). The transparent conductive glass (TCO) is placed on a hot plate ($\sim 125°C$) and the solution evenly spreads (with help of a dropper and a glass rod) on the surface and allowed to dry. Finally, when the dried gel (xerogel) is fired/autoclaved for particle sintering at 450°C, a mesoporous TiO_2 ceramic film results with high surface areas conductive for solar cell design [28, 29].

3.2. Spray Pyrolysis

In spray pyrolysis, the reaction takes place from the vapor phase at moderate high temperature, and can be performed in air for oxides. A solution of the dissolved precursor is sprayed in a vector gas as fine droplets onto a heated substrate. The solvent is evaporated or decomposes into gaseous products. The salt reacts to form a deposit generally based on oxide. The solution can also contain two reactive compounds and the temperature of the substrate allows the activation of the chemical reaction between the two compounds.

Such a technique has been used for depositing TiO_2 [30] and ZnO [31] compact layers on the surface of the transparent conductive. The compact layer was deposited before the nanostructured n-type transparent semiconductor in order to avoid any contact and subsequent short-circuiting between the SnO_2:F and p-type semiconductor layer (generally CuSCN). Porous TiO_2 having a duplex (double) structure (meso and macro-porous) [32] and $CuInS_2$ used in ETA-solar cell have also been deposited by spray pyrolysis [33]. For $CuInS_2$, excellent results were found in terms of infiltration when using a noncontinuously spray process [26].

3.3. Chemical Bath Deposition

Chemical bath deposition (CBD) technique has been used for more than 130 years. Yet, only the last decade or two have seen a resurgence of growth, largely due to CBD's success in thin-film photovoltaic cells and its ability to coat large areas in a reproducible and low cost process [34]. CBD is the most common solution method used for deposition of wide range of chalcogenide (e.g. CdS, ZnS, PbS, CdSe), $In(OH)_xS_y$ and chalcopyrite materials (e.g. $CuInS_2$ and $CuInSe_2$). The CBD process uses a controlled chemical reaction to achieve thin-film deposition by precipitation. A soluble salt of the required metal is dissolved in an aqueous solution to release cations. The nonmetallic element is provided by a suitable source compound that decomposes in the presence of hydroxide ions, releasing the anions. The anions and cations then react to form the compound which precipitates onto exposed surfaces. A complexing agent is often used to form metal complexes that limit rates of hydrolysis, imparting some stability to the bath, which would otherwise undergo rapid hydrolysis and precipitation. The homogeneity and stoichiometry of the product are maintained partly by the solubility product (K_{sp}) of the material in question. The deposition produces a conformal coating. The thickness of the CBD film tends to saturate between 50 and 200 nm. Thicker film can be deposited by multiple dip methods. However, the CBD technique can exhibit the risk of clogging the pores leading to voids in the structure. To avoid such problems, ion-layer gas reaction (ILGAR) and sequential methods like successive ion-layer adsorption and reaction (SILAR) have been developed and successfully applied.

3.4. Ion-Layer Gas Reaction

The method [35] has been very successful for depositing CdS, Cu_2S, In_2S_3, $CuInS_2$, and Al_2O_3 thin films. As an example the preparation of $CuInS_2$ by ILGAR is reported. It involves CBD of Cu(I) and In(I) salts by dipping the porous substrate in appropriate precursor solutions. In a subsequent gas-phase reaction with preheated H_2S vapor, the deposited ion salts are converted into $CuInS_2$ and annealed. The typical layer thickness after ILGAR cycle is 10–20 Å. Thicker layers can be obtained by repetitive cycling.

3.5. Successive Ion-Layer Absorption and Reaction

The method is also named sequential CBD (S-CBD) [36, 37]. It is well suited for deposition of sulfide, iodide or thiocyanate compounds. It consists in the sequential immersion of a substrate (the porous TiO_2 film for example) in two separate solutions containing a cation salt (Cd^{2+}, Cu^+) and an anion (S^{2-}, I^- or SCN^-) compound. A standard recipe is given here for CdS [21]. The substrate is successively immersed in four different beakers for about 30 s each: one contained diluted $Cd(NO_3)_2$ aqueous solution (0.05 M), another contains dilute Na_2S and the other two contain distillated water to rinse the sample from the excess of each precursor solution. The Cd^{2+} is adsorbed on the surface of the substrate and the sample is then dipped in water to remove the un-adsorbed ions. For the reaction with the anion the substrate is then dipped in the anionic precursor solution. The loosely bounded ions are removed by rinsing the sample by de-ionized water. One deposition SILAR cycle consists of the following sequence: Cd^{2+} adsorption/rinsing/S^{2-} reaction/rinsing. Such an immersion cycle is repeated several times until the required film thickness is reached (typically between 5 and 20 cycles). After several cycles the film became a dark yellow to orange.

3.6. Solution Casting: CuSCN

The method has been developed for infiltration of CuSCN in porous materials [27, 38, 39]. The CuSCN-deposition solution consists of CuSCN powder dissolved in n-propylsulfide $(CH_3CH_2CH_2)_2S$. Because CuSCN dissolves slowly in propylsulfide, the mixture is stirred overnight and then allowed to settle for at least one day. Solutions can also be doped with $Cu(II)(SCN)_2$ which is prepared by precipitation from a solution of LiSCN and $Cu(ClO_4)_2$ [40].

The final solution contains approximately 0.5% (v/v) CuSCN. The deposition of CuSCN into the pores of TiO_2 or between the ZnO nanowires is carried out on preheated $\sim 80°C$ substrate. The solution is allowed to flow slowly from a pipette while the pipette is moved in a rectangular pattern over the sample, giving a uniform layer. The process can be automatized using a custom-built apparatus designed for making multiple thin layers of low-solubility materials (Fig. 3). To fill a 4 μm porous TiO_2 film ~ 60 coating steps are used.

3.7. Electrodeposition

The method has been used for all the components of the ETA-solar cells: ZnO [11, 17, 41], CuSCN [24, 42, 43], CdTe [44], CdHgTe [45], CdSe [46], ZnTe [47]. In the process of electrochemical deposition, the conducting substrate is used as an electrode in an electrochemical cell and the deposition occurs by reduction or oxidation of the dissolved species. Electrodeposition (ECD) of semiconductor compound may also involve the co-deposition of two elements, which belong generally to II–VI and II–V families in the periodic table. For example reactions of electrodeposition of ZnO or CdSe can be summarized as the following:

$$Cd^{2+} + Se + 2e^- = CdSe \tag{1}$$

$$Zn^{2+} + 0.5O_2 + 2e^- = ZnO \tag{2}$$

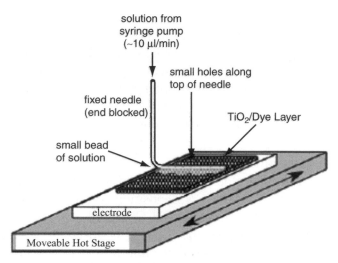

Fig. 3. In-house built machine for the automatic deposition of CuSCN (from O'Regan et al. [39]).

The conditions of deposition of various semiconductors used in ETA-solar cells are reported in Table 1. Thickness of the deposited layer depends on the amount of charge passed.

3.8. Gaz-Phase Deposition: Atomic Layer Deposition

The technique has been used for $CuInS_2$, Al_2O_3, In_2S_3 compounds [48]. The principle of chemical vapor deposition (CVD) is close to that of spray pyrolysis. The deposition is obtained thanks to a chemical reaction initiated by gaseous precursors in a reactor under vacuum. The reaction is activated by the temperature of the substrate generally above 500°C. For the gas-phase deposition of $CuInS_2$ thin films the metal precursors CuCl and $InCl_3$ are used, along with H_2S as the sulfur precursor. Alternating pulses of two different precursors, separated by a purge pulse of an inert gas, constitute a typical Al-CVD cycle. Multiple Al-CVD cycles are used to deposit a film. Between two pulses of different reactants, the Al-CVD reaction chamber is purged with argon. The metal precursors are evaporated from quartz boats inside the oven and the vapors are carried to the chamber by an argon flow. Evaporation temperatures are selected between 325 and 380°C for CuCl and between 300 and 350°C for $InCl_3$. The reaction temperature is between 350 and 500°C and the pressure is varied between 2 and 10 mbar. One deposition cycle consists of the following sequence: $CuCl$/purge/H_2S/purge/$InCl_3$/purge/H_2S/purge [48].

Table 1
Examples of semiconductors used in ETA-solar cells, obtained by electrochemical deposition

Materials	Precursors in solution	pH/temp.	Applied potential/current/charge	Morphology/film thickness	Post-treatment/anne aling	Ref.
ZnO	5×10^{-4} M ZnCl$_2$ + O$_2$ bubling + 0.1 M KCl	80°C	-1 V (SCE) 20 C/cm^2	Nanowires 1–2 μm		[17, 41]
CuSCN	0.05 M Cu(BF$_4$)$_2$ + 0.125 M KSCN	C.7–0.9 vs Pt				[42, 47]
ZnTe	0.02 M ZnSO$_4$ + 1×10^{-4} M TeO$_2$ + 0.5M LiNO$_3$	4.5/100°C	-1 V (Ag/AgCl)	Platelets/100 nm	30 min in Ar (to remove excess Te)	[13, 44]
CdTe	0.5 M CdSO$_4$ + 2.5×10^{-4} M TeO$_2$	1.6–2.2/90°C	-0.59 V (Ag/AgCl)	Polycrystalline/ 80–500 nm	3 min in sat. CdCl$_2$ in MeOH before boiling/400°C in Ar	[13, 47]
CdTe$_{1-x}$Hg$_x$						[45]
CdSe	0.05 M CdC$_4$H$_6$O$_4$ + 0.01 NTA + 0.05 selenosulfate with excess sulfite	8/20°C	-2.7 mA/cm^2/0.25 C/cm^2	Polycrystalline/ 20–30 nm	350°C in Ar for 1 h	[18, 46]
Se	0.5 M SeO$_3^{2-}$		2.7 mA/cm^2/3 min		150°C for 5 min	[27]

4. ETA-SOLAR CELL COMPONENTS

4.1. Nanostructured n-Type Semiconductors

Metal oxides such as TiO_2, ZnO and SnO_2 are promising electron acceptors. For example both ZnO and TiO_2 are n-type materials due to oxygen deficiency. They offer good electron transport properties, excellent physical and chemical stability and fabrication via facile techniques. Moreover they offer the possibility to control the microstructure in order to create a large interfacial area when coated with an absorber film.

4.1.1. TiO₂ Layer

The TiO_2 layer consists of anatase crystallites with a $\sim$30–50 nm size. Two kinds of layer with a different overall morphology are used in the ETA-solar cells.

The first kind corresponds to a very homogeneous mesoporous material (Fig. 4). Gas-phase measurements (BET) indicated that the inner surface area is 100–300 times larger than the geometric projection. Hence the thicker the film, the more surface area magnified. The mesoporous TiO_2 is obtained by a sol-gel method using Ti alkoxide precursors or adding TiO_2 powder (Degussa P25, for example) in ethanol. The porous films were prepared by

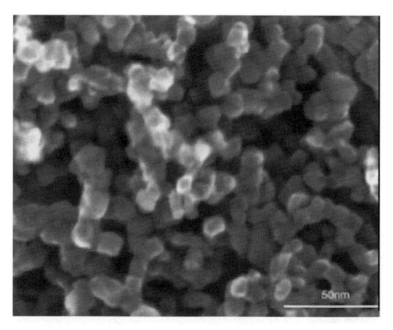

Fig. 4. SEM image of mesoporous TiO_2 doctor blading deposited film (Grätzel [49]).

doctor blading or screen printing and fired at 450°C in air for 30 min. The thickness of the mesoporous TiO$_2$ is ~2–5 μm. The film does not scatter the light very efficiently and optical bandgap is 3.2 eV.

The second kind corresponds to a macroporous TiO$_2$ and has a double structure. On the nanometer scale, the size of the TiO$_2$ crystallite and shape are similar to the mesoporous TiO$_2$, but on the micrometer scale the film is highly inhomogeneous exhibiting a widely open morphology (Fig. 5). Scanning Electron Microscopy (SEM) analysis shows that the surface enlargement on the micrometer scale is a factor ten. BET measurements however show that on the nanoscale the surface enlargement is similar to that of the mesoporous film [22]. The macroporous TiO$_2$ film is obtained by spray pyrolysis of 3.5 M titanium tetra-isopropylate in isopropanol at 160°C on a glass substrate (180°C). The macroporous morphology is favorized by the pyrolitic process, which agglomerates TiO$_2$ crystallites. After the spray deposition, a final annealing step at 450°C is applied. The thickness of the macroporous TiO$_2$ is ~3 μm. The visible light is strongly scattered in this type of film.

4.1.2. ZnO Layer

Another n-type semiconductor used in ETA-solar cells design is zinc oxide. Several electronic properties of ZnO (e.g. positions of the conduction

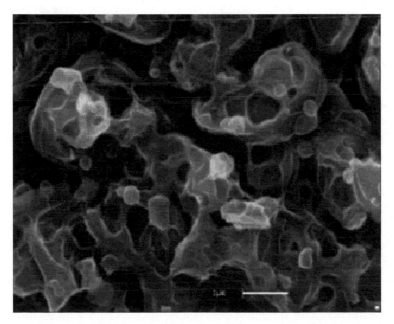

Fig. 5. SEM image of macroporous TiO$_2$ spray-pyrolyzed film (from Ernst et al. [13]).

Table 2
n-type wide bandgap oxides used in ETA-solar cells

n-type	Bandgap (eV)	Synthesis method	Surface enhancement times	Mobility (cm²/Vs)	Ref.
Mesoporous TiO$_2$ (anatase)	3.23	Sol-gel + doctor blading of an aqueous colloidal paste or screen printing	400	10^{-2} (Degussa P25)	[49–51]
Macroporous (μp) TiO$_2$		Spray pyrolysis	30–60		[32]
ZnO nanowire	3.31	Electrochemical	15	23	[17] [52a]

and valence bands, charge separation rates and bandgap) are comparable to TiO$_2$, but the electron mobility in ZnO is much greater than in TiO$_2$. Recently, a very low electron mobility measured by tetrahertz spectroscopy of 10^{-2} cm²/V s was reported [51] in nanocrystallite TiO$_2$–P25 which is two orders of magnitude lower than in ZnO nanowires [17]. ZnO is an inexpensive and environmental friendly material that can be synthesized in high purity and crystallinity at low temperature. In addition it can easily be grown with different morphologies (Table 2).

Vertically aligned ZnO nanowire arrays are easily grown electrochemically onto a transparent electrode composed of a F:SnO$_2$ layer on top of which is deposited by spray pyrolysis a compact ZnO layer [31]. Each nanowire is a single crystal with a wurtzite structure, oriented in the *c*-axis (002) direction that provides a highly efficient direct conduction path for electrons to the electrode (back contact) (Fig. 6). The ZnO nanowire diameter is between 20 and 150 nm and the length is 0.5–3 μm [11, 52a]. The surface enlargement is a factor 10–30 from that of a geometric projection.

4.2. Absorbers: CdTe, CuInS$_2$, a-Si, CdSe, CdS, Cu(In,Ga)S$_2$

The choice of the absorber is guided by their ability to absorb visible light ($1.1 < E_g < 1.8$ eV) and transfer electrons to large bandgap semiconductors. Photoelectrochemical experiments showed that CdS [53–55], PbS [56, 57], Bi$_2$S$_3$ [55, 58], CdSe [59] and InP [60] fulfill those requirements. CdTe and CuInS$_2$ with a bandgap around 1.5 eV are also able to inject electrons into macroporous TiO$_2$ and ZnO nanowires when used in solid-state devices [10, 20, 22, 25]. Semiconductor quantum dots (QDs) such as PbS with its tunable band

Fig. 6. SEM image of a ZnO nanowire array obtained by electrodeposition exhibiting a 25 surface enlargement factor.

Table 3
Various absorbers used in ETA-solar cells

Absorber	Bandgap (eV)	Absorption coefficient (cm^{-1})	Synthesis method	Ref.
$Cu_{1.8}S$	1.75	2.3×10^4 ($\lambda = 500$ nm)	AL-CVD	[61]
$CuInS_2$	1.55	10^5 ($\lambda = 500$ nm)	ILGAR	[35, 62]
			AL-CVD	[48]
CdS	2.5		SILAR	[21]
CdSe	1.75		ECD	[11, 18]
CdTe	1.54	1.56×10^4	EC and OMCVD	[13, 22]
	1.45			[20]
$Cd_{1-x}Hg_xTe$	1.07		ECD	[23, 24]
PbS Qdots	0.85		CBD	[56, 63]
	($\sim$1.1)		SILAR	[64]
a-Si-H$_x$	1.6		CVD	[17]

edge offer new possibility for harvesting light energy in the visible [56]. Several of the above cited semiconductors have been used in ETA-solar cells (Table 3).

The major task is to obtain uniform deposition of the absorber on the nano-structured or porous n-type materials. As already mentioned above, the techniques based on liquid phase are particularly well suited, because they are not expensive. However good results have been obtained with more expensive gas-phase techniques such as metal organic CVD (MOCVD) (CdTe, [20]), plasma-assisted CVD (a-Si:H [17]) and Al-CVD (CuInS$_2$ [48]).

Chalcosite Cu$_2$S and digenite Cu$_{1.8}$S are possibly interesting semiconductors for photovoltaic applications [61]. Their electronic structure is poorly understood because their crystal structure is complex. It consists of a close-packed lattice of S with mobile Cu occupying various types of interstitial sites with a statistical distribution depending on temperature. The experimental evidence for a semiconducting gap of about 1 eV.

4.3. Transparent p-Type Semiconductor (Hole Collector)

Candidate materials to be used as transparent hole conductors can be inorganic semiconductor such as CuI [65–67], CuSCN [4, 27, 38, 67–69], CuAlO$_2$ [70], NiO [71], or organic polymer such as spiro-OmeTAD, PEDOT: PSS (Poly (3,4-ethylenedioxythiophene) doped with polystyrene sulfonic (PSS) acid. In ETA-solar cell CuSCN and PEDOT:PSS have been used (Table 4).

4.3.1. CuSCN

CuSCN can be obtained by electrochemical deposition, solution casting and SILAR. The ETA-solar cells take advantage of the research done on

Table 4
p-type semiconductors used as hole collectors in ETA-solar cells

p-type	Bandgap (eV)	Synthesis method	Ref.
		Precipitation	[65]
CuI (γ-phase)	3.1	ECD + I$_2$ gas reaction	[66]
		SILAR	[67]
		ECD	[4, 68]
CuSCN	3.4	Solution casting	[27, 38, 69]
		SILAR	[67]
CuAlO$_2$	3.54	Ion exchange	[70]
NiO	3.55	Dipping in solution + firing at 400°C	[71]
ZnTe	2.1	Electrodeposition	[13]
PEDOT:PSS	5.2 (work function)	Spin casting from undiluted solution	[73]

solid-state DSSC. It was found that CuSCN electrodeposited was highly crystallographically oriented and physical contact with porous TiO$_2$ was not excellent. The solution casting method was reported to produce good impregnation inside TiO$_2$. In ETA-solar cell, undoped CuSCN deposited by solution casting has led to satisfactory results.

4.3.2. PEDOT:PSS

Because of its excellent electronic properties (electrical conductivity, electrochromic properties) and high stability, PEDOT stands out as one of the most studied and commercially most successful conducting polymer. Once doped with PSS acid oligomer, it becomes highly conductive [72]. Heavily doped PEDOT:PSS was chosen based on the value of its work functions (5.2 eV) [73] (Au work function is 5.1 eV close to the 4.85 eV work function of the I$^-$/I$_3^-$ couple. Generally its penetration in the sensitized n-type semiconductor is obtained by spin casting from undiluted aqueous solution of PEDOT:PSS (Baytron P™).

4.4. Intermediate Layers

4.4.1. Compact Oxide Layer

The hole conductor can form an ohmic contact with the SnO$_2$:F deposited on the glass substrate allowing charges to recombine at this interface. In order to avoid direct contact between the two materials which would short-circuit the solar cell, it is necessary to deposit a dense hole-blocking layer onto the SnO$_2$:F layer. The compact layer is an oxide, which currently has the same chemical composition as the n-type semiconductor. A 100 nm compact TiO$_2$ or ZnO layer is deposited by spray pyrolysis between the con-ducting glass and the porous TiO$_2$ or ZnO nanowires array, respectively. Spray pyrolysis of a solution contained titanium tetra-isopropoxide and acetylacetone in ethanol or zinc acetate dihydrate and acetic acid in water/ethanol mixture is performed at a temperature of 450°C. In the case of ZnO, the spraying process consists of cycle of 2 s with spray and 10 s without spray. These cycles avoid an abrupt decrease of the substrate temperature and, there-fore, nonuniform ZnO deposition due to the condensation of the spray.

4.4.2. Passivation of Interface Recombination: Electronic Engineering of the n-Type Semiconductor/Absorber Interface

A general finding is that for TiO$_2$-based ETA-solar cells new materials have to be introduced that act as buffer layer between the ETA light absorber and the TiO$_2$ substrate. While spontaneous decay of charge carrier

Table 5
Insulating oxide coating (blocking layer) and non-insulating buffer layer

Intermediary layer	Bandgap (eV)	Synthesis method	Ref.
Al_2O_3	9	Al-CVD	[10, 84]
		ILGAR	[85]
MgO	8	ED	[74]
CdS	2.4	CBD	[76]
		ILGAR	[35]
In_2S_3	2.1	Al-CVD	[10]
$In(OH)_xS_y$	2.8	Wet chemical precipitation	[89]
ZnO	3.4	ED	[74]

recombination within the absorber may be efficiently suppressed due to the ETA-solar cell configuration, the substantially enlarged surface area is expected to lead to a considerable increase of interface recombination probability (as compared to a flat surface of the same optical density). Therefore the passivation of interface recombination is of fundamental importance in these systems. The research done on the electronic engineering of solid-state DSSC has been successfully transferred to ETA-solar cells. The TiO_2 surface has been modified by introducing an insulating metal-oxide overlayer through which electron tunneling can occur (Al_2O_3, ZnO, MgO) [74] or buffer layer that relies on suitable band positions (CdS, In_2S_3) [75, 76] (Table 5).

4.4.2.1. Insulating oxide coating (blocking layer). Ultra-thin (≤ 2 nm) conformal insulating coatings as a mean to control/slow down the active interface recombination kinetics have been introduced in the field of DSSC [77–79]. The strategy has been particularly successful in the liquid junction dye solar cells based on mesoporous SnO_2/Dye/I^-/I_3^-. The open circuit (V_{oc}) and overall light-electricity conversion efficiency of the cell could be more than doubled in the presence of Al_2O_3 barrier coating [77]. Influence of insertion of various very thin insulating oxide coatings including Al_2O_3, MgO, ZnO, ZrO_2, Y_2O_3 has been recently summarized showing that some of them led to significant improvement of solid-state dye solar cell characteristics [74]. In TiO_2-based solid-state dye cells, TiO_2/dye/CuI [8, 80, 81] and TiO_2/dye/ CuSCN [82] considerable improvements in V_{oc} and fill factor (FF) were obtained when coating TiO_2 with Al_2O_3. The coating of Al_2O_3 was based on sequential chemisorption of a metal alkoxide on hydroxylated mesoporous

TiO$_2$ thin films and hydrolysis (often with a thermal annealing step) of the precursor in order to obtain the conformal insulator layer.

There are different theories for the physical function of ultra-thin Al$_2$O$_3$ coating, but the consensus is that the beneficial effect is related to a decrease of the interface recombination rate [77, 79, 83]. This has been confirmed also for ETA-solar cells based on mesoporous TiO$_2$/CuInS$_2$ [10, 14], in which the Al$_2$O$_3$ tunnel barrier coating was applied by the atomic-layer deposition (Al-CVD) technique using AlCl$_3$ and O$_2$ as precursor gases [84]. Passivation of sol-gel TiO$_2$ by ultra-thin layers of Al-oxide has been investigated using transient and spectral photovoltage (PV) techniques. The ultra-thin layer of Al-oxide in this analysis was prepared by the ILGAR technique and modified by thermal treatments in air, vacuum or Ar/H$_2$S atmosphere [85]. The samples were characterized by elastic recoil detection analysis (ERDA), X-ray photoelectron spectroscopy (XPS) and contact potential difference (CPD) technique. This study showed that without an Al-oxide surface layer atmosphere, electronic states in the forbidden gap of TiO$_2$ were formed during thermal treatments in vacuum and Ar/H$_2$S. The trap density was strongly reduced at the TiO$_2$/Al-oxide interface and the formation of electronic defects was prevented by a uniform ultra-thin layer of Al-oxide [85]. The effect of the intermediate layer has also been theoretically studied in CuInS$_2$-based ETA-solar cells. Physical reasons were suggested for the observed cell improvement, such as minimization of interface recombination (creation of an extrinsic interface, reduction of interface state density, interface band diagram tailoring) and chemical and electrostatic interactions at the interface (chemical buffer layer, stopping unwanted interdiffusion of species at interface) [86].

4.4.2.2. Noninsulating buffer layer. Improvement of cell characteristics on applying an intermediate noninsulating buffer layer (CdS, In$_2$S$_3$) in a TiO$_2$/CuInS$_2$ ETA system was also reported [10, 76]. In(OH)$_x$S$_y$ *** and In$_2$S$_3$ are known as buffers in conventional CuInS$_2$ thin-film solar cells (CIS, Mo/CuInS$_2$/In(OH)$_x$S$_y$/ZnO/ZnO:Al) where they improve the junction properties between CuInS$_2$ and ZnO [87]. The In$_2$S$_3$ layer was obtained by Al-CVD [88]. The In(OH)$_x$S$_y$ layers were prepared by a wet chemical precipitation process in a solution of InCl$_3$, thioacetamide and acetic acid [89]. In(OH)$_x$S$_y$ has proved to be a good buffer for TiO$_2$ solar cells. One of the requirements of a buffer layer is to have a wide bandgap in order to allow light to reach the absorber. A study of In(OH)$_x$S$_y$ deposited on conductive glass has then be carried out. Values between 2.4 and 3.4 eV have been measured when pH of the solution varies between 2.6 and 3. This change in bandgap was related to the stoichiometry of $In(OH)_xS_y$: higher content of S is expected for lower pH value while the contrary is observed for higher pH values (higher hydroxide or oxide contents). After annealing in argon at 300°C, E_g decreased from 2.4 to 2.2 eV indicating that hydroxide and oxide ions left the buffer layer. The interfacial $In(OH)_xS_y$ layer did improve the overall performance of ETA-solar cells using $CuInS_2$ as adsorber and hole conductor. It was shown that its presence decreases the electron–hole recombination [75].

Deposition of an interfacial CdS layer between TiO_2 and CdSe also improved the performance of the photoelectrochemical cell. This was attributed to passivation effect of the CdS on the TiO_2 surface rather than connected with energy level alignments [76].

5. VARIOUS TYPES OF ETA-SOLAR CELLS

The ETA-solar cells in the literature have two distinct conformations of materials sandwiched between the two conductive electrical ohmic contacts. One is a two-component system, which consists of a binary matrix of mesoporous TiO_2 (anatase) filled with a high purity smaller bandgap p-type material, which generates electron–hole pairs and plays the role of a majority hole carrier. The other one is a three component system which consists of a mesoporous TiO_2 or ZnO nanowire matrix coated by an extremely thin metal chalcogenide absorber and a wide bandgap p-type material (e.g. CuSCN or PEDOT:PSS). It is very important to note that not only the bandgap values must be taken into consideration for ETA-solar cell design, but the energy levels of the electronic bands have to be included too. The conduction band of the p-type component must be well above that of the n-type material, which should be below that of the absorber. Furthermore the processing conditions for the deposition of the components must be compatible. Finally cheap and nontoxic materials are required.

5.1. Two-Component ETA-Solar Cells or 3D Solar Cells

CdTe, after subsequent dip in saturated $CdCl_2$ solution in methanol and annealing in air at 400°C [90] and $CuInS_2$ are p-type semiconductors. It is well known from CdTe and $CuInS_2$ thin-film solar cells that they are able to conduct charge carriers very well, even up to a layer thickness of 1 μm. The resulting two-component solar cells are based on TiO_2 (Table 6).

Table 6
Characteristics of various two-component ETA-solar cells

Solar cells	Surface (cm²)	I_{sc} (mA/cm²)	V_{oc} (V)	FF	η (%)	Ref.
TiO₂/CdTe/Au	<1	8.9	0.67		2	[22, 23, 24]
TiO₂/Cd$_{1-x}$Hg$_x$Te/Au		15	0.57			
TiO₂/Al₂O₃/In₂S₃/ CuInS₂	0.03					
1. Made by Al-CVD		18	0.49	0.44	4	[10]
2. Made by spray pyrolysis		17	0.53	0.55	5	[26]

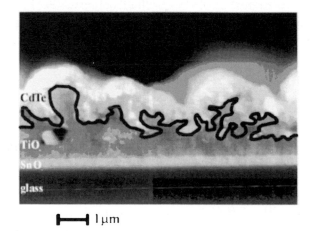

Fig. 7. SEM image of the cross section of the electrodeposited CdTe on macroporous TiO₂. The interface is marked with a black line (from Belaïdi et al. [23]).

5.1.1. *Macroporous TiO₂/CdTe Solar Cell [22–24]*

An early approach to the ETA-solar cell concept was developed at Hahn Meitner Institute-Berlin (Germany). The solar cell involved a 150 nm thick CdTe absorber deposited by electrochemical deposition on spray-pyrolyzed macroporous TiO₂ substrates supported on a conductive glass. The electrical back contact was a gold layer deposited on CdTe. The CdTe layer filled the TiO₂ macropores (Fig. 7). The enhancement of the TiO₂/CdTe interface was around 10 and the local thickness of CdTe layer was approximately one order of magnitude thinner than what is typical in conventional solid-state thin-film solar cells. The macroporous Ti₂O/CdTe/Au device gave 0.67 V as open-circuit voltage and 8.9 mA/cm² short-circuit current for 100 mW/cm² illumination [22, 24] (Table 6).

The energy bands of the TiO_2 and CdTe bands were measured by XPS and UPS [91]. Using the bandgap value of these compounds taken from the literature, a conduction band offset of 0.7 eV was found (Fig. 8). Numerical modeling for the $CdTe/TiO_2$ solar cell based on the simulation program SCAPS indicated that this large band offset was responsible for the low FF. To decrease this offset, alloying with mercury (Hg) appeared to be a solution, as it is known to reduce the bandgap by lowering the conduction band edge while maintaining the valence band at the same energy. Absorber layers of CdHgTe with different contents of Hg were made by electrochemical deposition. Solar cells composed of macroporous $TiO_2/Cd_{1-x}Hg_xTe$, exhibiting open-circuit voltage of 0.57 V and improved short-circuit current of 15 mA/cm², were obtained for an absorber with bandgap of 1.2 eV [23, 24]. However, the FF lied between 20 and 30% and remained a challenging problem to be solved in these cells.

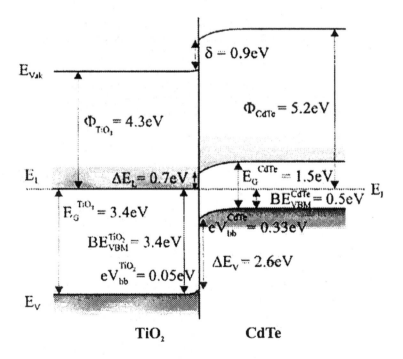

Fig. 8. Band alignment in the TiO_2/CdTe system. All data are from XPS and UPS measurements. TiO_2 values are taken from the literature (from Tiefenbacher et al. [91]).

For the passivation, a CdS buffer layer has been introduced between TiO$_2$ and CdTe [22]. A higher FF was observed but the short-circuit current was reduced progressively for thicker CdS layers indicating a series resistance effect and adsorption losses in the CdS layer. The open-circuit voltage decreased by approximately 100 mV and was independent of the method of preparation or thickness of CdS and post-treatment.

5.1.2. *Mesoporous TiO$_2$/Al$_2$O$_3$/In$_2$O$_3$/CuInS$_2$ Solar Cell [10, 26, 88]*

The solar cell was developed at Delft University of Technology (Netherlands). It was composed of a 2 μm thick mesoporous TiO$_2$ anatase layer whose pores were filled with high quality p-type CuInS$_2$ (Fig. 9) deposited by Al-CVD method [10, 84]. The internal surface of TiO$_2$ was about 500 times the geometrical area. By careful tuning of all the relevant Al-CVD parameters, the pores of the TiO$_2$ were completely filled with CuInS$_2$ (as seen by TEM analysis [10]). The results also showed that the lattices of TiO$_2$ and CuInS$_2$ were in intimate contact on the atomic scale, prerequisite for fast electron transfer across the interface. To get a good photoresponse, a decrease of sulfur vacancies was necessary, obtained by postdeposition annealing at 500°C in sulfur vapor followed by an annealing at 200°C in oxygen. And to suppress surface recombination a conformal 10 nm thick In$_2$S$_3$ buffer layer was deposited on TiO$_2$ by Al-CVD. In$_2$S$_3$ bandgap increases up to 2.8 eV due to contamination with oxygen.

Conduction band electrons can easily cross the interface between CuInS$_2$ and In$_2$S$_3$ as their conduction band energies are similar (Fig. 10),

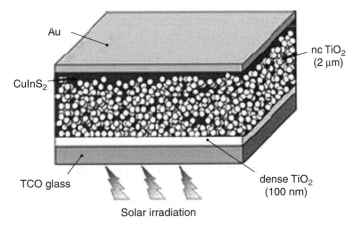

Fig. 9. Structure of the two-component macroporous TiO$_2$/CuInS$_2$ solar cell (from Nanu et al. [10]).

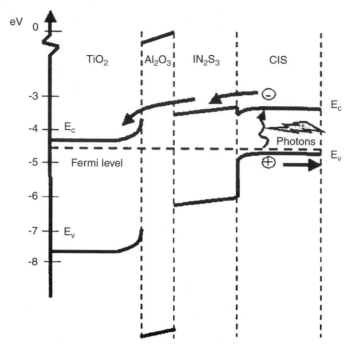

Fig. 10. Energy band diagram of a TiO$_2$/Al$_2$O$_3$/In$_2$S$_3$/CuInS solar cell (from Nanu et al. [10]).

where they are injected into TiO$_2$ thanks to the 1 eV conduction band offset when compared to anatase conduction band energy. Additionally a 1–2 nm amorphous Al$_2$O$_3$ tunnel barrier made by Al-CVD had to be included in the device between the mesoporous TiO$_2$ and In$_2$S$_3$. This prevented TiO$_2$ surface-modification during the Al-CVD and annealing process, due either to oxide-ion vacancies or copper migration from CuInS$_2$ through In$_2$S$_3$ buffer into the TiO$_2$ lattice. This research led to a breakthrough in the two-component ETA-solar cell study. The conversion efficiency of the solar cell was 4% under AM1.5 irradiation with an I_{sc} of 18 mA/cm^2, a V_{oc} of 0.49 V and a 0.44 FF between 360 and 900 nm with a maximum monochromatic incident photon-to-current conversion efficiency (external quantum efficiency, EQE) of 80% (Fig. 11).

Similar good results have been obtained by depositing CuInS$_2$ films with 300–400 nm thickness by spray deposition (300°C) into 2 μm thick mesoporous TiO$_2$(100 nm nanocrystallites and 50% porosity) [26]. The inter-penetrating network of CuInS$_2$ and TiO$_2$ was obtained by noncontinuously

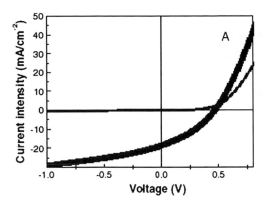

Fig. 11. Current voltage characteristics of a $TiO_2/Al_2O_3/In_2S_3/CuInS$ solar cell made by Al-CVD (from Nanu et al. [10]).

spraying process. According to XRD, well-crystallized $CuInS_2$ films are formed on colloidal and compressed mesoporous TiO_2 underlayers. The introduction of the buffer layer $In_x(OH)_yS_z$ was made by aerosol spray. An energy-conversion efficiency of 5% under AM1.5 irradiation was reported [26]. Hence, this new 3D solar-cell design has the potential to lower the price barrier and open up new production technologies for low-cost photovoltaic solar cells.

5.2. Three-Component ETA-Solar Cells

Initial reports on three-component ETA-solar cells were not very encouraging. Tennakone et al. [27] reported the $TiO_2/Sc/CuSCN$ heterostructure as a photovoltaic cell with an energy conversion efficiency ~0.13% in 1998. Three years later, the Könenkamp group [25] proposed the $TiO_2/CuInS_2/CuSCN$ heterostructure as a good candidate to act as an ETA-solar cell. They studied the dependence of the photocurrent as a function of the absorber layer thickness and showed the rectifying behavior of their samples under dark conditions, but did not report its energy conversion efficiency. Recently, several ETA-solar cells with different configurations were reported with conversion efficiency larger than 1% giving a clear demonstration on the potentiality of such a solar-cell concept (Table 7).

5.2.1. TiO₂/CdS/CuSCN ETA-Solar Cell [21]

Mesoporous $TiO_2/CdS/CuSCN$ ETA-solar cells were produced at IMRA-Europe (France). They were composed of a transparent conductive glass

Table 7
Characteristics of various three-component ETA-solar cells

ETA-solar cell	Surface (cm^2)	I_{sc} (mA/cm^2)	V_{oc} (V)	FF	η (%)	Ref.
TiO$_2$/CdS/CuSCN	0.5	>2	>0.8	0.6	1	[21]
TiO$_2$/In(OH)$_x$S$_y$/PbS/ PEDOT:PSS	0.03	8	0.3		1	[93, 94]
ZnO/CdTe/CuSCN	1	0.03	0.2	0.28		[20]
ZnO/CdSe/CuSCN	1	3.9^a	0.49	0.42	2.3	[11, 18]

[a]Value obtained under 1/3 sun illumination.

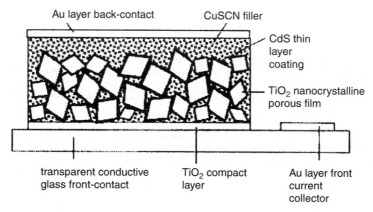

Fig. 12. Schematic of the TiO$_2$/CdS/CuSCN ETA-solar cell (from Larramona et al. [21]).

substrate, a hole blocking compact layer of TiO$_2$ of ~50 nm deposited by spray pyrolysis, a mesoporous TiO$_2$ film, an ultra-thin coating of CdS absorber, a transparent n-type CuSCN filling the pores and a gold back contact. As the absorber CdS is an almost wide bandgap ($E_g = 2.4$ eV), it was chosen as a model material for the proof of the ETA-solar cell concept (see Fig. 12).

The energy band diagram of the three components is reported in Fig. 13. Valence band energies of CuSCN and CdS coating have been measured as the ionization potential by means of ultraviolet photoelectron spectroscopy (UPS). Conduction band of CdS coating was calculated by subtracting the optical absorption onset from the measured valence band value. Data from the literature were used for the conduction band of TiO$_2$. The diagram shows that the bands are well positioned to get an efficient photogenerated charge separation and transfer. The conduction band offset between TiO$_2$ and CdS is low, favoring a good electron injection.

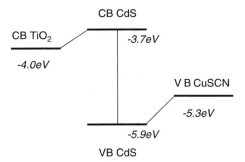

Fig. 13. Energy band levels of the components of TiO_2/CdS/CuSCN ETA-solar cell (IMRA courtesy).

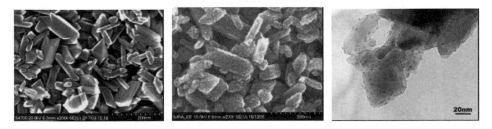

Fig. 14. SEM images of (left) uncoated nanocrystalline TiO_2 film and (middle) CdS-coated TiO_2 film (15 cycles of CdS coating by S-CBD). TEM image of TiO_2/CdS sample (<10 cycles of CdS coating) (right) (from Larramona et al. [21]).

The mesoporous films were made with anatase TiO_2 nanocrystals with an average particle size of ~40–50 nm, prepared by a sol-gel method (Fig. 14). The TiO_2 film thickness was typically 3–6 μm, and the film roughness factor (internal area over projected one) was 150–300 for a 3 μm thick film. Deposition of CdS coating was made by SILAR and a thin 10–20 nm CdS absorber layer was sufficient to obtain efficient light absorption. CuSCN filling was done by impregnation and evaporation following the solution casting and a TiO_2 pore filling volume of at least 50% was deduced from quantitative Energy Dispersive X-ray Microanalysis (EDX) measurements.

The CdS coating was homogeneously distributed onto the TiO_2 nanocrystals and all through the TiO_2 film thickness (Fig. 14 middle). The coating consisted on two types of coverage. First, CdS was present as flattened particles or dots (Fig. 14 right). Second, the whole TiO_2 crystal surface was in addition covered by a very thin film of CdS (<1 nm), whose presence was confirmed by EDX by focusing the beam in regions of the TiO_2 surface

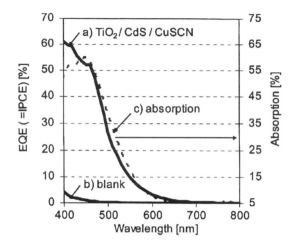

Fig. 15. (a) EQE of mesoporous TiO$_2$/CdS/CuSCN ETA-solar cell compared to the adsorption spectrum of mesoporous TiO$_2$/CdS film; (b) EQE of TiO$_2$/CuSCN hetero-junction is shown in (c) (from Larramona et al. [21]).

where dots were not present. XRD measurements confirmed presence of the cubic CdS phase.

Absorption spectra of TiO$_2$/CdS films showed large light absorption over the ~400–500 nm wavelength region (Fig. 15). Absorption attained 100% for the thicker TiO$_2$ films (≤6 μm). The absorption onset was at ~2.2 eV, a slightly lower photon energy than the bandgap of bulk CdS (2.4–2.5 eV).

Best cell efficiencies were obtained for 15–20 cycles of CdS deposition. This could be due to a balance between larger light absorption and less volume available for CuSCN filling. Buffer layers made as a thin layer of Al$_2$O$_3$ covering the TiO$_2$ nanocrystalline film (previous to CdS coating) did not improve the cell efficiency. Typical cell area was ~0.5 cm^2. The EQE (IPCE attained maximum values of 60% in Fig. 15, the integrated photocurrent being >3 mA/cm^2 (using the standard sun spectrum AM1.5). The maximum current that can be expected from this narrow absorber is ~6 mA/cm^2. For some cells of ~3 μm thickness, the IPCE was nearly coincident with the absorption spectrum, which means that the internal quantum efficiency (or APCE, absorbed photon to current conversion efficiency) was ~100%. Annealing at <200°C did not change cell efficiency. At higher annealing temperatures the cell efficiency decreased considerably.

The conversion efficiency of the best cell was ~1.3% at 1 sun with $J_{sc} > 2$ mA/cm^2, high $V_{oc} > 0.8$ V and good FF > 0.6 (Fig. 16). Unsealed

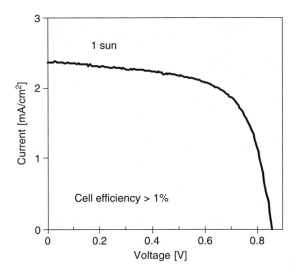

Fig. 16. *I–V* curve of mesoporous TiO$_2$/CdS/CuSCN ETA-solar cell (IMRA courtesy).

cells were stable for several months when left in the lab atmosphere, the loss being about 20%.

5.2.3. *TiO$_2$/In(OH)$_x$S$_y$/PbS/PEDOT:PSS ETA-Solar Cell [92–95]*

The macroporous TiO$_2$/In(OH)$_x$S$_y$/PbS/PEDOT:PSS ETA-solar cells were produced at the Hahn Meitner Institute (Germany). The PEDOT:PSS gel polymer has some advantages including high ionic conductivity. The choice of PbS was motivated by the fact that QDs can be obtained increasing the optical bandgap from 0.35 to 1.1 eV depending on the quantization strength.

Compact and macroporous layers of TiO$_2$ were prepared by several dipping in a solution containing titanium isopropoxide onto glass substrates coated with conductive SnO$_2$:F and final firing at 500°C for 30 min. The In(OH)$_x$S$_y$ and PbS layers were deposited by the SILAR technique from diluted InCl$_3$ containing a small amount of HCl to reach pH = 2.4, Pb(AcO$_2$) and diluted Na$_2$S (pH = 7–8) precursor salt solutions and annealing in air at 200°C and 120°C, respectively. During the SILAR process, the samples were rinsed after each dipping step while the water was exchanged after each 5 dips. PEDOT:PSS layers were spin coated. The contact areas were defined by carbon ink back contacts (Fig. 17).

A bandgap of 0.85 eV was determined for the PbS (bulk PbS E_g value is 0.37 eV) in the absorber layer. This was attributed to the formation of a

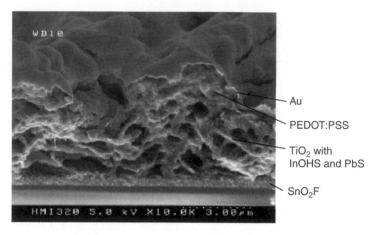

Fig. 17. SEM cross section of the ETA-solar cell $TiO_2/In(OH)_xS_y/PbS/PEDOT:PSS/Au$ (from Bayon et al. [92]).

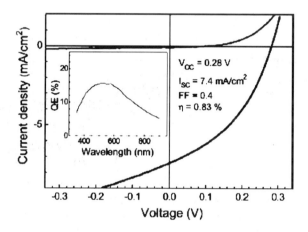

Fig. 18. Current-voltage characteristics of $TiO_2/In(OH)_xS_y$ PbS/PEDOT:PSS/Au in the dark and under illumination ($100\,mW/cm^2$) (inset: External quantum efficiency, from Bayon et al. [92]).

mixed $InPb_x(OH)_yS_z$ phase due to an ion-exchange mechanism. The ETA-solar cells have reached up to 1% conversion efficiency with $J_{sc} = 8\,mA/cm^2$ and $V_{oc} = 0.3\,V$ (Fig. 18). These low values were attributed to the low penetration of the materials (especially PEDOT:PSS) inside the TiO_2 pores (Fig. 17). The optimization of the entire cell assembly in terms of deposition techniques of all the components (especially for improving penetration

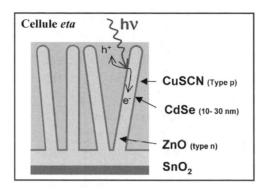

Fig. 19. Schema of the ZnO nanowire/CdSe/CuSCN ETA-solar cell.

into the TiO_2 pores) needs to be improved if higher efficiencies want to be achieved.

5.2.2. ZnO/CdSe/CuSCN ETA-Solar Cell [11, 18]

The ZnO nanowire/CdSe/CuSCN ETA-solar cell was produced at CNRS-Thiais (France). The schema of the three-component cell is shown in Fig. 19. CdSe is a good candidate to act as absorber in ZnO/absorber/CuSCN ETA-solar cells, because its conduction and valence bands are in a favorable position with respect to the n- and p-type materials, as can be seen from the band diagram of the ZnO/CdSe/CuSCN heterostructure (using the reported values for electron affinities of ZnO, CdSe and CuSCN) (Fig. 20). The heterostructure promotes electron transfer from the absorber to ZnO and hole transfer from the absorber to CuSCN.

The ZnO nanowire array was electrodeposited on conductive glass substrates that were first coated with a 100 nm compact ZnO layer by spray pyrolysis. As a second step, a 30 nm thin CdSe coating was electrodeposited on ZnO nanowires, giving a ZnO/CdSe core-shell nanowire array (Fig. 21). The samples were annealed in air at temperatures in the range of 350–400°C during 1 h. The empty space of the nanostructure was filled with CuSCN, deposited with the solution casting method. To finish the ETA-solar cell, a gold contact was vacuum evaporated on the CuSCN.

The electrochemically deposited ZnO consisted of vertically standing single crystal nanowires oriented along the c-axis. The nanowires were ~100–150 nm across and had a length within the range of 1–2 μm (Fig. 21a). The surface area was calculated to be enlarged by a factor of 10–15 over an unstructured layer. In addition to interesting geometric properties favorable

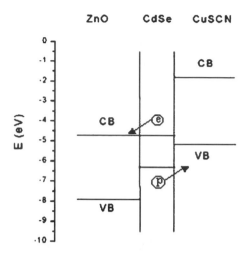

Fig. 20. Band diagram of /CdSe/CuSCN heterostructure. CB and VB are the conduction and valence bands, respectively (from Tena-Zaera et al. [18]).

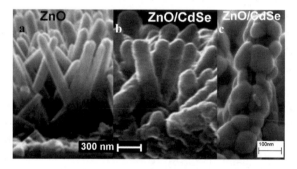

Fig.21. SEM images of ZnO nanowires (a) and ZnO/CdSe core-shell nanowires (as deposited (b) and annealed at 400°C (c)).

to light trapping, the n-type ZnO nanowire film exhibited unique electronic properties with electron transport occurring only along the axis of the wires. Carrier concentration and electron mobility were found to be equal to 10^{20} cm^{-3} and 23 cm^2/V s, respectively [17].

The CdSe ultra-thin layer was electrochemically deposited on the ZnO leading to the formation of an array of core-shell ZnO/CdSe nanowires (Fig. 21b). Annealing the ZnO/CdSe nanowires at high temperature up to 400°C influenced the morphology of the CdSe layer. Annealing below ≤350°C did not induce changes in CdSe morphology while, several "near-isolated" CdSe grains were observed after annealing at 400°C (Fig. 21c).

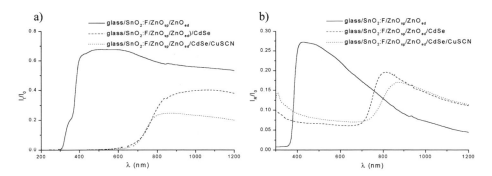

Fig. 22. Transmittance (a) and reflectance (b) of ZnO/CdSe/CuSCN ETA-solar cell, partial and total assembly (from Tena-Zaera et al. [18]).

The ZnO wurtzite phase and the sphalerite cubic and hexagonal CdSe phases were detected by XRD. Annealing temperatures higher than 250°C promoted a phase transition from CdSe cubic phase to hexagonal one. Additionally, a decrease in the CdSe peaks width was observed when the annealing temperature increases, indicating an increase of the CdSe grain size, which could be estimated by the Scherrer's formula, from 3 nm in as deposited ZnO/CdSe nanowires to 60 nm after annealing at 400°C [52b].

The annealing treatments also induced small changes in CdSe bandgap. From optical transmission measurements, the estimated bandgap value varied from 1.88 eV for as-deposited CdSe to 1.74 eV for samples annealed at temperatures ≤350°C, reaching the value of bulk CdSe. These results showed that as-deposited CdSe might have small quantum confinement, which was lost after annealing due to the crystal size increase. The small bandgap shift did not drastically affect the high solar-light trapping effect exhibited by the ZnO/CdSe nanowire array (Fig. 22). A similar shift was observed in the onset of EQE of the solid/liquid junction ZnO/CdSe nanowires/ $Fe(CN)_6^{-3/-4}$ solution (Fig. 23), in agreement with the estimated CdSe bandgap. Thus, it was infered that the CdSe band energies were well situated regarding the electron and hole collectors (ZnO and electrolyte, respectively). The EQE increased with the annealing temperature and, therefore, with the CdSe grain size, attaining a very good value after annealing at 400°C (i.e. 75%).

The ZnO/CdSe/CuSCN nanostructure showed efficient light trapping with R_E ~8% and A_E ~89% (Fig. 22a and b). The high A_E value obtained with less than 40 nm of CdSe thickness was similar to that obtained in crystalline Si with a thickness of ~100 μm.

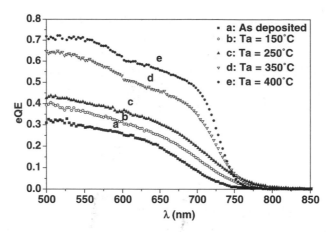

Fig. 23. EQE of ZnO/CdSe nanowires in presence of an electrolyte containing a redox couple $Fe(CN)_6^{-3/-4}$. Influence of the annealing.

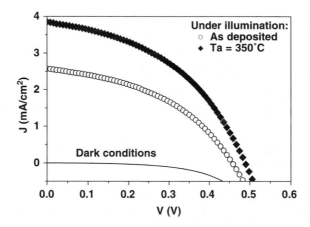

Fig. 24. Current–voltage curve of ZnO/CdSe/CuSCN ETA-solar cell (illumination ~1/3 sun) (from Tena-Zaera et al. [18]).

An energy conversion efficiency <1% was obtained for ZnO/CdSe/ CuSCN ETA-solar cell with as deposited CdSe layer (Fig. 24). However, an enhancement of the electrical properties (transport and injection) of ZnO/CdSe nanowires was also observed in *I–V* curves of complete ZnO/CdSe/CuSCN ETA-solar cells after annealing. Promising values in the range of 2.3% are obtained for samples annealed at 350°C in air for 1 h with an enhancement of open-circuit voltage (V_{oc}). However, the EQE dropped for ZnO/CdSe/

CuSCN ETA-solar cells, in comparison to ZnO/CdSe/Fe(CN)$_6^{-3/-4}$ photo-electrochemical cells. This drop revealed recombination losses at CdSe/CuSCN interfaces or in the CuSCN layer. The electron lifetime, measured from transient PV decay under bias light of one sun, was 14 μs. The decay time of the transient photocurrent at short circuit was also 14 μs, implying that recombination was limiting charge collection in these devices [96]. This in turn implied that reducing recombination would give significant increases in photocurrent.

6. CONCLUSION

The discipline of ETA-solar cell research is very new, essentially it is still in its infancy. We do not yet have a robust theory as to why one system works favorably while similar systems seem to exhibit very poor results. The energy conversion efficiencies of ETA-solar cells are typically lower (by a factor ~2–5) at present compared to devices based on liquid electrolytes. The lower performances can be attributed to the faster interfacial recombination kinetics, difficulty in achieving perfect infiltration of the p-type semiconductor into porous TiO$_2$ or nanostructured ZnO, and insufficient charge carrier mobility inside the p-type material. Hence, we are currently in a period of exploration and more focused work needs to follow. To advance the development of ETA-solar cells, we require a better fundamental understanding of the nanostructured interface that is responsible for the charge separation process. As the development of nanostructured semiconductors is expected to open new horizons in yet unknown areas of science and technology, further study of these materials is a challenging task. Among other obvious paths to develop, we can underline a few in particular. There is the problem of forming an intimate contact between the various layers in order to obtain good carrier injection. More research needs to be invested in the intermediate layer, which looks similar to the problems encountered in thin film solar cells (e.g. CIS, CIGS etc). There is also the problem of obtaining a good ohmic contact with the hole collector and the metal film, as the materials used thus far exhibit a very rough surface. Finally, one should ask whether reducing the absorber thickness is a safe strategy to achieve good efficiencies. A recent model by Taretto and Rau [97, 98] predicts that tunneling-enhanced recombination process in the absorber becomes significant as the film is reduced in thickness beyond a certain threshold. This model provides that a suggested optimal film thickness can be achieved by considering the built-in voltage as a variable. In their model they consider the

absorber as a p–i–n junction with tunneling assisted recombination and predict that 15% energy conversion efficiencies are possible for ETA-solar cells
with CdTe and CuInS$_2$ absorbers having a minimum thickness of around 15
and 20 nm. The results from all these exciting challenges for future research
will undoubtedly yield guiding principles in the way we approach advanced
photovoltaic devices. In addition, they may have a broader influence on
interpreting charge-carrier transport and the problem of the interfaces in
other applications such as computer processing chips.

ACKNOWLEDGMENTS

I would like to thank my co-workers J.R.S. Brownson and R. Tena-Zaera for
very fruitful discussions and review of the manuscript.

REFERENCES

[1]   B. O'Regan and M. Grätzel, Nature, 353 (1991) 737–740.
[2]   M. Grätzel, J. Photochem. Photobiol. A, 164 (2004) 3–14.
[3]   A. Hinsch, J.M. Kroon, R. Kern, I. Uhlendorf, J. Holzbock, A. Meyer and J. Ferber,
      Prog. Photovoltaics, 9 (2001) 425–438.
[4]   B. O'Regan and D.T. Schwartz, Chem. Mater., 7 (1995) 1349–1354.
[5]   K. Tennakone, G.R.R.A. Kumara, A.R. Kumarasinghe, K.G.U. Wijayantha and
      P.M. Sirimanne, Semicond. Sci. Technol., 10 (1995) 1689–1693.
[6]   U. Bach, D. Lupo, P. Conte, J.-E. Moser, F. Weissörtel, J. Salbeck, H. Spreizer and
      M. Grätzel, Nature, 395 (1998) 583–585.
[7]   L. Schimdt-Mende, S.M. Zakeeruddin and M. Grätzel, Appl. Phys. Lett., 86 (2005)
      13504–13506.
[8]   T. Taguchi, X. Zhang, I. Sutanto, K.I. Tokuhiro, T.N. Rao, H. Watanabe, T. Nakamori,
      M. Uragami and A. Fujishima, Chem. Commun., 19 (2003) 2480–2481.
[9]   R. Könenkamp, P. Hoyer and A. Wahi, J. Appl. Phys., 79 (1996) 7029–7035.
[10] M. Nanu, J. Schoonman and A. Goossens, Adv. Mater., 16 (2004) 453–456.
[11] C. Lévy-Clément, R. Tena-Zaera, M.A. Ryan, A. Katty and G. Hodes, Adv. Mater.,
      17 (2005) 1512–1515.
[12] H.W. Deckman, C.R. Wronski, H. Witzke and E. Yablonovitch, Appl. Phys. Lett.,
      42 (1983) 968–970.
[13] K. Ernst, I. Sieber, M. Neumann-Spallart, M.-C. Lux-Steiner and R. Könenkamp,
      Thin Solid Films, 361 (2000) 213–217.
[14] F. Lenzmann, M. Nanu, O. Kijatkina and A. Belaïdi, Thin Solid Films, 451–452
      (2004) 639–643.
[15] R. Könenkamp, K. Ernst, C.H. Fisher, M.C. Lux-Steiner and C. Rost, Phys. Stat.
      Sol. A, 182 (2000) 151–155.
[16] R. Könenkamp, L. Dloczik, K. Ernst and C. Olesch, Physica E, 14 (2002) 219–221.

[17] R. Könenkamp, K. Boedecker, M.C. Lux-Steiner, M. Poschenrieder, F. Zenia, C. Lévy-Clément and S. Wagner, Appl. Phys. Lett., 77 (2000) 2575–2577.

[18] R. Tena-Zaera, M.A. Ryan, A. Katty, G. Hodes, S. Bastide and C. Lévy-Clément, Compte Rendus Chim., 9 (2006) 717–729.

[19] C. Lévy-Clément, A. Katty, S. Bastide, F. Zenia, I. Mora and V. Muñoz-Sanjosé, Physica E, 14 (2002) 229–232.

[20] R. Tena-Zaera, A. Katty, S. Bastide, C. Lévy-Clément, B. O'Regan and V. Muñoz-Sanjosé, Thin Solid Films, 483 (2005) 372–377.

[21] G. Larramona, C. Choné, A. Jacob, D. Sakakura, B. Delatouche, D. Péré, X. Cieren, N. Magino and R. Bayon, Chem. Mater., 18 (2006) 1688–1696.

[22] K. Ernst, R. Engelhardt, K. Ellmer, C. Kelch, H.-J. Muffler, M.-C. Lux-Steiner and R. Könenkamp, Thin Solid Films, 387 (2001) 26–28.

[23] A. Belaïdi, R. Bayon, L. Dloczik, K. Ernst, M.C. Lux-Steiner and R. Könenkamp, Thin Solid Films, 431–432 (2003) 488–491.

[24] K. Ernst, A. Belaïdi and R. Könenkamp, Semic. Sci. Technol., 18 (2003) 475–479.

[25] I. Kaiser, K. Ernst, C.H. Fischer, M.C. Lux-Steiner and R. Könenkamp, Sol. Energy Mat. Sol. Cells, 67 (2001) 89–96.

[26] M. Nanu, J. Schoonman and A. Goossens, Nano Letters, 5 (2005) 1716–1719.

[27] K. Tennakone, G.R.R.A. Kumara, I.R. Kottegoda, V.P.S. Perera and G. M.L. Aponsu, J. Phys. D: Appl. Phys., 31 (1998) 2326–2330.

[28] C.J. Brinker, and G.W. Scherer, The Physics and Chemistry of Sol-Gel Processing, Academic Press, Boston, MA, 1990.

[29] Ch.-J. Barbé, F. Arendse, P. Comte, M. Jirousek, F. Lenzmann, V. Shklover and M. Grätzel, J. Am. Ceram. Soc., 80 (1997) 3157–3171.

[30] L. Kavan and M. Grätzel, Electrochim. Acta, 40 (1995) 643–652.

[31] B. O'Regan, D.T. Schwartz, S.M. Zakeeruddin and M. Grätzel, Adv. Mater., 12 (2000) 1263.

[32] S. Siebentritt, K. Ernst, Ch.-H. Fischer, R. Könenkamp and M.-Ch. Lux-Steiner, Proc. 14th European Photovoltaic Solar Energy Conf., pp. 1823–1826, H.S. Stephen & Associates Pubs., Bedford, UK, 1997.

[33] O. Kijatkina, M. Krunks, A. Mere, B. Mahrov and L. Dloczic, Thin Solid Films, 431–432 (2003) 105–109.

[34] G. Hodes (ed.), Chemical Solution Deposition of Semiconductor Films, Marcel Dekker, New York, 2003.

[35] J. Moller, C.-H. Fischer, H.J. Muffler, R. Konenkamp, I. Kaiser, C. Kelch and M.-Ch. Lux-Steiner, Thin Solid Films, 361 (2000) 113–117.

[36] Y.F. Nicolau and M. Dupuy, J. Electrochem. Soc., 137 (1990) 2915–2924.

[37] Y.F. Nicolau and J.C. Menard, J. Appl. Electrochem., 20 (1990) 1063–1066.

[38] B. O'Regan, F. Lenzmann, R. Muis and J. Wienke, Chem. Mater., 14 (2002) 5023–5029.

[39] B. O'Regan and F. Lenzmann, J. Phys. Chem. B, 108 (2004) 4342–4350.

[40] V.P.S. Perera, M.K.I. Senevirathna, P.K.D.D.P. Pitigala and K. Tennakone, Sol. Energy Mater. Sol. Cells, 86 (2005) 443–450.

[41] S. Peulon and D. Lincot, J. Electrochem. Soc., 145 (1998) 864–874.

[42] K. Tennakone, A.R. Kumarasinghe, P.M. Sirimanne, G.R.R.A. Kumara, Thin Solid Films, 261 (1995) 307–310.

[43] B. O'Regan and D.T. Schwartz, Chem. Mater., 10 (1998) 1501–1509.

[44] M.P.R. Panicker, M. Knaster and F.A. Kröger, J. Electrochem. Soc., 125 (1978) 566–572.

[45] M. Neumann-Spallart, G. Tamizhmani, A. Boutry-Forveille and C. Lévy-Clément, Thin Solid Films, 169 (1988) 315–322.

[46] M. Skyllas-Kazavos and B. Miller, J. Electrochem Soc., 127 (1980) 869.

[47] M. Neumann-Spallart and Ch. Königstein, Thin Solid Films, 265 (1995) 33–37.

[48] M. Nanu, J. Schoonman and A. Goossens, Chem. Vap. Deposition, 10 (2004) 45–49.

[49] M. Gratzel, Nature, 414 (2001) 338–344.

[50] V.B. Gusen, L.M. Genev and I.I. Kalinichenko, Zh. Prikl. Spekstrosk., 34 (1981) 939.

[51] E. Hendry, M. Koeberg, B. O'Regan and M. Bonn, Nano Letters, 6 (2006) 7755–7759.

[52] a) C. Lévy-Clément, J. Elais, R. Tena-Zaera, Proc. SPIE, Vol 6340, Solar Hydrogen and Nanotechnology, V. Vayssièred Ed. (2006) 63 400R.
      b) R. Tena-Zaera, A. Katty, S. Bastide and C. Lévy-Clément, Chem Mat. Submitted.

[53] H. Gerischer and M. Luebke, J. Electroanal. Chem., 204 (1986) 225–227.

[54] R. Vogel, K. Pohl and H. Weller, Chem. Phys. Lett., 174 (1990) 241–246.

[55] S. Kohtami, A. Kudo and T. Sakata, Chem. Phys. Lett., 206 (1993) 166–170.

[56] R. Vogel, P. Hoyer and H. Weller, Chem. J. Phys. Chem., 98 (1994) 3183–3188.

[57] R. Plass, S. Pelet, J. Krueger, M. Grätzel and U. Bach, J. Phys. Chem. B, 106 (2002) 7578–7580.

[58] L.M. Peter, K.G.U. Wijayantha, D. Riley and J.P. Waggett, J. Phys. Chem. B, 107 (2003) 8378–8381.

[59] D. Liu and P.V. Kamat, J. Phys. Chem., 97 (1993) 10769–10773.

[60] A. Zaban, O.I. Micic, B.A. Gregg and A. Nozik, Langmuir, 14 (1998) 3153–3156.

[61] L. Reijnen, B. Meester, J. Schoonman and A. Goossens, Mater. Sci. Eng. C, 19 (2002) 311–314.

[62] J. Möller, Ch.-H. Fischer, S. Siebentritt, R. Könenkamp and M.-Ch. Lux-Steiner, Proc. 2nd World Conf. Exhibition on Photovoltaic Solar Energy Conversion, pp. 209–211, H.S. Stephen & Associates Publishers, Bedford, UK, 1998.

[63] R. Bayon, R. Musembi, A. Belaidi, M. Bär, T. Guminskaya, M.-Ch. Lux-Steiner and Th. Dittrich, Sol. Energy Mater. Sol. Cells, 89 (2005) 13–25.

[64] I. Oja, A. Belaidi, L. Dloczik, M.-Ch. Lux-Steiner and Th. Dittrich, Semicond. Sci. Technol., 21 (2006) 520–526.

[65] K. Tennakone, G.R.R.A. Kumara, I.R.M. Kottegoda, V.P.S. Perera, G.M.L. Aponsu and K.G.U. Wijayantha, Sol. Energy Mater. Sol. Cells, 55 (1998) 283–289.

[66] C. Rost, I. Sieber, C. Fischer, M.-Ch. Lux-Steiner and R. Könenkamp, Mater. Sci. Eng., 69–70 (2000) 570–573.

[67] B.R. Sankapal, E. Goncalves, A. Ennaoui and M.C. Lux-Steiner, Thin Solid Films, 451–452 (2004) 128–132.

[68] C. Rost, I. Sieber, K. Ernst, S. Siebentritt, M.-C. Lux-Steiner and R. Könenkamp, Appl. Phys. Lett., 45 (1999) 692–694.

[69] G.R.R.A. Kumara, A. Konno. G.K.R. Senadeera, P.V.V. Jayaweera, D.B.R.A. De Silva and K. Tennakone, Sol. Ener. Mat. Sol. Cells, 69 (2001) 195–199.

[70] L. Dloczik, Y. Tomm, R. Könenkamp, M.-C. Lux-Steiner and Th. Dittrich, Thin Solid Films, 451–452 (2004) 116–119.

[71] J. Bandara and H. Weerasinghe, Sol. Energy Mater. Sol. Cells, 85 (2005) 385–390.

[72] L.B. Groenendaal, F. Jonas, D. Freitag, H. Pielartzik and J.R. Reynolds, Adv. Mater., 12 (2000) 481–494.

[73] E. Arici, N.S. Sariciftci and D. Meisner, Adv. Funct. Mater., 13 (2003) 165–171.

[74] J.-H. Yum, S. Nakade, D.-Y. Kim and S. Yanagida, J. Phys. Chem. B, 110 (2006) 3215–3219.

[75] J. Wienke, M. Krunks and F. Lenzmann, Semicond. Sci. Technol., 18 (2003) 867–888.

[76] O. Niitsoo, S.K. Sarkar, C. Pejoux, S. Rühle, D. Cahen and G. Hodes, J. Photochem. and Photobiology A: Chemistry, 181 (2006) 306–313.

[77] G.R.R.A. Kumara, K. Tennakone, V.P.S. Perera, A. Konno, A. Kaneko and M. Okuya, J. Phys. D: Appl. Phys., 34 (2001) 868–873.

[78] E. Palomares, J.N. Clifford, S.A. Haque, T. Lutz and J.R. Durrant, Chem. Commun., 14 (2002) 1464–1465.

[79] A. Kay and M. Grätzel, Chem. Mater., 14 (2002) 2930–2935.

[80] K. Tennakone, J. Bandara, P.K.M. Bandaranayake, G.R.A. Kumara and A. Konno, Jap. J. Appl. Phys, 40 (2001) L732–L734.

[81] X.-T. Zhang, H.-W. Liu, T. Taguchi, Q.-B. Meng, O. Sato and A. Fujishima, Sol. Energy Mater. Sol. Cells, 81 (2004) 197–203.

[82] B.C. O'Regan, S. Scully, A.C. Mayer, E. Palomares and J.R. Durrant, J. Phys. Chem. B, 109 (2005) 4616–4623.

[83] E. Palomares, J.N. Clifford, S.A. Haque, T. Lutz and J.R. Durrant, J. Am. Chem. Soc., 125 (2003) 475–482.

[84] M. Nanu, L. Reijnen, B. Meester, A. Goossens and J. Schoonman, Thin Solid Films, 431–432 (2003) 492–496.

[85] Th. Dittrich, H.-J. Muffler, M. Vogel, T. Gurinskaya, A. Ogacho, A. Belaidi, E. Strub, W. Bohne, J. Röhrich, O. Hilt and M.-Ch. Lux-Steiner, Appl. Surf. Sci., 240 (2005) 236–243.

[86] C. Grasso and M. Burgelman, Thin Solid Films, 451–452 (2004) 156–159.

[87] D. Braunger, D. Hariskos, T. walker and H.W. Schock, Sol. Energy Mater. Sol. Cells, 40 (1996) 97–102.

[88] M. Nanu, J. Schoonman and A. Goossens, Adv. Funct. Mater., 15 (2005) 95–100.

[89] R. Bayon, C. Guillen, M.A. Martinez, M.T. Guitterez and J. Herreroo, J. Electrochem. Soc., 145 (1998) 2775–2779.

[90] G. Fulop, M. Doty, P. Meyers, J. Betz and C.H. Liu, Appl. Phys. Lett., 40 (1982) 327–328.

[91] S. Tiefenbacher, C. Pettenkofer and W. Jägermann, J. Appl. Phys., 91 (2002) 1984–1987.

[92] R. Bayon, R. Musembi, A. Beladi, M. Bär, T. Guminskaya, M.-Ch. Lux-Steiner and Th. Dittrich, Sol. Energy Mater. Sol. Cells, 89 (2005) 13–25.

[93] R. Bayon, R. Musembi, A. Belaidi, M. Bär, T. Guminskaya, Ch.-H. Fischer, M.-Ch. Lux-Steiner and Th. Dittrich, C.R. Chimie., 9 (2006) 730–734.

[94] I. Oja, A. Belaïdi, M.-Ch. Lux-Steiner and Th. Dittrich, Semicon. Sci. Technol., 31 (2006) 520–526.

[95] S. Gravilov, I. Oja, B. Lim, A. Belaïdi, W. Bohne, E. Strub, J. Röhrich, M.-Ch. Lux-Steiner and Th. Dittrich, Phys. Stat. Sol. A, 203 (2006) 1024–1029.
[96] B. O'Regan, private communication.
[97] K. Taretto and U. Rau, Prog. Photovolt: Res. Appl., 12 (2004) 573–591.
[98] K. Taretto and U. Rau, Thin Solid Film, 480 (2005) 447–451.

Nanostructured Materials for Solar Energy Conversion
T. Soga (editor)

Chapter 15

Quantum Structured Solar Cells

A.J. Nozik[a,b]

[a]Center for Basic Sciences, National Renewable Energy Laboratory, 1617 Cole Blvd., Golden, CO 80401, USA
[b]Department of Chemistry, University of Colorado, Boulder, CO 80309, USA

1. INTRODUCTION

The cost of delivered photovoltaic (PV) power is determined by the PV module conversion efficiency and the capital cost of the PV system per unit area. To achieve very low cost PV power, it is necessary to develop cells that have very high conversion efficiency and moderate cost. Toward this end, we have been investigating the possibility of achieving high conversion efficiency in single-bandgap solar cells by capturing the excess energy of electron – hole pairs created by the absorption of solar photons larger than the bandgap to do useful work before these high-energy electron – hole pairs convert their excess kinetic energy (equal to the difference between the photogenerated electron energy and the conduction band energy) to heat through phonon emission [1–4]. These highly excited electrons and holes are termed hot electrons and hot holes (or hot carriers); in semiconductor nanocrystals, the photogenerated electron – hole pairs are correlated and are termed excitons. Semiconductor nanocrystals (also called quantum dots, QDs) have discrete electronic states, and the absorption of photons with energies greater than the energy difference between the highest hole state (1 S_h) and the lowest electron state (1 S_e) (also termed the HOMO-LUMO transition) produces excited excitons.

The extraction of useful work from hot electron – hole pairs (hot carriers) is difficult in bulk semiconductors because the cooling process that occurs through inelastic carrier–phonon scattering and subsequent hot-carrier cooling is very fast (sub-ps). However, the formation of discrete quantized levels in QDs affects the relaxation and cooling dynamics of high-energy excitons and this could enhance the power conversion efficiency by either using the excess energy of excited excitons to create additional excitons (a process termed "multiple exciton generation, MEG" [5]), or allowing electrical or chemical free energy to be extracted from the excited excitons through charge separation before the excitons relax and produce heat.

As is well known, the maximum thermodynamic efficiency for the conversion of unconcentrated solar irradiance into electrical free energy in the radiative limit, assuming detailed balance, a single threshold absorber, a maximum yield of one electron–hole pair per photon, and thermal equilibrium between electrons and phonons, was calculated by Shockley and Queisser in 1961 [6] to be about 31%; this analysis is also valid for the conversion to chemical free energy [7, 8]. This efficiency is attainable in semiconductors with bandgaps ranging from about 1.2 to 1.4 eV.

However, the solar spectrum contains photons with energies ranging from about 0.5 to 3.5 eV. Photons with energies below the semiconductor bandgap are not absorbed, while those with energies above the bandgap create electrons and holes (charge carriers) with a total excess kinetic energy equal to the difference between the photon energy and the bandgap. This excess kinetic energy creates an effective temperature for an ensemble of photogenerated carriers that can be much higher than the lattice temperature; such carriers are called "hot electrons and hot holes," and their initial temperature upon photon absorption can be as high as 3000°K with the lattice temperature at 300°K. In bulk semiconductors, the division of this kinetic energy between electrons and holes is determined by their effective masses, with the carrier having the lower effective mass receiving more of the excess energy [1]. Thus,

$$\Delta E_e = (h\nu - E_g)[1 + m_e^*/m_h^*]^{-1} \tag{1}$$

$$\Delta E_h = (h\nu - E_g) - \Delta E_e \tag{2}$$

where E_e is the energy difference between the conduction band and the initial energy of the photogenerated electron, and E_h the energy difference between the valence band and the photogenerated hole (see Fig. 1). However, in QDs, the distribution of excess energy is determined by the quantized energy level structure in the QDs and the associated selection rules for the optical transitions between the hole and electron levels [9].

In the Shockley–Queisser analysis, a major factor limiting the conversion efficiency to 32% is that the absorbed photon energy above the semiconductor bandgap is lost as heat through electron–phonon scattering and subsequent phonon emission, as the carriers relax to their respective band edges (bottom of conduction band for electrons and top of valence band for holes) (see Fig. 1) and equilibrate with the phonons. The main approach to reduce

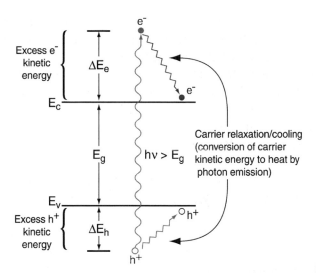

Fig. 1. Hot carrier relaxation/cooling dynamics in semiconductors (From Reference 1).

this loss and increase efficiency above the 32% limit has been to use a stack of cascaded multiple p–n junctions in the absorber with bandgaps better matched to the solar spectrum; in this way higher-energy photons are absorbed in the higher-bandgap semiconductors and lower-energy photons in the lower-bandgap semiconductors, thus reducing the overall heat loss due to carrier relaxation via phonon emission. In the limit of an infinite stack of bandgaps perfectly matched to the solar spectrum, the ultimate conversion efficiency at one-sun intensity increases to about 66%. For practical purposes, the stacks have been limited to two or three p–n junctions; actual efficiencies of about 32% have been reported in PV cells with two cascaded p–n junctions. Other approaches to exceed the Shockley–Queisser limit include hot carrier solar cells [1–3], solar cells producing multiple electron–hole pairs per photon [10–14], multiband and impurity solar cells [12, 15], and thermo-PV/thermophotonic cells [12]. Here, we will only discuss hot carrier and MEG solar cells, and the effects of size quantization in semiconductor QDs on the carrier dynamics that control the probability of these processes.

There are two fundamental ways to utilize hot carriers or hot excitons for enhancing the efficiency of photon conversion. One way produces an enhanced photovoltage, and the other way produces an enhanced photocurrent. The former requires that the carriers be extracted from the photoconverter before they cool [2, 3], while the latter requires the energetic hot carriers to produce a second (or more) electron–hole pair through MEG – a process that is the inverse of an Auger process whereby two electron–hole pairs recombine to produce a single highly energetic electron–hole pair. In order to achieve the former, the rates of photogenerated-carrier separation, transport, and interfacial transfer across the semiconductor interface must all be fast compared to the rate of carrier cooling [3, 4, 16, 17]. The latter requires that the rate of exciton multiplication is greater than the rate of carrier cooling and forward Auger processes.

Hot electrons and hot holes generally cool at different rates because they generally have different effective masses; for most inorganic semiconductors electrons have effective masses that are significantly lighter than holes and consequently cool more slowly. Another important factor is that hot-carrier cooling rates are dependent upon the density of the photogenerated-hot carriers (viz, the absorbed light intensity) [18–20]. Here, most of the dynamical effects we will discuss are dominated by electrons rather than holes; therefore, we will restrict our subsequent discussion primarily to the relaxation dynamics of photogenerated electrons.

Finally, in recent years it has been proposed [3, 4, 16, 21–24] and experimentally verified in some cases [1, 25–27], that the relaxation dynamics of photogenerated carriers may be markedly affected by quantization effects in the semiconductor (i.e., in semiconductor quantum wells (QWs), quantum wires, QDs, superlattices, and nanostructures). That is, when the carriers in the semiconductor are confined by potential barriers to regions of space that are smaller than or comparable to their deBroglie wavelength or to the Bohr radius of excitons in the semiconductor bulk, the relaxation dynamics can be dramatically altered; specifically the hot-carrier cooling rates may be dramatically reduced, and the rate of producing multiple excitons per photon could become competitive with the rate of carrier cooling [1] (see Fig. 2).

2. RELAXATION DYNAMICS OF HOT CARRIERS IN BULK SEMICONDUCTORS

Upon photoexcitation with a laser pulse, the initial carrier distributions are usually not Boltzmann-like, and the first step toward establishing equilibrium is for the hot carriers to interact separately among themselves and with the initial population of cold carriers through their respective carrier–carrier collisions and inter-valley scattering to form separate Boltzmann distributions of electrons and holes. These two Boltzmann distributions can then be separately assigned an electron and hole temperature that reflects the distributions of kinetic energy in the respective charge carrier populations. If photon absorption produces electrons and holes with initial excess kinetic energies at least kT above the

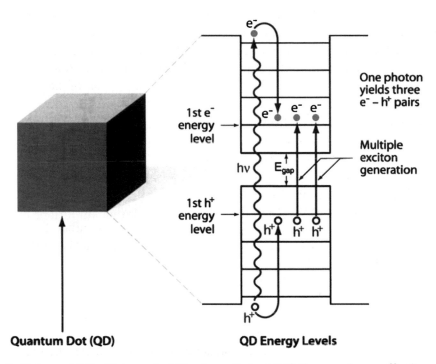

Quantum Dot (QD)   **QD Energy Levels**

Fig. 2. Enhanced PV efficiency in QD solar cells by MEG (inverse Auger effect).

conduction and valence bands, respectively, then both initial carrier temperatures are always above the lattice temperature and the carriers are called hot carriers. This first stage of relaxation or equilibration occurs very rapidly (<100 fs)
[18, 19], and this process is often referred to as carrier thermalization (i.e.,
formation of a thermal distribution described by Boltzmann statistics).

After the separate electron and hole populations come to equilibrium
among themselves in less than 100 fs, they are still not yet in equilibrium
with the lattice. The next step of equilibration is for the hot electrons and hot
holes to equilibrate with the semiconductor lattice. The lattice temperature is
the ambient temperature and is lower than the initial hot electron and hot hole
temperatures. Equilibration of the hot carriers with the lattice is achieved
through carrier–phonon interactions (phonon emission) whereby the excess
kinetic energy of the carriers is transferred from the carriers to the phonons; the
phonons involved in this process are the longitudinal optical (LO) phonons.
This may occur by each carrier undergoing separate interactions with the
phonons, or in an Auger process where the excess energy of one carrier type
is transferred to the other type, which then undergoes the phonon interaction.

The phonon emission results in cooling of the carriers and heating of the lattice until the carrier and lattice temperatures become equal. This process is termed carrier cooling, but some researchers also refer to it as thermalization; however, this latter terminology can cause confusion with the first stage of equilibration that just establishes the Boltzmann distribution among the carriers. Here, we will restrict the term thermalization to the first stage of carrier relaxation, and we will refer to the second stage as carrier cooling (or carrier relaxation) through carrier–phonon interactions.

The final stage of equilibration results in complete relaxation of the system; the electrons and holes can recombine, either radiatively or nonradiatively, to produce the final electron and hole populations that existed in equilibrium in the dark before photoexcitation. Another important possible pathway following photoexcitation of semiconductors is for the photogenerated electrons and holes to undergo spatial separation. Separated photogenerated carriers can subsequently produce a photovoltage and a photocurrent (PV effect) [28–30]; alternatively, the separated carriers can drive electrochemical oxidation and reduction reactions (generally labeled redox reactions) at the semiconductor surface (photoelectrochemical energy conversion) [31]. These two processes form the basis for devices/cells that convert radiant energy (e.g., solar energy) into electrical [28–30] or chemical-free energy (PV cells and photoelectrochemical cells, respectively) [31].

2.1. Quantum Wells and Superlattices

Semiconductors show dramatic quantization effects when charge carriers are confined by potential barriers to small regions of space where the dimensions of the confinement are less than their deBroglie wavelength; the length scale at which these effects begin to occur range from about 10 to 50 nm for typical semiconductors (Groups IV, III–V, II–VI). In general, charge carriers in semiconductors can be confined by potential barriers in one spatial dimension, two spatial dimensions, or in three spatial dimensions. These regimes are termed quantum films, quantum wires, and QDs, respectively. Quantum films are also more commonly referred to simply as QWs.

One-dimensional QWs, hereafter called quantum films or just QWs, are usually formed through epitaxial growth of alternating layers of semiconductor materials with different bandgaps. A single QW is formed from one semiconductor sandwiched between two layers of a second semiconductor having a larger bandgap; the center layer with the smaller bandgap semiconductor forms the QW while the two layers sandwiching the center layer create the potential barriers. Two potential wells are actually formed in the QW structure; one well

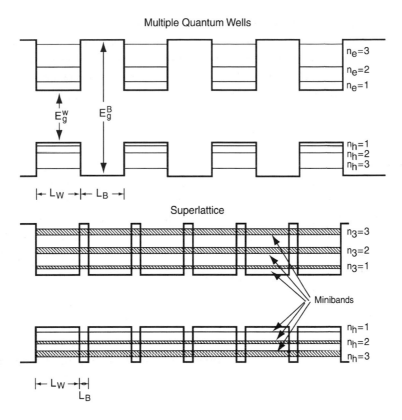

Fig. 3. Difference in electronic states between MQW structures (barriers >40 Å) and superlattices (barriers <40 Å); miniband formation occurs in the superlattice structure, which permits carrier delocalization (From Reference 1)

is for conduction-band electrons, the other for valence-band holes. The well depth for electrons is the difference (i.e., the offset) between the conduction-band edges of the well and barrier semiconductors, while the well depth for holes is the corresponding valence band offset. If the offset for either the conduction or valence bands is zero, then only one carrier will be confined in a well.

Multiple QW (MQW) structures consist of a series of QWs (i.e., a series of alternating layers of wells and barriers). If the barrier thickness between adjacent wells prevents significant electronic coupling between the wells, then each well is electronically isolated; this type of structure is termed a MQW. On the other hand, if the barrier thickness is sufficiently thin to allow electronic coupling between wells (i.e., there is significant overlap of the electronic wavefunctions between wells), then the electronic charge distribution can become delocalized along the direction normal to the well layers. This coupling also leads to a broadening of the quantized electronic states of the wells; the new broadened and delocalized quantized states are termed *minibands* (see Fig. 3). A MQW structure that exhibits strong electronic coupling between the wells is termed a *superlattice*. The critical thickness at which miniband formation just begins to occur is about 40 Å [32, 33]; the electronic coupling increases rapidly with decreasing thickness and miniband formation is very strong below 20 Å [32]. Superlattice structures yield efficient charge transport normal to the layers because the charge carriers can move through the minibands; the narrower the barrier, the wider the miniband and the higher the carrier mobility. Normal transport in MQW structures (thick barriers) require thermionic emission of carriers over the barriers, or if electric fields are applied, field-assisted tunneling through the barriers [34].

2.1.1. Measurements of Hot Electron Cooling Dynamics in QWs and Superlattices

Hot-electron cooling times can be determined from several types of time-resolved PL experiments. One technique involves hot luminescence non-linear correlation [35–37], which is a symmetrized pump-probe type of experiment. Fig. 2 of Ref. [35] compares the hot-electron relaxation times as a function of the electron energy level in the well for bulk GaAs and a 20-period MQW of GaAs/$Al_{0.38}Ga_{0.62}As$ containing 250 Å GaAs wells and 250 Å $Al_{0.38}Ga_{0.62}As$ barriers. For bulk GaAs, the hot-electron relaxation time varies from about 5 ps near the top of the well to 35 ps near the bottom of the well. For the MQW, the corresponding hot-electron relaxation times are 40 ps and 350 ps.

Another method uses time-correlated single-photon counting to measure PL lifetimes of hot electrons. Fig. 4 shows 3-D plots of PL intensity as a function of energy and time for bulk GaAs and a 250 Å/250 Å GaAs/$Al_{0.38}Ga_{0.62}As$ MQW [20]. It is clear from these plots that the MQW sample exhibits much longer-lived hot luminescence (i.e., luminescence above the lowest $n = 1$ electron to heavy-hole transition at 1.565 eV) than bulk GaAs. Depending upon the emitted photon energy, the hot PL for the MQW is seen to exist beyond times ranging from hundreds to several thousand ps. On the other hand, the hot PL intensity above the bandgap (1.514 eV) for bulk GaAs is negligible over most of the plot; it is only seen at the very earliest times and at relatively low photon energies.

Calculations were performed [20] on the PL intensity versus time and energy data to determine the time dependence of the quasi-Fermi-level, electron temperature, electronic specific heat, and ultimately the dependence of the characteristic hot-electron cooling time on electron temperature.

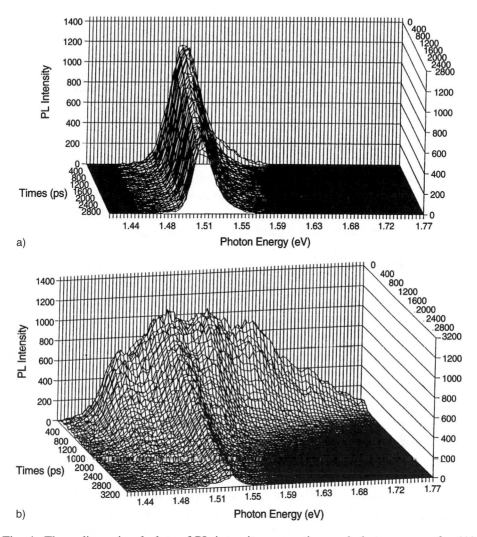

Fig. 4. Three-dimensional plots of PL intensity versus time and photon energy for (A) bulk GaAs and (B) 250 Å GaAs/250 Å Al$_{0.38}$Ga$_{0.62}$As MQW (From Reference 1).

The cooling, or energy-loss, rate for hot electrons is determined by LO phonon emission through electron–LO–phonon interactions. The time constant characterizing this process can be described by the following expression [38–40]:

$$P_{e} = -\frac{dE}{dt} = \frac{\hbar\omega_{LO}}{\tau_{avg}} \exp(-\hbar\omega_{LO}/kT_{e}) \tag{3}$$

where P_e is the power loss of electrons (i.e., the energy-loss rate), hT_{LO} the LO phonon energy (36 meV in GaAs), T_e the electron temperature, and τ_{avg} the time constant characterizing the energy-loss rate.

The electron energy-loss rate is related to the electron temperature decay rate through the electronic specific heat. Since at high light intensity the electron distribution becomes degenerate, the classical specific heat is no longer valid. Hence, the temperature and density-dependent specific heat for both the QW and bulk samples need to be calculated as a function of time in each experiment so that τ_{avg} can be determined.

The results of such calculations (presented in Fig. 2 of Reference [20]) show a plot of τ_{avg} versus electron temperature for bulk and MQW GaAs at high and low carrier densities. These results show that at a high carrier density [$n \sim (2\text{–}4) \times 10^{18}\,\text{cm}^{-3}$], the τ_{avg} values for the MQW are much higher ($\tau_{avg} = 350\text{–}550$ ps for T_e between 440 and 400 K) compared to bulk GaAs ($\tau_{avg} = 10\text{–}15$ ps over the same T_e interval). On the other hand, at a low carrier density [$n \sim (3\text{–}5) \times 10^{17}\,\text{cm}^{-3}$] the differences between the τ_{avg} values for bulk and MQW GaAs are much smaller.

A third technique to measure cooling dynamics is PL upconversion [20]. Time resolved luminescence spectra were recorded at room temperature for a 4000 Å bulk GaAs sample at the incident pump powers of 25, 12.5, and 5 mW. The electron temperatures were determined by fitting the high-energy tails of the spectra; only the region which is linear on a semilogarithmic plot was chosen for the fit. The carrier densities for the sample were 1×10^{19}, 5×10^{18}, and $2 \times 10^{18}\,\text{cm}^{-3}$, corresponding to the incident excitation powers of 25, 12.5, and 5 mV, respectively. Similarly, spectra for the MQW sample were recorded at the same pump powers as the bulk. Fig. 5 shows τ_{avg} for bulk and MQW GaAs at the 3 light intensities, again showing the much slower cooling in MQWs (by up to two orders of magnitude).

The difference in hot-electron relaxation rates between bulk and quantized GaAs structures is also reflected in time-integrated PL spectra. Typical results are shown in Fig. 6 for single photon counting data taken with 13 spec pulses of 600 nm light at 800 kHz focused to about 100 μm with an average power of 25 mW [41]. The time-averaged electron temperatures obtained from fitting the tails of these PL spectra to the Boltzmann function show that the electron temperature varies from 860 K for the 250 Å/250 Å MQW to 650 K for the 250 Å/17 Å superlattice, while bulk GaAs has an electron temperature of 94 K, which is close to the lattice temperature (77 K). The variation in the electron temperatures between the quantized structures can be attributed to differences in electron delocalization between

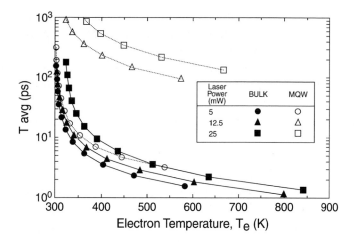

Fig. 5. Time constant for hot-electron cooling (τ_{avg}) vs electron temperature for bulk GaAs and GaAs MQWs at three excitation intensities (From Reference 1).

MQWs and SLs, and the associated non-radiative quenching of hot-electron emission.

As shown above, the hot carrier cooling rates depend upon photogenerated carrier density; the higher the electron density the slower the cooling rate. This effect is also found for bulk GaAs, but it is much weaker compared to quantized GaAs. The most generally accepted mechanism for the decreased cooling rates in GaAs QWs is an enhanced "hot phonon bottleneck" [42–44]. In this mechanism a large population of hot carriers produces a non-equilibrium distribution of phonons (in particular, optical phonons which are the type involved in the electron–phonon interactions at high carrier energies) because the optical phonons cannot equilibrate fast enough with the crystal bath; these hot phonons can be re-absorbed by the electron plasma to keep it hot. In QWs the phonons are confined in the well and they exhibit slab modes [43], which enhance the "hot phonon bottleneck" effect.

An investigation of PV cells that are based on a p–i–n structure with the i-region consisting of a superlattice has been reported [45]. The concept is to use a superlattice region with a low value of the lowest energy transition to absorb a large fraction of the solar photons, create a hot-electron distribution within the superlattice layer that cools slowly because of the miniband formation, separate the hot electrons and holes and transport them to the higher bandgap n- and p-contacts using the electric field produced by the p–i–n structure. The results show that the concept for higher efficiency hot carrier

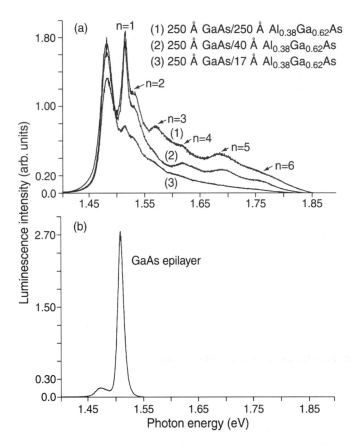

Fig. 6. (a) Time-integrated PL spectra for MQWs and SLs showing hot luminescence tails and high-energy peaks arising from hot-electron radiative recombination form upper quantum levels. (b) Equivalent spectrum for bulk GaAs showing no hot luminescence (From Reference 1).

production and transport is undermined by the fact that under operating conditions of forward bias for the cell, cold carriers from the contacts are injected in the superlattice region and lower the hot carrier temperature. This effect could perhaps be alleviated by using selective contacts; the use of solar concentration may also help to improve conversion efficiency.

2.2. Relaxation Dynamics of Hot Excitons in Quantum Dots

As discussed above, slowed hot-electron cooling in QWs and superlattices that is produced by a *hot* phonon bottleneck requires very high light intensities in order to create the required photogenerated carrier density of greater than about $1 \times 10^{18} \, \mathrm{cm}^{-3}$. This required intensity, possible with laser excitation, is many orders of magnitude greater than that provided by solar radiation at the earth's surface (maximum solar photon flux is about $10^{18} \, \mathrm{cm}^{-2} \mathrm{s}^{-1}$; assuming a carrier lifetime of 1 ns and an absorption coefficient of $1 \times 10^{5} \, \mathrm{cm}^{-1}$, this translates into a photoinduced electron density of about $10^{14} \, \mathrm{cm}^{-3}$ at steady state). Hence, it is not possible to obtain slowed hot carrier cooling in semiconductor QWs and superlattices with solar irradiation via a *hot* phonon bottleneck effect; solar concentration ratios greater than 10^{4} would be required, resulting in severe practical problems.

However, the situation with three-dimensional confinement in QDs is potentially more favorable. In the QD case, slowed hot-electron cooling is theoretically possible even at arbitrarily low light intensity; this effect is simply called a "phonon bottleneck," without the qualification of requiring hot phonons (i.e., a non-equilibrium distribution of phonons). Furthermore, it is also anticipated that the slowed cooling could make the rate of exciton multiplication (inverse Auger effect) an important process in QDs [1, 13, 46, 47]. PL blinking in QDs (intermittent PL as a function of time) has been explained [48, 49] by an Auger process whereby if two electron–holes pairs are photogenerated in a QD, one pair recombines and transfers its recombination energy to one of the remaining charge carriers, ionizing it over the potential barrier at the surface into the surface region. This creates a charged QD that quenches radiative emission after subsequent photon absorption; after some time, the ionized electron can return to the QD core and the PL is turned on again. Since this Auger process can occur in QDs, the inverse Auger process, whereby one high-energy electron–hole pair (created from a photon with $h\nu > E_g$) can generate two electron–hole pairs, can also occur in QDs [47]. The following discussion will present a discussion of the hot carrier cooling dynamics.

2.2.1. *Phonon Bottleneck and Slowed Hot-Electron Cooling in Quantum Dots*

The first prediction of slowed cooling at low light intensities in quantized structures was made by Boudreaux, Williams and Nozik [3]. They anticipated that cooling of carriers would require multi-phonon processes when the quantized levels are separated in energy by more than phonon energies. They analyzed the expected slowed cooling time for hot holes at the surface of highly-doped n-type TiO_2 semiconductors, where quantized energy levels arise because of the narrow space charge layer (i.e., depletion layer) produced by the high doping level. The carrier confinement in this case is produced by the band bending at the surface; for a doping level of $1 \times 10^{19} \, \mathrm{cm}^{-3}$ the potential well can be approximated as a triangular well extending 200 Å from the semiconductor bulk to the surface and with a depth of 1 eV at the surface barrier. The multiphonon relaxation time was estimated from

$$\tau_c \sim T^{-1} \exp(E/kT) \tag{4}$$

where τ_c is the hot carrier cooling time, T is the phonon frequency, and E is the energy separation between quantized levels. For strongly quantized electron levels, with E > 0.2 eV, ϑ_c could be >100 ps according to Equation (4).

However, carriers in the space charge layer at the surface of a heavily doped semiconductor are only confined in one dimension, as in a quantum film. This quantization regime leads to discrete energy states which have dispersion in k-space [50]. This means the hot carriers can cool by undergoing inter-state transitions that require only one emitted phonon followed by a cascade of single phonon intrastate transitions; the bottom of each quantum state is reached by intrastate relaxation before an interstate transition occurs. Thus, the simultaneous and slow multiphonon relaxation pathway can be bypassed by single phonon events, and the cooling rate increases correspondingly.

More complete theoretical models for slowed cooling in QDs have been proposed by Bockelmann and co-workers [23, 51] and Benisty and co-workers [22, 24]. The proposed Benisty mechanism [22, 24] for slowed hot carrier cooling and phonon bottleneck in QDs requires that cooling only occurs via LO phonon emission. However, there are several other mechanisms by which hot electrons can cool in QDs. Most prominent among these is the Auger mechanism [52]. Here, the excess energy of the electron is transferred via an Auger process to the hole, which then cools rapidly because of its larger effective mass and smaller energy level spacing. Thus, an Auger mechanism for hot-electron cooling can break the phonon bottleneck [52]. Other possible mechanisms for breaking the phonon bottleneck include electron–hole scattering [53], deep level trapping [54], and acoustical–optical phonon interactions [55, 56].

2.2.2. Experimental Determination of Relaxation/Cooling Dynamics and a Phonon Bottleneck in Quantum Dots

Over the past several years, many investigations have been published that explore hot-electron cooling/relaxation dynamics in QDs and the issue of a phonon bottleneck in QDs. The results are controversial, and there are many reports that both support [25–27, 57–74] and contradict [27, 54, 75–88] the prediction of slowed hot-electron cooling in QDs and the existence of a phonon bottleneck. One element of confusion that is specific to the focus of this manuscript is that while some of these publications report relatively long hot-electron relaxation times (tens of ps) compared to what is observed in bulk semiconductors, the results are reported as being not indicative of a phonon bottleneck because the relaxation times are not excessively long and PL is observed [89–91] (theory predicts very long relaxation lifetimes (hundreds of ns to μs) of excited carriers for the extreme, limiting condition of a full phonon bottleneck; thus, the carrier lifetime would be determined by non-radiative processes and PL would be absent). However, since the interest here is on the rate of relaxation/cooling compared to the rate of electron separation and transfer, and MEG, we consider that slowed relaxation/cooling of carriers has occurred in QDs if the relaxation/cooling times are greater than 3–5 ps (about an order of magnitude greater than that for bulk semiconductors). This is because electron separation and transport and MEG can be very fast (sub-ps). For solar fuel production, previous work that measured the time of electron transfer from bulk III–V semiconductors to redox molecules (met-allocenium cations) adsorbed on the surface found that ET times can also be sub-ps to several ps [92–95]; hence photoinduced hot carrier separation, transport, and transfer can be competitive with electron cooling and relaxation if the latter is greater than about 10 ps. MEG rates can also be in the sub-ps regime [5, 96].

In a series of papers, Sugawara et al. [62, 63, 65] have reported slow hot-electron cooling in self-assembled InGaAs QDs produced by Stranski-Krastinow (SK) growth on lattice-mismatched GaAs substrates. Using time-resolved PL measurements, the excitation-power dependence of PL, and the current dependence of electroluminescence spectra, these researchers report cooling times ranging from 10 ps to 1 ns. The relaxation time increased with electron energy up to the 5th electronic state. Also, Mukai and Sugawara [97] have recently published an extensive review of phonon bottleneck effects in QDs, which concludes that the phonon bottleneck effect is indeed present in QDs.

Gfroerer et al. report slowed cooling of up to 1 ns in strain-induced GaAs QDs formed by depositing tungsten stressor islands on a GaAs QW with AlGaAs barriers [74]. A magnetic field was applied in these experiments to sharpen and further separate the PL peaks from the excited state transitions, and thereby determine the dependence of the relaxation time on level separation. The authors observed hot PL from excited states in the QD, which could only be attributed to slow relaxation of excited (i.e., hot) electrons. Since the radiative recombination time is about 2 ns, the hot-electron relaxation time was found to be of the same order of magnitude (about 1 ns). With higher excitation intensity sufficient to produce more than one electron–hole pair per dot the relaxation rate increased.

A lifetime of 500 ps for excited electronic states in self-assembled InAs/GaAs QDs under conditions of high injection was reported by Yu et al. [69]. PL from a single GaAs/AlGaAs QD [72] showed intense high-energy PL transitions, which were attributed to slowed electron relaxation in this QD system. Kamath et al. [73] also reported slow electron cooling in InAs/GaAs QDs.

QDs produced by applying a magnetic field along the growth direction of a doped InAs/AlSb QW showed a reduction in the electron relaxation rate from $10^{12}\,\mathrm{s}^{-1}$ to $10^{10}\,\mathrm{s}^{-1}$ [64, 98].

In addition to slow electron cooling, slow hole cooling was reported by Adler et al. [70, 71] in SK InAs/GaAs QDs. The hole relaxation time was determined to be 400 ps based on PL rise times, while the electron relaxation time was estimated to be less than 50 ps. These QDs only contained one electron state, but several hole states; this explained the faster electron cooling time since a quantized transition from a higher quantized electron state to the ground electron state was not present. Heitz et al. [66] also report relaxation times for holes of about 40 ps for stacked layers of SK InAs QDs deposited on GaAs; the InAs QDs are overgrown with GaAs and the QDs in each layer self-assemble into an ordered column. Carrier cooling in this system is about two orders of magnitude slower than in higher-dimensional structures.

All of the above studies on slowed carrier cooling were conducted on self-assembled SK type of QDs. Studies of carrier cooling and relaxation have also been performed on II–VI CdSe colloidal QDs by Klimov et al. [57, 81], Guyot-Sionnest et al. [60], Ellingson et al. [26], and Blackburn et al. [25]. The Klimov group first studied electron relaxation dynamics from the first-excited 1P to the ground 1S state using interband pump-probe spectroscopy [81]. The CdSe QDs were pumped with 100 fs pulses at 3.1 eV to create high-energy electron and holes in their respective band states, and then probed with fs white light continuum pulses. The dynamics of the interband bleaching and induced absorption caused by state filling was monitored to determine the electron relaxation time from the 1P to the 1S state. The results showed very fast 1P to 1S relaxation, on the order of 300 fs, and were attributed to an Auger process for electron relaxation which bypassed the phonon bottleneck. However, this experiment cannot separate the electron and hole dynamics from each other. Guyot-Sionnest et al. [60]

followed up these experiments using fs infrared pump-probe spectroscopy. A visible pump beam creates electrons and holes in the respective band states and a subsequent IR beam is split into an IR pump and an IR probe beam; the IR beams can be tuned to monitor only the intraband transitions of the electrons in the electron states, and thus can separate electron dynamics from hole dynamics. The experiments were conducted with CdSe QDs that were coated with different capping molecules (TOPO, thiocresol, and pyridine), which exhibit different hole-trapping kinetics. The rate of hole trapping increased in the order: TOPO, thiocresol, and pyridine. The results generally show a fast relaxation component (1–2 ps) and a slow relaxation component (≈200 ps). The relaxation times follow the hole-trapping ability of the different capping molecules, and are longest for the QD systems having the fastest hole-trapping caps; the slow component dominates the data for the pyridine cap, which is attributed to its faster hole-trapping kinetics.

These results [60] support the Auger mechanism for electron relaxation, whereby the excess electron energy is rapidly transferred to the hole which then relaxes rapidly through its dense spectrum of states. When the hole is rapidly removed and trapped at the QD surface, the Auger mechanism for hot-electron relaxation is inhibited and the relaxation time increases. Thus, in the above experiments, the slow 200 ps component is attributed to the phonon bottleneck, most prominent in pyridine-capped CdSe QDs, while the fast 1–2 ps component reflects the Auger relaxation process. The relative weight of these two processes in a given QD system depends upon the hole-trapping dynamics of the molecules surrounding the QD.

Klimov et al. further studied carrier relaxation dynamics in CdSe QDs and published a series of papers on the results [57, 58]; a review of this work was also recently published [59]. These studies also strongly support the presence of the Auger mechanism for carrier relaxation in QDs. The experiments were done using ultrafast pump-probe spectroscopy with either 2 beams or 3 beams. In the former, the QDs were pumped with visible light across its bandgap (hole states to electron states) to produce excited state (i.e., hot) electrons; the electron relaxation was monitored by probing the bleaching dynamics of the resonant HOMO to LUMO transition with visible light, or by probing the transient IR absorption of the 1S to 1P intraband transition, which reflects the dynamics of electron occupancy in the LUMO state of the QD. The 3 beam experiment was similar to that of Guyot-Sionnest et al. [60] except that the probe in the experiments of Klimov et al. is a white light continuum. The first pump beam is at 3 eV and creates electrons and holes across the QD bandgap. The second beam is in the IR and is delayed with respect to the optical pump; this beam re-pumps electrons that have relaxed to the LUMO backup in energy. Finally, the third beam is a broad band white light continuum probe that monitors photoinduced interband absorption changes over the range of 1.2 to 3 eV. The experiments were done with two different caps on the QDs: a ZnS cap and a pyridine cap. The results showed that with the ZnS-capped CdSe the relaxation time from the 1P to 1S state was about 250 fs, while for the pyridine-capped CdSe, the relaxation time increased to 3 ps. The increase in the latter experiment was attributed to a phonon bottleneck produced by rapid hole trapping by the pyridine, as also proposed by Guyot-Sionnest et al. [60]. However, the time scale of the phonon bottleneck induced by hole trapping by pyridine caps on CdSe that were reported by Klimov et al. was not as great as that reported by Guyot-Sionnest et al. [60].

Recent results were reported [25, 26] for the electron cooling dynamics in InP QDs where the QD surface was modified to affect hole trapping and also where only electrons were injected into the QD from an external redox molecule (sodium biphenyl) so that holes necessary for the Auger cooling mechanism were not present in the QD [25]. For InP, HF etching was found to passivate electronic surface states but not hole surface states [99, 100]; thus holes can become localized at the surface in both etched and unetched TOPO-capped QDs, and the dynamics associated with these two samples will not deviate significantly. The relaxation was found to be bi-exponential and suggests the presence of two subsets of QDs within the sample [25, 26]. Since etching has been shown to inefficiently passivate hole traps, it is proposed that two subsets of QDs are probed in the experiment: one subset in which the hole and electron are efficiently confined to the interior of the nanocrystal (hole trap absent; exciton confined to the QD core), and one subset in which the hole is localized at the surface of the QD on a phosphorous dangling bond (hole trap present; charge-separated QD) [25, 26].

With the electron and hole confined to the QD core, strong electron–hole interaction leads to efficient, fast relaxation via the Auger mechanism, and in QDs where the hole is localized at the surface the increased spatial separation inhibits the Auger process and results in slower relaxation. The data imply that hole trapping at the intrinsic surface state occurs in less than 75 fs [25].

To further investigate the mechanisms involved in the intraband relaxation, experiments were conducted in which only electrons are present in the QDs and holes are absent. Sodium biphenyl is a very strong reducing agent which has been shown to successfully inject electrons into the conduction band of CdSe QDs [101, 102], effectively bleaching the 1S transition and allowing an IR-induced transition to the $1P_e$ level. Sodium biphenyl was therefore

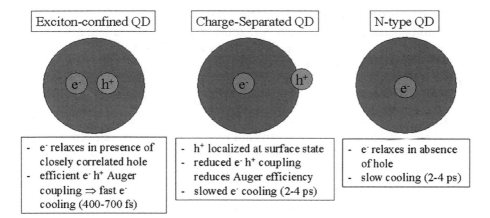

Fig. 7. Different electron–hole configurations in a QD and the resulting relaxation/cooling dynamics (From Reference 46).

used to inject electrons into the 1S electron level in InP QDs [25]. This $1S_e$ electron may be excited to the $1P_e$ level with an IR pump and its relaxation dynamics back to the ground 1S state monitored. Time-resolved, IR-induced transitions in n-type (electron injected) InP QDs show that the relaxation of the excited electrons from the 1P to the 1S level can be fit to a single exponential, with an average time constant of 3.0 ps, corresponding to a relaxation rate of $0.092 \, eV \, ps^{-1}$; in neutral 50 Å TOP/TOPO-capped InP QDs, the relaxation shows a large 400 fs component indicative of fast electron cooling. Similar conclusions were reported for electrons injected into ZnO and CdSe colloidal QDs [27]. These experiments confirm that in the absence of a core-confined hole, electronic relaxation is slowed by about an order of magnitude. However, it should be noted that the relaxation rate in the absence of a hole is close to the relaxation rate with the hole localized at the surface. This is surprising and raises the question of why electron cooling in the absence of a hole is not longer. Possible explanations have been proposed [103] including that (1) positive counter ions of the oxidized sodium biphenyl are adsorbed on the QD surface and behave like a trapped hole in producing a significant Coulomb interaction with the electron to permit Auger cooling and (2) an enhanced Huang-Rhees parameter occurs in charged QDs and enhances multi-phonon relaxation. A summary of these experiments investigating the effects of electron cooling on electron–hole separation is shown in Fig. 7.

Recent results by Guyot-Sionnest et al. show that the nature of the surface ligands has a major effect on the relaxation dynamics [27]. Depending upon the surface ligand stabilizing the QDs, the relaxation or cooling dynamics of hot excitons could be varied from 3.8 ps for tetradecylphosphonic acid, 8 ps for oleic acid, 10 ps for octadecylamine, to 27 ps for dodecanethiol ligands. The later cooling rate is nearly 2 orders of magnitude slower than that for naked QDs or QDs capped with TOP-TOPO.

In contradiction to the results showing slowed cooling in QDs, many other investigations exist in the literature in which a phonon bottleneck was apparently not observed. These results were reported for both self-organized SK QDs [54, 75–88] and II–VI colloidal QDs [81, 83, 85]. However, in several cases [66, 89, 91], hot-electron relaxation was found to be slowed, but not sufficiently for the authors to conclude that this was evidence of a phonon bottleneck.

3. MULTIPLE EXCITON GENERATION IN QUANTUM DOTS

The formation of multiple electron–hole pairs per absorbed photon in photoexcited bulk semiconductors is a process typically explained by impact ionization (I.I.). In this process, an electron or hole with kinetic energy greater than the semiconductor bandgap produces one or more additional electron–hole pairs. The kinetic energy can be created either by applying an electric field or by absorbing a photon with energy above the semiconductor bandgap energy. The former is well studied and understood [104–106]. The latter process is less well studied, but has been observed in photoexcited p–n junctions of Si, Ge, and InSb [107–110].

However, impact ionization has not contributed meaningfully to improved quantum yield in working solar cells, primarily because the I.I. efficiency does not reach significant values until photon energies reach the ultraviolet region of the spectrum. In bulk semiconductors, the threshold photon energy for I.I. exceeds that required for energy conservation alone because, in addition to conserving energy, crystal momentum must be conserved. Additionally, the rate of I.I. must compete with the rate of energy relaxation by electron–phonon scattering. It has been shown that the rate of I.I. becomes competitive with phonon scattering rates only when the kinetic energy of the electron is many times the bandgap energy (E_g) [104–106]. The observed transition between inefficient and efficient I.I. occurs slowly; for example, in Si the I.I. efficiency was found to be only 5% (*i.e.*, total quantum yield = 105%) at $h\nu \approx 4\,\text{eV}(3.6E_g)$, and 25% at $h\nu \approx 4.8\,\text{eV}$ $(4.4E_g)$ [110, 111]. This large blue-shift of the threshold photon energy for I.I. in semiconductors prevents materials such as bulk Si and GaAs from yielding improved solar conversion efficiencies [11, 111].

However, in QDs the rate of electron relaxation through electron–phonon interactions can be significantly reduced because of the discrete character of the electron–hole spectra, and the rate of Auger processes, including the inverse Auger process of exciton multiplication, is greatly enhanced due to carrier confinement and the concomitantly increased electron–hole Coulomb interaction. Furthermore, crystal momentum need not be conserved because momentum is not a good quantum number for three-dimensionally-confined carriers. Indeed, very efficient multiple electron–hole pair (multi-exciton) creation by one photon was reported recently in PbSe nanocrystals by Schaller and Klimov [14]. They reported an excitation energy threshold for the formation of two excitons per photon at $3E_g$, where E_g is the absorption energy gap of the nanocrystal (HOMO-LUMO transition energy. Schaller and Klimov reported a QY value of 218% (118% I.I. efficiency) at $3.8E_g$; QYs above 200% indicate the formation of more than two excitons per absorbed photon. Other researchers have recently reported [5] a QY value of 300% for 3.9 nm diameter PbSe QDs at a photon energy of $4E_g$, indicating the formation of three excitons per photon for every photoexcited QD in the sample. Evidence was also provided that showed the threshold for MEG by optical excitation is $2E_g$, not $3E_g$ as reported previously for PbSe QDs [14], and it was also shown that comparably efficient MEG occurs also in PbS nanocrystals. A new possible mechanism for MEG was introduced [14] that invokes a coherent superposition of multiple-excitonic states, meaning that multiple excitons are essentially created instantly upon absorption of high-energy photons. Most recently, MEG has been reported in CdSe QDs [112], and in PbTe QDs [113] and seven excitons per photon were reported in PbSe QDs at 7 times the bandgap [112].

Multiexcitons are detected by monitoring the signature of multiexciton decay dynamics using transient absorption (TA) spectroscopy [5, 14, 112]. The magnitude of the photoinduced absorption change at the band edge is proportional to the number of electron–hole pairs created in the sample. The transients are detected by probing either with a band edge (energy gap or HOMO-LUMO transition energy $\equiv E_g$) probe pulse, or with a mid-IR probe pulse that monitors intraband transitions in the newly created excitons. Although both the band-edge and mid-IR probe signals would incorporate components from excitons with energy above the $1S_h–1S_e$ exciton, multiple-exciton Auger recombination analysis relies only on data for delays >5 ps, by which time carrier multiplication and cooling are complete.

The dependence of the MEG QY on the ratio of the pump photon energy to the bandgap ($E_{h\nu}/E_g$) is shown in Fig. 8 for PbSe, PbS, and PbTe QDs.

A.J. Nozik

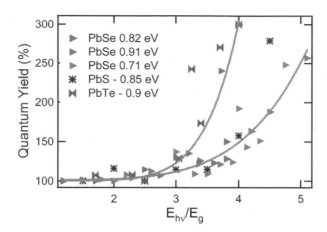

Fig. 8. MEG QYs for PbS, PbSe, PbTe, the solid lines are guides to the eye.

For the 3 PbSe QD samples, E_g = 0.72 eV (dia. = 5.7 nm), E_g = 0.82 eV (dia. = 4.7 nm), and E_g = 0.91 eV (dia. = 3.9 nm). For all three samples, the sharp rise in QY begins at about three times the energy gap, a result in agreement with that reported. The data show that for the 3.9 nm QD (E_g = 0.91 eV), the QY reaches a value of 300% at $E_{h\nu}/E_g$ = 4.0, indicating that the QDs produce three excitons per absorbed photon. For the other two PbSe samples (E_g = 0.82 eV (4.7 nm dia.) and 0.72 eV (5.7 nm dia.)), it is estimated that a QY of 300% is reached at an $E_{h\nu}/E_g$ value of 5.5. It was noted that the $2P_h$–$2P_e$ transition in the QDs is resonant with the $3E_g$ excitation, corresponding to the sharp onset of increased MEG efficiency. If this symmetric transition ($2P_h$–$2P_e$) dominates the absorption at $\sim 3E_g$, the resulting excited state provides both the electron and the hole with excess energy of $1E_g$, in resonance with the lowest exciton absorption (at $1E_g$). Our data also showed that the QY begins to surpass 100% at E_h/E_g values greater than 2.0 (see Fig. 3). In ref. [5], 16 QY values were carefully measured between $2.1E_g$ and $2.9E_g$ (mean value = 109.8%) and 11 QY values between $1.2E_g$ and $2.0E_g$ (mean value = 101.3%). Application of statistical t-tests show that the QY values for photon energies between $1E_g$ and $2E_g$ were not statistically different from 100% (P value = 0.105), while the difference in QYs between $1.2E_g$–$2.0E_g$ and $2.1E_g$–$2.9E_g$ were very statistically significant with a P value of 0.001. Also, simple visual inspection of Fig. 3 indicated a significant difference between the QY values between $1E_g$–$2E_g$ and $2E_g$–$3E_g$. For PbS and PbTe QDs, the bandgaps were 0.85 and 0.90 eV, respectively, corresponding to diameters of 5.5 nm and 4.2 nm.

4. QUANTUM DOT SOLAR CELL CONFIGURATIONS

The two fundamental pathways for enhancing the conversion efficiency (increased photovoltage [2, 3] or increased photocurrent [10, 11] can be accessed, in principle, in at least three different QD solar cell configurations; these configurations are shown in Fig. 9 and they are described below. However, it is emphasized that these potential high efficiency configurations are conceptual and there is no experimental evidence yet that demonstrates actual enhanced conversion efficiencies in any of these systems.

4.1. Photoelectrodes Composed of Quantum Dot Arrays

In this configuration, the QDs are formed into an ordered 3-D array with inter-QD spacing sufficiently small such that strong electronic coupling occurs and minibands are formed to allow long-range electron transport (see Fig. 9A). The system is a 3-D analog to a 1-D superlattice and the miniband structures formed therein [1] (see Fig. 3). The delocalized quantized 3-D miniband states could be expected to slow the carrier cooling and permit the transport and collection of hot carriers to produce a higher photopotential in a PV cell or in a photoelectrochemical cell where the 3-D QD array is the photoelectrode [114]. Also, MEG might be expected to occur in the QD arrays, enhancing the photocurrent (see Fig. 2). However, hot-electron transport/collection and MEG cannot occur simultaneously; they are mutually exclusive and only one of these processes can be present in a given system.

Significant progress has been made in forming 3-D arrays of both colloidal [115–117] and epitaxial [118] II–VI and III–V QDs. The former have been formed via evaporation and crystallization of colloidal QD solutions containing a uniform QD size distribution; crystallization of QD solids from broader size distributions lead to close-packed QD solids, but with a high degree of disorder. Concerning the latter, arrays of epitaxial QDs have been formed by successive epitaxial deposition of epitaxial QD layers; after the first layer of epitaxial QDs is formed, successive layers tend to form with the QDs in each layer aligned on top of each other [118, 119]. Theoretical and experimental studies of the properties of QD arrays are currently under way. Major issues are the nature of the electronic states as a function of inter-dot distance, array order vs disorder, QD orientation and shape, surface states, surface structure/passivation, and surface chemistry. Transport properties of QD arrays are also of critical importance, and they are under investigation.

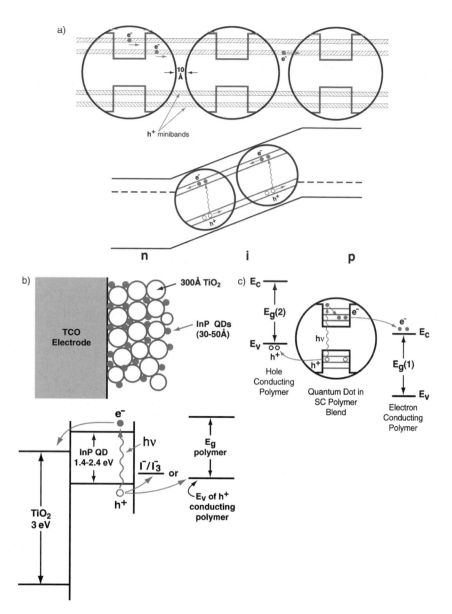

Fig. 9. Configurations for QD solar cells. (a) a QD array used as a photoelectrode for a photoelectrochemical or as the i-region of a p–i–n PV cell; (b) QDs used to sensitize a nanocrystalline film of a wide bandgap oxide semiconductor (viz. TiO$_2$) to visible light. This configuration is analogous to the dye-sensitized solar cell where the dye is replaced by QDs; (C) QDs dispersed in a blend of electron- and hole-conducting polymers. In configurations A, B, C, the occurrence of impact ionization could produce higher photocurrents and higher conversion efficiency. In A, enhanced efficiency could be achieved either through impact ionization or hot carrier transport through the minibands of the QD array resulting in a higher photopotential (From Reference 13).

4.2. Quantum Dot-Sensitized Nanocrystalline TiO₂ Solar Cells

This configuration is a variation of a recent promising new type of PV cell that is based on dye-sensitization of nanocrystalline TiO_2 layers [120–122]. In this latter PV cell, dye molecules are chemisorbed onto the surface of 10–30 nm-size TiO_2 particles that have been sintered into a highly porous nanocrystalline 10–20 μm TiO_2 film. Upon photoexcitation of the dye molecules, electrons are very efficiently injected from the excited state of the dye into the conduction band of the TiO_2, affecting charge separation and producing a PV effect. The cell circuit is completed using a non-aqueous redox electrolyte that contains I^-/I_3^- and a Pt counter electrode to allow reduction of the adsorbed photooxidized dye back to its initial non-oxidized state (via I_3^- produced at the Pt cathode by reduction of I^-).

For the QD-sensitized cell, QDs are substituted for the dye molecules; they can be adsorbed from a colloidal QD solution [123] or produced in situ [124–127] (see Fig. 9B). Successful PV effects in such cells have been reported for several semiconductor QDs including InP, CdSe, CdS, and PbS [123–127]. Possible advantages of QDs over dye molecules are the tunability of optical properties with size and better heterojunction formation with solid hole conductors. Also, as discussed here, a unique potential capability of the QD-sensitized solar cell is the production of quantum yields greater than one by MEG (inverse Auger effect) [47]. Dye molecules cannot undergo this process. Efficient inverse Auger effects in QD-sensitized solar cells could produce much higher conversion efficiencies than are possible with dye-sensitized solar cells.

4.3. Quantum Dots Dispersed in Organic Semiconductor Polymer Matrices

Recently, PV effects have been reported in structures consisting of QDs forming junctions with organic semiconductor polymers. In one configuration, a disordered array of CdSe QDs is formed in a hole-conducting polymer—MEH-PPV (poly(2-methoxy, 5-(2′-ethyl)-hexyloxy-p-phenylenevinylene) [128]. Upon photoexcitation of the QDs, the photogenerated holes are injected into the MEH-PPV polymer phase, and are collected via an electrical contact to the polymer phase. The electrons remain in the CdSe QDs and are collected through diffusion and percolation in the nanocrystalline phase to an electrical contact to the QD network. Initial results show relatively low conversion efficiencies [128, 129] but improvements have been reported with rod-like CdSe QD shapes [130] embedded in poly(3-hexylthiophene) (the rod-like shape enhances electron transport through the nanocrystalline

QD phase). In another configuration [131], a polycrystalline TiO_2 layer is used as the electron conducting phase, and MEH-PPV is used to conduct the holes; the electron and holes are injected into their respective transport mediums upon photoexcitation of the QDs.

A variation of these configurations is to disperse the QDs into a blend of electron and hole-conducting polymers (see Fig. 9C). This scheme is the inverse of light emitting diode structures based on QDs [132–136]. In the PV cell, each type of carrier–transporting polymer would have a selective electrical contact to remove the respective charge carriers. A critical factor for success is to prevent electron–hole recombination at the interfaces of the two polymer blends; prevention of electron–hole recombination is also critical for the other QD configurations mentioned above.

All of the possible QD-organic polymer PV cell configurations would benefit greatly if the QDs can be coaxed into producing multiple electron–hole pairs by the inverse Auger/MEG process [47]. This is also true for all the QD solar cell systems described above. The various cell configurations simply represent different modes of collecting and transporting the photo-generated carriers produced in the QDs.

CONCLUSION

The relaxation dynamics of photoexcited electrons in semiconductor QDs can be greatly modified compared to the bulk form of the semiconductor. Specifically, the cooling dynamics of highly energetic (hot) electrons created by absorption of supra-bandgap photons can be slowed by at least one order of magnitude (4–7 ps vs 400–700 fs). This slowed cooling is caused by a so-called "phonon bottleneck" when the energy spacing between quantized levels in the QD is greater than the LO-phonon energy, thus inhibiting hot-electron relaxation (cooling) by electron–phonon interactions. In order to produce the slowed hot-electron cooling via the phonon bottleneck, it is necessary to block an Auger process that could bypass the phonon bottleneck and allow fast electron cooling. The Auger cooling process involves the transfer of the excess electron energy to a hole, which then cools rapidly because of its higher effective mass and closely-spaced energy levels. Blocking the Auger cooling is achieved by rapidly removing the photogenerated hole before it undergoes Auger scattering with the photogenerated electron, or by injecting electrons into the LUMO level (conduction band) of the QD from an external electron donating chemical species and then exciting these electrons with an IR pulse. Slowed electron cooling in QDs offers the potential to use QDs in solar cells to enhance their conversion efficiency. In bulk semiconductors, the hot electrons (and holes) created by absorption of supra-bandgap photons cool so rapidly to the band edges that the excess kinetic energy of the photogenerated carriers is converted to heat and limits the theoretical Shockley-Queisser thermodynamic conversion efficiency to about 32% (at one sun). Slowed cooling in QDs could lead to their use in solar cell configurations wherein impact ionization (the formation of two or more electron–hole pairs per absorbed photon) or hot-electron separation, transport, and transfer can become significant, thus producing enhanced photocurrents or photovoltages and corresponding enhanced conversion efficiencies with thermodynamics limits of 66% (one sun). Three configurations for QD solar cells have been described here that would produce either enhanced photocurrent or photovoltage.

ACKNOWLEDGMENTS

The author is supported by the U.S. Department of Energy, Office of Science, Office of Basic Energy Sciences, Division of Chemical Sciences, Geosciences and Biosciences. Vital contributions to the work reviewed here have been made by Olga Micic, Randy Ellingson, Matt Beard, Jim Murphy, Jeff Blackburn, Phil Ahrenkiel, Justin Johnson, Pingrong Yu, Andrew Shabaev, and Alexander Efros.

REFERENCES

[1]   A.J. Nozik, Annu. Rev. Phys. Chem., 52 (2001) 193*.
[2]   R.T. Ross and A.J. Nozik, J. Appl. Phys., 53 (1982) 3813.
[3]   D.S. Boudreaux and F. Williams, A.J. Nozik, J. Appl. Phys., 51 (1980) 2158.
[4]   F.E. Williams and A.J. Nozik, Nature, 311 (1984) 21.
[5]   R.J. Ellingson, M.C. Beard, J.C. Johnson, P. Yu, O.I. Micic, A.J. Nozik, A. Shabaev, and A.L. Efros, Nano Lett., 5 (2005) 865.
[6]   W. Shockley and H. J. Queisser, J. Appl. Phys., 32 (1961) 510.
[7]   R.T. Ross, J. Chem. Phys., 45 (1966) 1.
[8]   R.T. Ross, J. Chem. Phys., 46 (1967) 4590.
[9]   R.J. Ellingson, J.L. Blackburn, P. Yu, G. Rumbles, O.I. Micic and A.J. Nozik, J. Phys. Chem. B, 106 (2002) 7758.
[10] P.T. Landsberg, H. Nussbaumer, G. Willeke, J. Appl. Phys., 74 (1993) 1451.

---

*This chapter is derived from previously published reviews and manuscripts by the author in references [1] and [46].

[11] S. Kolodinski, J.H. Werner, T. Wittchen, H.J. Queisser, Appl. Phys. Lett., 63 (1993) 2405.

[12] M.A. Green, Third Generation Photovoltaics, Bridge Printery, Sydney, 2001.

[13] A.J. Nozik, Physica E, 14 (2002) 115.

[14] R. Schaller, V. Klimov, Phys. Rev. Lett., 92 (2004) 186601.

[15] A. Luque, A. Marti, Phys. Rev. Lett., 78 (1997) 5014.

[16] A.J. Nozik, D.S. Boudreaux, R.R. Chance, F. Williams, Charge Transfer at Illuminated Semiconductor-Electrolyte Interfaces, in: M. Wrighton (Ed.) Advances in Chemistry, vol 184, ACS, New York, 1980, p. 162.

[17] A.J. Nozik, Philos. Trans. R. Soc. London. Ser. A, A295 (1980) 453.

[18] W.S. Pelouch, R.J. Ellingson, P.E. Powers, C.L. Tang, D.M. Szmyd, A.J. Nozik, Phys. Rev. B, 45 (1992) 1450.

[19] W.S. Pelouch, R.J. Ellingson, P.E. Powers, C.L. Tang, D.M. Szmyd, A.J. Nozik, Semicond. Sci. Technol., 7 (1992) B337.

[20] Y. Rosenwaks, M.C. Hanna, D.H. Levi, D.M. Szmyd, R.K. Ahrenkiel, A.J. Nozik, Phys. Rev. B, 48 (1993) 14675.

[21] F. Williams, A.J. Nozik, Nature, 271 (1978) 137.

[22] H. Benisty, C.M. Sotomayor-Torres, C. Weisbuch, Phys. Rev. B, 44 (1991) 10945.

[23] U. Bockelmann, G. Bastard, Phys. Rev. B, 42 (1990) 8947.

[24] H. Benisty, Phys. Rev. B, 51 (1995) 13281.

[25] J.L. Blackburn, R.J. Ellingson, O.I. Micic, A.J. Nozik, J. Phys. Chem. B, 107 (2003) 102.

[26] R.J. Ellingson, J.L. Blackburn, J.M. Nedeljkovic, G. Rumbles, M. Jones, H. Fu, A.J. Nozik, Phys. Rev. B, 67 (2003) 075308.

[27] P. Guyot-Sionnest, B. Wehrenberg, D. Yu, J. Chem. Phys., 123 (2005) 074709.

[28] J.I. Pankove, Optical Processes in Semiconductors, Dover, New York, 1975.

[29] S. Sze, Physics of Semiconductor Devices, Wiley, New York, 1981.

[30] M.A. Green, Solar Cells, The University of New South Wales, Kensington, Aust., 1992.

[31] A.J. Nozik, Annu. Rev. Phys. Chem., 29 (1978) 189.

[32] R. Dingle (Ed.) Applications of Multiquantum Wells, Selective Doping, and Superlattices, Academic Press, New York, 1987.

[33] M.W. Peterson, J.A. Turner, C.A. Parsons, A.J. Nozik, D.J. Arent, C. Van Hoof, G. Borghs, R. Houdre, H. Morkoc, Appl. Phys. Lett., 53 (1988) 2666.

[34] C.A. Parsons, B.R. Thacker, D.M. Szmyd, M.W. Peterson, W.E. McMahon, A.J. Nozik, J. Chem. Phys., 93 (1990) 7706.

[35] D.C. Edelstein, C.L. Tang, A.J. Nozik, Appl. Phys. Lett., 51 (1987) 48.

[36] Z.Y. Xu, C.L. Tang, Appl. Phys. Lett., 44 (1984) 692.

[37] M.J. Rosker, F.W. Wise, C.L. Tang, Appl. Phys. Lett., 49 (1986) 1726.

[38] J.F. Ryan, R.A. Taylor, A.J. Tuberfield, A. Maciel, J.M. Worlock, A.C. Gossard, W. Wiegmann, Phys. Rev. Lett., 53 (1984) 1841.

[39] J. Christen, D. Bimberg, Phys. Rev. B, 42 (1990) 7213.

[40] W. Cai, M.C. Marchetti, M. Lax, Phys. Rev. B, 34 (1986) 8573.

[41] A.J. Nozik, C.A. Parsons, D.J. Dunlavy, B.M. Keyes, R.K. Ahrenkiel, Solid State Comm. 75 (1990) 297.

[42] P. Lugli, S.M. Goodnick, Phys. Rev. Lett., 59 (1987) 716.

[43] V.B. Campos, S. Das Sarma, M.A. Stroscio, Phys. Rev. B, 46 (1992) 3849.

[44] R.P. Joshi, D.K. Ferry, Phys. Rev. B, 39 (1989) 1180.

[45] M.C. Hanna, Z.W. Lu, A.J. Nozik, Future Generation Photovoltaic Technologies, in: R.D. McConnell (Ed.) AIP Conference Proceedings 404, American Institute of Physics, Woodbury, NY, 1997, pp. 309–317.

[46] A.J. Nozik, Next Generation Photovoltaics, in: A. Marti, M. Green (Eds.), Series in Optics and Optoelectronics, Institute of Physics, 2004, pp. 196–218*.

[47] A.J. Nozik, (1997) unpublished manuscript.

[48] M. Nirmal, B.O. Dabbousi, M.G. Bawendi, J.J. Macklin, J.K. Trautman, T.D. Harris, L.E. Brus, Nature, 383 (1996) 802.

[49] A.L. Efros, M. Rosen, Phys. Rev. Lett., 78 (1997) 1110.

[50] M. Jaros, Quantum Wells, Superlattices, Quantum Wires, and Dots, in: A.N. Broers, C. Hilsum, R.A. Stradling (Eds.), Physics and Applications of Semiconductor Microstructures, Oxford University Press, New York, 1989, pp. 83–106.

[51] U. Bockelmann, T. Egeler, Phys. Rev. B, 46 (1992) 15574.

[52] A.L. Efros, V.A. Kharchenko, M. Rosen, Solid State Commun. 93 (1995) 281.

[53] I. Vurgaftman, J. Singh, Appl. Phys. Lett., 64 (1994) 232.

[54] P.C. Sercel, Phys. Rev. B, 51 (1995) 14532.

[55] T. Inoshita, H. Sakaki, Phys. Rev. B, 46 (1992) 7260.

[56] T. Inoshita, H. Sakaki, Phys. Rev. B, 56 (1997) R4355.

[57] V.I. Klimov, A.A. Mikhailovsky, D.W. McBranch, C.A. Leatherdale, M.G. Bawendi, Phys. Rev. B, 61 (2000) R13349.

[58] V.I. Klimov, D.W. McBranch, C.A. Leatherdale, M.G. Bawendi, Phys. Rev. B, 60 (1999) 13740.

[59] V.I. Klimov, J. Phys. Chem. B, 104 (2000) 6112.

[60] P. Guyot-Sionnest, M. Shim, C. Matranga, M. Hines, Phys. Rev. B, 60 (1999) R2181.

[61] P.D. Wang, C.M. Sotomayor-Torres, H. McLelland, S. Thoms, M. Holland, C.R. Stanley, Surf. Sci., 305 (1994) 585.

[62] K. Mukai, M. Sugawara, Jpn. J. Appl. Phys., 37 (1998) 5451.

[63] K. Mukai, N. Ohtsuka, H. Shoji, M. Sugawara, Appl. Phys. Lett., 68 (1996) 3013.

[64] B.N. Murdin, A.R. Hollingworth, M. Kamal Saadi, R.T. Kotitschke, C.M. Ciesla, C.R. Pidgeon, P.C. Findlay, H.P.M. Pellemans, C.J.G.M. Langerak, A.C. Rowe, R.A. Stradling, E. Gornik, Phys. Rev. B, 59 (1999) R7817.

[65] M. Sugawara, K. Mukai, H. Shoji, Appl. Phys. Lett., 71 (1997) 2791.

[66] R. Heitz, M. Veit, N.N. Ledentsov, A. Hoffmann, D. Bimberg, V.M. Ustinov, P.S. Kop'ev, Z. I. Alferov, Phys. Rev. B, 56 (1997) 10435.

[67] R. Heitz, A. Kalburge, Q. Xie, M. Grundmann, P. Chen, A. Hoffmann, A. Madhukar, D. Bimberg, Phys. Rev. B, 57 (1998) 9050.

[68] K. Mukai, N. Ohtsuka, H. Shoji, M. Sugawara, Phys. Rev. B, 54 (1996) R5243.

[69] H. Yu, S. Lycett, C. Roberts, R. Murray, Appl. Phys. Lett., 69 (1996) 4087.

[70] F. Adler, M. Geiger, A. Bauknecht, F. Scholz, H. Schweizer, M.H. Pilkuhn, B. Ohnesorge, A. Forchel, Appl. Phys., 80 (1996) 4019.

[71] F. Adler, M. Geiger, A. Bauknecht, D. Haase, P. Ernst, A. Dörnen, F. Scholz, H. Schweizer, J. Appl. Phys., 83 (1998) 1631.

[72] K. Brunner, U. Bockelmann, G. Abstreiter, M. Walther, G. Böhm, G. Tränkle, G. Weimann, Phys. Rev. Lett., 69 (1992) 3216.

[73] K. Kamath, H. Jiang, D. Klotzkin, J. Phillips, T. Sosnowski, T. Norris, J. Singh, P. Bhattacharya, Inst. Phys. Conf. Ser., 156 (1998) 525.

[74] T.H. Gfroerer, M.D. Sturge, K. Kash, J.A. Yater, A.S. Plaut, P.S.D. Lin, L.T. Florez, J.P. Harbison, S.R. Das, L. Lebrun, Phys. Rev. B, 53 (1996) 16474.

[75] X.-Q. Li, H. Nakayama, Y. Arakawa, Phonon Decay and Its Impact on Carrier Relaxation in Semiconductor Quantum Dots, in: D. Gershoni (Ed.) Proc. 24th Int. Conf. Phys. Semicond., World Sci., Singapore, 1998, pp. 845–848.

[76] J. Bellessa, V. Voliotis, R. Grousson, D. Roditchev, C. Gourdon, X.L. Wang, M. Ogura, H. Matsuhata, Relaxation and Radiative Lifetime of Excitons in a Quantum Dot and a Quantum Wire, in: D. Gershoni (Ed.) Proc. 24th Int. Conf. Phys. Semicond., World Scientific, Singapore, 1998, pp. 763–766.

[77] M. Lowisch, M. Rabe, F. Kreller, F. Henneberger, Appl. Phys. Lett., 74 (1999) 2489.

[78] I. Gontijo, G.S. Buller, J.S. Massa, A.C. Walker, S.V. Zaitsev, N.Y. Gordeev, V.M. Ustinov, P.S. Kop'ev, Jpn. J. Appl. Phys., 38 (1999) 674.

[79] X.-Q. Li, H. Nakayama, Y. Arakawa, Jpn. J. Appl. Phys., 38 (1999) 473.

[80] K. Kral, Z. Khas, Phys. Status Solidi. B, 208 (1998) R5.

[81] V.I. Klimov, D.W. McBranch, Phys. Rev. Lett., 80 (1998) 4028.

[82] D. Bimberg, N.N. Ledentsov, M. Grundmann, R. Heitz, J. Boehrer, V.M. Ustinov, P.S. Kop'ev, Z.I. Alferov, J. Lumin., 72–74 (1997) 34.

[83] U. Woggon, H. Giessen, F. Gindele, O. Wind, B. Fluegel, N. Peyghambarian, Phys. Rev. B, 54 (1996) 17681.

[84] M. Grundmann, R. Heitz, N. Ledentsov, O. Stier, D. Bimberg, V.M. Ustinov, P.S. Kop'ev, Z.I. Alferov, S.S. Ruvimov, P. Werner, U. Gösele, J. Heydenreich, Superlattices Microstruct. 19 (1996) 81.

[85] V.S. Williams, G.R. Olbright, B.D. Fluegel, S.W. Koch, N. Peyghambarian, J. Modern Optics, 35 (1988) 1979.

[86] B. Ohnesorge, M. Albrecht, J. Oshinowo, A. Forchel, Y. Arakawa, Phys. Rev. B, 54 (1996) 11532.

[87] G. Wang, S. Fafard, D. Leonard, J.E. Bowers, J.L. Merz, P.M. Petroff, Appl. Phys. Lett., 64 (1994) 2815.

[88] J.H.H. Sandmann, S. Grosse, G. von Plessen, J. Feldmann, G. Hayes, R. Phillips, H. Lipsanen, M. Sopanen, J. Ahopelto, Phys. Status Solidi. B, 204 (1997) 251.

[89] R. Heitz, M. Veit, A. Kalburge, Q. Xie, M. Grundmann, P. Chen, N.N. Ledentsov, A. Hoffmann, A. Madhukar, D. Bimberg, V.M. Ustinov, P.S. Kop'ev, Z.I. Alferov, Physica E (Amsterdam)., 2 (1998) 578.

[90] X.-Q. Li, Y. Arakawa, Phys. Rev. B, 57 (1998) 12285.

[91] T.S. Sosnowski, T.B. Norris, H. Jiang, J. Singh, K. Kamath, P. Bhattacharya, Phys. Rev. B, 57 (1998) R9423.

[92] R.D.J. Miller, G. McLendon, A.J. Nozik, W. Schmickler, F. Willig, Surface Electron Transfer Processes, VCH Publishers, New York, 1995.

[93] A. Meier, D.C. Selmarten, K. Siemoneit, B.B. Smith, A.J. Nozik, J. Phys. Chem. B, 103 (1999) 2122.

[94] A. Meier, S.S. Kocha, M.C. Hanna, A.J. Nozik, K. Siemoneit, R. Reineke-Koch, R. Memming, J. Phys. Chem. B, 101 (1997) 7038.

[95] S.J. Diol, E. Poles, Y. Rosenwaks, R.J.D. Miller, J. Phys. Chem. B, 102 (1998) 6193.

[96] R.D. Schaller, V.M. Agranovich, V.I. Klimov, Nat. Phys., 1 (2005) 189.

[97] K. Mukai, M. Sugawara, The Phonon Bottleneck Effect in Quantum Dots, in: R.K. Willardson, E.R. Weber (Eds.), Semiconductors and Semimetals, Vol 60, Academic Press, San Diego, 1999, p. 209.

[98] B.N. Murdin, A.R. Hollingworth, M.Kamal-Saadi, R.T. Kotitsche, C.M. Ciesla, C.R. Pidgeon, P.C. Findlay, H.A. Pellemans, C.J.G.M. Langerak, A.C. Rowe, R.A. Stradling, E. Gornik, Suppression of LO phonon scattering in "quasi" quantum dots, in: D. Gershoni (Ed.) Proc. 24th Int. Conf. Phys. Semicond., World Scientific, Singapore, 1998, pp. 1867–1870.

[99] L. Langof, E. Ehrenfreund, E. Lifshitz, O.I. Micic, A.J. Nozik, J. Phys. Chem. B, 106 (2002) 1606.

[100] O.I. Micic, A.J. Nozik, E. Lifshitz, T. Rajh, O.G. Poluektov, M.C. Thurnauer, J. Phys. Chem., 106 (2002) 4390.

[101] M. Shim, P. Guyot-Sionnest, Nature, 407 (2000) 981.

[102] M. Shim, C. Wang, P.J. Guyot-Sionnest, J. Phys. Chem., 105 (2001) 2369.

[103] A. Shabaev, A.L. Efros, A.J. Nozik, Nanoletters. (2006), in press.

[104] H.K. Jung, K. Taniguchi, C. Hamaguchi, J. Appl. Phys., 79 (1996) 2473.

[105] D. Harrison, R.A. Abram, S. Brand, J. Appl. Phys., 85 (1999) 8186.

[106] J. Bude, K. Hess, J. Appl. Phys., 72 (1992) 3554.

[107] V.S. Vavilov, J. Phys. Chem. Solids, 8 (1959) 223.

[108] J. Tauc, J. Phys. Chem. Solids, 8 (1959) 219.

[109] R.J. Hodgkinson, Proc. Phys. Soc., 82 (1963) 1010.

[110] O. Christensen, J. Appl. Phys. 47 (1976) 690.

[111] M. Wolf, R. Brendel, J.H. Werner, H.J. Queisser, J. Appl. Phys., 83 (1998) 4213.

[112] R.D. Schaller, M.A. Petruska, V.I. Klimov, Appl. Phys. Lett., 87 (2005) 1.

[113] J.E. Murphy, M.C. Beard, A.G. Norman, S.P. Ahrenkiel, J.C. Johnson, P. Yu, O.I. Micic, R.J. Ellingson, A.J. Nozik, J. Am. Chem. Soc., 28 (2006) 3241.

[114] A.J. Nozik, (1996) unpublished manuscript.

[115] C.B. Murray, C.R. Kagan, M.G. Bawendi, Annu. Rev. Mater. Sci., 30 (2000) 545.

[116] O.I. Micic, S.P. Ahrenkiel, A.J. Nozik, Appl. Phys. Lett., 78 (2001) 4022.

[117] O.I. Micic, K.M. Jones, A. Cahill, A.J. Nozik, J. Phys. Chem. B, 102 (1998) 9791.

[118] M. Sugawara, Self-Assembled InGaAs/GaAs Quantum Dots, in: M. Sugawara (Ed.) Semiconductors and Semimetals, Vol 60, Academic Press, San Diego, 1999, pp. 1–350.

[119] Y. Nakata, Y. Sugiyama, M. Sugawara, in: M. Sugawara (Ed.) Semiconductors and Semimetals, Vol 60, Academic Press, San Diego, 1999, pp. 117–152.

[120] A. Hagfeldt, M. Grätzel, Acc. Chem. Res., 33 (2000) 269.

[121] J. Moser, P. Bonnote, M. Grätzel, Coord. Chem. Rev., 171 (1998) 245.

[122] M. Grätzel, Prog. Photovoltaics, 8 (2000) 171.

[123] A. Zaban, O.I. Micic, B.A. Gregg, A.J. Nozik, Langmuir, 14 (1998) 3153.

[124] R. Vogel, H. Weller, J. Phys. Chem., 98 (1994) 3183.

[125] H. Weller, Ber. Bunsen-Ges. Phys. Chem., 95 (1991) 1361.

[126] D. Liu, P.V. Kamat, J. Phys. Chem., 97 (1993) 10769.

[127] P. Hoyer, R. Könenkamp, Appl. Phys. Lett., 66 (1995) 349.

[128] N.C. Greenham, X. Peng, A.P. Alivisatos, Phys. Rev. B, 54 (1996) 17628.

[129] N.C. Greenham, X. Peng, A.P. Alivisatos, A CdSe Nanocrystal/MEH-PPV Polymer Composite Photovoltaic, in: R. McConnell (Ed.) Future Generation Photovoltaic Technologies: First NREL Conference, Am. Inst. Phys., 1997, p. 295.

[130] W.U. Huynh, X. Peng, P. Alivisatos, Adv. Mater., 11 (1999) 923.

[131] A.C. Arango, S.A. Carter, P.J. Brock, Appl. Phys. Lett., 74 (1999) 1698.

[132] B.O. Dabbousi, M.G. Bawendi, O. Onitsuka, M.F. Rubner, Appl. Phys. Lett., 66 (1995) 1316.

[133] V. Colvin, M. Schlamp, A.P. Alivisatos, Nature, 370 (1994) 354.

[134] M.C. Schlamp, X. Peng, A.P. Alivisatos, J. Appl. Phys., 82 (1997) 5837.

[135] H. Mattoussi, L.H. Radzilowski, B.O. Dabbousi, D.E. Fogg, R.R. Schrock, E.L. Thomas, M.F. Rubner, M.G. Bawendi, J. Appl. Phys., 86 (1999) 4390.

[136] H. Mattoussi, L.H. Radzilowski, B.O. Dabbousi, E.L. Thomas, M.G. Bawendi, M.F. Rubner, J. Appl. Phys., 83 (1998) 7965.

Nanostructured Materials for Solar Energy Conversion
T. Soga (editor)

Chapter 16

Quantum Well Solar Cells and Quantum Dot Concentrators

K.W.J. Barnham[a], I. Ballard[a], A. Bessière[a], A.J. Chatten[a], J.P. Connolly[a], N.J. Ekins-Daukes[a], D.C. Johnson[a], M.C. Lynch[a], M. Mazzer[a,d], T.N.D. Tibbits[a], G. Hill[b], J.S. Roberts[b] and M.A. Malik[c]

[a]Blackett Laboratory, Physics Department, Imperial College London, London, SW7 2BW, UK
[b]EPSRC National Centre for III-V Technologies, Sheffield S1 3JD, UK
[c]Chemistry Department, University of Manchester, Manchester M13 9PL, UK
[d]CNR-IMM, University Campus Lecce, Italy

1. INTRODUCTION

This chapter reviews the development over the past half a decade of the quantum well solar cell (QWSC) and the quantum dot concentrator (QDC). The study of nanostructures such as quantum wells (QWs) and quantum dots (QDs) has dominated opto-electronic research and development for the past two decades. Photovoltaic (PV) applications of these nanostructures have been less extensively studied. Our group has pioneered the study of both QWs and QDs for PV applications. Others groups have made significant contributions, particularly to the study of lattice matched QWSCs and the theory of the cells, particularly in the ideal limit. These earlier contributions have been acknowledged in the two reviews, which were written at the turn of the new century [1,2]. In this chapter we will review recent advances since then, concentrating in particular on studies of the strain-balanced quantum well solar cell (SB-QWSC) as a concentrator cell [3] and the thermodynamic modelling of the QDC [4].

The SB-QWSC offers a way to extend the spectral range of the highest efficiency single-junction cell, the GaAs cell. We will discuss how this can in principle lead to higher efficiency in both single-junction and multi-junction cells and offers particular advantages in high-concentration systems. The high efficiency, wide spectral range and small cell size make these systems particularly attractive for high concentration, building-integrated applications using direct sunlight.

The QDC is a novel, non-tracking concentrator. It is a promising approach for concentrating diffuse sunlight and is therefore complementary to the SB-QWSC in building integrated concentrator photovoltaics (BICPV).

2. STRAIN-BALANCED QUANTUM WELL SOLAR CELLS

The highest efficiency single-junction solar cells under both 1-sun conditions and concentration are GaAs cells. Both single-junction records have held for a decade and a half [5]. The GaAs bandgap (1.42 eV) is, however, rather high as optimal efficiency requires a bandgap around 1.1 eV at both 1 sun and high concentration [6]. Since those records were established, the main effort to raise efficiency has gone into developing tandem cells, which can now achieve 37.4%. These have a high-bandgap GaInP top cell grown lattice matched to the GaAs bottom cell with electrical connection between the cells provided by a tunnel junction. As the cells are in series, the same current passes through both cells and is limited by the current generated in the poorer cell. This turns out to be the GaAs bottom cell in most spectra, again because of the relatively high GaAs bandgap. The utility of the tandem approach has yet to be demonstrated at the very high-concentration levels required for cost reduction and in the varied spectral conditions of building-integrated applications. A tandem or multi-junction cell can achieve high efficiencies when optimised at the series current level of specific spectral conditions and temperature. In the varying spectral conditions of BICPV, with the cell temperature varying, plus the difficulty of maintaining good tunnel junction performance at high-current levels, a highly efficient, single-junction cell with wide spectral range may well, over a year, harvest comparable electrical energy to a tandem cell with nominally higher efficiency.

In summary, for terrestrial concentrator applications of "Third Generation" GaAs-based cells, either single- or multi-junction, it is important to be able to lower the bandgap of the conventional GaAs cell. The problem in doing so, which III–V cell designers have had to face for at least two decades, is that though there are *higher-bandgap* alloys such as GaInP and AlGaAs that can be grown lattice matched to GaAs, there is no lattice matched binary or ternary III–V compound with *lower-bandgap* than GaAs. To extend the GaInP/GaAs tandem cell efficiency, considerable effort is going into studying the quaternary GaInNAs which can be grown lattice matched to GaAs, but demonstrates poor minority carrier lifetimes resulting in insufficient current to avoid limiting the multi-junction performance [7].

The addition of lattice-matched QWs to AlGaAs, InP and GaAsP cells has been shown to extend the spectral range and enhance the efficiency [1,2], but when it comes to enhancing the GaAs cell, the QW approach also suffers from the absence of a lattice matched lower-bandgap ternary alloy. A number

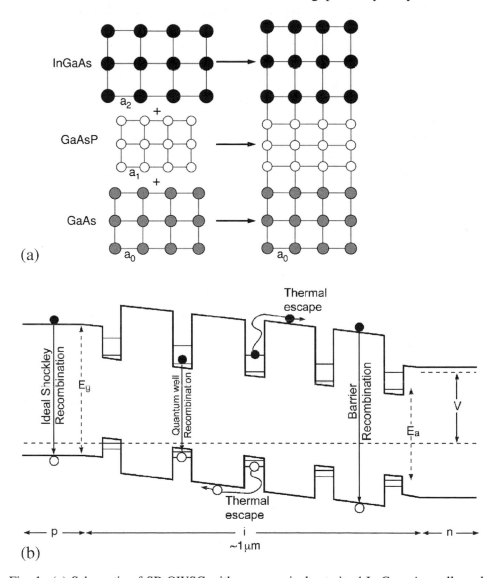

Fig. 1. (a) Schematic of SB-QWSC with compressively strained $In_xGa_{1-x}As$ wells and tensile-strained $GaAs_{1-y}P_y$ barriers. (b) Energy band-edge diagram of p–i–n SB-QWSC. Note that the $GaAs_{1-y}P_y$ barriers in the i-region are higher than the bulk GaAs in p and n regions.

of strained InGaAs wells with lower-bandgap than GaAs and larger lattice constant can be grown compressively strained to GaAs. However, only a limited number of wells can be grown before relaxation occurs and these give insufficient current enhancement to overcome the loss of voltage [8].

In SB-QWSC, illustrated schematically in Fig. 1a, the low-bandgap, higher-lattice-constant (a_2) alloy $In_xGa_{1-x}As$ wells (In composition $x \sim 0.1$–0.2) are compressively strained. The higher-bandgap, lower-lattice-constant (a_1) alloy $GaAs_{1-y}P_y$ barriers (P composition $y \sim 0.1$) are in tensile strain [3]. Of the many possible balance conditions we find that the *zero-stress* gives the best material quality [9] and enables at least 65 wells to be grown without relaxation [10].

Fig. 2 shows the spectral response of a 50 shallow well SB-QWSC. The cell is a p–i–n diode with an i-region containing 50 QWs 7 nm wide of compressively strained $In_xGa_{1-x}As$ with $x \sim 0.1$ inserted into tensile strained $GaAs_{1-y}P_y$ barrier regions. The QWs extend absorption from bulk bandgap E_g to threshold energy E_a determined by the confinement energy as in Fig. 1b, giving the extra absorption and wider spectral response demonstrated in Fig. 2.

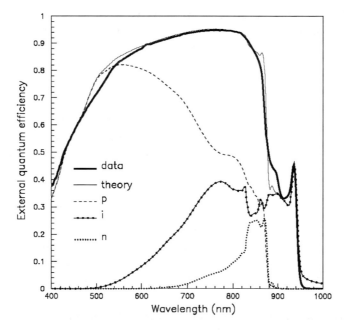

Fig. 2. Spectral response (external quantum efficiency at zero bias) of a 50-well SB-QWSC. The fit shows separate contributions of p, i and n regions as discussed in the text.

Alternative approaches to lowering the GaAs involve the growth of InGaAs on a relaxed virtual substrate [11, 12]. However, this approach inevitably results in material with a residual dislocation density. Dislocations are completely absent in a SB-QWSC [3]. This is important if the recombination is to be minimal, as discussed in the next section. It should be noted in Fig. 1 that the GaAsP barriers in the i-region have higher-bandgaps than GaAs, which also helps reduce the unwanted recombination.

3. IDEAL DARK-CURRENT BEHAVIOUR AT CONCENTRATOR CURRENT LEVELS

An important observation has been made when studying the dark currents of a range of SB-QWSC with differing well number and depth. We observe that, at current levels corresponding to 200× concentration and above, the dark currents of SB-QWSCs have ideality $n = 1$ as observed for the best GaAs p–n cells. The importance of this observation is that one can expect minimum recombination and maximum efficiency at high concentration. Concentration levels ~400× are generally accepted to be necessary if GaAs cells are to be cost competitive in terrestrial systems.

When we studied a range of SB-QWSCs with different well numbers and depth we observed a further, in this case unexpected, result. The reverse bias saturation current of the $n = 1$ contribution does not show the absorption threshold energy dependence expected of an ideal Shockley diode. We explain this behaviour in terms of two contributions to the $n = 1$ current as discussed below.

A range of SB-QWSCs were grown by metal-organic vapour phase epitaxy (MOVPE) at the EPSRC National Centre for III–V Technologies, Sheffield. Further growth details can be found in Refs. [3, 13]. Growth included a series of structures with P fraction $y = 0.08$ and a varying number (10–65) of shallow wells ($x = 0.1$), a 50 QW device with P fraction $y = 0.08$ and an intermediate depth well of In ($x = 0.13$) and a second series with P fraction $y = 0.08$ and 20, 30 or 40 deeper wells ($x = 0.17$). Details of the characterisation can be found in Refs. [5, 11]. Dark currents were measured at 25°C on fully metallised test diodes. A typical result is shown in Fig. 3. At currents corresponding to 200× concentration and above the ideality $n = 1$ contribution dominates.

We have measured the dark-current densities of between 8 and 18 fully metallised devices for each wafer and have fitted with two exponentials, one with ideality $n \sim 2$ and the other with ideality n fixed at 1.

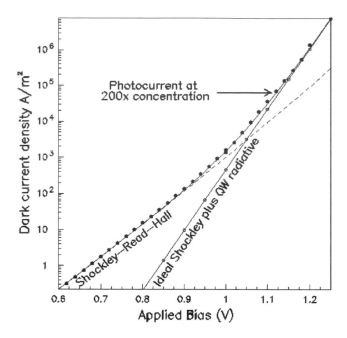

Fig. 3. Measured dark currents at 25°C of the device in Fig. 2 compared in the $n \sim 2$ region and the $n = 1$ region with the models discussed in the text.

$$J_d = J_{01}(e^{eV/kT} - 1) + J_{02}(e^{eV/nkT} - 1) \qquad (1)$$

The *reverse bias saturation current density* of the $n = 1$ contribution (J_{01}), which we estimate from the zero bias intercept of the ideality $n = 1$ fit, is plotted in Fig. 4 against the threshold energy E_a. The latter we take to be the energy of the e1–h1 exciton for the SB-QWSCs and the GaAs bulk band-edge for the homostructure p–i–n control cell. For each wafer in Fig. 4 we show the extrapolated intercept for the device with lowest dark current. In Ref. [4] we argue that the lowest dark current is the one most representative of the structure. The experimental increase in intercept with well number at fixed band edge is not very marked, as discussed later.

The absorption edge dependence of the intercept J_{01} of the $n = 1$ contribution is expected to be determined by the square of the intrinsic carrier density and therefore to vary exponentially with band edge as indicated by the line based on the control homostructure p–i–n cell. In fact, the trend of the data is significantly below this expectation, particularly as the wells get deeper.

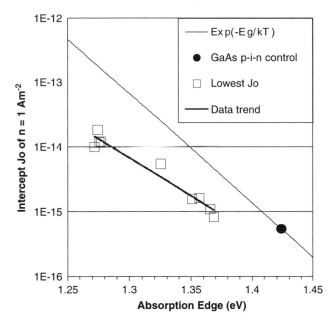

Fig. 4. Reverse bias saturation current density (J_{01}) from the intercept of the ideality $n = 1$ fits to the lowest measured dark current, plotted against exciton position.

4. MODELLING THE $n \sim 2$ AND $n = 1$ DARK-CURRENT BEHAVIOUR

To explain this unexpected behaviour, as the wells get deeper, we describe the data in the $n \sim 2$ region with a model we have developed for QWSCs in two lattice-matched material systems [14, 15]. The model solves for the variation in the carrier distributions $n(x)$ and $p(x)$ with position x through the i-region using the known QW density of states, assuming the depletion approximation holds. This approach gives similar results to an exact self-consistent calculation up to the voltages at which the $n = 1$ contribution dominates. From carrier densities, a recombination rate is determined assuming the Shockley–Hall–Read (SHR) approach [16]. This requires the non-radiative lifetimes of the carriers. The evidence suggests we can equate the electron and hole lifetimes [14] so we are left with two parameters which depend on material quality, the carrier lifetimes τ_B and τ_W in the barrier and well, respectively. For lattice-matched QWSCs, we determine these two parameters by fitting homostructure or double heterostructure controls which have i-regions formed from the material of the barrier or the well, respectively. In the case of SB-QWSC, bulk material of comparable bandgap to the QW and of similar material quality does not exist. We assume that, for these low P and In fractions, the well and barrier quality are sufficiently similar, "so that the lifetimes can be considered to be equal" (i.e., $\tau_B = \tau_W = \tau$). We find reasonable fits can be obtained to the $n \sim 2$ region of the dark current of these variable well samples with the single parameter τ.

We anticipate that there are two distinct contributions to the $n = 1$ current. Firstly the standard, ideal Shockley diode current [17]. This assumes no recombination in the depletion region but does assume the radiative and non-radiative recombination of injected minority carriers with majority carriers in the field-free regions. This contribution depends in a standard way on the minority-carrier diffusion lengths, doping levels and the surface recombination in the neutral regions [17]. We can estimate this current from the minority-carrier parameters obtained when fitting the spectral response in the neutral regions as in Fig. 2. Further examples are given in Ref. [15] where details can be found including the effect of strain on QW depth and barrier height.

The second contribution to the $n = 1$ current results from the recombination of carriers injected into the QWs and barriers in the depletion region. Like the ideal Shockley current, this is expected to have both radiative and non-radiative contributions. However, we assume that the non-radiative contribution in the i-region is described by the SHR $n \sim 2$ model discussed above. The radiative contribution to the QW recombination can be estimated by a detailed balance argument. This relates the photons absorbed to the photons radiated, as discussed in Ref. [18] and references cited therein. The radiated spectrum as a function of photon energy E in an electrostatic field F, $L(E,F)\,dE$, is determined by the *generalised Planck equation*:

$$L(E,F)\,dE = \frac{2\pi n^2 L_W}{h^3 c^2} \frac{\alpha(E,F)E^2}{e^{(E-\Delta E_F)/k_B T} - 1}\,dE \tag{2}$$

We integrate this spectrum over the energy and the cell geometry as described in Ref. [18] to give a total radiative current. This will depend on the *quasi-Fermi level separation* ΔE_F and the absorption coefficient $\alpha(E,F)$ as a function of energy and field. For this study, we assume that $\Delta E_F = eV$ where V is the diode bias. In Section 5, we discuss evidence from single- and 5-well SB-QWSCs that $\Delta E_F < eV$ though we have yet to observe this effect in the high well-number samples reported here [19].

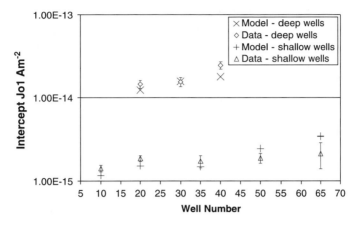

Fig. 5. Mean of measurements of the intercept of the $n = 1$ dark current from fits to 8–18 devices with $n \sim 2$ and $n = 1$ exponential terms. Data are plotted against number of wells and compared with the radiative plus ideal model described in the text.

The absorption coefficient $\alpha(E,F)$ is calculated from first principles [18] in the programme used to fit the spectral response of the QWs assuming unity quantum efficiency for escape from the wells. It should be noted that the important parameters for both the ideal Shockley (minority-carrier diffusion lengths) and the QW radiative current levels (absorption coefficient $\alpha(E,F)$) are therefore determined by the spectral response fits in the bulk and QW regions, respectively as in Fig. 2.

The sum of the two $n = 1$ terms is compared with the typical data in Fig. 3. In Fig. 5 the measured mean intercept from the experimental double-exponential fits is compared with the sum of the Shockley ideal and QW-radiative terms. Reasonable agreement is observed even though there are essentially no free parameters for the model. Note also that for a given well depth the intercept is only weakly dependent on the well number.

Fig. 6 shows the absorption threshold energy dependence of the ratio of the QW-radiative current intercept to the sum of the intercepts of the ideal Shockley and the QW radiative currents. It can be clearly seen that the dark current is becoming increasingly radiatively dominated as the threshold moves to lower energies and the wells get deeper. For a given absorption edge the ratio is not strongly dependent on the number of wells.

We conclude that at concentrator current levels the ideality $n = 1$ behaviour becomes increasingly dominated by radiative recombination in the wells

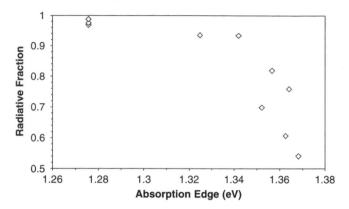

Fig. 6. Ratio of QW-radiative well current to sum of QW-radiative current plus ideal Shockley current against QW absorption edge given by first exciton position.

rather than non-radiative and radiative recombination in the GaAs p and n regions as the wells get deeper. This is important not only because deeper wells take the cell to absorption edges corresponding to higher efficiency [6] but also because the recombination can be further reduced by photon recycling techniques to be discussed in Section 6. However, to get PV power form these devices we must first consider if this low-recombination behaviour persists under light illumination as will be discussed in the next section.

5. TESTING ADDITIVITY IN SB-QWSCS

In the last section the dark current J_d behaviour of SB-QWSC, described by Eq. (1), was discussed at some length. Under illumination, the current–voltage curve $J(V)$ of the highest efficiency III–V cells can generally be described by

$$J(V) = J_{sc} - J_d \tag{3}$$

where $J_{sc} = J(0)$ is the *short-circuit current density*.

It is important to check if the *additivity condition* in Eq. (3) holds in SB-QWSCs for three reasons:

(i)   The ideal dark-current behaviour discussed in the last section will only translate into high efficiency if it does hold.

(ii)   There are theoretical claims based on a thermodynamic model that the suppressed radiative recombination we have observed in the dark in low well number QWSCs [18,19] will result in increased recombination in the light [20].

(iii)   QWSCs are p–i–n devices and the built-in field must be maintained across the i-region at the operating bias to ensure efficient collection of photo-generated carriers. Background doping in the i-region or the build-up of one or other of the charge carriers in the wells could lead to the failure of additivity and a resulting loss of efficiency.

We have tested additivity for the 1-, 5- and 10-well samples used to confirm that the suppressed radiative recombination observed in low-well number strained and lattice-matched QWSCs also occurs in the SB devices [19]. The SB-QWSCs were light-biased to the zero-current, open-circuit voltage V_{oc} by illumination with a laser exciting on the continuum in the QW and the photoluminescence (PL) spectrum measured with a spectrometer and charged coupled device (CCD) system. The laser intensity is equivalent to around 1 sun. The PL spectrum is then compared with the electroluminescence spectrum (EL) obtained when the laser is turned off and the cell biased externally to the same voltage (0.88 V in the example in Fig. 7) at the same temperature. Significant radiative recombination suppression is observed in the 1- and 5-well SB-QWSC, consistent with the earlier strained- and lattice-matched QWSCs [19]. The suppression is smaller in the 5-well devices, consistent with the zero suppression observed in the 10-well case. However, in the 1-, 5- and 10-well samples the PL and EL spectra are virtually identical, as can be seen for the single-well example in Fig. 7. At V_{oc} the $J(V)$ in Eq. (3) is zero and the equality of the PL and EL spectra suggest that the photo-generated recombination (PL) and the radiative-dark current (EL) cancel. This result is not consistent with the model in Ref. [20].

It is important to extend these tests to higher-current levels corresponding to the high concentrations in which these cells must operate for cost-effective terrestrial applications and to current levels where good dark-current performance is observed. Here the main problem becomes the challenge of contact fabrication such that resistance effects are minimised. We are currently limited by device processing. This can be seen in Fig. 8 where we show the efficiency of a 50-well SB-QWSC as a function of concentration. The cell parameters were measured in a 3000 K black-body spectrum in a shuttered measurement system. The 1-sun intensity was set by calculating the short-circuit current level I_{sc} to be expected by integrating the measured spectral response of the cell over an AM1.5D spectrum. The cell was then moved

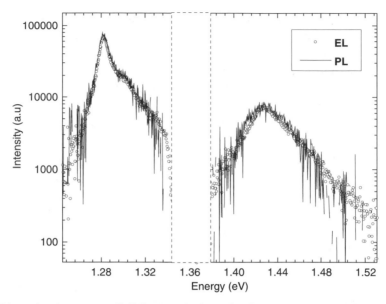

Fig. 7. Photo-luminescence (full line) and electroluminescence (scattered points) spectra at 290.9 K of the single-well SB-QWSC biased at +0.88 V. Data points in the filter window have been removed.

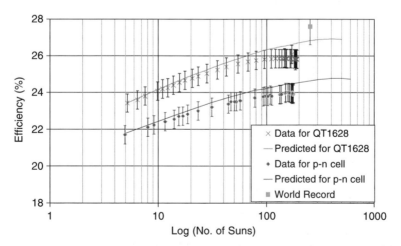

Fig. 8. Efficiency of 50-well SB-QWSC measured in a shuttered concentrator system compared with p–n control cell. The full lines show the predictions of additivity. If additivity holds at high concentration, the QW cell could achieve an efficiency close to the World single-junction record [5].

closer to the illumination with its temperature maintained by a Peltier control system. The cell position was optimised at each distance, and the effective concentration was assumed to be proportional to I_{sc}.

The first observation to be noted from Fig. 8 is that the 50 shallow-well SB-QWSC has a significantly higher efficiency than that of the p–n cell, which has comparable material quality. The prediction of additivity in Eq. (3) is shown for both cells by the full lines. These were generated assuming J_d was given by fits to the measured dark current, which included the cell series resistance measured in the dark. The data fall significantly below this prediction at ~200 suns. This could be the result of an extra resistive term under light-generation or the failure of additivity at high concentration. We believe the former explanation at present as the effect is also observed in the p–n cell, which has no i-region. Furthermore, EL studies indicate that the deviation from additivity is accompanied by non-uniform current behaviour.

We are currently studying improvements to our masks and contacting procedures in order to extend the additivity studies to higher-concentration levels and to challenge the World single-junction efficiency limit.

6. PHOTON RECYCLING IN SB-QWSCs

As mentioned in Section 5, the dominance of radiative recombination at high-current levels suggests that the dark current could be further reduced by photon recycling schemes. The analysis presented there indicated that radiative recombination in the wells dominates over radiative and non-radiative recombination in the GaAs regions. A SB-QWSC is ideal therefore to observe photon recycling effects as the *generalised-Planck equation* in Eq. (2) favours the emission of photons well below the GaAs bandgap energy that can be reflected by a distributed Bragg reflector (DBR) at the back of the cell. These will only be re-absorbed in the QWs rather than regions of the cell where non-radiative recombination dominates as occurs in conventional homostructure cells.

A number of 50-well SB-QWSCs have been grown at the EPSRC National Centre for III–V Technologies in the Quantax reactor, which will take two substrates. One, the upstream wafer (U) was a conventional GaAs wafer. The second, GaAs substrate, the downstream one (D), had a 20.5 period DBR optimised for high reflectivity within the QW wavelengths, which had been grown in an earlier run. A SB-QWSC was then grown on both wafers so that the overgrown device was as similar as possible and comparison of the U and D wafer would enable the effect of the DBR to be extracted.

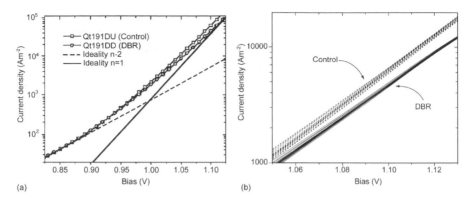

Fig. 9. (a) Typical dark current of a 50-well SB-QWSC control (squares) and DBR (circles) cell showing the ideality $n = 1$ (solid) and ideality $n \sim 2$ (dotted) dark-current components. (b) High-bias results plotted separately for the 20 devices in each case. Ideality $n = 1$ dark currents for QT1910 control devices (broken lines) and DBR devices (full lines) clearly have different reverse saturation currents J_{01}.

The wafers were first processed as fully metallised diodes to minimise resistive effects such as those discussed in the last section. The dark currents of 20 different devices were measured from both the DBR (D) and control (U) wafers. Typical dark currents for a wafer QT1910 with 50 relatively deep (In ~ 0.17) wells are shown in Fig. 9a. The ideality $n \sim 2$ and $n = 1$ components from a fit of the dark current to Eq. (1) are also shown. The dark currents of all the QT1910 devices shown in Fig. 9b clearly display a separation between the DBR and non-DBR devices at high voltage when radiative recombination dominates. On the other hand, the Shockley–Read–Hall ideality $n \sim 2$ dark-current component is not significantly changed between the DBR and non-DBR devices. The average overall DBR devices of the intercepts J_{01} of the $n = 1$ component is $(6.7 \pm 0.1) \times 10^{-21}$ Am^{-2}, which is significantly less than that of the control devices $(8.9 \pm 0.3) \times 10^{-21}$ Am^{-2}. This reduction is consistent with the expectations of a model for the electroluminescence based on Eq. (1) as described in Ref. [21].

We believe this is the first direct observation of the effect of photon recycling on the dark current of a solar cell.

7. QWSCs FOR TANDEM AND THERMOPHOTOVOLTAIC APPLICATIONS

The ability to tailor the bandgap of a GaAs-based cell without introduction of dislocations gives the SB-QWSC an advantage as the lower cell in a

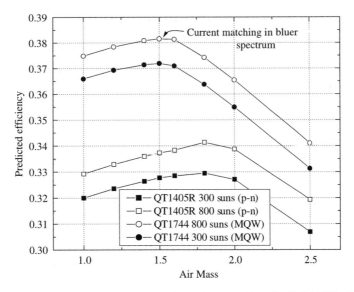

Fig. 10. The predicted performance of tandem cells of GaInP/SB-QWSC and GaInP/GaAs tandems using measured dark currents at 300 and 800 suns.

GaInP/GaAs monolithic tandem-cell arrangement, where it is well known that the bottom GaAs cell limits the series current in most solar spectra. This advantage has been demonstrated theoretically and a typical result is shown in Fig. 10 [22]. The Quantax reactor has been unable to grow the GaInP top cell but its replacement, a Thomas Swan reactor, will have this facility and be able to grow tandem cells based on SB-QWSCs.

The QWSC approach can be used in the GaInAsP/InP system and both lattice-matched and SB-QW cells have been demonstrated. The ability to tailor the bandgap and to operate at a higher voltage for a given output power has advantages for thermophotovoltaics, particularly in the case of narrow-band, rare-earth, radiant emitters. There is insufficient space in this chapter to discuss this work, but this application is extensively discussed in Ref. [23].

8. THE QUANTUM DOT CONCENTRATOR

The luminescent concentrator was originally proposed in the late 1970s [24]. It consisted of a transparent sheet (thickness, D) doped with appropriate organic dyes. Sunlight incident on the top surface of the sheet (width, W, and length, L) is absorbed by the dye and then re-radiated isotropically. Ideally the luminescence has high quantum efficiency (QE) and much of the flux is trapped in the sheet by total internal reflection. The luminescence propagates by waveguide action along the length (L) of the waveguide to a cell (area = $D \cdot W$) mounted at the edge of the sheet. This gives an upper limit to the concentration ratio of L/D if all the light is absorbed in the depth (D) and no luminescence is lost through the escape-cone or re-absorption. A stack of sheets doped with different dyes can separate the light, and solar cells can be chosen to match the different luminescent wavelengths to convert the trapped light at the edge of the module. The advantages over geometric concentrators include that solar tracking is unnecessary and that both direct and diffuse radiation can be collected. However, the development of this promising concentrator was limited by the stringent requirements on the luminescent dyes, particularly suitable red-shifts and stability under illumination [25].

The QDC is an updated version of the luminescent concentrator in which the dyes are replaced by QDs [4]. The first advantage of the QDs over dyes is the ability to tune the absorption threshold simply by choice of dot diameter. Secondly, since they are composed of crystalline semiconductor, the dots should be inherently more stable than dyes. Thirdly, the red-shift may be tuned by the choice of spread of QD sizes [4].

We have developed a series of thermodynamic models for single concentrator slabs and modules [26–28], which are comprised of a slab with a solar cell bonded to one edge. The models were developed by applying detailed balance arguments to relate the absorbed light to the spontaneous emission using self-consistent three-dimensional (3D) fluxes. The models were derived by applying the method of Schwarzschild and Milne [29], in which the angular dependence of the radiative intensity described by Chandrasekhar's [30] general three dimensional transfer equation is ignored and the radiation is considered as consisting simply of forward (+) and backward (−) streams. We have extended this approach to streams parallel to the x, y and z axes of a concentrator slab and apply appropriate reflection boundary conditions to the radiation depending on whether it falls within the escape cone or the solid angle of total internal reflection.

In addition to the forward and backward radiation streams in each co-ordinate direction, we also distinguish what happens when the direction of propagation, θ is greater or less than the critical angle θ_c. Escaping photons with $\theta < \theta_c$ and trapped photons with $\theta > \theta_c$ are treated as separate streams [26–28].

The models allow for:

(i) a significant fraction of the incident flux to be absorbed by the concentrator

(ii) spectral overlap of the incident radiation with the luminescence
(iii) re-absorption of radiation emitted into the escape cone
(iv) losses due to absorption in the host material,

all of which could not be accounted for in earlier models [31].

The model has recently been extended [28] to stacks of slabs doped with QD of different diameters so as to achieve higher efficiency matching slabs of different absorption thresholds to cells of appropriate bandgaps, as envisaged for the original dye-doped approach [25].

Full details of all the thermodynamic models can be found in Refs [26–28]. Here we will report some typical comparisons of the model with data. The detailed balance condition ensures that the trapped flux depends crucially on the absorption coefficient of the dots near threshold. This is analogous to the importance of the QW absorption coefficient $\alpha(E,F)$ in Eq. (3). However, the problem is more complex as the *photon chemical potential* $\mu(x,y,z)$ varies much more strongly with position in the slab than the equivalent *quasi-Fermi level separation* ΔE_F in Eq. (3).

For QDs with δ-function density of states and Gaussian distributed diameters the dependence of the QD absorption near threshold $\alpha(E_\gamma)$ is expected to be Gaussian in the photon energy E_γ. The experimental absorption can often be fitted by a Gaussian down to threshold and this was the situation with the CdSe/CdS core-shell QDs in acrylic studied in Fig. 11. Given this $\alpha(E_\gamma)$ the special variation of $\mu(x,y,z)$ is fitted with a Newton–Raphson approach. Following convergence the fluxes escaping the surfaces of the slab can be calculated as shown in Fig. 11. The calculation agrees well with the shape and position of the flux collected by a solar cell mounted at the right surface. The figure also shows the gain in escaping flux at the edge compared with the flux escaping from the top surface. The luminescence in the flux at the edge is red-shifted due to re-absorption when compared to the flux escaping the top. This is also reproduced by the models.

Short-circuit currents, J_{sc}, resulting from the radiation escaping the right surfaces of various slabs and modules were measured and are compared with the predicted values in Table 1. These measurements were performed on a slab of CdSe/CdS QDs in acrylic, a slab of red dye in Perspex, a mirrored slab of red dye in Perspex and a module comprising a slab of red dye in Perspex. In each case J_{sc} was measured using silicon solar cells, one of which was bonded to the edge of a slab of red dye in Perspex with sil-gel to form the module. The mirrored slab had aluminium evaporated onto the left, near, far and bottom surfaces.

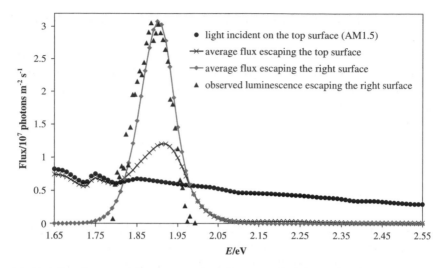

Fig. 11. Predicted average fluxes escaping the top and right surfaces together with the normalised observed luminescence escaping the right surface for a $L \times W \times D = 42 \times 10 \times 5$ mm slab of CdSe/CdS QDs with QE = 0.5 in acrylic. The slab has perfect mirrors on the bottom, near, far and left surfaces and is illuminated by AM1.5 at normal incidence.

Table 1
Comparison of short-circuit currents with predictions of Thermodynamic Model

Slab/module	Slab size (mm)	QE	$J_{sc}/$mA m^{-2} at $x = L$	
			Exp	Pred
CdSe/CdS QD slab	$42 \times 10 \times 5$	0.50	11.1 ± 2.0	10.0 ± 1.4
Red dye slab	$40 \times 15 \times 3$	0.95	20.1 ± 2.0	22.1 ± 1.7
Mirrored red dye slab	$40 \times 15 \times 3$	0.95	26.0 ± 2.0	26.2 ± 2.6
Red dye module	$40 \times 15 \times 3$	0.95	31.1 ± 2.0	29.3 ± 2.8

The slabs and module were positioned on a matt black stage with a matt black background to avoid unwanted reflections and were illuminated at normal incidence by a calibrated tungsten halogen lamp. The uncertainty in the measurements is due to current generated by coupling of the incident light into the edges of the solar cell. Allowance has been made for this by background measurements. The uncertainty in the predictions is mainly due to uncertainty in the low-absorption coefficient of the host material. In these calculations, we assumed that the reflectivity of the evaporated Al mirrors was 0.9 and that the Si concentrator cell, which has an anti-reflection coating and a textured surface, gives a reflectivity of 0.05 for the slab/sil-gel/cell interface in the module.

It is encouraging for the 3D thermodynamic model that the measurements in Fig. 11 and Table 1 agree well with the predictions, in particular given that the materials have very different losses owing to the high QE of the dyes and relatively low QE of the QDs. Our confidence in the models is also increased by the agreement with experiment for the slab with mirrored surfaces and the further agreement for the module, which has a solar cell bonded to the exit face.

9. CONCLUSION

We have demonstrated the performance of the SB-QWSC that makes it a good candidate for use in terrestrial high-concentration systems, particularly in the varying spectral conditions of a BICPV system. The advantages are:

   (i)   Higher efficiency than comparable homostructure GaAs cells.
  (ii)   Lower bandgap than GaAs, without dislocations, suitable for concentrator and multi-junction applications.
 (iii)   Wider-spectral range than single-junction GaAs cells.
 (iv)   Radiative-recombination dominated dark current for deep wells at concentrator current levels.
  (v)   Dark currents reduced by photon-recycling demonstrated for the first time.

We have also demonstrated a 3D thermodynamic model capable of describing the performance of dye-doped and QD-doped slabs of luminescent concentrators. The model is a powerful tool for analysing the performance of the luminescent concentrators. The fits show that concentrator performance is currently limited by the QE of the QDs dispersed in the plastics.

ACKNOWLEDGMENTS

We have benefited from the financial support of the EPSRC, EU Framework VI, Ashden Trust and Imperial BP Strategic Alliance.

REFERENCES

[1]   K.W.J. Barnham, P. Abbott, I. Ballard, D.B. Bushnell, A.J. Chatten, J.P. Connolly, N.J. Ekins-Daukes, B.G. Kluftinger, J. Nelson, C. Rohr, M. Mazzer, G. Hill, J.S. Roberts, M.A. Malik and P. O'Brien, Electrochemical Society Proc., 2001–10 (2001) 31–45.

[2]   K.W.J. Barnham, I. Ballard, J.P. Connolly, N.J. Ekins-Daukes, B.G. Kluftinger J. Nelson and C. Rohr, J. Mater. Sci.: Mater. Electron., 11 (2000) 531–536.

[3]   N.J. Ekins-Daukes, K.W.J. Barnham, J.G. Connolly, J.S. Roberts, J.C. Clark, G. Hill and M. Mazzer, Appl. Phys. Lett., 75 (1999) 4195–4198.

[4]   K. Barnham, J.L. Marques, J. Hassard and P. O'Brien, Appl. Phys. Lett., 76 (2000) 1197–1200.

[5]   M.A. Green, K. Emery, D.L. King, S. Igori and W. Warta, Prog. Photovolt.: Res. Appl., 13 (2005) 387–392.

[6]   J.S. Ward, M.W. Wanlass, K.A. Emery and T.J. Coutts, Proc. 23rd IEEE Photovoltaic Specialists Conf., pp. 650–654, Louisville, IEEE, 1993.

[7]   D.J. Friedman, J.F. Geisz, S.R. Kurtz and J.M. Olson, 2nd World Conference and Exhibition on Photovoltaic Energy Conversion, pp. 3–7, Vienna, 1998.

[8]   P.R. Griffin, J. Barnes, K.W.J. Barnham, G. Haarpaintner, M. Mazzer, C. Zanotti-Fregonara, E. Grunbaum, C. Olson, C. Rohr, J.P.R. David, J.S. Roberts, R. Grey and M.A. Pate, J. Appl. Phys., 80 (1996) 5815–5820.

[9]   N.J. Ekins-Daukes, K. Kawaguchi and J. Zhang, Cryst. Growth Des., 2 (2002) 287–292.

[10] M.C. Lynch, I.M. Ballard, D.B. Bushnell, J.P. Connolly, D.C. Johnson, T.N.D. Tibbits, K.W.J. Barnham, N.J. Ekins-Daukes, J.S. Roberts, G. Hill, R. Airey and M. Mazzer, J. Mater. Sci., 40 (2005) 1445–1449.

[11] F. Dimroth, P. Lanyi, U. Schubert and A.W. Bett, J. Electron. Mater., 29 (2000) 42–46.

[12] R.R. King, C.M. Fetzer, P.C. Colter, K.M. Edmondson, D.C. Law, A.P. Stavrides, H. Yoon, G.S. Kinsey, H.L. Cotal, J.H. Ermer, R.A. Sherif, K. Emery, W. Metzger, R.K. Ahrenkiel and N.H. Karam, Proc. 3rd World Conf. Photovoltaic Energy Conversion, pp. 622–625, Osaka, 2003.

[13] D.B. Bushnell, K.W.J. Barnham, J.P. Connolly, M. Mazzer, N.J. Ekins-Daukes, J.S. Roberts, G. Hill, R. Airey and L. Nasi, J. Appl. Phys., 97 (2005) 124908.

[14] J.P. Connolly, J. Nelson, I. Ballard, K.W.J. Barnham, C. Rohr, C. Button, J.S. Roberts and C.T. Foxon, Proc. 17th European Photovoltaic Solar Energy Conf., Munich, Germany, pp. 204–207, 2001.

[15] J.P. Connolly, I.M. Ballard, K.W.J. Barnham, D.B. Bushnell, T.N.D. Tibbits and J.S. Roberts, Proc. 19th European Photovoltaic Solar Energy Conf., pp. 355–358, Paris, 2004.

[16] W. Shockley and W.T. Read, Phys. Rev., 87 (1952) 835; R.N. Hall, Phys. Rev., 87 (1952) 387.

[17] W.T. Shockley, Bell Syst. Tech. J., 28 (1949) 435.

[18] J. Nelson, J. Barnes, N. Ekins-Daukes, B. Kluftinger, E. Tsui, K. Barnham, C.T. Foxon, T. Cheng and J.S. Roberts, J. Appl. Phys., 82 (1997) 6240.

[19] A. Bessière, J.P. Connolly, K.W.J. Barnham, I.M. Ballard, D.C. Johnson, M. Mazzer and G. Hill, J.S. Roberts, Proc. 31st IEEE Photovoltaic Specialists Conf., pp. 679–683, Orlando, IEEE, 2005.

[20] A. Luque, A. Marti and L. Cuadra, IEEE Trans. Elect. Devices, 48 (2001) 2118–2124.

[21] D.C. Johnson, I.M. Ballard, K.W.J. Barnham, A. Bessière, D.B. Bushnell, J.P. Connolly, J.S. Roberts, G. Hill, C. Calder, Proc. 31st IEEE Photovoltaic Specialists Conf., Orlando, pp. 699–703, IEEE, 2005.

[22] T.N. Tibbits, I.M. Ballard, K.W.J. Barnham, D.B. Bushnell, N.J. Ekins-Daukes, R. Airey, G. Hill and J.S. Roberts, Proc. 3rd World Conf. Photovoltaic Energy Conversion, pp. 2718–2721, Osaka, 2003.

[23] J.P. Connolly and C. Rohr, Semicon. Sci. Technol., 18 (2003) 216–220.

[24] A. Goetzberger and W. Greubel, Appl. Phys., 14 (1977) 123–139.

[25] A. Goetzberger, W. Stahl and V. Wittwer, Proc. 6th European Photovoltaic Solar Energy Conf., pp. 209–215, Reidel, Dordrecht, 1985.

[26] A.J. Chatten, K.W.J. Barnham, B.F. Buxton, N.J. Ekins-Daukes and M.A. Malik, Proc. 3rd World Conf. Photovoltaic Energy Conversion, pp. 2657–2660, Osaka, 2003.

[27] A.J. Chatten, K.W.J. Barnham, B.F. Buxton, N.J. Ekins-Daukes and M.A. Malik, Proc. 19th European Photovoltaic Solar Energy Conf. Exhibition, pp. 109–112, Paris, (2004) 82–85.

[28] A.J. Chatten, D. Farrell, C. Jermyn, P. Thomas, B.F. Buxton, A. Büchtemann, R. Danz and K.W.J. Barnham, Proc. 31st IEEE Photovoltaic Specialists Conf., Orlando, IEEE, 2005.

[29] E.A. Milne, Monthly Notices Roy. Astron. Soc. Lond., 81 (1921) 361.

[30] S. Chandrasekhar, Radiative Transfer, Clarendon, Oxford, UK, 1950.

[31] E. Yablonovitch, J. Opt. Soc. Am., 70 (1980), p. 1362.

Nanostructured Materials for Solar Energy Conversion
T. Soga (editor)

Chapter 17

Intermediate Band Solar Cells (IBSC) Using Nanotechnology

A. Martí[a], C.R. Stanley[b] and A. Luque[a]

[a]Instituto de Energía Solar, Universidad Politécnica de Madrid, E.T.S.I. Telecomunicación, Ciudad Universitaria s/n Madrid, Madrid 28040, Spain
[b]Department of Electronics and Electrical Engineering, University of Glasgow, Glasgow G12 8QQ, UK

1. INTRODUCTION

The operation of the intermediate band solar cell (IBSC) relies on the electrical and optical properties of the intermediate band (IB) materials. These are characterized by the existence of an electronic band located between the conduction and valence bands of an otherwise conventional semiconductor band gap (Fig. 1). We use "semiconductor", only to restrict ourselves to the IB materials created from host semiconductors into which impurities are

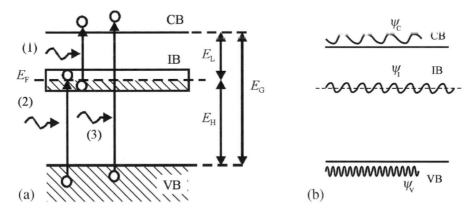

Fig. 1. (a) Simplified band structure of an intermediate band material in equilibrium. E_F is the Fermi level. The photon absorption processes involved (marked as 1, 2 and 3) are also shown. (b) The formation of an intermediate *band* from a collection of deep impurity levels arises from delocalization of the electron wavefunctions Ψ_I.

inserted in one way or the other, which might turn out in the future to be a very limited vision.

For ideal photovoltaic performance, the gap that separates the IB from the conduction band (CB), E_L, has to be in the range of 0.71 eV, while the gap that separates the IB from the valence band (VB), E_H has to be in the range of 1.24 eV [1, 2]. These gaps can be swapped, i.e., the IB can be closer to the VB rather than to the CB. On the other hand, the IB has to be half-filled with electrons (i.e. it has to be "metallic"), as it has to provide both empty states to accommodate electrons from the VB and occupied states to supply electrons to the CB.

At the outset of research into IBSCs, it was speculated that the materials with an IB might exist. At the time of writing, however, several studies have identified numerous candidates [3–8], sometimes even providing experimental support [9]. These materials would exhibit the IB as a bulk property. Quantum dots were earlier proposed [10] as a means to create a semiconductor with the necessary IB characteristics; therefore, their application to the realization of IBSCs will be a major theme of this chapter.

The IB could be regarded as a collection of energy levels within the semiconductor band gap. However, such levels usually lead to nonradiative recombination that would be detrimental for solar cell performance. Therefore, this raises the question as to why an IB material should be different.

The answer is that the wavefunctions of the electrons in these intermediate energy levels must be delocalised. When this phenomenon takes place, we really speak of a "band" rather than just of a "collection of deep impurity levels" [11] (Fig. 1b). Delocalisation is already a property of the electron wavefunction in the conduction and valence bands; so, in this regard, the electronic properties of the IB will be similar. Therefore, when an electron recombines from the CB to the IB, for example, the electric charge carried by the electron moves from one extended state to another. No element in the lattice structure of the IB material gets strongly perturbed because of this transition [12] because no sudden interaction with the electron's electric field occurs (no "breathing mode" or strong vibration can exist). Under these circumstances, the energy from the recombining electron cannot be transferred to the lattice (nonradiative recombination through an intermediate state has been prevented), but only to a photon (radiative recombination) or to another electron (Auger recombination). Auger recombination is less detrimental for the IBSC compared with conventional solar cells [13, 14], since it is possible for the energy lost by an electron recombining from, say, the IB to the VB that is used to pump an electron from the

IB to the CB. Indeed, the energy cannot be transferred to the lattice via a multiple collision process with lattice phonons owing to the large value of the gap involved in the system (0.71 eV). A multiple collision process would imply the simultaneous emission of several phonons (each may be with an energy of 10–50 meV), which is considered unlikely.

The delocalization of the electron wavefunction in the IB can be achieved inducing a Mott transition [11], which involves a change of the wavefunction from a localized to a delocalized state as the number of impurities introduced [15] is increased. As the density of impurity increases, the accompanying electrons screen the charge of the impure nuclei to the extent that they stop sensing the interaction with these nuclei and begin to act as free electrons, characterized by an extended wavefunction.

When only radiative recombination occurs, a solar cell is said to operate in its radiative limit by which its limiting efficiency is calculated [16, 17]. Radiative losses are inevitable once we accept that a photon can also be absorbed, since photon absorption and radiative recombination are dual processes (Roosbroeck–Shockley relation [18]). In the radiative limit, the IBSC exhibits a limiting efficiency of 63.2% [1, 2], substantially higher than the figures for single gap (40.7%) and two-junction (55.4%) solar cells operated at their radiative limits.

The assumption that radiative recombination is the only loss mechanism is sufficient for determining the limiting efficiency of single gap solar cells. However, in the case of an IBSC, an additional hypothesis is required, i.e., the absorption of photons is *selective*. This means that if a photon of energy E can be absorbed through an optical transition from, say, the VB to the IB, this has to be the only absorption process by which such a photon is absorbed. Initially, this might appear to be a stringent requirement, but in practice [19] more relaxed conditions apply requiring only that the absorption coefficient, α_{CI}, associated with optical transitions involving the lowest of the band gaps (E_L) is weaker than the one associated with the absorption coefficient, α_{IV}, of the next band gap, E_H, which in turn is lower than the absorption coefficient, α_{CV}, related to the total band gap, E_G. Some photon selectivity is also implied – but usually not recognized – when computing the limiting efficiency of multijunction solar cells where the cells are located in a stack. The cell at the top of this stack must be perfectly transparent to the photons absorbed by the next cell(s) in the stack.

While manufacturing a practical cell, the IB material has to be sandwiched between two conventional semiconductors (Fig. 2) one p type (p emitter) and the other n type (n emitter), which serve to isolate the IB

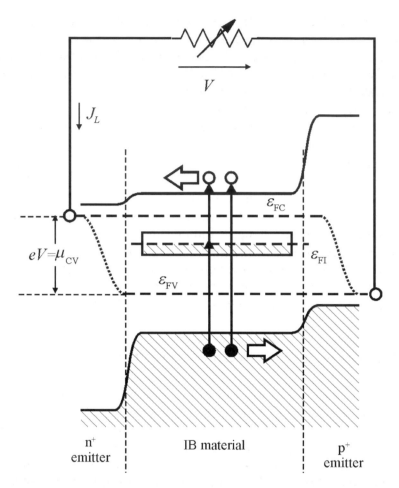

Fig. 2. Structure of an intermediate band solar cell showing the intermediate band material sandwiched between conventional *p* and *n* semiconductors, and the mechanisms of
current extraction and voltage preservation.

from the contacts [2]. In this configuration, the CB electrons generated from
the solar radiation are extracted through the *n* emitter, their chemical energy
$\mu_{CV} = eV$, is delivered to the external load (*V* being the voltage drop at the
load) and returned to cell as VB electrons through the *p* emitter.

When a single-gap semiconductor is used, only photons with energy
above the band gap threshold E_G can be absorbed and, therefore, contribute
to the photogenerated current J_L. However, when a half-filled IB exists, two
photons with energy below the band gap threshold can also pump a single

electron from the VB to the CB. The first photon, with energy above E_H will pump an electron from the VB to the IB (which has empty states to receive this electron) and the second, with energy above E_L, will pump an electron from the IB (that has also filled states) to the CB. In this way the photocurrent generated, J_L, is higher than the current that would have been generated by a conventional semiconductor of band gap E_G, without an IB.

However, this higher current alone is not sufficient for the IBSC efficiency to exceed that of its single-gap counterpart if the output voltage is not preserved. In this regard, each band must be described by its own quasi-Fermi (ε_{FC}, ε_{FV} and ε_{FI} for the CB, VB and IB, respectively). Physically, this means that the time an electron takes to relax within each band is much smaller than the time it takes to recombine between bands. The standard IBSC theory also assumes that the IB quasi-Fermi level, ε_{FI}, is fixed to its equilibrium position. For high illumination densities, this is achieved when the density of states for electrons in the IB is comparable to the densities of states in the CB and VB [20]. Under these circumstances, it is clear from Fig. 2 that the output voltage is limited by the high band gap E_G and not by either of the lower gaps, E_L and E_H.

A few comments are needed on the transport properties of the electrons in the IB. In an ideal IBSC, no carrier transport through the IB is required since this band is isolated from the external contacts by the p and n emitters. Consequently, the effective mass of electrons in the IB can be very high, and since the carrier effective mass is related to the energy bandwidth of the IB (the bandwidth narrows as the effective mass increases), the IB can be very narrow in terms of energy. In practical devices, however, some internal electron transport through the IB might be required to equilibrate the generation of carriers between regions with strong and weak absorption of photons [21].

Finally, we should mention that the study of the operation of the IBSC has been extended to the case in which more than one IB exists [22, 23]. The use of IB materials are proposed to manufacture up and down converters [24, 25].

2. CIRCUIT MODEL FOR THE INTERMEDIATE BAND SOLAR CELL

The basic operation of the IBSC has been described by Luque and Martí [1]. We will briefly review this model here, emphasizing some aspects not previously discussed in detail.

In the ideal model, the IBSC is considered to operate in its radiative limit. In addition, ideal photon selectivity is considered so that a photon with energy E can only be absorbed by a unique type of transition (from the VB to the IB, from the IB to the CB or from the VB to the CB). Complete photon absorption is assumed (absorptivity equal to unity).

In the model, the electron current density, J_C, extracted from the CB is given by:

$$J_C = J_{L,CI} - J_{CI}(\mu_{CI}) + J_{L,CV} - J_{CV}(\mu_{CV}) \qquad (1)$$

Similarly, the hole current density, J_V, extracted from the VB is given by:

$$J_V = J_{L,IV} - J_{IV}(\mu_{CI}) + J_{L,CV} - J_{CV}(\mu_{CV}) \qquad (2)$$

In these equations, $J_{L,XY}$ is the photogenerated current originating from the absorption of photons that cause transitions from band Y to band X; J_{XY} is the net recombination current caused by the radiative recombination of electrons from band X to band Y; $\mu_{CV} = \varepsilon_{FC} - \varepsilon_{FV}$, $\mu_{CV} = \varepsilon_{FC} - \varepsilon_{FV}$ and $\mu_{CV} = \varepsilon_{FC} - \varepsilon_{FV}$ are the quasi-Fermi level splits and also the photon chemical potentials. The word *net* here means that the reabsorption of photons emitted by radiative recombination is taken into account in the calculation of J_{XY} [26]. Under this interpretation, J_{XY} is calculated as the number of photons (multiplied by the electron charge) per unit of time and area that finally escape from the cell due to the radiative recombinations that take place between band X and Y.

The equations to calculate $J_{L,XY}$ (for the particular case of a cell illuminated by the sun assumed to be a black body source at temperature T_S) and J_{XY} in this ideal case are collected in Table 1. The current density–voltage (J–V) characteristic of the IBSC is obtained by solving simultaneously the set of equations:

$$J = J_C = J_V \qquad (3)$$

$$eV = \mu_{CV} = \mu_{CI} + \mu_{IV} \qquad (4)$$

Eqs. 1 and 2 can be represented in equivalent circuit form by the circuit shown in Fig. 3, where the voltage between nodes X and Y corresponds to the quasi-Fermi level split μ_{XY}.

If we now relax the photon selectivity condition and allow for overlap of the absorption coefficients, the IBSC circuit model must incorporate new elements [27], as represented in Fig. 4. To understand the physical origin of these elements, it is necessary to describe the different recombination and photon recycling processes that can take place. A photon generated through

Table 1
Current parameters used to describe the operation of the IBSC in the ideal case

Definitions:

$$F(T,\mu,E_A,E_B,H) = \frac{2eH}{h^3c^2} \int_{E_A}^{E_B} \frac{\varepsilon^2}{\exp\left(\dfrac{\varepsilon-\mu}{kT}\right)-1} d\varepsilon \tag{I-1}$$

H: Radiation view factor (or étendue or Lagrange invariant [28] per unit of area).
h: Planck's constant.
c: speed of light in vacuum.
e: electron charge.

Photogenerated currents $J_{L,XY}$ [(1)]:

$$J_{L,CI} = F(T_S,0,E_L,E_H,H_S) + F(T_C,0,E_L,E_H,\pi-H_S) \tag{I-2}$$

$$J_{L,IV} = F(T_S,0,E_H,E_G,H_S) + F(T_C,0,E_H,E_G,\pi-H_S) \tag{I-3}$$

$$J_{L,CV} = F(T_S,0,E_G,\infty,H_S) + F(T_C,0,E_G,\infty,\pi-H_S) \tag{I-4}$$

Net radiative recombination currents [(2)]:

$$J_{CI} = F(T_C,\mu_{CI},E_L,E_H,\pi) \tag{I-5}$$

$$J_{IV} = F(T_C,\mu_{IV},E_I,E_V,\pi) \tag{I-6}$$

$$J_{CV} = F(T_C,\mu_{CV},E_G,\infty,\pi) \tag{I-7}$$

[(1)] The illumination spectrum is that of the sun, considered as a black-body at temperature T_S. The term $F(T_S,0,E_A,E_B,H_S)$ must be substituted by the appropriate photon irradiance when other sources are used. Under dark conditions, $F(T_S,0,E_A,E_B,H_S)$ must be substituted by $F(T_C,0,E_A,E_B,H_S)$ to account for thermal generation.
[(2)] The emission of photons takes place only through the front surface (a back reflector is assumed). No restriction on the angle of emission is applied so that the view factor for photon emission per unit of area is π.

radiative recombination between the CB and the IB can follow various paths as illustrated in Fig. 5, and summarized as follows:

(1) escape from the cell. The rate at which this process takes place is given by $J_{CI,CI}^{\text{ext}}$;
(2) promotion of an electron from the VB to the CB at a rate $J_{CV,CI}^{\text{int}}$;
(3) a fraction of the electrons generated in (2) will recombine with the VB at a rate $J_{CV,CI}^{\text{ext}}$, emitting a photon that escapes from the cell;
(4) promotion of an electron from the VB to the IB at a rate $J_{IV,CI}^{\text{int}}$; and

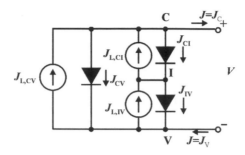

Fig. 3. Equivalent circuit for an ideal IBSC. Under normal conditions (non-degeneracy), the currents J_{XY} follow the Shockley's diode equation quite accurately.

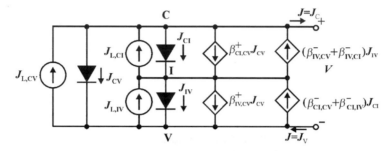

Fig. 4. Equivalent circuit for the ideal IBSC when band absorption coefficients overlap.

(5)   some of the electrons generated in (4) will recombine with the VB at a rate $J_{IV,CI}^{ext}$, emitting a photon that escapes from the cell.

The description above suggests the introduction of the general notation $J_{b,a}^{z}$ for each of the processes described, where a is the initial transition generating the photon (CI, if it is from the CB to the IB; IV for the IB to the VB and CV for the CB to the VB) and b is the other transition involved in the process. If this "other transition" consists of the promotion of an electron to an upper band, we set $z \equiv$ int standing for *internal* reabsorption; if it consists of the emission of a photon that escapes from the cell, we write $z \equiv$ ext for *external* emission. This notation can also be applied to the description of the remaining transitions, as illustrated in Fig. 5b where the chain starts with a transition from the IB to the VB, and in Fig. 5c where the process is started by a transition from the CB to the VB. The dependence of these current densities as a function of different variables and, in particular, as a function of the absorption coefficients α_{XY} is given in Table 2.

Mapping the above processes to the circuit elements is now straightforward. For example, let us focus in the processes in Fig. 5a. For convenience

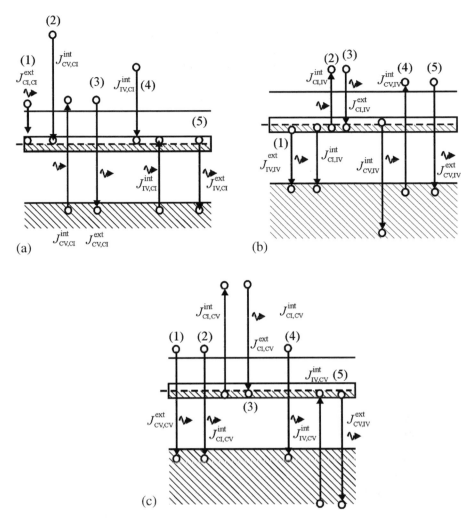

Fig. 5. (a) Possible paths for a photon that is emitted in a radiative recombination process between the CB and the IB. (b) Paths for a recombination process between the IB and the VB. (c) Paths for a recombination process from the CB to the VB.

(somewhat arbitrarily), we note first that process (3), from the CB to the VB is equivalent to one involving a step from the CB to the IB (we will refer to this process as "above") and followed by from the IB to the VB (we will refer to this process as "below"), both occurring at the same rate $J_{CV,CI}^{ext}$ (divided by the electron charge). Then, the net electron generation rate from the VB to the IB is given by $J_{IV,CI}^{int} - J_{CV,CI}^{ext}$ (below) = $\beta_{IV,CI}^{-} J_{CI}$ and leads to the corresponding current-dependent element. Process (2), that combines the recombination of electrons from the CB to the IB followed by the generation of electrons from the VB to the CB, is equivalent to a net electron generation rate from the VB to the IB given by $J_{CV,CI}^{int}$. Together with the recombination rate $J_{IV,CI}^{ext}$ they lead to a net generation rate from the VB to the IB occurring at rate $J_{CV,CI}^{int} - J_{IV,CI}^{ext} = \beta_{CV,CI} J_{CI}$ that leads to the corresponding current-dependent

Table 2
Current parameters used to describe the operation of the IBSC for the case in which the band absorption coefficients overlap.

Definitions:

$$f(T,\mu,H,a) = \frac{2eH}{h^3 c^2} \int_0^\infty \frac{\alpha_a}{\alpha_{CV} + \alpha_{IV} + \alpha_{CI}} \frac{A(\varepsilon)\varepsilon^2}{\exp\left(\dfrac{\varepsilon - \mu}{kT}\right) - 1} d\varepsilon \tag{II-1}$$

$$J_{a,b}^{int}(\mu) = \frac{8e\pi n_r^2 W}{h^3 c^2} \int_0^\infty \frac{\alpha_a \alpha_b}{\alpha_{CV} + \alpha_{IV} + \alpha_{CI}} \frac{\varepsilon^2}{\exp\left(\dfrac{\varepsilon - \mu}{kT_C}\right) - 1} d\varepsilon \tag{II-2}$$

$$J_{a,b}^{ext} = \frac{2e\pi(H_f + H_r)}{h^3 c^2} \int_0^\infty \frac{A(\varepsilon)\alpha_a \alpha_b}{\alpha_{CV} + \alpha_{IV} + \alpha_{CI}} \frac{\varepsilon^2}{\exp\left(\dfrac{\varepsilon - \mu_b}{kT}\right) - 1} d\varepsilon \tag{II-3}$$

H_S: Radiation view factor (or étendue or Lagrange invariant [28]) per unit of area.
H_f: Étendue of the photons emitted from the front (for emission to air without angular restriction $H_f = \pi$).
H_r: Étendue of the photons emitted from the rear (for emission to a substrate with the same refraction index than the cell active layers, $H_r = \pi n_r^2$).
α_a: Absorption coefficient for transitions between the bands indicated by suffix a.
n_r : cell refraction index.
h : Planck's constant.
c : speed of light in vacuum.
e : electron charge.
$A(\varepsilon)$: absorptivity. $[A(\varepsilon) = 1 - \exp[-(\alpha_{CV} + \alpha_{CI} + \alpha_{IV})W]$ for a non-textured, planar structure without back reflector].
a, b: suffixes that can take the values *CV*, *CI* and *IV*.
Photogenerated currents $J_{L,XY}^{(1)}$:

$$J_{L,CI} = f(T_S,0,H_S,CI) + f(T_C,0,H_f - H_S,CI) + f(T_C,0,H_r,CI) \tag{II-4}$$

$$J_{L,IV} = f(T_S,0,H_S,IV) + f(T_C,0,H_f - H_S,IV) + f(T_C,0,H_r,IV) \tag{II-5}$$

$$J_{L,CV} = f(T_S,0,H_S,CV) + f(T_C,0,H_f - H_S,E_G,CV) + f(T_C,0,H_r,CV) \tag{II-6}$$

Table 2 (*continued*)

Diode currents and current gain factors:

$$J_{CV} = J_{CV,CV}^{ext} \tag{II-7}$$

$$J_{IV} = J_{CV,IV}^{ext} + J_{IV,IV}^{ext} + J_{CI,IV}^{int} \tag{II-8}$$

$$J_{IV} = J_{CV,IV}^{ext} + J_{IV,IV}^{ext} + J_{CI,IV}^{int} \tag{II-9}$$

$$\beta_{a,b}^{-} = \frac{J_{a,b}^{int} - J_{c,b}^{ext}}{J_b} \qquad\qquad \beta_{a,b}^{+} = \frac{J_{a,b}^{int} + J_{c,b}^{ext}}{J_b} \tag{II-10}$$

---

[1] For illumination sources other than the sun, the term $2H_S\varepsilon^2/(h^3c^2)\left[\exp\left(\dfrac{\varepsilon-\mu}{kT_S}\right)-1\right]^{-1}$

must be substituted by the appropriate spectral photon irradiance. Under dark conditions, T_S is set equal to T_C. For most practical situations under illumination, the terms $f(T_C...$ can be neglected.

element in the circuit in Fig. 4. Finally, as far as the processes in Fig. 5a are concerned, the rates $J_{CI,CI}^{ext} + J_{CV,CI}^{ext}$ (above) $+ J_{IV,CI}^{int}$ lead to diode element crossed by current J_{CI}. The mapping of the rest of the processes illustrated in Fig. 5b and Fig. 5c into circuit elements is carried out in a similar way. The value of the parameters $\beta_{a,b}^{-}$ and $\beta_{a,b}^{+}$ appearing in the circuit is also given in Table 2.

A few words of clarification on the "diode models" used are appropriate. The diode currents described in Table 2 are not proportional to $\exp[e\mu_{XY}/(mkT)] - 1$ (with $m = 1$ because we are in a radiative-recombination-only framework), as is the case in a conventional diode model. This is because the model in Table 2 is valid even for the degenerate case when the quasi-Fermi levels approach the band edges. This is a useful approach when carrying out theoretical studies. If degeneracy is included, it automatically prevents the quasi-Fermi levels from entering into the bands, and hence the open-circuit voltage from exceeding the band gap. However, when the bands are not degenerated, the following approximation can be made:

$$\int_0^\infty \frac{\varepsilon^2}{\exp\left(\dfrac{\varepsilon-\mu}{kT_C}\right)-1}\,d\varepsilon \approx \exp\left(\frac{\mu}{kT_C}\right)\int_0^\infty \varepsilon^2 \exp\left(-\frac{\varepsilon}{kT_C}\right)d\varepsilon \tag{5}$$

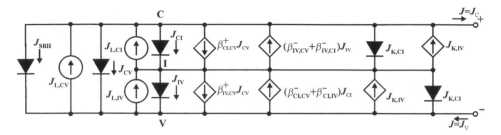

Fig. 6. IBSC equivalent circuit including Auger effects and nonradiative recombination between bands.

resulting in

$$J_{b,a}^z \propto \exp\frac{\mu_a}{kT_C} \tag{6}$$

If the factor "−1" is incorporated, we have to assume that the photocurrents $J_{L,a}$ are caused only by photons in excess of the background thermal radiation, so that $J_{L,a} = 0$ in dark conditions. Under this approach, the conventional diode model is restored.

The circuit model can be completed by including Auger [14] and other nonradiative recombination mechanisms. Hence, the Auger recombination of an electron from the CB to the IB (current component $J_{K,CI}$) can pump an electron from the VB to the IB. Similarly, the Auger recombination of an electron from the IB to the VB (current component $J_{K,IV}$) can pump an electron from the IB to the CB. Other nonradiative recombination mechanisms, such as Shockley–Read–Hall (SRH) recombination from the CB to the VB through defect levels, can be modelled by one (or more) diodes connected between the CB and VB (current component J_{SRH}). The resulting equivalent circuit is represented in Fig. 6.

3. THE QUANTUM DOT INTERMEDIATE BAND SOLAR CELL (QD–IBSC)

As mentioned in the introduction, quantum dots can provide a means to engineer IB materials and solar cells. The energy levels that lead to the IB arise from the states of electrons confined in the dots (Fig. 7). Quantum dots are preferred [10] over other low dimensional structures such as quantum

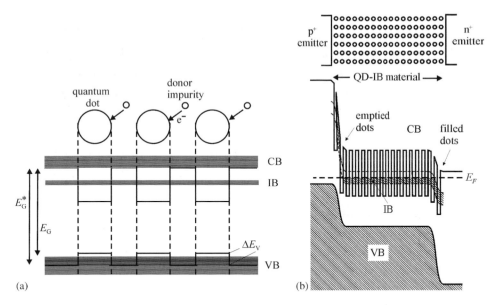

Fig. 7. (a) Formation of an intermediate band from the confined electronic states in an array of quantum dots [31]. (b) Simplified energy band diagram of a QD–IBSC in equilibrium (modified after Ref. 32).

wells or wires, because quantum wells do not provide a true zero density of states between the IB and the CB. Also, the optical transitions between the IB and the CB are forbidden in quantum wells for illumination perpendicular to the plane of growth [29, 30].

When using quantum dots, the zero density of states between the CB and the IB impedes the recombination of an electron (or, in equivalent terms, the capture of a CB electron by the dot). The phenomenon is known as the *phonon bottleneck* effect although, as Nozik [33] has pointed out, it has as many defenders as detractors. According to the phonon–bottleneck theory, nonradiative recombination between the CB and the IB is prevented because the energy of a single phonon is not sufficient to account for the energy difference between the CB from the IB, if it is accepted that a process involving multiple phonon emission is unlikely. At this point, we face a paradox. On the one hand, it is accepted in the context of quantum dots that an electron in the CB cannot be captured easily by the dot because the energy of a phonon is too low. On the other hand, the low energy of phonons does not seem to be an obstacle to effective recombination through levels in the center of the gap of bulk semiconductors (SRH recombination [34, 35]), where they are usually located much further in terms of energy (0.5–0.7 eV) from the CB than is the case for the QDs (0.2–0.4 eV).

The explanation of this paradox comes once more from considering the extended nature or otherwise of the wavefunction of electrons confined in the dots. In the case of bulk-semiconductors, we have already mentioned that states located in the middle of the gap act as effective recombination centers because their wavefunctions are localized (with a range extending, may be, over only a couple of atoms) making it possible for a breathing mode to exist to account for the conservation of energy during the transition. However, the wavefunctions of the electrons in the confined states of the dots are, by comparison, more extended although confined within the quantum dot, they range typically over 30–60 atoms, depending on the dot size dot. Therefore, the low energy available from a single phonon together with the difficulty in creating a breathing mode might be the reason for the existence of the phonon–bottleneck effect. This might also explain why the phonon–bottleneck effect is encountered in some cases (no breathing mode is created) and not in others (the breathing mode is created). At this point, it should be mentioned that the phonon–bottleneck effect should be prevented when manufacturing a quantum dot laser, but is favored in an IBSC.

Either way, by using quantum dots, we have the technological prospect of creating an IB isolated from the CB and the VB through a zero density of states. In addition, the capture times, in particular those related to the CB to IB transitions, might be sufficiently high in comparison with the relaxation times for electrons within the bands, so that the three distinct quasi-Fermi levels required by IBSC theory can coexist.

Half-filling of the IB can be achieved by modulation doping, i.e., by inserting impurity donors in the barrier region at a concentration of approximately one impurity atom per quantum dot [36]. However, to provide the charge density to sustain the electric field required at the emitter junctions, the QDs close to the *p* emitter will be emptied of electrons while those close to the *n* emitter will be filled (see Fig. 7b). Under these circumstances, these layers of quantum dots will fail either to provide empty states for electrons from the VB, or to act as a supplier of electrons to the CB. To solve this potential problem, important when the number of QD layers that can be grown is small, it has been proposed [37] that *p* and *n* layers of tailored thickness and doping are inserted between the QD layers and the *n* and *p* emitters, respectively.

We now turn our attention to the absorption coefficients. Those relating to transitions from the VB to the CB do not present a particular problem since they correspond to the VB to the CB transitions of the host semiconductor in which the QD system is embedded. The absorption of photons from the VB to the IB is also possible [38] as will be shown from experimental data in the next section. In fact, due to the Roosbroeck–Shockley relation [18], since quantum dot lasers show that the emission of photons due to transitions from the IB to the VB is feasible, the reverse process, the absorption of a photon due to a transition from the VB to the IB is also possible.

The absorption of photons from the IB to the CB is more problematic. For a single dot, photon absorption would imply a transition between a confined level within the dot and an extended state in the CB, with poor wavefunction overlapping between the two (IB and E_2 in Fig. 8a), except possibly for those states close to the CB-band edge (IB and E_1, Fig. 8a) [31]. Still, overlapping might not be sufficient to make possible photon absorption since it is only a necessary condition (the matrix element for an optical transition from the IB to the CB could still be zero if the initial and final states are orthogonal through the Hamiltonian describing the optical interaction). Therefore, photon absorption by means of transitions from the IB to the CB will be weak but detectable [39]. Increasing the overlap between wavefunctions by, for example, preserving the translational symmetry of the dot array (this could arise from increasing their density) might strengthen the absorption coefficient (Fig. 8b).

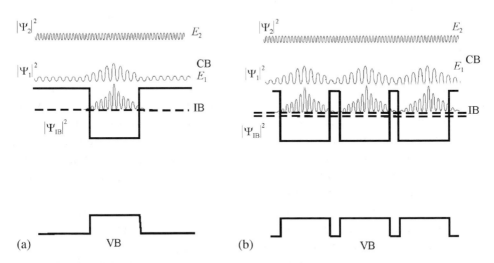

Fig. 8. (a) Qualitative form of the wavefunctions of electrons in the confined level of a quantum dot (IB) and in the conduction band. Level E_1 is supposed to be located close to the conduction band edge while level E_2 is far from the conduction band edge and approaches the wavefunction of a free electron. (b) The wavefunctions in an array of QDs.

Weak absorption from the IB to the CB is not a disadvantage as perceived by IBSC theory. The maximum performance of an IBSC is achieved when the absorption coefficients do not overlap. In the scheme presented, photon-absorbing transitions are only important for a restricted range of energies limited to the lowest energy transitions. This is favorable because their absorption coefficients will not overlap with the ones for VB to IB transitions, and even less so for VB to CB transitions.

Several material systems such as $InAs_{0.85}Sb_{0.15}/AlAs_{0.56}Sb_{0.44}/InP$, with theoretical gaps close to the optimum and negligible VB offsets, have been proposed [40] for implementation of the QD–IBSC concept. Nevertheless, practical experimentation still remains restricted to the InAs/GaAs system, even though the gaps are not optimum. It has been our purpose to test the principles of operation of the IBSC with InAs/GaAs rather than to produce the optimum device. The technology for realizing InAs quantum dots is relatively mature; this allows us to concentrate on the underlying principles of QD–IBSC operation.

4. OPTICAL AND ELECTRICAL CHARACTERIZATION OF QUANTUM-DOT INTERMEDIATE BAND SOLAR CELLS

In this section, we will describe the experimental work that has been carried on QD–IBSCs. The QD–IBSCs have been based on the incorporation of 10 or more layers of InAs QDs between the p and n type emitter layers of an otherwise conventional GaAs solar cell structure. The cells have been produced at the University of Glasgow within the framework of the European-founded project FULLSPECTRUM [41].

Three types of structure are grown by molecular beam epitaxy and fabricated into solar cells. In the first structure, the doping that is required to half-fill the dots with electrons was introduced into the barrier regions in the form of planar or modulation doping (δ-doping) as illustrated in Fig. 9. We refer to the cells made from this wafer as "δ–QD–IBSC". The second structure (named "no–δ–QD–IBSC") was identical to the first, but lacked the δ-doping layers. Finally, the third structure was a control sample in which both the δ-doping and quantum dots were omitted ("GaAs-ref"). Details concerning the rest of the layers within the cells are also shown in Fig. 9.

The QD devices performance as solar cells is shown in [42] Fig. 10, but their efficiency does not exceed that of the GaAs reference cell. This is due primarily to a small increase in short-circuit current associated with weak absorption of sub-band-gap light by the QDs, and to the degradation of their open-circuit voltage, degradation that the IBSC concept is intended to

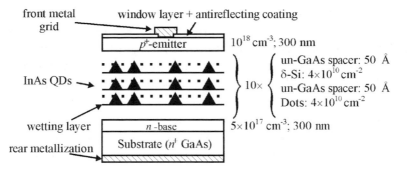

Fig. 9. Basic epitaxial layer structure of the quantum dot intermediate solar cells used in the experiments.

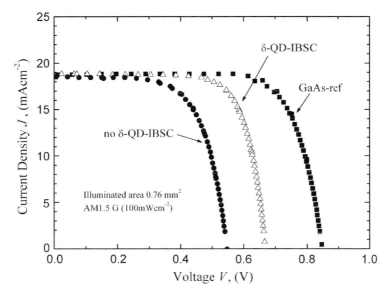

Fig. 10. Current density–voltage characteristics of quantum-dot intermediate band and GaAs reference solar cells (modified after Ref. 42).

prevent. Similar results are obtained by [43] in samples grown by MOVPE. There are, however, several reasons to explain this degradation, without implying that the operational principles of the IBSC do not apply.

If we recall the simplified band gap diagram of the quantum dot system plotted in Fig. 7a, we observe a VB offset, ΔE_V, which provides quantum confinement for the holes. However, since the effective mass of the holes is much greater than that of the electrons, it is expected that the confined energy levels in the VB will be closely grouped and become an extension of the VB. Therefore, the total band gap of the QD–IBSC is reduced when compared to that of the host semiconductor without QDs. A comparison of the performance of the cells based only on the open-circuit voltage is therefore not straightforward. For example, even in the case where the QD–IBSC operates under ideal conditions, its open-circuit voltage is limited by E_G and will not exceed the open-circuit voltage of the cell manufactured from the host semiconductor with the higher gap $E_G{}^*$.

For the InAs/GaAs system, the VB offset is estimated to be 72 meV [44], translating to a minimum reduction in open-circuit voltage of 72 mV. We will have to add to this reduction the impact of additional nonradiative recombination mechanisms induced by the possible presence of a higher density of defects in the QD structures, when compared with that of the bulk GaAs cell.

On the other hand, the short-circuit current of the three types of cell appears to be approximately the same, but a closer analysis of their quantum efficiencies shows that this is not the case (Fig. 11).

The QD samples, both with and without the δ-doping layers, show extended quantum efficiency response towards the infrared when compared with their GaAs counterpart. It must be recalled here that the quantum efficiency of a solar cell at energy E is simply the ratio between the electrons that circulate through the cell per unit of time under short-circuit conditions divided by the number of incident photons at that energy, also per unit of time. Therefore, the extended quantum efficiency response proves the production of photocurrent for below band gap energy photons, one of the principles sustaining the operation of the IBSC.

To understand how this current is extracted, we need to return to the fundamental circuit model depicted in Fig. 3. For the cell operated in short-circuit conditions and illuminated with below band gap energy photons, the result is $J_{CV} = 0$ and $J_{L,CV} = 0$. The modified equivalent circuit is shown in Fig. 12.

Since $J_{L,CI} \neq J_{L,IV}$, in general, one of the diodes has to be forward-biased while the other is reverse-biased. Furthermore, from the discussion in Section 3 relating to the strength of the absorption coefficients, we can assume $J_{L,CI} < J_{L,IV}$, and it then follows that diode D_{IV} becomes forward-biased (this, obviously, will also be the case when we later assume $J_{L,CI} = 0$). Two arguments are now possible to explain the extraction of photocurrent for incident photons of energy E lower than E_G:

(a) If the absorption coefficient $\alpha_{CI}(E) \neq 0$ (in reality, we mean "significantly different from zero"), photons with energy E will also pump electrons from the IB to the CB ($J_{L,CI} \neq 0$). This photogeneration enables the extraction of current, as is apparent from the circuit in Fig. 12b where the

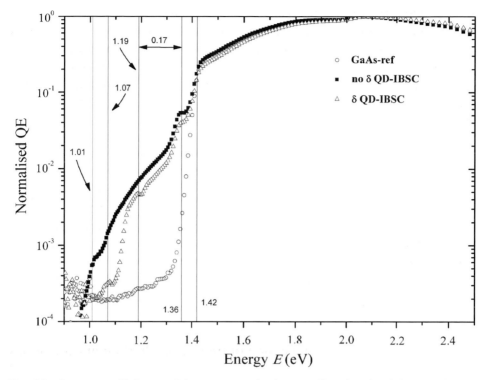

Fig. 11. Quantum efficiency of the quantum-dot intermediate band and GaAs reference solar cells.

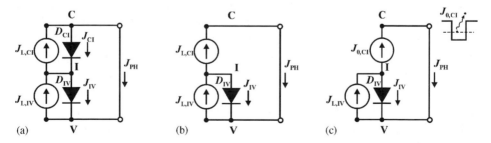

Fig. 12. (a) Equivalent circuit of the IBSC operated in short-circuit conditions and illuminated with below bandgap energy photons. (b) The circuit for a significant current $J_{L,CI} \neq 0$, but still with $J_{L,CI} < J_{L,IV}$. (c) The circuit for negligible current $J_{L,CI} \approx 0$ and with the current extracted by the thermal generation of carriers between the IB and the CB.

(ideal) diode D_{CI} is reverse-biased (and has therefore "disappeared" from the circuit). Consequently, it is current $J_{L,CI}$ that limits the extraction of the photocurrent I_{PH}. Nevertheless, the cell still operates in accordance with basic IBSC principles.

(b) If the absorption of photons from the IB to the CB is negligible ($J_{L,CI} \approx 0$), we have to rely on thermal pumping of electrons from the IB to the CB to explain the extraction of photocurrent. This thermal pumping is represented schematically by the inset in Fig. 12c; it is nothing other than the current circulating through diode D_{CI} in reverse, and limited by the reverse saturation current of the diode, $J_{0,CI}$ (here represented as a current source). However, "in reverse" implies a negative quasi-Fermi level split between the IB and the CB, placing the CB quasi-Fermi level below that of the IB. The appearance of this split is also a working hypothesis of the IBSC model.

To illustrate better the situation, it is instructive to calculate the value of $J_{0,CI}$, even for a cell operated in the radiative limit. At $T_C = 300\,\mathrm{K}$ and for a gap $E_L = 0.2\,\mathrm{eV}$ (the one estimated for the QD–IBSC based on InAs/GaAs), this is given by

$$J_{0,CI} \approx \frac{2\pi}{h^3 c^2} kT_C(E_L^2 + kT_C E_L + k^2 T_C^2) \exp\frac{-E_L}{kT_C} \approx 9\,\mathrm{mAcm}^{-2} \tag{7}$$

The result is $9\,\mathrm{mAcm}^{-2}$, a very high current density. In fact, it is around one third of the short-circuit current density of a standard GaAs solar cell operated under one sun illumination. This causes the diode D_{CI} to behave almost as a short-circuit, unable to block the photogenerated current $J_{L,IV}$. (A split of $-3\,\mu\mathrm{V}$ would suffice to sustain a current density of $1\,\mu\mathrm{Acm}^{-2}$).

We believe case b (thermal escape) rather than case a, which offers the more likely explanation for the operation of the QD–IBSC (a combination of both mechanisms could also be possible) under quantum efficiency measurments. For example, the assumption under case b allows us to explain why an extended quantum efficiency response has also been obtained for the no–δ–QD–IBSC samples (only empty states in the IB are required for $J_{L,IV}$). In fact, extended quantum efficiency response has also been measured in multiple quantum well solar cells [45], devices with essentially the same structure as the ones studied here, but incorporating quantum wells instead of quantum dots. In the case of the quantum wells, the existence of a continuous two-dimensional density of states between the confined energy levels and the CB would cause the current density $J_{0,CI}$ to be even higher and therefore, to require a negligible quasi-Fermi level split for the same given current density.

The analysis above lends support to IBSC theory through the existence of a quasi-Fermi level split between the IB and the CB, although this split is negative. To proceed further in the description of the experimental operation of the IBSC, we need to show that this split can be positive. Ideally, a split higher than 25 mV at room temperature must be demonstrated for the value of $J_{O,CI}$ given above, which would require a photocurrent $J_{L,CI}$ of at least 15 mAcm^{-2}. This has proved to be extremely difficult given the low value of the absorption coefficient α_{CI}. The required current would even be higher if other recombination mechanisms between the CB and the IB exist.

As an alternative to light excitation, a forward bias applied to the cells in the dark can also produce the required split. With this in mind, electroluminescence measurements have been carried out and the results are shown in Fig. 13. The peaks observed have a correspondence with the ones shown in the quantum efficiency curves although the electroluminescent peaks appear red-shifted (30–70 meV) when compared to the equivalent peaks revealed by the quantum efficiency measurements. This kind of shift is generally known as a Stokes-shift [46], and in this experiment energy differences could arise between confined levels in the dot with the cell operated

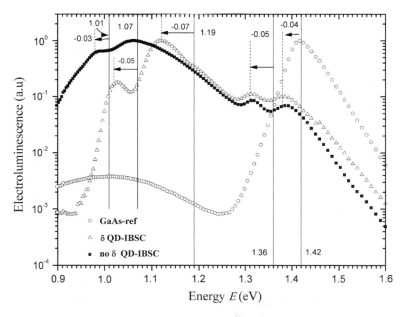

Fig. 13. Electroluminescence corresponding to QD–IBSC and GaAs reference cells. The forward bias current is 5 Acm^{-2}. The peaks observed in the quantum efficiency measurement are shown by connected vertical lines for reference.

in short-circuit conditions (quantum efficiency measurements) and in forward bias (electroluminescence measurements).

According to the equations and the model in Table 2, the electroluminescence (number of photons emitted by the cell per unit of time and area), N_{ph}^{XY}, corresponding to radiative recombination processes between band X and Y is given by

$$N_{\text{ph}}^{XY} = \frac{1}{e}(J_{\text{CV,XY}}^{\text{ext}} + J_{\text{CI,XY}}^{\text{ext}} + J_{\text{IV,XY}}^{\text{ext}}) \tag{8}$$

Eq. (8) can be used to model the electroluminescence. When combined with quantum-efficiency data, we have shown [27, 32] that the electroluminescence spectra are better explained if a split of about 150 meV between the IB and the CB quasi-Fermi levels is assumed.

On the other hand, capacitive vs. voltage measurements have indicated [42] that the approximate band gap diagram of these QD–IBSCs (both δ–QD–IBSC and no–δ–QD–IBSC), consisting of 10 layers of quantum dots only, is better represented by Fig. 14, i.e., with all the quantum dots embedded in the space charge region rather located in a flat band potential region as depicted for the ideal case in Fig. 7b. Under these circumstances, most of the QD layers adjacent to the p emitter will be empty of electrons, while others close to the n emitter will be filled. Only a few remaining layers of QDs (those whose confined energy level is crossed by the quasi-Fermi level) will have empty states to receive electrons from the VB and filled states to supply electrons to the CB for the sub-band-gap photon absorption processes involved in the operation of the IBSC.

The operation of the QD–IBSC under such nonideal space charge region conditions has also been studied [32], including removing the requirement for the IB quasi-Fermi to be fixed at its equilibrium position. A multiple fit to the experimental results confirms again the quasi-Fermi level split between the CB and the IB (Fig. 15), even under the more realistic conditions. This multiple fit that involves the dark current–voltage characteristic of the cells, also allows to capture cross sections for electrons ($2.35 \times 10^{-12}\,\text{cm}^{-2}$) and holes ($3 \times 10^{-14}\,\text{cm}^{-2}$) by the IB to be determined. In this respect, Fig. 16 shows the dark current density–voltage characteristic of the cells according to this fit, together with the contributions of the currents J_{SRH} and $J_{\text{CI}} = J_{\text{IV}}$.

For a volumetric dot density of $4 \times 10^{16}\,\text{cm}^{-3}$, derived from a planar density of $4 \times 10^{10}\,\text{cm}^{-2}$ and a QD layer spacing of $10^{-6}\,\text{cm}$, these capture cross sections imply lifetimes for the CB to the IB and for the IB to VB

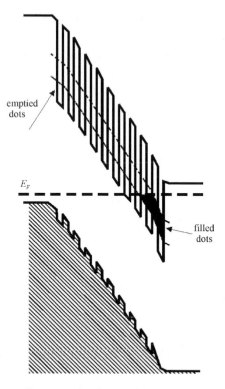

Fig. 14. Simplified energy diagram of a QD–IBSC in equilibrium when all the dots are located in the space charge region.

recombination process of 0.5 and 40 ps, respectively. These times, although low, imply a relatively inefficient recombination rate since the energy level involved (the IB) is not located in the center of the gap.

Through these experiments, the QD–IBSC has proved itself to be a valuable tool for testing the principles of operation of the IBSC, as well as a means to take the concept into practice. We hope to have encouraged others to undertake research on the topic, maybe with other material systems and experiments. In addition, the related theory has also provided a new framework under which the operation of other QD-based devices might be studied.

The strategic importance of research on novel concepts to increase the efficiency of photovoltaic energy conversion must not be underestimated. A breakthrough in this field [47] will contribute to a reduction in the cost of producing photovoltaic energy, supply a new economic market, and provide a significant source of inexhaustible energy to a world demanding clean energy resources.

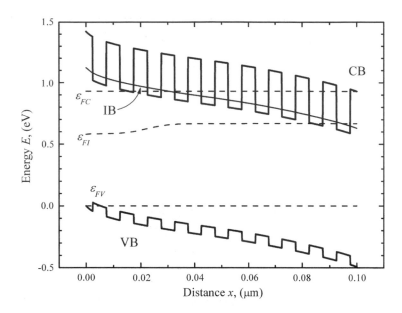

Fig. 15. Energy bandgap diagram corresponding to the δ–QD–IBSC samples that best fits the experimental results. The plot represents the dots fully immersed in the space charge region for a 5 Acm^{-2} forward current. (modified after Ref. 32).

5. SUMMARY

The IBSC is a novel solar cell with a potential limiting efficiency that exceeds that of single gap solar cells. The higher limiting efficiency results from an increase in the photogenerated current while preserving its open-circuit voltage. Its performance is based on the so-called IB materials that are characterized by the existence of an "intermediate band" within the conventional band gap of a host semiconductor. It is this band which is responsible for the increased photogenerated current by assisting the absorption of below band gap photons in a two-step process via the IB. The preservation of the open-circuit voltage associated with the wider gap of the host semiconductor is a consequence of the coexistence of separate quasi-Fermi levels, one for each of the CB, the VB and the IB.

In this chapter, we have described the practical implementation of IBSCs through the use of InAs quantum dot technology, and have outlined the underlying theory on which the operation of IBSCs is founded. In essence, the IB in the InAs/GaAs quantum dot system arises from the confined energy states of the electrons in the InAs dots, and the existence of three distinct

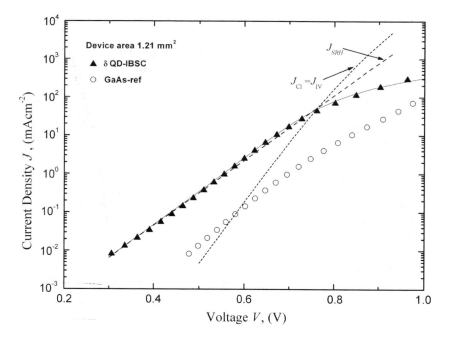

Fig. 16. Dark current density–voltage characteristics of the δ–QD–IBSC and GaAs reference cells (modified after Ref. 32).

quasi-Fermi levels is possible due to the isolation of the IB from the CB by a true zero density of states.

Characterization of our QD–IBSCs has revealed the production of photogenerated current by below band gap energy photons. Analysis of a range of experimental results, reinforced with equivalent circuit modelling, has shown that the behavior of the QD–IBSCs is best described in terms of distinct quasi-Fermi levels, one for each of the CB, the VB and the IB. Two of the fundamental theoretical requirements for IBSCs have therefore been demonstrated.

ACKNOWLEDGMENTS

We acknowledge our coworkers N. López, C. Farmer, E. Antolín, and E. Cánovas for their valuable comments and assistance with experiments and cell processing. This work has been supported by the European Commission under contract No. SES6-CT-2003-502620.

REFERENCES

[1] A. Luque and A. Martí, Increasing the efficiency of ideal solar cells by photon induced transitions at intermediate levels, Phys. Rev. Lett., 78 (26) (1997) 5014–5017.

[2] A. Luque and A. Martí, A metallic intermediate band high efficiency solar cell, Prog. Photovoltaic Res. Appl., 9 (2) (2001) 73–86.

[3] C. Tablero, Survey of intermediate band materials candidates, Solid State Commun., 133 (2005) 97–101.

[4] C. Tablero, Survey of intermediate band materials based on ZnS and ZnTe semiconductors, Sol. Energ. Mater. Sol. Cells, 90 (5) (2006) 588–596.

[5] J. Fernández, C. Tablero and P. Wahnón, Application of the exact exchange potential method to half metallic intermediate band alloy semiconductor, J. Chem. Phys., 120 (2004) 10780–10785.

[6] C. Tablero and P. Wahnón, Analysis of metallic intermediate-band formation in photovoltaic materials, Appl. Phys. Lett., 82 (2003) 151–153.

[7] P. Wahnón and C. Tablero, Ab-initio electronic structure calculations for metallic intermediate band formation in photovoltaic materials, Phys. Rev. B Condens. Matter, 65 (165115) (2002) 1–10.

[8] C. Tablero, P. Palacios, J. Fernández and P. Wahnón, Properties of intermediate band materials, Sol. Energ. Mater. Sol. Cells, 87 (2005) 323–331.

[9] K.M. Yu, W. Walukiewicz, J. Wu, W. Shan, J.W. Beeman, M.A. Scarpulla, D. Dubon and P. Becla, Diluted II–VI oxide semiconductors with multiple band gaps, Phys. Rev. Lett., 91 (24) (2003) 24603-1–24603-4.

[10] A. Martí, L. Cuadra and A. Luque, Quantum dot intermediate band solar cell, in Proceedings of the 28th IEEE Photovoltaics Specialists Conference, pp. 940–943, IEEE, New York, 2000.

[11] A. Luque, A. Martí and L. Cuadra, High efficiency solar cell with metallic intermediate band, in Proceedings of the 16th European Photovoltaic Solar Energy Conference, pp. 59–61, James & James, London, 2000.

[12] A. Luque, A. Martí, E. Antolín and C. Tablero, Intermediate bands versus levels in non-radiative recombination, Physica B, 382 (2006) 320–327.

[13] S.P. Bremner, C.B. Honsberg and R. Corkish, Non-ideal recombination and transport mechanisms in multiple band gap solar cells, in Proceedings of the 28th IEEE Photovoltaics Specialists Conference, pp. 1206–1208, IEEE, New York, 2000.

[14] A. Luque, A. Martí and L. Cuadra, Impact-ionization-assisted intermediate band solar cell, IEEE Trans. Elec. Dev., 50 (2) (2003) 447–454.

[15] N.F. Mott, Metal-insulator transition, Rev. Mod. Phys., 40 (4) (1968) 677–683.

[16] W. Schockley and H.J. Queisser, Detailed balance limit of efficiency of p–n junction solar cells, J. Appl. Phys., 32 (1961) 510–519.

[17] G.L. Araújo and A. Martí, Absolute limiting efficiencies for photovoltaic energy conversion, Sol. Energ. Mater. Sol. Cells, 33 (1994) 213–240.

[18] W. van Roosbroeck and W. Shockley, Photon-radiative recombination of electrons and holes in germanium, Phys. Rev., 94 (6) (1954) 1558–1560.

[19] L. Cuadra, A. Martí and A. Luque, Influence of the overlap between the absorption coefficients on the efficiency of the intermediate band solar cell, IEEE Trans. Elec. Dev., 51 (6) (2004) 1002–1007.

[20] A. Martí, L. Cuadra and A. Luque, Quantum dot analysis of the space charge region of intermediate band solar cell, in Proceedings of the 199th Electrochemical Society Meeting, (R.D. McConnell and V.K. Kapur Eds.), pp. 46–60, The Electrochemical Society, Pennington, 2001.

[21] A. Martí, L. Cuadra and A. Luque, Quasi drift-diffusion model for the quantum dot intermediate band solar cell, IEEE Trans. Elec. Dev., 49 (9) (2002) 1632–1639.

[22] A. Brown and M.A. Green, Intermediate band solar cell with many bands: Ideal performance, J. Appl. Phys., 94 (2003) 6150–6158.

[23] A.S. Brown, M.A. Green and R. Corkish, Limiting efficiency for multiband solar cells containing three and four bands, Phys. E, 14 (2002) 121–125.

[24] T. Trupke, M.A. Green and P. Wuerfel, Improving solar cells by downconversion of high energy photons, J. Appl. Phys. 92 (3) (2002) 1668–1674.

[25] T. Trupke, M.A. Green and P. Wuerfel, Improving solar cells by the upconversion of sub-band-gap light, J. Appl. Phys., 92 (7) (2002) 4117–4122.

[26] A. Martí, J.L. Balenzategui and R.F. Reyna, Photon recycling and Shockley S-diode equation, J. Appl. Phys., 82 (1997) 4067–4075.

[27] A. Luque, A. Martí, C. Stanley, N. López, L. Cuadra, D. Zhou and A. McKee, General equivalent circuit for intermediate band devices: Potentials, currents and electroluminescence, J. Appl. Phys., 96 (2004) 903–909.

[28] A. Luque, Solar Cells and Optics for Photovoltaic Concentration, Non-imagining Optics and Static Concentration, Adam Hilguer, Bristol, 1989.

[29] P. Harrison, Quantum Wells Wires and Dots, Wiley, New York, 2000.

[30] J.P. Loehr and M.O. Manasreh, Semiconductor Quantum Wells and Superlattices for Long-Wavelength Infrared Detectors, Artech House, Boston, 1993.

[31] A. Martí, L. Cuadra and A. Luque, Intermediate Band Solar Cells, in Next Generation Photovoltaics: High Efficiency through Full Spectrum Utilization, Optics and Optoelectronics (A. Martí and A. Luque, eds.), pp. 140–162, Institute of Physics Publishing, Bristol, 2003.

[32] A. Luque, A. Martí, N. López, E. Antolín and E. Cánovas, Operation of the intermediate band solar cell under nonideal space charge region conditions and half filling of the intermediate band, J. Appl. Phys., 99 (2006) 094503-1–094503-9.

[33] A. Nozik, Quantum dot solar cells, in Next Generation Photovoltaics: High Efficiency through Full Spectrum Utilization, Optics and Optoelectronics (A. Martí and A. Luque, eds.), Institute of Physics Publishing, Bristol, 2003.

[34] W. Shockley and W.T. Read, Statistics of the recombination of holes and electrons, Phys. Rev., 87 (1952) 835–842.

[35] R.N. Hall, Electron-hole recombination in germanium, Phys. Rev., 87 (1952) 387.

[36] A. Martí, L. Cuadra and A. Luque, Partial filling of a quantum dot intermediate band for solar cells, IEEE Trans. Electron Devices, 48 (2001) 2394–2399.

[37] A. Martí, L. Cuadra and A. Luque, Quasi drift-diffusion model for the quantum dot intermediate band solar cell, IEEE Trans. Electron Devices, 49 (2002) 1632–1639.

[38] M. Sugawara, Self-Assembled InGaAs/GaAs Quantum Dots. Vol. 60, Academic Press, San Diego, 1999.

[39] J. Phillips, K. Kamath, X. Zhou, N. Chervela and P. Bhattacharya, Intersubband absorption and photoluminescence in Si-doped self-organized InAs/GaAlAs quantum dots, J. Vac. Sci. Technol. B, 16 (1997) 1243–1346.

[40] M.Y. Levy, C. Honsberg, A. Martí and A. Luque, Quantum dot intermediate band solar cell material systems with negligible valence band offsets, in 31st IEEE Photovoltaics Specialists Conference, pp. 90–93, IEEE, New York, 2005.

[41] Fullspectrum: A new PV wave making more efficient use of the solar spectrum. Project sponsored by the European Commission. Ref. SES6-CT-2003-502620. www.fullspectrum-eu.org.

[42] A. Martí, N. López, E. Antolín, E. Cánovas, C. Stanley, C. Farmer, L. Cuadra and A. Luque, Novel semiconductor solar cell structures: the quantum dot intermediate band solar cell, Thin Solid Films, 512–513 (2006) 638–644.

[43] A.G. Norman, M.C. Hanna, P. Dippo, D.H. Levi, R.C. Reedy and J.S. Ward, InGaAs/GaAs QD superlattices: MOVPE growth, structural and optical characterization, and application in intermediate-band solar cells, in 31st IEEE Photovoltaic Specialists Conference, pp. 43–48, IEEE, New York, 2005.

[44] L. Cuadra, A. Martí, A. Luque, C.R. Stanley and A. McKee, Strain considerations for the design of the quantum dot intermediate band solar cell in the $In_xGa_{1-x}As/Al_yGa_{1-y}As$ material system, in Proceedings of the 17th European Photovoltaic Solar Energy Conference, pp. 98–101, WIP-Renewable Energies and ETA, Munich, Florence, 2001.

[45] N.J. Ekins-Daukes, K.W.J. Barnham, J.P. Connolly, J.S. Roberts, J.C. Clark, G. Hill and M. Mazzer, Strain-balanced GaAsP/InGaAs quantum well solar cells, Appl. Phys. Lett., 75 (1999) 4195–4197.

[46] J.I. Pankove, in Optical Processes in Semiconductors, pp. 113, Dover Publications, New York, 1971.

[47] A. Luque, Photovoltaic markets and costs forecast based on a demand elasticity model, Prog. in Photovoltaics: Res. Appl., 9 (2001) 303–312.

Nanostructured Materials for Solar Energy Conversion
T. Soga (editor)

Chapter 18

Nanostructured Photovoltaics Materials Fabrication and Characterization

Ryne P. Raffaelle

Rochester Institute of Technology, 85 Lomb Memorial Drive, Rochester, NY 14623-5608, USA

1. INTRODUCTION

The developmental efforts in regards to conventional Si- and GaAs-based cells have resulted in nearly eliminating the efficiency differences between actual devices and realistic theoretical estimates. Over the past decade much of the cell-efficiency improvements have resulted from the move toward multi-junction devices. However, as researchers continue to push the envelope, they are looking toward new approaches, such as the use of nanotechnology, in improving device efficiencies. Theoretical results of Luque and Marti [1] have shown that an intermediate-band solar cell, which gets its name from an intermediate band of states resulting from the introduction of quantum dots (QDs), can exceed the Shockley and Queisser model efficiency of not only a single junction but also a tandem cell device.

A QD is a granule of a semiconductor material whose size is on the nanometer scale. These nanocrystallites behave essentially as a 3-dimensional potential well for electrons (i.e., the quantum mechanical "particle in a box"). By introducing a single sized dot into an ordered array within the intrinsic region of a p–i–n solar cell (see Fig. 1), Luque and Marti calculated a theo-retical efficiency of 63%. QDs have already been used successfully to improve the performance of devices such as lasers, light-emitting diodes, and photodetectors.

The idea of improving photovoltaic performance through the introduc-tion of electronic states at or near a photovoltaic junction was proposed as far back as 1960 by Wolf [2]. However, Schockley and Queisser [3] argued against this idea citing the problem of recombination losses that would make

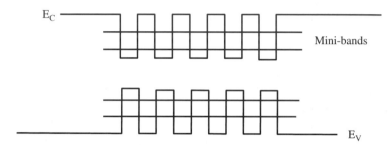

Fig. 1. Exploded view of an intermediate-band solar cell (from Luque and Marti [1]) showing the ordered array of QDs in the intrinsic region of a p–i–n solar cell.

Fig. 2. Quantized mini-band states resulting from QD array. The position in energy of the mini-bands will depend upon dot size (well width) and separation.

the approach impractical. This recombination problem has hindered other quantum mechanical approaches to improving efficiencies, such as through the use of multiple quantum well devices. Barnham and Duggan [4] showed that a photovoltaic device incorporating multiple quantum wells theoretically could achieve efficiencies of 40%. In this model, the electronic states in the potential "wells" are "quantized" and therefore have discrete energy levels. The overlap of the quantum mechanical states from well-to-well results in mini-band levels that will depend on the width of the wells (see Fig. 2). This dependence of the energy levels to well width can then be used to "tune" the device to the solar spectrum. Unfortunately, the problems associated with growing defect-free structures of this type have prohibited the anticipated efficiency improvements.

It was theoretically shown that a deep-level electronic state near the junction could be used to exceed the theoretical single-junction efficiency [5]. A claim of a 35% efficient solar cell owing to the impurity photovoltaic effect was made. This claim was later retracted after the comments published by Luque, Werner, and others. However, it has been shown that the impurity photovoltaic effect can indeed improve efficiencies as was demonstrated with the doping of a Si-based devices with indium [6].

It has also been shown that a quantum efficiency of greater than 1 can be achieved by the introduction of impurities, creating an inverse Auger mechanism that has a theoretical efficiency limit of 43% [7]. It was recently demonstrated that seven excitons could be created from a single photon [8]. In addition to an intermediate-band solar cell, several other approaches to improving photovoltaic device efficiency through the use of quantum-confined materials and structures have been proposed [9, 10].

The basic principle behind the efficiency increases offered by a QD intermediate-band solar cell is that the discrete states that result from the inclusion of the dots allow for absorption of sub-bandgap energies. The reason that this approach can exceed that of an ordinary dual-junction cell is that when the current it extracted it is limited by the host bandgap and not the individual photon energies. In a dual-junction solar cell, the current must be "matched" between the two the junctions. This means that the same amount of current must be passed through both junctions, and therefore the overall device efficiency is limited by the current generating ability of the weaker of the two junctions. In addition, if a dual-junction or other multi-junction device is grown monolithically, in which the junctions are connected in series, it is required that there be tunnel junctions grown in between the various active regions. Problems with lattice mismatch and the increased number of interfaces (and therefore interfacial defects) are impediments to this approach in photovoltaic development. Although, tremendous achievements have been made in developing multi-junction solar cells, this problem will be compounded as we continue to try and increase the number of junctions. The problems with lattice mismatch and interfacial defects are the same problems that have plagued the development in multi-quantum well structures as well.

Much like the energy dependence with multiple quantum wells, the energy states of the QDs are inversely proportional to their size. The ground state absorption energy of a QD is determined by

$$\hbar\omega = E_g + \frac{\hbar^2}{2m_e^* r^2}\pi^2 \tag{1}$$

where E_g is the bulk semiconductor bandgap, h the Planck's constant, m_e^* the electron effective mass, and r the QD radius. Fig. 3 shows the theoretical increase in the QD effective bandgap with decreasing particle size in comparison to its bulk value for CdSe.

If QDs are produced in an ordered array within an insulating medium, the wavefunctions associated with the discrete electronic states of the QDs

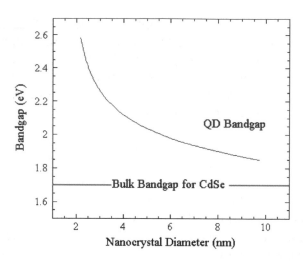

Fig. 3. The theoretical effective bandgap for spherical CdSe QDs as a function of their diameter.

will overlap creating "mini-bands" within the insulating region. A simple approximation of the effective energy gap of these mini-bands would be the ground state absorption of the QDs which make up the array. This then puts a constraint on the range of QDs sizes, in addition to their Bohr exciton radius, which can produce and intermediate bands of consequence.

As QD sizes become too small, their effective bandgap will approach that of the host material and negate any sub-gap absorption of any consequence. For a real material system, the differential between the electron affinities of the host and QDs must also be taken into consideration. For example, in determining the size of InAs QDs which could produce an intermediate band in GaAs, we note that there is a 0.83 eV difference in the conduction band edge between GaAs and InAs based on their electron affinities of 4.03 and 4.9 eV, respectively (see Fig. 4). To achieve absorption much below the GaAs band edge, the QDs size needs to be >4 nm, assuming an InAs bandgap of 0.38 eV and an effective mass of 0.023 me. Therefore, using the Bohr exciton radius in InAs of 28.2 nm as the upper bound, InAs QDs in GaAs need to be somewhere between 4 and 28 nm to achieve an intermediate band. If we assume dot radii in the range of 4.0–10.0 nm that is necessary for strong quantum confinement and yet be intermediate, the dots will need to be spaced at over twice their radii or at approximately 8.0–20.0 nm. This will provide a density of states for the intermediate band to be on the order of 10^{17}–10^{18} cm^{-3}. This range is typical for most solar cell materials and will provide the cell with a

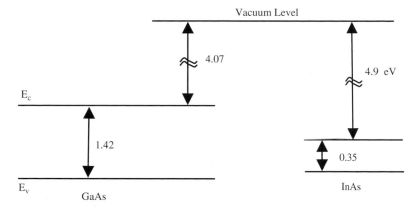

Fig. 4. An idealized energy band diagram depicting the bulk bandgaps and electron affinities for GaAs and InAs.

strong absorption coefficient. The width of the mini-bands under these conditions is well below the point at which stimulated emission becomes a problem.

There are a number of other considerations in fabricating QDs for photovoltaic applications. For example, asymmetry in the QD shape, regularity in the QD size, doping of the QDs, and degree of crystalline perfection are examples of properties which could dramatically affect the utility of QDs in real devices [11]. It should definitely be noted that the importance of these various considerations can vary dramatically based upon the role of the QDs in a given photovoltaic device application [4, 9, 10, 12]. Furthermore, although they share much in the way of common physics, the QDs or other quantum confined materials used in dye-sensitized or polymeric solar cells have many other properties that must be considered due to their unique roles in these types of devices [13–16].

The possibility of making large area, inexpensive, lightweight, and flexible solar cells has attracted many researchers to organic thin films based on soluble conducting polymers for some time (see Fig. 5) [17]. However, it was not until recently when researchers began to look into nanomaterial/conjugated polymer complexes that devices with reasonable efficiencies became possible [18–23]. Photon absorption in the organic-based composites produces bound-state excitons. Dissociation of these charge pairs can be accomplished by the potential difference across a polymer-metal junction provided the excitons are near the interface. However, the dissociation can also be accomplished via electron-accepting impurities. Thus, under illumination, a preferential transfer of electron to the acceptors leaves

Fig. 5. Lightweight and flexible nanomaterial/polymer thin film solar cell.

holes to be preferentially transported through the conjugated polymer. This process is known as photoinduced charge transfer.

Since the discovery of photoinduced charge transfer, a variety of acceptor materials have been introduced into donating conjugated polymers to produce photovoltaic devices (i.e., buckminster fullerenes [24, 25], CdSe QDs [15, 16], rods [13], and single-wall carbon nanotubes (SWCNTs) [26–29]). The devices are produced by placing the doped polymeric films between a transparent conductive oxide (TCO) top contact and a metallic back contact. There has also been recent work on thin-film polymeric solar cells which incorporate QDs as well as QD/SWCNT complexes [30]. Typically, very high loading levels of semiconductor nanoparticles are required to overcome the poor electron transport properties of the polymer. Inclusion of SWCNTs to the polymer/QD mixture can improve the transport problem [31]. SWCNTs are excellent electron-transport materials and their length far exceeds the longest quantum-rod materials. Therefore, improved electron conductivity can be achieved with much lower-weight percent loading levels [32].

2. III–V QUANTUM DOT SYNTHESIS

The majority of III–V solar cell materials are grown using an ultra-high vacuum technique, such as molecular beam epitaxy (MBE) or organometallic vapor phase epitaxy (OMVPE), which deposit sequential monolayers of materials which are lattice-matched. The techniques can be used to grow QDs using the Stranski–Krastanow (S–K) mode [33, 34]. This type of growth occurs by depositing a material with a different lattice parameter, but low interfacial energy. This growth results in the spontaneous formation of islands after a small initial stage of layer-by-layer growth. A wide variety of III–V QDs have been developed for electronic applications using this method [35, 36]. An array of QDs may be produced by alternating intrinsic cladding layers and QD layers. The QDs will self-order in the vertical direction due to the fact that the QDs in a given layer tend to grow directly above those in the previous layers due to the lattice strain [37].

InAs QDs are typically grown on GaAs using metallorganics such as trimethyl gallim (TMGa) and trimethyl indium (TMIn) as precursor materials, along with hydrides phosphine (PH_3), arsine (AsH_3), and 1% AsH_3 in hydrogen. Disilane (10^2 ppm Si_2H_6) and dimethyl zinc (10^3 ppm DMZn) diluted with hydrogen were used as sources of n- and p-dopants, respectively. The typical growth temperatures are between 500 and 700°C and a pressure of around 600 Torr. Fig. 6 shows atomic force microscopy (AFM) images of InAs grown on GaAs by OMVPE under similar conditions. These S–K QDs have an aspect ratio of approximately 4 to 1 in diameter to height as is seen in a cross-section of the data (see Fig. 7). This will produce a directional anisotropy in the absorption. The lowest energy states of the QDs would be those that correspond to their smallest dimension (i.e., height). The heights shown in Fig. 7 range from approximately 5–10 nm. This would yield an effective bandgap range of 1.0–0.5 eV based on Eq. (1).

The self-ordering that can occur when sequential layers of S–K QDs (grown as above) are sequentially deposited with strain-compensating barrier layers in between are shown in the cross-sectional transmission electron microscopy (TEM) image in Fig. 8. The contrast in these images results from both lattice strain and the compositional differences. The InAs QDs are the black regions indicated by the arrows in the figure and the growth direction is toward the top of the page.

Photoluminescence spectroscopy of the InAs QD array shown in Fig. 9(a) demonstrates that the QD array does have energy states with and effective bandgap that are consistent with the theoretical predictions based upon their size. Fig. 9(b) shows photoluminescence emission from the same QD array as a function of excitation wavelength. The discrete mini-band absorption is manifested as peaks on the resulting topographic graph where height corresponds to the measured luminescent intensity.

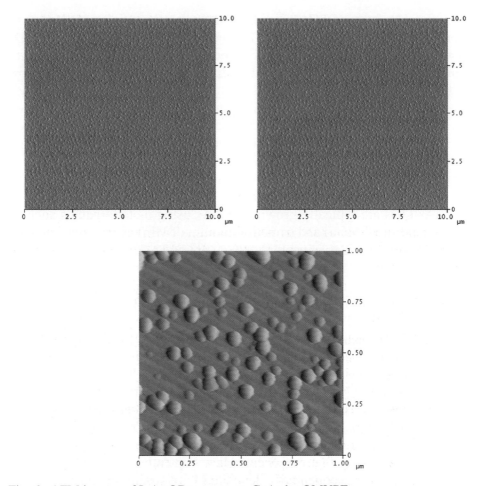

Fig. 6. AFM images of InAs QDs grown on GaAs by OMVPE.

In terms of their possible application as an intermediate-band materials, it is necessary to demonstrate sub-host bandgap absorption by the QD array materials. Fig. 10 shows the absorbance as a function of illumination energy for the 20-layer InAs QD array grown on GaAs. A substantial increase in sub-GaAs bandgap (i.e., 1.42 eV) absorption is seen with the introduction of the QDs.

3. COLLOIDAL QUANTUM DOT SYNTHESIS

The majority of the colloidal QD synthesis performed to date has focused on CdSe (see Fig. 11) [38–41]. However, new reports on other II–VIs,

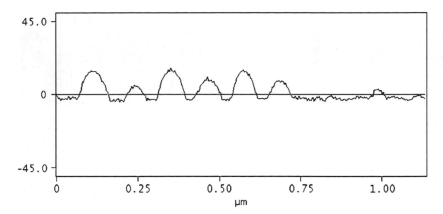

Fig. 7. A representative cross-section of the data shown in Fig. 6 of the InAs QDs grown on GaAs by OMVPE. The *y*-axis is given in nanometers and yields an average diameter to height ratio of 4:1.

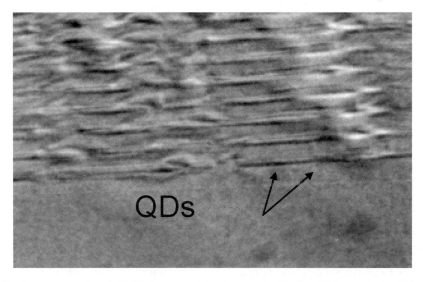

Fig. 8. A cross-sectional TEM image of the bottom half of a 20-layer InAs QD array grown on GaAs.

III–Vs, and ternary are increasing in frequency. Such as the nanocrystal-polymer composites that have been produced using CdTe with polyaniline, polypyrrole, or poly(3,4-ethylenedioxythiophene) (PEDT) [42], and the synthesis of $CuInS_2$ and $CuInSe_2$ chalcopyrite QDs for space photovoltaic applications [43]. Colloidal synthesis of nanoparticles offers greater control

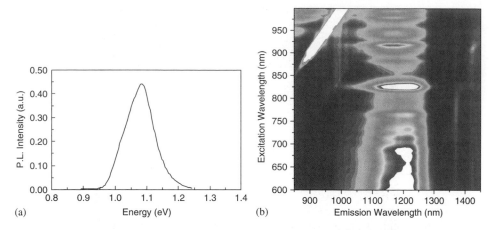

(a)

(b)

Fig. 9. (a) Photoluminescence spectroscopy from a 20-layer InAs QD array grown on GaAs under monochromatic 433 nm laser excitation and (b) photoluminescent topographic map resulting from spectroscopic excitation.

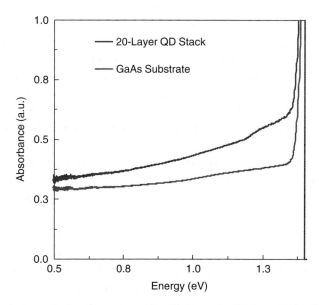

Fig. 10. Optical absorption versus energy for GaAs and a 20-layer InAs QD layer grown on GaAs.

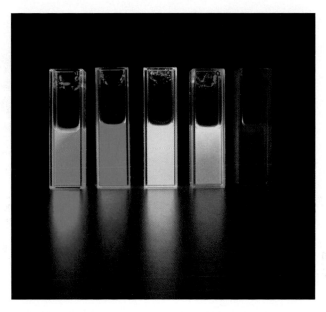

Fig. 11. Photograph of CdSe colloidal QDs under UV illumination.

over morphology and size than do Stranski–Krastanov methods, particularly for very small particles. In the past decade, synthetic control over colloidal nanoparticles has greatly improved, especially for the II–VI and III–V semiconductors [40, 44]. Recent work with single-source precursors has shown them to be viable starting materials for the synthesis of high-quality colloidal QDs [45–47]. Single-source precursors are discrete molecules that include all the elements required in the final material. These precursors can be designed with many properties in mind, including stoichiometry, solubility, and volatility.

CdSe QDs are conventionally synthesized using a CdO precursor [48]. A typical synthesis will combine CdO, stearic acid, and 1-octadecene with heat (i.e., 200°C) under $Ar_{(g)}$. After cooling to room temperature, trioctylphosphine oxide (TOPO) and octadeclyamine (ODA) will be added, and this mixture will be brought to 280°C. At this point a 1 M selenium-trioctylphosphine solution is injected. The reaction is held at 250°C for 30 min and then cooled to room temperature. Extraction of the CdSe-TOPO QDs occurs via sequential washes using a separatory funnel with a 10:1 mixture by volume of methanol: hexanes until a distinct interfacial separation is observed. The CdSe-TOPO QDs are precipitated with acetone and re-suspended in chloroform for subsequent characterization and/or utilization [49].

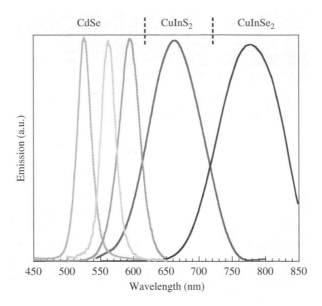

Fig. 12. Photoluminescence emission from 3 different sized CdSe, a $CuInS_2$, and a $CuInSe_2$ QD suspension. The lines represent the approximate ranges in bandgap tunability demonstrated with these systems.

A typical colloidal synthesis using a single-source precursor would start with something like $(PPh_3)_2CuIn(SEt)_4$, which is a charge-neutral molecule which decomposes in solution at 200°C to form QDs of $CuInS_2$. By modification of the surface with alkyl groups during synthesis, soluble colloids can be formed. The surface ligands can be exchanged in a postprocessing step with any Lewis base. Modification of the nanocrystal surface with functionalized ligands can be used to facilitate binding to functional groups on other materials or surfaces such as carbon nanotubes [50].

Colloidal QDs are generally much more spheroidal than their S–K counterparts and posses a larger range of bandgap tunability. Fig. 12 shows the photoluminescence emission spectrum for 3 different size CdSe dispersion, as well as one for $CuInS_2$ and $CuInSe_2$ QD sample. The dashed lines represent the ranges in bandgaps that have been demonstrated by varying the size of the QDs in these systems [46, 51].

In addition to OMVPE and colloidal growth techniques, there are a wide variety of other methods currently being employed to produce QD and other quantum confined inorganic materials. Examples include thermal sublimation, electrochemical synthesis, pulsed laser synthesis, and chemical

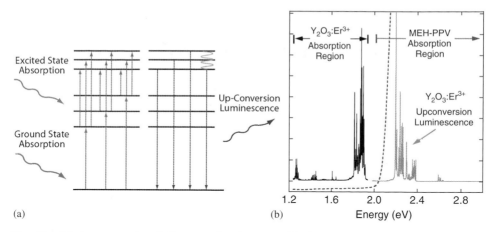

Fig. 13. (a) An energy band diagram showing a multi-photon absorption and re-emission process and (b) absorption of erbium doped yttria nanoparticles and subsequent re-emission at higher energy overlayed with the absorption spectrum of the conducting polymer MEH–PPV.

bath techniques such as the EDTA assisted UREA precipitation technique. These techniques are being employed to produce QDs of a wide variety of materials in addition to those mentioned above for potential photovoltaic applications such as PbSe, TiO_2, $Y_2O_3:Er^{3+}$, $Y_2O_3:Eu^{3+}$, ZnS:Ag, ZnS:Cl, $Y_2O_3:Er^{3+}$, $YAG:Tb^{3+}$ among many others. Many of these materials are particularly attractive for some of the more novel photovoltaic applications mentioned above (i.e., up-conversion, down-conversion, etc.) Fig. 13(a) shows a schematic of how a multi-photon process can be achieved using the discrete-like states of a quantum confined material. Fig. 13(b) shows a real demonstration of this phenomena using erbium doped yttria nanoparticles which absorb below the band-edge of their host polymer (MEH-PPV) and re-emits the energy above it.

4. SINGLE-WALL CARBON NANOTUBES SYNTHESIS

Since they were first imaged by Iijima and co-workers [52] in 1991, carbon nanotubes have generated an interest among scientists and engineers that surpasses almost any material known to man. The unique mechanical and electronic properties of both the single-wall and multi-walled varieties of carbon nanotubes have proven to be a rich source of new physics and have lead to applications in a wide variety of materials and devices [53]. Carbon nanotubes can be envisioned as a rolled up graphene sheet, or multiple sheets as in the case of a multi-walled nanotube. The role-up vector will determine the so-called "chirality" of the SWNT, which relates to whether the structure will be metallic or semiconducting. The optoelectronic properties of a SWNT will depend directly on this chiral angle, as well as the diameter of the nanotube. In some cases, the π-orbital overlap can lead to a metallic SWNT which acts as a 1-dimenionsal ballistic conductor. Other chiral angles exhibit the electronic energy transitions similar to a traditional semiconductor with a bandgap energy [53].

The synthesis of carbon nanotubes can be accomplished in a wide variety of methods that involve the catalytic decomposition of a carbon containing gas or solid. Some of the most common techniques are chemical vapor deposition (CVD), arc-discharge, and laser vaporization synthesis [53, 54]. The synthesis conditions (temperature, pressure, carrier gas, etc.), metal catalyst type (most commonly iron, nickel, cobalt, or yttrium), and carbon source (graphite or hydrocarbon) have all been shown to influence the properties of the resulting carbon nanotubes [53, 55, 56]. In general, the by-products are the principal component of the as-produced materials or raw SWNT "soot." By-products such as graphitic and amorphous carbon phases, metal catalysts, fullerenes, and carbonaceous coatings on the SWNTs may not only dominate the physical characteristics of the raw soot, but they also pose significant challenges in any subsequent purification [57–63]. Further development of SWNT-based applications is expected to require material standardization, specifically with respect to electronic type and degree of purity.

SWNT diameters typically range from 0.4 to 2 nm and recently lengths up to 1.5 cm have been reported [64]. Such dimensions give rise to aspect ratios (length/diameter) of over ten million, which is extremely advantageous for establishing low percolation thresholds in polymer or ceramic composites. In fact, the extraordinary conductivity of carbon nanotubes (electrical $= 10^4$ S/cm [65] and thermal $= 6600$ W/mK for a (10, 10) SWNT [66]) coupled with their high aspect ratio has contributed to dramatic enhancement in composite conductivity at low-weight percent doping levels [67, 68]. It is this remarkable conductivity along with the extremely high specific surface area (up to ~ 1600 m^2/g [69]) of carbon nanotubes which has generated so much interest in their use for solar cells. Fig. 14(a) shows a schematic of a pulsed laser (Alexandrite 755 nm) vaporization reactor that can be used to fabricate SWNTs. Fig. 14(b) shows a scanning electron micrograph of as-produced materials. The laser pulse used to generate these particular materials was rastered (corner to corner over 1 cm^2 with 50%

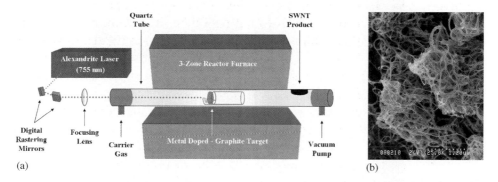

Fig. 14. (a) Schematic of a pulsed laser vaporization reactor used to fabricate SWNTs and (b) a scanning electron micrograph of the as-produced material.

overlap of $100 \, \mu s$ pulses at a repetition rate of $10 \, Hz$) over the surface of a graphite (1–$2 \, \mu m$) target doped with 2% w/w Ni (sub-micron) and 2% w/w Co ($<2 \, \mu m$), at an average power density of $100 \, W/cm^2$. The reaction furnace temperature was maintained at $1150°C$, with a chamber pressure of $700 \, Torr$ under $100 \, sccm$ flowing $Ar_{(g)}$ [70].

Purification of SWNT "raw soots" can be performed using an established protocol [70]. In summary, 50–$100 \, mg$ of raw SWNT soot is refluxed at $120°C$ in $3 \, M$ nitric acid for $16 \, h$, and then filtered over a $1 \, \mu m$ PTFE membrane filter with copious amounts of water. The membrane filter was dried at $70°C$ in vacuo to release the resulting SWNT paper from the filter paper. The L-SWNT paper was thermally oxidized in air at $450°C$ for $1 \, h$ in a Thermolyne 1300 furnace, followed by a $6 \, M$ hydrochloric acid wash for $30 \, min$ using magnetic stirring, equivalent filtering steps, and a final oxidation step at $550°C$ for $1 \, h$. Fig. 15(a) shows a scanning electron micrograph of the subsequent "purified" SWNTs. In addition, these SWNTs can be shortened or "cut" using a combination of etching in $H_2SO_4:H_2O_2$ and ultra-sonication (see Fig. 15(b)).

As mentioned previously, we can envision a nanotube as a rolled-up graphene sheet. Depending on its particular roll up vector (i.e., direction with respect to the benzene rings and length), it will be either metallic or semiconducting with a diameter dependent bandgap. Fig. 16 shows the theoretical band energies as a function of SWNT diameter for the different possible chiralities. This is commonly referred to as a Katura plot [71]. $^{s}E_{11}$ and $^{s}E_{22}$ correspond to the first and second Van Hove singularities for the semiconducting SWNTs, respectively.

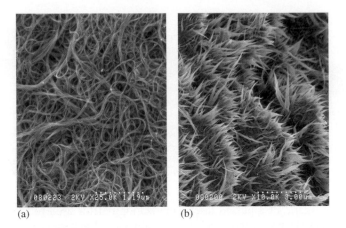

(a)   (b)

Fig. 15. (a) Purified and (b) cut SWNTs.

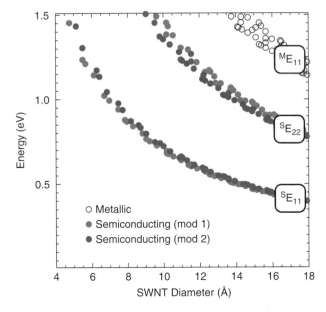

Fig. 16. Theoretical energy states for the various SWNT chiralities as a function of their diameter.

In a given SWNT sample, there is normally a distribution of diameters and chiralities regardless of the synthesis method employed. It is possible to determine the various chiralities present using photoluminescence energy mapping (see Fig. 17). Unfortunately, due to the various transition probabilities

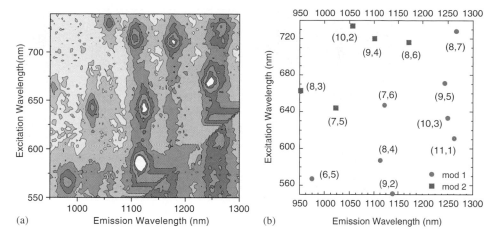

Fig. 17. (a) Photoluminescent map of the SWNT chiralities present in a laser generated SWNT sample and (b) the corresponding theoretically generated positions based upon the roll-up vector components.

of the various states responsible for the absorption and emission, more work is required before precise quantification of the various chiralities represented in a given will be possible. Also, although there are many groups currently pursuing chirality separations, no current efficient large-scale method has been developed.

5. MULTI-WALL CARBON NANOTUBES SYNTHESIS

The area of multi-wall carbon nanotubes (MWNTs) has also developed rapidly during the past decade since their discovery [72]. MWNTs are the concentric graphene layers spaced 0.34 nm apart, with diameters ranging from 10 to 200 nm and lengths up to hundreds of microns [53].

These materials have been produced by several techniques including arc discharge [72, 73], laser ablation [74], flame synthesis [75], and a variety of CVD methods [76–80]. Two of the most promising methods for depositing commercial quantities of aligned MWCNs are the "floating catalyst" CVD method [81–84] and the injection CVD method [85–93]. For the injection method, an organic solvent containing a dissolved organometallic compound that decomposes to form the catalysts is injected into a two-zone furnace (see Fig. 18(a)). In the first zone, both the solvent and the catalyst vaporize. A carrier gas sweeps the vapors into the second zone where the

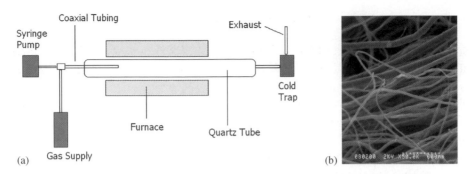

Fig. 18. (a) A schematic of a spray chemical vapor reactor for fabrication MWNT and (b) a scanning electron micrograph of MWNT.

organometallic compound decomposes to yield nanoparticles of the metal catalyst. Solvent and ligand molecules serve as the carbon source for nanotube growth at the catalyst sites. The organometallic compounds commonly used are metallocenes (Fe, Co, Ni, or Ru) [86–93] and iron pentacarbonyl, [Fe(CO)$_5$] [94, 95].

Fig. 18(b) shows MWNTs synthesized by and established low-pressure CVD procedure [96]. In summary, toluene solutions of the catalyst precursor (0.06 M) were injected through a 20 gauge needle into the first zone of a two zone furnace using a syringe pump. Solutions were injected at rates ranging from 1.5 to 2.0 mL/h, with typically 2.5–4.0 mL being delivered. The temperature of zone one was held at 200 ± 1°C, which is above the boiling point of the solvent and above the decomposition temperature of the precursor. A mixed carrier gas of 4% hydrogen in nitrogen (1.5 slpm) swept the hydrocarbon and catalyst gases into the hot zone (650–800°C) of the furnace. The MWNTs deposited on the walls of the 35 mm fused silica reactor tube, just inside the entrance of the hot zone.

MWNTs dispersed in poly(p-phenylene vinylene) (PPV) have been used for photovoltaic devices. Diode response from an MWNT/PPV/Al device that exhibited a doubling in percent external quantum efficiency compared to an ITO/PPV/Al device. The device was illuminated at 485 nm with an intensity of 37 μW/cm^2, producing a V_{oc} and I_{sc} equal to 0.90 V and 0.56 μA/cm^2, respectively. This open-circuit voltage coincides with the workfunctions of Al and MWNTs, namely 4.3 and 5.1 eV. Photoluminescence data reflect a proposed energy transfer from PPV to MWNTs, attributed to hole collection at the MWNT electrode [97].

6. NANOMATERIAL COMPLEXES

Semiconducting QD and SWNT complexes are being evaluated for a variety of optoelectronic applications [98]. Each material has shown a rich chemistry for functionalization, and several recent reports have evaluated the direct coupling of these two materials [50, 98–102]. The previous attempts at attachment of QDs to SWNTs, however, relied on covalent methods utilizing carboxylic acid groups, which alters the SWNT electronic properties by disrupting the sp^2 hybridized conjugation [98]. In comparison, noncovalent approaches which use π-orbital interactions between the functionalizing compound and SWNTs mitigate this concern [103–106], as previously demonstrated in the case of immobilizing proteins on SWNTs [103]. Noncovalent coupling of QDs to SWNTs is expected to produce a material which facilitates selective wavelength absorption, charge transfer to SWNTs, and efficient electron transport; all essential properties of additives for polymer photovoltaics [29, 50].

Synthesis of SWNT-QD heterostructures have been demonstrated which involve oxidized SWNTs attached to cadmium selenide (CdSe) and TiO_2 [107, 108]. Attachment of the QDs to the SWNTs occurred primarily at functionalized ends for shortened SWNTs (i.e., $<300\,nm$) and at both sidewall and end functionalized sites for longer SWNTs [108]. Based on optical absorption spectroscopy, depletion of the energy states for the CdSe nanocrystals results in a proposed charge transfer from the CdSe nanocrystals to the SWNTs [107]. Fig. 19(a) shows an example of a covalent attachment scheme for CdSe QDs and SWNTs. Fig. 19(b) shows a TEM image of the material complex which resulted using this scheme.

Fig. 20 shows a noncovalent attachment scheme for the attachment of CdSe QDs to SWNTs using a 4-aminothiophenol (ATP) ligand and an intermediary 1-pyrenebutyric acid *N*-hydroxy-succinimide ester (PBASE) [103] molecule. Presumably, this approach could be used for a wide variety of nanocrystal to SWNT attachments. Fig. 21 shows a TEM image of resulting noncovalent CdSe–SWNT complexes.

7. NANOSTRUCTURED MATERIAL CHARACTERIZATION

As demonstrated above, there are several commonly employed techniques used to characterize both epitaxially grown and colloidal QDs, as well as the various carbon nanotubes and other nanomaterials. These include photoluminescence spectroscopy, transmission spectroscopy, absorption spectroscopy, Raman spectroscopy and atomic force and scanning tunneling microscopy,

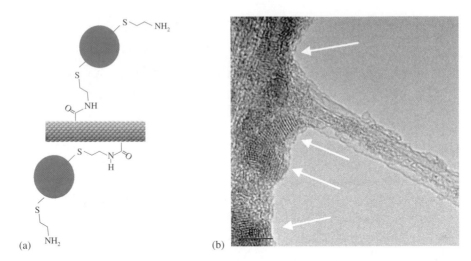

Fig. 19. (a) Schematic of a covalent attachment scheme for CdSe QDs and functionalized SWNTs and (b) a TEM micrograph of the resulting SWNT-QD complex (from Landi et al. [102]).

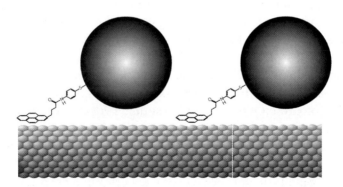

Fig. 20. Schematic of noncovalent attachment of CdSe QDs to SWNTs, using a 4-ATP ligand and an intermediary 1-PBASE (from Chen et al. [103]) molecule.

just to name a few. Some of these techniques are much more suited to certain types of materials. For example, AFM is much more straightforward when applied to the epitaxial QDs. Collodial dots or carbon nanotubes tend to agglomerate and move under the AFM probe when attempting to image as a dry powder.

The physical properties associated with QDs, carbon nanotubes etc. like diameter distributions (and length for nanotubes), defect content, and

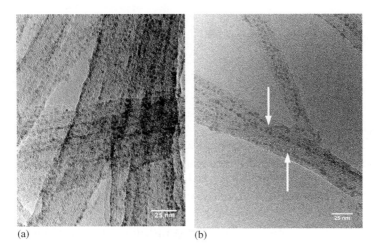

(a)   (b)

Fig. 21. TEM images of CdSe-MPB-SWNTs (a) showing the abundant coverage of QDs along the sidewalls of the SWNTs and (b) arrangement in an ordered array.

aggregation and bundling effects, are influential factors that may impact device performance. Attempts at characterizing these properties have strongly relied upon microscopy (SEM, TEM, and AFM), but due to the sample-limiting nature of these measurements it is difficult to estimate the bulk character [53]. Spectroscopic techniques like Raman, photoluminescent, and optical absorption can be used to accurately assess the size distributions.

SWNTs sizes can be on the diameter-dependent resonant Raman effects and the relation between optical bandgap and SWNT diameter [109, 110]. The characteristic radial breathing mode (RBM) at low energy ($100–300\,cm^{-1}$) for the SWNTs is used to calculate the diameter distribution based on Eq. (1). The following relationship between diameter (d, nm) and Raman shift (ω_{RBM}, cm^{-1}) has been reported for bundled SWNTs [110]:

$$\omega_{RBM} = (224/d) + 14 \tag{2}$$

For our typical laser vaporization conditions, the diameter range is between 1.2 and 1.4 nm. It is important to note that the RBM is not observed for MWNTs, preventing the diameter distribution to be calculated with Raman spectroscopy. Optical absorption spectroscopy is another technique for evaluating the diameter distribution of SWNTs, since the optical transitions are based on the discrete electronic transitions associated with the Van Hove singularities of the semiconducting and metallic types [109]. The ith pair of discrete electronic transition energies corresponding to these singularities is approximated by the following:

$$^{S,M}E_{ii} = \frac{2na_{c-c}\gamma_0}{d_{SWNT}}$$ (3)

Where n is an integer, having values of 1, 2, 4, 5, or 7 for semiconducting (S) SWNTs and $n = 3$ or 6 for metallic (M) SWNTs in the spectral range of interest [111], a_{c-c} is the carbon–carbon bond distance with a value of 0.142 nm, and d_{SWNT} is the SWNT diameter [112, 113]. The carbon–carbon overlap integral for SWNTs, γ_0, has been reported to range from 2.45 to 3.0 eV [114]. In general, the absorption peaks are broadened due to the superposition of absorbances for the overall diameter distribution, unless a debundling effect occurs which can be achieved through surfactant-stabilized or organic-solvent dispersions [115, 116].

Analysis of the length distribution for bulk samples of carbon nanotubes can further enable the aspect ratio (length/diameter) to be calculated. The aspect ratio is an important parameter for ceramic or polymer composite applications where structural, thermal, and/or electrical conductivity enhancement is desired. The ability to achieve a low percolation threshold in these composites is facilitated with dopant materials having a high aspect ratio, such as carbon nanotubes. Microscopic analysis has been the prominent method of estimating the length distribution with electron micrographs and AFM images being routinely used [117]. For the laser-generated SWNTs, the typical length scale is ~1–10 μm, giving an aspect ratio for individual SWNTs of 1000–10000. In comparison, the injection CVD MWNTs having a similar length scale would have an aspect ratio from 10 to 1000, since the diameters are significantly larger.

Qualitative estimation of the defect density in both SWNTs and MWNTs is performed using a combination of microscopy and Raman spectroscopy. The presence of kinks, vacancies, or other defective sites can be observed using TEM, although the degree of occurrence in the bulk sample is less conclusive. Raman spectroscopy has also been used for establishing the defect density based on the relative ratio of the D-Band (~1350 cm^{-1}) intensity to the G-Band (~1600 cm^{-1}) [118]. The D-Band is the Raman mode associated with a disorder-induced dispersion relation, which indicates the presence of vacancies or nonaromatic sites in the graphitic structure [119]. Evident from the Raman spectra in Fig. 3(a) are the corresponding D- and G-Bands for SWNTs and MWNTs. The relative peak ratios can be qualitatively assigned to the defect density in the samples, thus showing a higher defect density in the MWNTs compared to the SWNTs.

The degree of bundling in SWNT samples is an important property, especially for composite preparation where the exfoliation of bundles is imperative to reduction of the percolation threshold through lower doping levels. We have observed such debundling effects previously in SWNT-Nafion polymer composites where a factor of five reduction in average bundle size was observed by cross-sectional analysis with SEM [68]. Additionally, debundling effects can be probed using Raman spectroscopy by monitoring the Raman shifts in the RBM or changes in the lineshape associated with the Breit–Wigner–Fano (BWF) lineshape of the G-Band [110, 120].

The use of thermogravimetric analysis (TGA) to monitor the thermal decomposition profile of the carbonaceous components found in carbon nanotubes is commonly employed. In has been shown to be an outstanding tool for the evaluation of various SWNT purification techniques [121]. Determination of the actual metal impurity content can be performed by adjusting the high temperature residue based on the oxidation products of the metal catalysts. However, since both SWNTs and MWNTs are generally synthesized in quartz reaction vessels, the occasional presence of SiO_2 can influence these calculations with proper adjustment made by quantitative electron dispersive X-ray spectroscopy (EDS). Such TGA residue values are important since certain applications require ceramic or polymer composites containing carbon nanotubes, and evaluation of the relative mass of active material is a critical step toward establishing accurate doping levels.

REFERENCES

[1]   A. Luque and A. Marti, Phys. Rev. Lett., 78 (1997) 5014–5017.
[2]   M. Wolf, Proc. IRE, 48 (1960) 1246.
[3]   W. Shockley and H. Quiesser, J. Appl. Phys., 32 (1961) 510.
[4]   K. Barnham and G. Duggan, J. Appl. Phys., 67 (1990) 3490.
[5]   M.J. Keevers and M.A. Green, J. Appl. Phys., 75 (1994) 4022.
[6]   M.J. Keevers and M.A. Green, J. Appl. Phys., 78 (1994) 8.
[7]   J. Werner, S. Kolodinski and H.J. Quiesser, Phys. Rev. Lett., 72 (1994) 3851.
[8]   R.D. Schaller, M. Sykora, J.M. Pietryga and V.I. Klimov, Nano Lett., 6 (2006) 424–429.
[9]   K. Barnham, J.L. Marques, J. Hassard and P. O'Brien, Appl. Phys. Lett., 76 (2000) 1197.
[10] A.J. Nozik, Proc. Int. Symp., (2001) 61–68.
[11] A. Marti, L. Cuadra and A. Luque, Physica E, 14 (2002) 150–157.
[12] V. Aroutiounian, S. Petrosyan, A. Khachatryan and K. Touryan, J. Appl. Phys., 89 (2001) 2268–2271.

[13] W.U. Huynh, J.J. Dittmer and A.P. Alivisatos, Science, 295 (2002) 2425–2427.

[14] W.U. Huynh, X. Peng and A.P. Alivisatos, Adv. Mater, 11 (1999) 923.

[15] L. Kronik, N. Ashkenasy, M. Leibovitch, E. Fefer, Y. Shapira, S. Gorer and G. Hodes, J. Electrochem. Soc., 145 (1998) 1748.

[16] D.S. Ginger and N.C. Greenham, J. Appl. Phys., 87 (2000) 1361.

[17] G.A. Chamberlain, Solar Cells, 8 (1983) 47.

[18] G. Yu, J. Gao, J.C. Hummelen, F. Wudl and A.J. Heeger, Science, 270 (1995) 1789–1791.

[19] J. Nelson, Curr. Opin. Solid State Mater. Sci., 6 (2002) 87.

[20] H.H.A.N.S. Sariciftci, J. Mater. Res., 17 (2004) 1924–1945.

[21] J.-M. Nunzi, C.R. Phys., 3 (2002) 523.

[22] P. Peumans, A. Yakimov and S.R. Forrest, J. Appl. Phys., 93 (2003) 3693.

[23] M. Granstrom, K. Petritsch, A.C. Arias, A. Lux, M.R. Andersson and R.H. Friend, Nature, 395 (1998) 257–360.

[24] N. Camaioni, G. Ridolfi, G. Casalbore-Miceli, G. Possamai, L. Garlaschelli and M. Maggini, Sol. Energy. Mater. Sol. Cells, 76 (2002) 107–113.

[25] J.J.M. Halls, K. Pichler, R.H. Friend, S.C. Moratti and A.B. Holmes, Appl. Phys. Lett., 68 (1996) 3120–3122.

[26] E. Kymakis and G.A.J. Amaratunga, Appl. Phys. Lett., 80 (2002) 112–114.

[27] E. Kymakis, I. Alexandrou and G.A.J. Amaratunga, Synth. Met., 127 (2002) 59–62.

[28] E. Kymakis, I. Alexandrou and G.A.J. Amaratunga, J. Appl. Phys., 93 (2003) 1764.

[29] B.J. Landi, R.P. Raffaelle, S.L. Castro and S.G. Bailey, Prog. Photovot: Res. Appl., 13 (2005) 1–8.

[30] B.J. Landi, H.J. Ruf, C.M. Evans, S.G. Bailey, S.L. Castro and R.P. Raffaelle, Sol. Energy. Mater. Sol. Cells, 87 (2005) 733–746.

[31] R.P. Raffaelle, B.J. Landi, J.D. Harris, S.G. Bailey and A.F. Hepp, Mater. Sci. Eng. B, 116 (2005) 233–243.

[32] M.J. Biercuk, M.C. Llaguno, M. Radosavljevic, J.K. Hyun, A.T. Johnson, J.E. Fischer, Appl. Phys. Lett., 80 (2002) 2767–2769.

[33] A. Stintz, G.T. Liu, H. Li, L.F. Lester and K.J. Malloy, IEEE Photon. Technol. Lett., 12 (2000) 591.

[34] W. Seifert, N. Carlsson, J. Johansson, M.-E. Pistol and L. Samuelson, J. Cryst. Growth, 170 (1997) 39.

[35] S.M. Kim, Y. Wang, M. Keever and J.S. Harris, IEEE Photon. Technol. Lett., 16 (2004) 377.

[36] N. Nuntawong, S. Huang, Y.B. Jiang, C.P. Hains and D.L. Huffaker, Appl. Phys. Lett., 87 (2005) 113105.

[37] Y.Q.A.D. Uhl, J. Cryst. Growth, 257 (2003) 225–230.

[38] C.B. Murray, D.J. Norris and M.G. Bawendi, J. Am. Chem. Soc., 115 (1993) 8706.

[39] X. Peng, M.C. Schlamp, A.V. Kadavanich and A.P. Alivisatos, J. Am. Chem. Soc., 119 (1997) 7019–7029.

[40] X. Peng, J. Wickham and A.P. Alivisatos, J. Am. Chem. Soc., 120 (1998) 5343–5344.

[41] X. Peng, M.C. Schlamp, A. Kadavanich and A.P. Alivisatos, J. Am. Chem. Soc., 119 (1997) 7019.

[42] D.V. Talapin, S.K. Poznyak, N.P. Gaponik, A.L. Rogach and A. Eychmuller, Phys. E, 14 (2002) 237–241.

[43] S.L. Castro, B.J. Landi, R.P. Raffaelle and S.G. Bailey, 2nd International Energy Conversion Engineering Conference, Providence, RI 2004.

[44] R.L.Wells. Gladfelter, J. Clust. Sci., 8 (1997) 217–238.

[45] S.L. Castro, S.G. Bailey, R.P. Raffaelle, K.K. Banger and A.F. Hepp, J. Phys. Chem. B, 108 (2004) 12429–12435.

[46] S.L. Castro, S.G. Bailey, R.P. Raffaelle, K.K. Banger and A.F. Hepp, Chem. Mater, 15 (2003) 3142–3147.

[47] E.E. Foos, R.J. Jouet, R.L. Wells, A.L. Rheingold and L.M. Liable-Sands, J. Organomet. Chem., 582 (1999) 45–52.

[48] J.J. Li, A. Wang, W. Guo, J.C. Keay, T.D. Mishima, M.B. Johnson and X. Peng, J. Am. Chem. Soc., 125 (2003) 12567–12575.

[49] W.W. Yu, L. Qu, W. Guo and X. Peng, Chem. Mater, 15 (2003) 2854–2860.

[50] B.J. Landi, H.J. Ruf, C.M. Evans, S.G. Bailey, S.L. Castro and R.P. Raffaelle, Sol. Ener. Mat. Sol. Cells, 80 (2005) 733–736.

[51] Z.A. Peng and X. Peng, J. Am. Chem. Soc., 123 (2001) 183–184.

[52] S. Iijima, Nature, 354 (1991) 56.

[53] H. Dai, Surf. Sci., 500 (2002) 218–241.

[54] T. Guo, P. Nikolaev, A.G. Rinzler, D. Tomanek, D.T. Colbert and R.E. Smalley, J. Phys. Chem., 99 (1995) 10694–10697.

[55] A.C. Dillon, P.A. Parilla, J.L. Alleman, J.D. Perkins and M.J. Heben, Chem. Phys. Lett., 316 (2000) 13–18.

[56] E. Munoz, W.K. Maser, A.M. Benito, M.T. Martinez, G.F. de la Fuente, A. Righi, E. Anglaret and J.L. Sauvajol, Synth. Met., 121 (2001) 1193–1194.

[57] I.W. Chiang, B.E. Brinson, R.E. Smalley, J.L. Margrave and R.H. Hauge, J. Phys. Chem. B, 105 (2001) 1157–1161.

[58] I.W. Chiang, B.E. Brinson, A.Y. Huang, P.A. Willis, M.J. Bronikowski, J.L. Margrave, R.E. Smalley and R.H. Hauge, J. Phys. Chem. B, 105 (2001) 8297–8301.

[59] A.C. Dillon, T. Gennett, K.M. Jones, J.L. Alleman, P.A. Parilla and M.J. Heben, Adv. Mater., 11 (1999) 1354–1358.

[60] A.C. Dillon, T. Gennett, P.A. Parilla, J.L. Alleman, K.M. Jones and M.J. Heben, Mater. Res. Soc. Symp. Proc., 633 (2001) A5.2.1–A5.2.6.

[61] A.R. Harutyunyan, B.K. Pradhan, J. Chang, G. Chen and P.C. Eklund, J. Phys. Chem. B, 106 (2002) 8671–8675.

[62] J.-M. Moon, K.H. An, Y.H. Lee, Y.S. Park, D.J. Bae and G.-S. Park, J. Phys. Chem. B, 105 (2001) 5677–5681.

[63] K.L. Strong, D.P. Anderson, K. Lafdi and J.N. Kuhn, Carbon, 41 (2003) 1477–1488.

[64] S. Huang, M. Woodson, R. Smalley and J. Liu, Nano Letters, 4 (2004) 1025–1028.

[65] A. Thess, R. Lee, P. Nikolaev, H. Dai, P. Petit, J. Robert, C. Xu, Y.H. Lee, S.G. Kim, A. Rinzler, D.T. Colbert, G. Scuseria, D. Tomanek, J.E. Fischer and R. Smalley, Science, 273 (1996) 483–487.

[66] S. Berber, Y.-K. Kwon and D. Tomanek, Phys. Rev. Lett., 84 (2000) 4613–4616.

[67] M.J. Biercuk, M.C. Llaguno, M. Radosavljevic, J.K. Hyun, A.T. Johnson and J.E. Fischer, Appl. Phys. Lett., 80 (2002) 2767–2769.

[68] B.J. Landi, R.P. Raffaelle, M.J. Heben, J.L. Alleman, W. VanDerveer and T. Gennett, Nano Letters, 2 (2002) 1329–1332.

[69] R. Baughman, C. Cui, A.A. Zakhidov, Z.B. Iqbal, J.N. Barisci G.M. Spinks, G.G. Wallace, A. Mazzoldi, D. De Rossi, A.G. Rinzler, O. Jaschinski, S. Roth and M. Kertesz, Science, 284 (1999) 1340–1344.

[70] B.J. Landi, H.J. Ruf, J.J. Worman and R.P. Raffaelle, J. Phys. Chem. B, (2004) ASAP published on web 10/11/2004.

[71] H. Kataura, Y. Kumazawa, Y. Maniwa, I. Umezu, S. Suzuki, Y. Ohtsuka and Y. Achiba, Synth. Met., 103 (1999) 2555.

[72] S. Iijima, Nature, 354 (1991) 56–58.

[73] H. Huang, H. Kajiura, S. Tsutsui, Y. Hirano, M. Miyakoshi, A. Yamada and M. Ata, Chem. Phys. Lett., 343 (2001) 7–14.

[74] S. Bandow, S. Asaka, Y. Saito, A.M. Rao, L. Grigorian, E. Richter and P.C. Eklund, Phys. Rev. Lett., 80 (1998) 3779–3782.

[75] R.L. Vander Wal, T.M. Ticich and V.E. Curtis, Chem. Phys. Lett., 323 (2000) 217–223.

[76] J. Geng, C. Singh, D.S. Shephard, M.S.P. Shaffer, B.F.G. Johnson and A.H. Windle, Chem. Commun., (2002) 2666–2667.

[77] R. Kurt, C. Klinke, J.-M. Bonard, K. Kern and A. Karimi, Carbon, 39 (2001) 2163–2172.

[78] M. Cinke, J. Li, B. Chen, A. Cassell, L. Delzeit, J. Han and M. Meyyappan, Chem. Phys. Lett., 365 (2002) 69–74.

[79] P. Mauron, C. Emmenegger, A. Züttle, C. Nützenadel, P. Sudan and L. Schlapbach, Carbon, 40 (2002) 1339–1344.

[80] M. Nath, B.C. Satishkumar, A. Govindaraj, C.P. Vinod and C.N.R. Rao, Chem. Phys. Lett., 322 (2000) 333–340.

[81] C. Singh, T. Quested, C.B. Boothroyd, P. Thomas, I.A. Kinloch, A.I. Abou-Kandil and A.H. Windle, J. Phys. Chem. B, 106 (2002) 10915–10922.

[82] Z. Zhou, L. Ci, X. Chen, D. Tang, X. Yan, D. Liu, Y. Liang, H. Yuan, W. Zhou, G. Wang and S. Xie, Carbon, 41 (2003) 337–342.

[83] B.C. Satishkumar, A. Govindaraj and C.N.R. Rao, Chem. Phys. Lett., 307 (1999) 158–162.

[84] S. Huang, L. Dai and A.W.H. Mau, J. Phys. Chem. B, 103 (1999) 4223–4227.

[85] H.W. Zhu, C.L. Xu, D.H. Wu, B.Q. Wei, R. Vajtai and P.M. Ajayan, Science, 296 (2002) 884–886.

[86] R. Andrews, D. Jacques, A.M. Rao, F. Derbyshire, D. Qian, X. Fan, E.C. Dickey and J. Chen, Chem. Phys. Lett., 303 (1999) 467–474.

[87] R. Kamalakaran, M. Terrones, T. Seeger, P. Kohler-Redlich, M. Rühle, Y.A. Kim, T. Hayashi and M. Endo, Appl. Phys. Lett., 77 (2000) 3385–3387.

[88] A. Cao, L. Ci, G. Wu, B. Wei, C. Xu, J. Liang and D. Wu, Carbon, 39 (2001) 152–155.

[89] E.C. Dickey, C.A. Grimes, M.K. Jain, K.G. Ong, D. Qian, P.D. Kichambare, R. Andrews and D. Jacques, Appl. Phys. Lett., 79 (2001) 4022–4024.

[90] M. Mayne, N. Grobert, M. Terrones, R. Kamalakaran, M. Rühle, H.W. Kroto and D.R.M. Walton, Chem. Phys. Lett., 338 (2001) 101–107.

[91] C. Singh, M. Shaffer, I. Kinloch and A. Windle, Phys. B, 323 (2002) 339–340.

[92] B. Wei, R. Vajtai, Y.Y. Choi and P.M. Ajayan, Nano Letters, 2 (2002) 1105–1107.

[93]   C. Singh, M.S.P. Shaffer and A.H. Windle, Carbon, 41 (2003) 359–368.

[94]   F. Rohmund, L.K.L. Falk and E.E.B. Campbell, Chem. Phys. Lett., 328 (2000) 364–373.

[95]   X.Y. Liu, B.C. Huang and N.J. Coville, Carbon, 40 (2002) 2791–2799.

[96]   J.D. Harris, A.F. Hepp, R.P. Raffaelle, T. Gennett, R.V. Wal, B.J. Landi, Y. Luo and D.A. Scherson, 1st International Energy Conversion Engineering Conference, Portsmouth, VA, 2003.

[97]   H. Ago, K. Petritsch, M.S.P. Shaffer, A.H. Windle and R.H. Friend, Adv. Mater, 11 (1999) 1281–1285.

[98]   L. Sheeney-Haj-Ichia, B. Basnar and I. Willner, Angew. Chem. Int. Ed., 44 (2005) 78–83.

[99]   S. Banerjee and S.S. Wong, Nano Letters, 2 (2002) 195–200.

[100]   S. Banerjee and S.S. Wong, J. Am. Chem. Soc., 125 (2003) 10342.

[101]   J.M. Haremza, M.A. Hahn, T.D. Krauss, S. Chen and J. Calcines, Nano Letters, 2 (2002) 1253–1258.

[102]   B.J. Landi, S.L. Castro, C.M. Evans, H.J. Ruf, S.G. Bailey and R.P. Raffaelle, Mat. Res. Soc. Symp. Proc., 836 (2005) L2.8.

[103]   R.J. Chen, Y. Zhang, D. Wang and H. Dai, J. Am. Chem. Soc., 123 (2001) 3838–3839.

[104]   D.M. Guldi, G.M.A. Rahman, N. Jux, N. Tagmatarchis and M. Prato, Angew. Chem. Int. Ed., 43 (2004) 5526–5530.

[105]   S.-P. Han, T. Cagin and W.A. Goddard III, Mat. Res. Soc. Symp. Proc., 772 (2003) M6.3.1.

[106]   S.G. Stepanian, V.A. Karachevtsev, A.Y. Glamazda, U. Dettlaff-Weglikowska and L. Adamowicz, Mol. Phys., 101 (2003) 2609–2614.

[107]   S. Banerjee and S.S. Wong, Nano Letters, 2 (2002) 195–200.

[108]   J.M. Haremza, M.A. Hahn, T.D. Krauss, S. Chen and J. Calcines, Nano Letters, 2 (2002) 1253–1258.

[109]   H. Kataura, Y. Kumazawa, Y. Maniwa, I. Umezu, S. Suzuki, Y. Ohtsuka and Y. Achiba, Synth. Met., 103 (1999) 2555–2558.

[110]   A.M. Rao, J. Chen, E. Richter, U. Schlecht, P.C. Eklund, R.C. Haddon, U.D. Venkateswaran, Y.-K. Kwon and D. Tomanek, Phys. Rev. Lett., 86 (2001) 3895–3898.

[111]   Y. Lian, Y. Maeda, T. Wakahara, T. Akasaka, S. Kazaoui, N. Minami, N. Choi and H. Tokumoto, J. Phys. Chem. B, 107 (2003) 12082–12087.

[112]   T.W. Odom, J.-L. Huang, P. Kim and C.M. Lieber, Nature, 391 (1998) 62–64.

[113]   J.W.G. Wilder, L.C. Venema, A.G. Rinzler, R.E. Smalley and C. Dekker, Nature, 391 (1998) 59–62.

[114]   A. Hagen and T. Hertel, Nano Letters, 3 (2003) 383–388.

[115]   B.J. Landi, H.J. Ruf, J.J. Worman and R.P. Raffaelle, J. Phys. Chem. B, 76 (2004) 107–113.

[116]   M.J. O'Connell, S.M. Bachilo, C.B. Huffman, V.C. Moore, M.S. Strano, E.H. Haroz, K.L. Rialon, P.J. Boul, W.H. Noon, C. Kittrell, J. Ma, R.H. Hauge, R.B. Weisman and R.E. Smalley, Science, 297 (2002) 593–596.

[117]   C.A. Furtado, U.J. Kim, H.R. Gutierrez, L. Pan, E.C. Dickey and P.C. Eklund, J. Am. Chem. Soc., 126 (2004) 6095–6105.

[118] M. Endo, Y.A. Kim, Y. Fukai, T. Hayashi, M. Terrones, H. Terrones and M.S. Dresselhaus, Appl. Phys. Lett., 79 (2001) 1531–1533.

[119] M.S. Dresselhaus, G. Dresselhaus, A. Jorio, A.G.S. Filho and R. Saito, Carbon, 40 (2002) 2043–2061.

[120] N. Bendiab, R. Almairac, M. Paillet and J.L. Sauvajol, Chem. Phys. Lett., 372 (2003) 210–215.

[121] B.J. Landi, H.J. Ruf, C.M. Evans, C.D. Cress and R.P. Raffaelle, J. Phys. Chem. B, 109 (2005) 9952–9965.

Index